**some common conversions**

| | |
|---|---|
| Mass | $1\,u = 1.66 \times 10^{-27}\,kg$ |
| Length | $1\,m = 39.4\,in. = 3.28\,ft$ |
| | $1\,mi = 1.61\,km = 5280\,ft$ |
| | $1\,in. = 2.54\,cm$ (exact) |
| Volume | $1\,l = 10^3\,cm^3 = 0.0353\,ft^3$ |
| | $1\,ft^3 = 0.0283\,m^3$ |
| Time | $1\,y = 3.15 \times 10^7\,s$ |
| Speed | $1\,mi/h = 1.47\,ft/s = 0.447\,m/s = 1.61\,km/h$ |
| Force | $1\,lb = 4.45\,N$ |
| Energy | $1\,J = 0.239\,cal = 0{,}738\,ft\text{-}lb$ |
| | $1\,kcal = 4186\,J$ |
| | $1\,kWh = 3.6 \times 10^6\,J$ |
| | $1\,MeV = 10^6\,eV = 1.602 \times 10^{-13}\,J$ |
| Power | $1\,W = 0.738\,ft\text{-}lb/s$ |
| | $1\,hp = 550\,ft\text{-}lb/s = 746\,W$ |
| Pressure | $1\,mm\,Hg = 133\,N/m^2 = 1.36\,cm\,H_2O$ |
| | $1\,lb/in.^2 = 6895\,N/m^2 = 51.7\,mm\,Hg$ |
| | $1\,atm = 760\,mm\,Hg = 1033\,cm\,H_2O =$ |
| | $\quad 1.01 \times 10^5\,N/m^2 = 14.7\,lb/in.^2$ |

**physics
for the
life
sciences**

# physics for the life sciences

ALAN H. CROMER
Professor of Physics
Northeastern University

New York   St. Louis   San Francisco   Düsseldorf   Johannesburg   **McGRAW-HILL**
Kuala Lumpur   London   Mexico   Montreal   New Delhi   **BOOK COMPANY**
Panama   Rio de Janeiro   Singapore   Sydney   Toronto

This book was set in Imperial by York Graphic Services, Inc.
The editors were Jack L. Farnsworth and Laura Warner;
the designer was Barbara Ellwood;
and the production supervisor was Joe Campanella.
The drawings were done by Felix Cooper.
The printer was the Murray Printing Company;
the binder, The Book Press, Inc.

**Library of Congress Cataloging in Publication Data**
Cromer, Alan H        date
   Physics for the life sciences.
   Includes bibliographies.
   1. Physics.  I. Title.
QC23.C917                530                              73-14683
ISBN 0-07-014430-3

2 3 4 5 6 7 8 9 0 MUBP 7 9 8 7 6 5 4

# contents

# preface

The purpose of this book is to give students in biology, pharmacy, premedicine, physical therapy, physical education, and the allied health sciences the physics background they need for their professional work. The selection of material has been made on the basis of its appropriateness to the life sciences and its suitability for an introductory physics course. These criteria have resulted in some shifts in the customary emphasis of topics, but they have not limited the broad scope expected of a general physics textbook.

Since life science students often do not understand why they are required to take a physics course, the pertinence of the material to life processes is stressed at every opportunity by use of realistic biological examples to illustrate each physical principle and by inclusion of many problems that relate the physics to the life sciences. I have found that this approach increases the interest of students who otherwise have no motivation to study physics.

The mathematics in this book is limited to elementary algebra because many students, even those who have studied calculus, are not sufficiently proficient in advanced mathematics to benefit from its use. The use of only simple mathematics enables the student to concentrate directly on the physics.

The book is divided into five sections: Mechanics, Properties of Matter (which includes thermodynamics), Wave Phenomena (which includes optics), Electricity and Magnetism, and Modern Physics.

Chapters 2, 4, and 5 in Part I (Mechanics) cover the basic concepts of force, acceleration, and energy that are required for the rest of the book. Chapters 2 and 3 constitute a self-contained unit on statics suitable for students of physical therapy and physical education. The concept of similarity scaling, first introduced in Chap. 1, is used again in Chaps. 5 and 6 to derive interesting relations between size and function in animals.

Chapter 7 in Part II (Properties of Matter) covers the static and dynamic properties of fluids and applies them to respiration and circulation. The specific properties of gases, liquids, and solids relevant to biology are given in Chaps. 8 to 10. The treatment of thermodynamics in Chap. 11 goes beyond the usual introductory discussion in order to develop the concepts of entropy and free energy, which are so important for understanding the dynamics of chemical reactions.

Sound, light, and optics are covered in Part III (Wave Phenomena). Since this section is largely independent of Part II, it can be studied before Part II.

Chapters 16 and 17 in Part IV (Electricity and Magnetism) cover the basic principles of electrostatics and electric currents needed for an understanding of bioelectric phenomena. Magnetism is included for completeness in Chap. 18. The emphasis here is on the role electromagnetism plays in the practical utilization of electricity.

The fundamentals of quantum physics and their significance for chemistry are covered in Chap. 19 of Part V (Modern Physics). Chapter 20 develops the concepts of nuclear physics needed for an understanding of modern developments in nuclear medicine.

The last section or two in most chapters contains advanced material or specialized applications that can be omitted without significant loss of continuity. By the inclusion or omission of these sections, the length and level of the course can be adjusted to the needs of the class.

I am very grateful to Nathan Alpert, Bernard Gottschalk, Edward Neighbor, and Eugene Saletan for reading portions of the book and for giving me very valuable criticism. I also want to thank Jim Sigel for his useful advice. Although every effort has been made to eliminate errors, I am afraid that there inevitably will be some. For these I apologize in advance, and hope that they will be reported to me as they are discovered so that they can be corrected.

**ALAN H. CROMER**

physics
for the
life
sciences

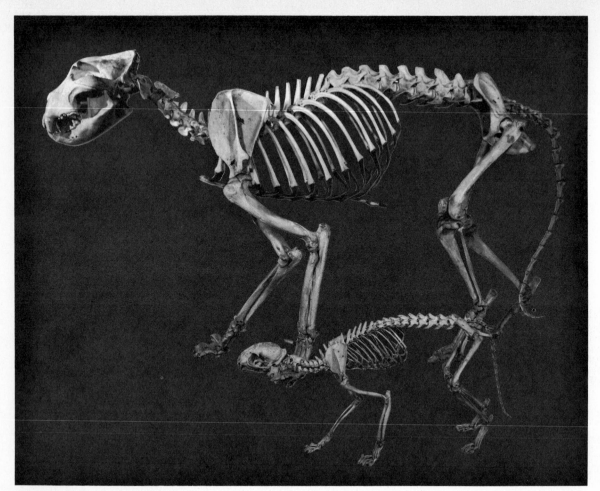

Skeletons of a large animal (lion) and a small animal (ocelot). Although these animals are anatomically very similar, the leg bones of the lion are disproportionately thicker than the leg bones of the ocelot, in accordance with the scaling laws discussed in Sec. 1.4. [*Courtesy T. McMahon, Harvard University.*]

# 1 mechanics

For thousands of years natural philosophers have speculated about the nature of the physical world. More than 2000 years ago ancient Greek scientists made considerable progress on some isolated problems, but it was not until the laws of mechanics were established by Galileo and Newton that a coherent understanding of all physical phenomena became possible. Mechanics is the study of the conditions under which objects remain at rest (*statics*) and the laws governing objects in motion (*dynamics*). These laws are universal in scope, applying as well to the motion of a satellite about the earth as to the motion of a runner about a track. More significant, mechanics is the foundation on which the rest of physics is built.

# CHAPTER 1 measurement

The human mind ascribes many different attributes to people and things, such as length, weight, color, beauty, and patriotism. Some attributes are clearly measurable, and others are not. Thus there exist well-defined procedures for measuring length and weight, but not beauty or patriotism. [Color is an intermediate case, for while a numerical value can be assigned to every color (Sec. 14.4), color cannot be ranked in order.] Physics is the study of the measurable attributes of things. The basic concepts of physics are defined in terms of measurements, and the purpose of physical theories is to correlate the results of measurements. A physical theory, no matter how abstractly it may be stated, is ultimately a statement about concrete operations performed in a laboratory.

## 1.1 THE INTELLECTUAL ORIGINS OF PHYSICS

Modern physics is the confluence of two rather different intellectual streams. One stream can be traced back to the early philosophers of Asia Minor, who were the first men in recorded history to ask questions about the fundamental nature (*physis*) of the material world. Their reasoning often tended to be speculative rather than scientific, but it was free of much of the mythology that beclouded the minds of most men in those days.

In the Greek colonies of Asia Minor (Ionia), especially the city of Miletus, the philosophers Thales (640?-546 B.C.), Anaximandros (610-545 B.C.), Anaximenes (?-525 B.C.), and others developed the concept of unity in the natural world. They believed that in spite of the apparent differences between material objects such as rocks, trees, clouds, and horses, there is an underlying sameness to all things. Each man had a different idea about the nature of this universal essence (Thales thought it was water, and Anaximenes thought it was air), but their great concept of the unity of matter remains a major doctrine of physics today.

The Ionian philosophers were also true scientists, and borrowing from the earlier work of the Egyptians and Babylonians, they made important advances in mathematics, astronomy, geology, and biology. Their work was continued on the Greek mainland in the fifth century B.C. and in the Hellenistic cities (especially Alexandria) in the fourth and third centuries B.C. The great Archimedes of Syracuse (287-212 B.C.) lived in this later period. His work in statics and hydrostatics is very modern in approach, and so little progress was made in the following 1800 years that he would have had no difficulty in understanding the work of Galileo (A.D. 1564–1642). In fact Archimedes might have started the second stream of ideas that, together with the early Greek search for the fundamental nature of things, constitutes modern physics. But unfortunately most of his written work was lost to Europe for a long time, and it was not until Galileo that the second stream clearly emerged.

It was Galileo who developed the modern method of studying simple systems by means of experimental measurement and mathematical analyses. He studied the motion of objects sliding down inclined planes and disentangled the relevant from the irrelevant features of the motion. The relevant feature was usually a measurable quantity such as the mass of the object or the time required to move a fixed distance. He tried to find the relation between these numerical measurements and to express the results in mathematical terms. It turned out that the results of an investigation could often be summarized in very simple terms, such as: The distance traveled down an incline is proportional to the square of the time. Galileo showed that the laws of nature (or some of them at least) obey simple mathematical equations, and ever since then physicists have continued the search for the mathematical relations connecting the results of their measurements.

So what is physics? It is a motivation and a method. The motivation is the same as that of the Greeks: to find the fundamental nature of things. But the method is that of Galileo: to investigate simple systems by means of experimentation and mathematical analysis. The problems under investigation sometimes seem childish and sometimes esoteric: Galileo rolls balls down an incline, Joule spins a paddle wheel in water, Rutherford experiments with rare radioactive elements. But time and again the results lead to deep and profound insights into the nature of things, fulfilling the aim of the Greek philosophers in ways they could only dream about.

In this book we shall study the method of physics and some of the results it has obtained in the 400 years since Galileo. These results are important for anyone interested in the nature of things because they apply universally to the whole material universe, including living organisms. Although it may be debated whether the mystery of life itself can be understood in terms only of the known physical laws, it is unquestionably true that life cannot be understood without reference to these laws. Indeed, as we shall see, physics gives important insights into the nature of living systems.

## 1.2  MEASUREMENT

Physics deals with the things that can be measured. What can be measured depends to a large degree on the current state of technology. For instance, the radiation emitted by radioactive substances could not be measured before the invention of devices to detect such radiation. The scope of physics continually increases as new inventions expand the range of possible measurements.

All sciences rely on measurements to some degree, but usually the measurement is auxiliary to the main purpose. Thus a zoologist might

a

b

c

d

e

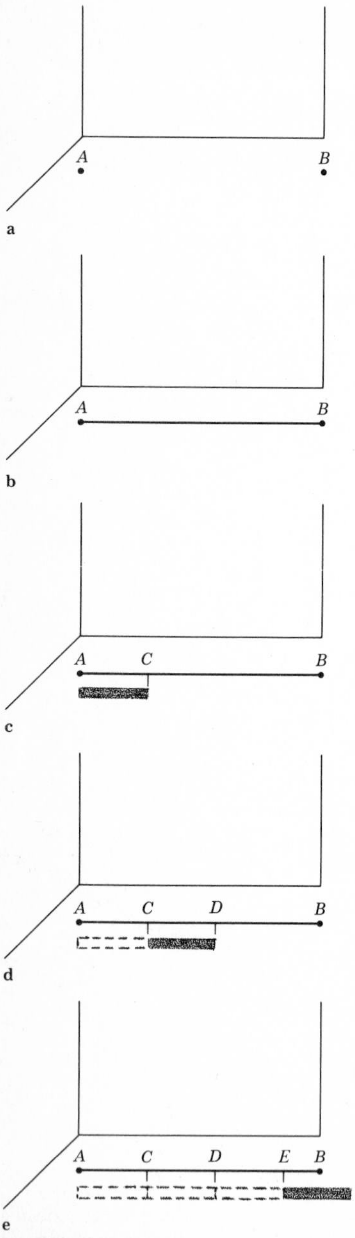

FIGURE 1.1
The steps required to measure the
length of a room.

carefully measure the weight of his experimental mice in order to determine the effect of a drug on their growth. This measurement is incidental to the problem of the drug's metabolic function. In physics, however, the measurement itself is a primary object of interest. This is because a particular concept, such as length, time, or temperature, is understood only in terms of the method used to measure it. This way of defining things is called *operationalism*, and its use avoids assigning unwarranted metaphysical meaning to a concept and introducing extraneous and possibly false connotations.

Consider for instance the concept of length. Operationally the length of something is defined as the number obtained when a specific set of operations, called *measuring its length*, is performed. These operations can be illustrated by examining the steps required to measure the length of a room:

*1* Mark two points, one at either end of the room, to define the exact interval to be measured (points *A* and *B* in Fig. 1.1*a*).
*2* Stretch a surveyor's chain (a piece of string will do) between these two points to define the straight line between them (Fig. 1.1*b*).
*3* Take a meterstick and lay it along the stretched chain with one end at point *A*. Mark where the other end of the stick touches the chain (point *C* in Fig. 1.1*c*) and move the stick along the chain until the front end is lined up with this mark (Fig. 1.1*d*). Repeat this step until point *B* is reached.
*4* When point *B* is reached, the stick will be in the position shown in Fig. 1.1*e*. Mark the point on the stick that touches *B*.
*5* The *length* (in meters) is the number of marks on the chain plus the fraction of the stick required to reach from the last mark (*E*) on the chain to *B*.

We see that by detailing the actual procedure used to measure a length we can avoid having to say anything about the essential nature of space or distance (which would be metaphysics). Length is what is measured with a ruler, and we never need to know any more about it in order to do physics.

Measurements are always made relative to some standard, called a *unit*. In the present example the unit is the meter, and the final result is given as so many meters, say 3.7 m. This can be converted into other units if the length of the meter is known in terms of these units. For instance, the conversion from meters to centimeters is

$$1 \text{ m} = 100 \text{ cm}$$

so the length of the room can be given as

$$3.7 \text{ m} \times 100 \text{ cm/m} = 370 \text{ cm}$$

The conversion from meters to feet is

$$1 \text{ m} = 3.28 \text{ ft}$$

so the length can also be given as

$$3.7 \text{ m} \times 3.28 \text{ ft/m} = 12.1 \text{ ft}$$

**REMARK** A common difficulty in converting units is knowing whether to divide or multiply by the conversion factor. For example, given that there is 0.197 wey in a hogshead (which is true), how many hogsheads are in 20 weys?

From the equation

$$1 \text{ hogshead} = 0.197 \text{ wey}$$

the number of hogshead in 1 wey is found by dividing the equation on both sides by 0.197 to obtain

$$\frac{1 \text{ hogshead}}{0.197} = \frac{0.197 \text{ wey}}{0.197}$$

or

$$1 \text{ wey} = 5.08 \text{ hogsheads}$$

Thus there are 5.08 hogsheads/wey so

$$20 \text{ weys} = 20 \text{ wey} \times 5.08 \text{ hogsheads/wey} = 101.6 \text{ hogsheads}$$

The trick is that if the units are treated as algebraic symbols, the units will come out properly in a correct calculation.

At one time the meter and the foot were the respective national standards of length of France and England. Today the meter is the scientific standard of length in all countries and the national standard in most countries except the United States. (Even the United Kingdom is converting to the metric system.) The metric system is described in Appendix VIII, and conversions between English and metric units are given inside the front cover.

The different units of length can be converted back and forth because they all characterize the same *dimension*. (The wey and hogshead are old English units with the dimension of volume.) To specify the dimension of a quantity, the symbol of the quantity is put in square brackets. Thus the dimension of length is written [$l$].
**REMARK** In physics the words *unit, dimension,* and *scale* have specific meanings that are more restrictive than their everyday meanings. Dimension refers to the quality of the concept being considered: length, volume, time, force, etc., all have different dimensions. It does not refer to amount or magnitude. (It is good English to say "She has generous dimensions" but bad physics.) Scale may be used to refer to size or magnitude. ("She is built on a large scale" is both good English and good physics.) A unit is the standard by which the magnitude (scale) of some dimension is measured. As we have just seen, different units can be used to measure the same dimension.

There are many cases of interest in which the direct measurement of a length using a meterstick is not possible, and indirect methods must be employed. However, even in an indirect measurement, a meterstick type of measurement must be made at some stage.

For example, to measure the distance $d$ between two points $A$ and $B$ on opposite sides of a river, a surveyor's transit is used (Fig. 1.2). A third point $C$ is picked on the surveyor's side of the river, and sightings are taken with the transit at $B$ and $C$ to measure the angles $\theta_1$ and $\theta_2$. The length $b$ of the base line $BC$ is measured with a chain and meterstick, as previously described. From these measurements the length $d$ can be found using the law of sines.

In this book, however, trigonometry will be avoided as much as possible, and problems like this will be solved *graphically* by making a scale drawing.

A scale drawing is begun by choosing a suitable scale. In this example, if $b = 0.5$ mile, the scale

$$10 \text{ cm} = 1 \text{ mi}$$

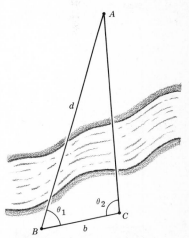

FIGURE 1.2
The distance $d$ between the two points $A$ and $B$ on opposite sides of a river is found by measuring the distance $b$ and the angles $\theta_1$ and $\theta_2$.

might be used. With this scale the line $BC$ is drawn 5 cm long. A protractor is used to draw lines at the angles $\theta_1$ and $\theta_2$ from points $B$ and $C$; they are extended until they intersect at point $A$. Finally, the distance $AB$ is measured with a ruler and converted back into the original units. For instance, suppose $AB$ is found to be 20.5 cm. According to our scale, the conversion factor is 0.1 mi/cm, so

$$20.5 \text{ cm} = 20.5 \text{ cm} \times 0.1 \text{ mi/cm} = 2.05 \text{ mi}$$

The graphical method is not as accurate as the trigonometric method, but it is conceptually much simpler and perfectly adequate for most of our purposes. Some surveying problems are given at the end of this chapter to prepare you for vectors in the next chapter.

Large distances on the earth are measured by a sequence of such *triangulations*, and the size of the earth itself is ultimately determined in this way. Once the size of the earth is known, the distance to the sun can be found by measuring the angle of the sun from two different points on the earth at the same time and using the known distance between these points as the base line (Fig. 1.3). The distance to the sun is then used to measure the distance to a (nearby) star by taking the diameter of the earth-sun orbit as the base line (Fig. 1.4). Thus even astronomical measurements are related, often through a large number of intermediate steps, to a direct measurement of distance with chain and meterstick.

Other concepts require other methods of measurement. Time is particularly subtle. To measure time a device is needed that continually repeats some event, so that the interval between two events can be taken as the unit of time. The rotation of the earth on its axis is very convenient for this purpose, and the unit of time might be one sidereal day, the time between the passing of a given star through the zenith on two successive nights. (This is not the same as the mean solar, or 24-hour, day, which is the time between two successive passings of the sun through the zenith averaged over 1 year.)

Clocks are mechanical devices constructed to repeat some event over and over. They are calibrated against the earth's rotation, and discrepancies between the clock and the earth are attributed to inaccuracies in one or the other. Atomic clocks are more accurate than the earth, which does not rotate with absolute uniformity because of tidal friction between the oceans and the ocean floor.

The standard unit of time is the *second* (s), which is 1/86,400 of the mean solar day in the year 1900. This is not the same as 1/86,400 of a current 24-hour day because the rate of the earth's rotation has changed since 1900. As a clock, the earth runs a little slow, and an extra second was added to July 30, 1972 to correct for this.†

Length and time are two fundamental dimensions in physics. Only three other fundamental dimensions (mass, temperature, and charge) will be introduced; all the other dimensions will be defined in terms of these. For instance, the (average) speed $v$ of a car in an auto race is the distance $d$ the car travels (obtained by measuring the racetrack and the number of circuits made) divided by the total elapsed time (measured with a stopwatch). Speed is an important concept, but

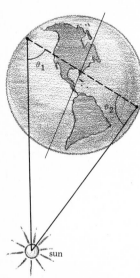

**FIGURE 1.3**
The distance to the sun is found by measuring the angles $\theta_1$ and $\theta_2$ from two points on the earth that are a known distance apart.

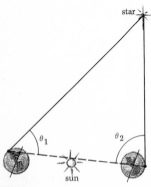

**FIGURE 1.4**
The distance to a star is found by using the diameter of the earth-sun orbit as a base line.

---

† Since 1967 the second has been defined by the cesium clock, an atomic clock controlled by electronic transitions in cesium.

its dimension $[v]$ can be related to the dimensions of length $[l]$ and time $[t]$ through its definition: the dimension $[v]$ is equal to $[l]$ divided by $[t]$, or $[v] = [l/t]$.

The dimensions of area and volume are also related to $[l]$. To measure the area (or volume) of something, certain measurements of length are required, plus some mathematical calculations. Thus the area of a circle requires the measurement of its radius $r$ (Fig. 1.5). If $r = 2.5$ ft, the area $A$ is

$$A = \pi r^2 = \pi \times (2.5 \text{ ft})^2 = 6.25\pi \text{ ft}^2 = 19.6 \text{ ft}^2$$

The unit in this case is the square foot (abbreviated ft²), and the dimension is $[l^2]$. Likewise a unit of volume is the cubic foot (abbreviated ft³), and the dimension is $[l^3]$.

For example, the volume of a sphere of radius 2.5 ft is

$$V = \frac{4\pi}{3}r^3 = \frac{4\pi}{3} \times (2.5 \text{ ft})^3$$

$$= \frac{4\pi}{3} \times 15.6 \text{ ft}^3 = 65.4 \text{ ft}^3$$

As we remarked before, various units can be used to measure area and volume, but the dimension of each of these quantities is always the same.

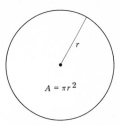

FIGURE 1.5
A circle of radius $r$.

## 1.3 ACCURACY, SIGNIFICANT FIGURES, AND EXPERIMENTAL ERROR

There is a limit to the accuracy of any measurement, depending on the equipment used and the skill of the experimenter. For example, if a meterstick marked in centimeters and tenths of centimeters, i.e., millimeters, were used to measure the size of this book, the result might be accurate to 0.1 cm, the smallest division on the stick. If the length of a room were measured, the accuracy would probably be less because the meterstick would have to be slid along the floor, as shown in Fig. 1.1. At each step the position of the end of the stick must be marked on the floor, which introduces an additional source of error into the measurement.

The determination of the accuracy of a measurement is as important as the measurement itself, and every experimenter should give both the result of his measurement and an estimate of its accuracy. A measurement of the size of this book might be quoted as $24.2 \pm 0.1$ cm. The symbol $\pm$ (read "plus or minus") means that the actual length is believed to lie between 24.1 and 24.3 cm. The quantity 0.1 cm is the *uncertainty* or *error* in the measurement.

There are two distinct kinds of error that can enter into a measurement, *random* errors and *systematic* errors. A random error is introduced each time the meterstick is slid across the floor in measuring the length of the room. The mark placed on the floor to indicate the end of the stick can never be put exactly at the end but will be put a bit to one side or the other. Once the stick is moved, another random error is introduced when the front of the stick is lined up with this mark. Again the stick will be put slightly to one side or the other of the mark. The characteristic of a random error is that it is as likely to make the final result too large as too small. Random errors are unavoidable, but they can be made very small with enough hard work. Thus repeating a measurement a large number of times and averaging the results will cancel out the random errors to a large degree.

A systematic error is usually the result of a flaw in either the equipment or the experimental procedure. The measurement of the length of the room will have a systematic error if the meterstick is slightly longer than 1 m (a faulty instrument) or if the mark is made in such a way that it is always in front of the stick (faulty procedure). In either case the measurement would come out with a length less than the true length of the room no matter how many times it was repeated. There is no way to detect and eliminate systematic errors in an experiment except to check and recheck instruments and procedure.

The error quoted by an experimentalist is usually his estimate of the random error only. There is no reasonable way to estimate a systematic error because if one is suspected, the experimenter would ordinarily go back and repeat his work to eliminate it.

It is often unnecessary to quote the error associated with a measured number explicitly since, by convention, the number of digits used to express the number roughly indicates the error. If the length of this book is quoted as 24.2 cm without the error estimate, one would assume that the true length is probably between 24.15 and 24.25 cm, i.e., that the error is about 0.05 cm. Although the error in this case is actually 0.1 cm, the difference in these two types of error estimate is unimportant for most purposes. It is usually sufficient to know the number of *significant figures* in a measurement.

The number of significant figures is the number of reliably known digits in the measurement. There are three significant figures in the length 24.2 cm, since even the last digit is known to be either 1, 2, or 3, and certainly not 5, 6, or 7. It would be incorrect to quote the result as 24.20 cm, because this would imply that the measurement was accurate to four significant figures. But in this case nothing is known about the last digit, which could as well be 3 (for example, 24.23) as 7 (for example, 24.27).

In dealing with large numbers, a problem can arise in indicating the number of significant figures correctly. For instance, the distance from the earth to the sun is about 92,900,000 mi. How many figures are significant? Certainly the distance is not known to the nearest half mile (so not all the zeros can be significant), but it might be known to within $\pm 5000$ mi (so that the first zero after the 9 would be significant). To avoid misunderstanding it is often convenient to use powers of 10 to express such a number. The number is given then as $92.90 \times 10^6$ mi. The first factor gives all the significant figures, and the second factor tells how many places to the right the decimal point should be moved. (Appendix I discusses the use of powers of 10 in more detail.)

In the course of a calculation, the number of digits can increase beyond the significance of the result. In such a case the final number should be rounded off to leave only the significant figures.

For example, suppose again that the length of this book is found to be $24.2 \pm 0.1$ cm and the width is found to be $19.5 \pm 0.1$ cm. From these measurements the area of the book is calculated to be

$$24.2 \text{ cm} \times 19.5 \text{ cm} = 471.90 \text{ cm}^2$$

But not all the digits in the final result can be significant. At one extreme the length may really be 24.3 cm and the width may be 19.6 cm, so that the area is about 476 cm$^2$. At the other extreme the length may be only 24.1 cm and the width only 19.4 cm, so the area is only about 468 cm$^2$.

As a general rule, the result of a calculation cannot have more significant figures than the number of digits in the least significant figure used in the calculation. In this last example the two numbers used in the calculation each had three significant figures, so the final result should be rounded off to three significant figures and quoted as 472 cm$^2$.

**REMARK** The use of significant figures to indicate the accuracy of a result is not as precise as giving the actual errors, but it is sufficient for most of our purposes. In the last example, for instance, it is questionable whether the last digit in 472 is really significant. It is not necessary to go into this here. The important point is that the last two digits in 471.90 certainly are not significant.

Some quantities that are fundamental to physics are measured over again every few years, as better techniques for improving the accuracy become available. For example, the speed of light in empty space (vacuum) is a universal constant of nature, and its exact determination is important in many areas of physics and astronomy. The speed of light is usually expressed in kilometers per second (km/s) and is found by measuring the distance a pulse of light travels in 1 s. In 1862 Foucault found the speed to be $298,000 \pm 500$ km/s, or, as we should write it, $(2.980 \pm 0.005) \times 10^5$ km/s. Note that the first zero after the 8 is significant and the others are not.

**REMARK** To convert kilometers to meters or miles, the conversion factor must be known. If the converted number is to have as many significant figures as the original number, the conversion factor must be known to the same number of significant figures. A kilometer is, of course, exactly $10^3$ m. A mile is now also defined to be an exact multiple of the kilometer, the relation being

$$1 \text{ mi} = 1.609344 \text{ km}$$

For conversions requiring only three significant figures, the relation 1 mi = 1.61 km may be used.

Since 1862 the speed of light has been measured with increasing accuracy by a large number of investigators. Table 1.1 lists some of these measurements to show how the number of significant figures has increased over the years. The newer measurements generally do not correct the older measurements but only increase the accuracy of the result.

An interesting exception to this is the 1935 measurement of Michelson, Pease, and Pearson. Their result differs from the 1967 measurement of Grosse by 18 km/s, or 9 times their estimated error. They had repeated their measurement over 2000 times to reduce the random error to only 2 km/s. Unfortunately there was a large unsuspected systematic error in their experiment which became appar-

TABLE 1.1 **Speed of light measured by various investigators**

| Date | Investigator | Measurement, km/s |
|---|---|---|
| 1676 | Römer | $220,000 \pm ?$ |
| 1849 | Fizeau | $313,300 \pm ?$ |
| 1862 | Foucault | $298,000 \pm 500$ |
| 1875 | Cornu | $299,990 \pm 200$ |
| 1880 | Michelson | $299,910 \pm 50$ |
| 1883 | Newcomb | $299,860 \pm 30$ |
| 1926 | Michelson | $299,796 \pm 4$ |
| 1928 | Helstaedt | $299,778 \pm 10$ |
| 1935 | Michelson, Pease, and Pearson | $299,774 \pm 2$ |
| 1941 | Anderson | $299,776 \pm 6$ |
| 1949 | Aslakson | $299,792 \pm 3.5$ |
| 1950 | Essen | $299,792.5 \pm 1.0$ |
| 1952 | Froome | $299,792.6 \pm 0.7$ |
| 1953 | MacKenzie | $299,792.4 \pm 0.5$ |
| 1957 | Bergstrand | $299,792.85 \pm 0.16$ |
| 1958 | Froome | $299,792.50 \pm 0.10$ |
| 1967 | Grosse | $299,792.50 \pm 0.05$ |

ent only when still more accurate measurements were made in the 1950s.

Every experiment is haunted by the possibility of serious systematic error. A good researcher is aware of this and does everything he can to check over his experiment. But the final check comes only after the experiment is repeated independently by someone else.

## 1.4 SCALING: AN INTRODUCTION TO MATHEMATICAL ANALYSIS

An interesting biological problem that can be investigated by simple mathematical analysis is how the size of a structure is related to its function. For example, is it possible to have a single cell as large as an ant or an ant as large as a man, or are the sizes of these structures determined by their function? These are questions of *scaling*, i.e., questions about how the properties of structures depend on size.

The answer to such questions requires an understanding of how the basic geometrical quantities of length, area, and volume vary with the size of an object. Figure 1.6 shows two cubes $C$ and $C'$ of different size. The length of an edge of cube $C'$ is twice the length of an edge of cube $C$. We say that cube $C'$ is larger than cube $C$ by the *scale factor $L$*, where $L$ is equal to 2 in this case. The scale factor is the ratio of corresponding lengths in two similar figures.

If instead of lengths we compare areas, it is clear that a face of cube $C'$ has 4 times the area of a corresponding face of cube $C$: the ratio of these areas is $L^2 = 2^2 = 4$. On the other hand, the volume of cube $C'$ is 8 times the volume of cube $C$: the ratio of these volumes is $L^3 = 2^3 = 8$.

This result, obvious for a cube, is also true for any two similar figures, regardless of shape. Figure 1.7 shows two similar figures of

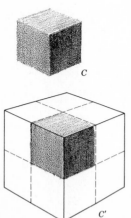

FIGURE 1.6
Two cubes of different size. An edge of $C'$ is twice the length of an edge of $C$.

different size. The scale factor is the ratio of corresponding lengths in the figures; for instance,

$$L = \frac{d'}{d}$$

(Since the two figures are similar, the scale factor $L$ is the same for any two corresponding lengths.) The ratio of any two corresponding areas, e.g., the ratio of the cross-sectional areas $A$ and $A'$ indicated in Fig. 1.7, is $A'/A = L^2$. The ratio of the volumes $V$ and $V'$ of the two figures is $V'/V = L^3$.

The importance of these geometrical relations comes from the fact that some physical properties of a body depend on volume and some depend on area. The ratio of such properties will therefore depend on the size of the body. For example, the weight of an animal depends on its volume. Figure 1.8 shows two ants that are identical in shape and material. The giant ant has a scale factor $L = d'/d$ relative to the normal ant, and consequently the giant ant weighs $L^3$ times as much as the normal ant:

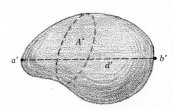

Weight of giant ant = $L^3 \times$ weight of normal ant

On the other hand, the strength of an ant (or any other organism) depends only on the cross-sectional area of its muscles. (Think of a weight lifter. The length of his arms is normal, but the cross-sectional area of his arms is extraordinarily large. All his increased strength comes from the increased cross-sectional area of his muscles.) Therefore the giant ant will be $L^2$ times stronger than the normal ant and will be able to lift $L^2$ times as much weight:

FIGURE 1.7
Two similar figures of different size.

Weight giant ant can lift = $L^2 \times$ weight normal ant can lift

The *relative strength* of an animal is defined as the ratio of the weight the animal can lift to the body weight of the animal. From the above relations for these quantities we get

$$\begin{aligned}
\text{Relative strength of giant ant} &= \frac{\text{weight giant ant can lift}}{\text{weight of giant ant}} \\
&= \frac{L^2 \times \text{weight normal ant can lift}}{L^3 \times \text{weight of normal ant}} \\
&= \frac{\text{weight normal ant can lift}}{L \times \text{weight of normal ant}} \\
&= \frac{1}{L} \times \text{relative strength of normal ant}
\end{aligned}$$

GIANT ANT

This means that the relative strength of the giant ant is smaller than the relative strength of the normal ant by the factor $1/L$.

It is often said that an ant is stronger than a man because it can lift 3 times its body weight whereas a man can lift only about half of his body weight; i.e., the relative strength of a normal ant is 3, and the relative strength of a man is 0.5. However, it is wrong to conclude from this that the ant is stronger; the ant should have a larger relative strength just because of its small size. To really compare the strengths of an ant and a man, one must take the difference in their size into account.

NORMAL ANT

FIGURE 1.8
Two ants of similar shape but different size.

A normal ant is about 0.5 in. long, and a man is about 72 in. long. A giant ant the size of a man would have the scale factor

$$L = \frac{72 \text{ in.}}{0.5 \text{ in.}} = 144$$

relative to a normal ant. The relative strength of this man-sized ant would be only

$$\frac{1}{L} \times \text{relative strength of normal ant} = \tfrac{1}{144} \times 3 = \tfrac{1}{48}$$

which is much less than the relative strength of a man. An ant, therefore, is intrinsically weaker than a man. In fact, a man-sized ant is not a biologically viable creature: since it could lift only one-forty-eighth of its body weight, it could not even lift its own legs to climb over small obstacles.

What we have said about the strength of muscles applies also to bones and other structural material. For an animal of a given shape the strength of its bones relative to its body weight depends on its size, and the larger the animal the smaller its relative strength. In nature we find that the shape of larger animals is very different from the shape of smaller animals. Figure 1.9 shows a dog and an elephant drawn the same size. The thickness of the elephant's legs is greater than that of the dog's. An animal the size of an elephant cannot have the shape of a dog because the ratio of the strength of the bones to body weight would be too small. The bones and muscles of large animals must be disproportionately thicker than the bones and muscles of small animals.

The scaling of other properties can be studied in the same way. In later chapters the dependence on size of such things as running speed, pulse rate, and power output will be considered. Now we shall apply the principles of scaling to one other problem: Why do cells divide when they reach a certain size?

For simplicity consider a spherical cell. Figure 1.10 shows two cells at different stages of growth. The scale factor of the older (larger) cell relative to the younger (smaller) cell is $L = R'/R$, where $R$ and $R'$ are the radii of the two cells, as shown. The volume of the older cell is $L^3$ times the volume of the younger cell. This means that the older cell has $L^3$ times as much metabolizing material as the younger cell and so requires $L^3$ times as much oxygen (and other vital substances) per minute:

Oxygen requirement/min of older cell
$$= L^3 \times \text{oxygen requirement/min of younger cell}$$

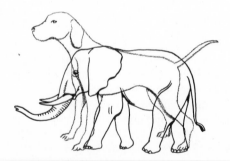

FIGURE 1.9
A dog and an elephant drawn the same size.

All the oxygen consumed by the cell must come through the cell wall, so that the maximum amount of oxygen the cell can obtain per minute is proportional to the area of the cell wall. Thus the older cell can obtain at most $L^2$ times as much oxygen per minute as the younger cell:

Maximum amount of oxygen obtainable/min by older cell
$= L^2 \times$ maximum amount of oxygen obtainable/min by younger cell

The ratio of the maximum amount of oxygen obtainable to the oxygen required is called the *viability factor*. It clearly must be greater than 1 for the cell to survive. From the above relations it is easy to see that

$$\text{Viability factor of older cell} = \frac{1}{L} \times \text{viability factor of younger cell}$$

A young cell has a viability factor greater than 1. The last relation shows that as the cell grows, its viability factor decreases and starts to approach 1. To avoid suffocation the cell must either stop growing or divide. By dividing, the one large cell with a small viability factor is replaced by two smaller cells, each with a larger viability factor.

This discussion of scaling is intended to show that a little knowledge of mathematics and physics is useful in understanding some general features of living systems. As the principles of physics are developed in this book, they will be applied to other problems in biology and the health sciences.

**FIGURE 1.10**
Two spherical cells at different stages of growth.

## PROBLEMS

*1* (*a*) Convert 30 ft to inches. (*b*) Convert 12 m to feet. (*c*) Convert 7.5 in. to centimeters. (*d*) Convert 150 weys to hogsheads.
*Ans.* (*a*) 360 in.; (*b*) 39.4 ft; (*c*) 19.0 cm; (*d*) 762 hogsheads

*2* (*a*) What is the conversion from square feet to square inches, i.e., how many square inches are in 1 ft²? (*b*) What is the conversion from cubic feet to cubic inches?

*3* (*a*) What is the conversion from square meters to square centimeters? (*b*) What is the conversion from cubic meters to cubic centimeters?
*Ans.* (*a*) 1 m² = $10^4$ cm²; (*b*) 1 m³ = $10^6$ cm³

*4* (*a*) What is the conversion from square meters to square feet? (*b*) What is the conversion from cubic meters to cubic feet?

*5* (*a*) What is the area of a circle 3.5 cm in diameter? (*b*) Convert the area to square meters.
*Ans.* (*a*) 9.6 cm²; (*b*) 9.6 × $10^{-4}$ m²

*6* (*a*) What is the volume of a sphere with a radius of 1.3 ft? (*b*) Convert the volume to cubic meters.

*7* A room is 14.5 ft long and 9.5 ft wide. Find the area of the room in (*a*) square feet, (*b*) square inches, and (*c*) square meters.
*Ans.* (*a*) 138 ft²; (*b*) 1.99 × $10^4$ in.²; (*c*) 12.8 m²

*8* A cubic foot of water weighs 62.3 lb. (*a*) How much does a cubic inch of water weigh? (*b*) How much does a cubic centimeter weigh?

*9* A gallon (U.S. liquid) of water weighs 8.33 lb. Find the volume of a gallon in (*a*) cubic feet and (*b*) cubic centimeters (see Prob. 8).
*Ans.* (*a*) 0.134 ft³; (*b*) 3.79 × $10^3$ cm³

*10* In hydrology (the science of water distribution on land) a common unit of volume is the *acre-foot*, which is the volume of water that could cover 1 acre to a uniform depth of 1 ft. (*a*) Given that 1 acre = 43,560 ft², how many gallons are in 1 acre-ft (see Prob. 9)? (*b*) A local thundershower drops 50 acre-ft of water. What is the weight of this water?

*11* A new unit, the *newton* (N), is introduced in Chap. 4. (*a*) Given that 1 N = 0.225 lb, what is the weight in newtons of a cubic foot of water (see Prob. 8)? (*b*) What is the weight in newtons of a cubic meter of water?
*Ans.* (*a*) 277 N; (*b*) 9.78 × $10^3$ N

*12* A car moving at a constant speed of 60 mi/h, travels 60 mi in 1 h. (*a*) How many feet does it travel in 1 h (remember, 1 mi = 5280 ft)? (b) How many feet does it travel in 1 s? (c) What is the conversion from miles per hour to feet per second?

13 (a) What is the conversion from miles per hour to kilometers per hour? (b) What is the conversion from miles per hour to meters per second?

Ans. (a) 1 mi/h = 1.61 km/h; (b) 1 mi/h = 0.448 m/s

14 A surveyor wishes to find the distance from point $B$ to point $A$ on the opposite side of a river (Fig. 1.2). He measures the base line $BC$ and the angles $\theta_1$ and $\theta_2$. Make a scale drawing and find the distance $BA$, given that $\theta_1 = 86°$, $\theta_2 = 83°$, and $BC = 0.15$ mi.

15 An obelisk (Fig. 1.11) casts a shadow of length $d = 35$ ft. At the same time the angle $\theta$ is found to be 75°. What is the height of the obelisk? Ans. 130 ft

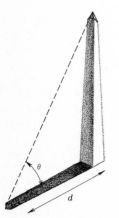

FIGURE 1.11 Problem 15.

16 A man walks north for 2 mi and then, turning 45°, he walks northeast for 3 mi. What is the straight-line distance between the beginning and end of the walk?

17 A man's height is measured to be 5 ft 11 in., with an uncertainty of 0.5 in. Express his height in feet, keeping only significant figures.

Ans. 5.92 ft

18 The length of a rectangular piece of land is found by surveying to be 1235 ± 25 ft, and its width is found to be 736 ± 15 ft. Calculate the area of the land in square feet, and express the result so that the number of significant figures is displayed properly.

19 The 1970 Census quotes the population of Boston as 627,776. This number may differ by as much as 2 percent from the true population at that time. Quote the population of Boston so that the correct number of significant figures is indicated properly. (Why do you think the Census Bureau includes insignificant digits in its census figures?) Ans. $6.3 \times 10^5$

20 In a drug study, 50 mice are fed a daily diet that includes 50 mg (1 mg = 1 milligram = $10^{-3}$ g) of drug X, and a control group of 50 mice is fed the same diet without the drug. Measurements are made daily of the weight of each mouse. List possible sources of error in this experiment, and note which errors are random and which are systematic. Take nothing for granted, because in practice any conceivable error may (and often does) occur.

21 A woman 5 ft 0 in. tall weighs 110 lb. How much would a 5 ft 5 in. woman of similar shape weigh? Ans. 140 lb

22 A spherical cell of radius $R'$ divides into two equal daughter cells, each of radius $R$. (a) Find the scale factor $L = R'/R$. (b) What is the ratio of the surface area of the two daughter cells to the surface area of the parent? (c) If the viability factor of the parent is 1, what is the viability factor of each daughter?

23 A brass cylinder 10 in. high and 2 in. in radius weighs 4 lb. It is glued to the underside of a horizontal board, as shown in Fig. 1.12. The glue has a maximum strength of 2 lb/in.². If a force greater than the maximum is applied to the glued surface, the glue will break. (a) What is the maximum strength of the glued surface in this case? (b) What is the weight of a similar brass cylinder 10 times as large (i.e., with the scale factor $L = 10$)? (c) Can the same glue hold the larger cylinder to the horizontal surface?

Ans. (a) 25.1 lb; (b) 2510 lb; (c) no

24 Land mammals range in size from the pigmy shrew (2 in. long) to the Indian elephant (130 in. long). Discuss factors that prevent mammals from being appreciably larger or smaller than these.

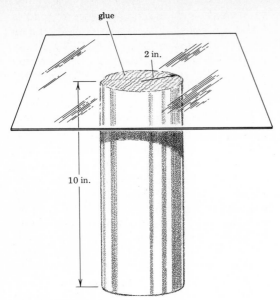

glue

2 in.

10 in.

FIGURE 1.12    Problem 23.

# BIBLIOGRAPHY

BRIDGMAN, P. W.: "The Nature of Physical Theory," Dover Publications, Inc., New York, 1964. The concept of operationalism is expounded and used as the basis of a positivist philosophy. This book gives a penetrating analysis of the nature of truth and the meaning of meaning.

GALILEI, GALILEO: "Dialogues concerning Two New Sciences," trans. by Henry Crew and Alfonso de Salvio, The Macmillan Company, New York, 1914, reprinted by Dover Publications, Inc., New York, 1954. The original treatment of the laws of scaling. Though first published in 1638, it remains a clear and delightful account of "the impossiblity of increasing the size of structures to vast dimensions either in art or in nature."

HOYLE, FRED: "Astronomy," Doubleday & Co., Inc., Garden City, N.Y., 1962. The various direct and indirect methods used by astonomers to measure distances are explained in chaps. 6, 9, and 10.

JEANS, JAMES: "The Growth of Physical Science," Fawcett Publications, Inc., New York, 1961. A very readable account of the history of physics from ancient times to the twentieth century.

SARTON, GEORGE: "A History of Science: Ancient Science through the Golden Age of Greece," Harvard University Press, Cambridge, Mass., 1960.

————: "A History of Science: Hellenistic Science and Culture in the Last Three Centuries B.C.," Harvard University Press, Cambridge, Mass., 1959. These two volumes give an authoritative account of science in the ancient world. Written shortly before Professor Sarton's death, these books are the distillation of the life's work of a great historian.

SMITH, J. MAYNARD: "Mathematical Ideas in Biology," Cambridge University Press, Cambridge, 1968. Chapter 1 of this brief but fascinating book applies the concept of scaling to a variety of biological problems.

# CHAPTER 2 force

The next two chapters develop the laws of *statics*, which are the conditions under which an object remains at rest. These laws have universal applicability and can be used to design the struts holding up a bridge as well as to understand the function of muscles in maintaining body posture. The study of statics also provides a good introduction to force, which is one of the most important concepts in physics.

## 2.1 PROPERTIES OF FORCE

It is possible to define force operationally by describing the operations required to measure it. This is the way the concepts of length and time were introduced in Chap. 1; but since such a procedure is more likely to obscure than clarify the properties of force we want to emphasize, we shall not bother to define force at all, taking it to be a primitive concept with the following properties.

**Property 1** *A force is a push or a pull,* such as we experience whenever we contract a muscle. *A force is always applied by one material object to another.* Thus the hand in Fig. 2.1 exerts a force **F** (by means of the rope) to the box.

**Property 2** *A force is characterized by both its magnitude and the direction in which it acts.* That is, both the magnitude and the direction are required to specify a force completely. The magnitude can be expressed in various units, and for now the English pound (lb) will be used. (The dimension $[f]$ of force is discussed in Sec. 4.4.) A force of 1 lb is the force required to support a 1-lb weight. The direction of a force is the direction that the force would tend to move the object to which it is applied, in the absence of other forces. The direction is indicated by an arrow like that in Fig. 2.1, which shows that the force exerted by the hand is along the rope. The direction of a force is not always evident from inspection, but flexible ropes always transmit forces along their length.

Quantities characterized by both a magnitude and a direction are called *vectors*. An arrow can be used to represent the magnitude of a vector at the same time that it gives its direction. In a given problem one simply adopts a scale, say 1 cm = 4 lb, and then a force of 8 lb is represented by an arrow 2 cm long. Everything one needs to know about the force is conveniently represented by such arrows. The vector **F** in Fig. 2.1 indicates a force of 6 lb with the scale 1 cm = 5 lb, since its length is 1.2 cm.

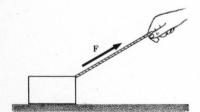

**FIGURE 2.1**
A force **F** is applied by the hand to the box.

The symbol for a force is written in boldface type **(F)** to emphasize that it is a vector quantity. The same symbol written in italics ($F$) refers to the magnitude of the force alone.

**Property 3 (Newton's third law of motion)**   *Forces always come in pairs.* When one object (say the hand in Fig. 2.1) applies a force to another (the box), the second object *simultaneously* applies a force of *equal magnitude* and in the *opposite direction* to the first object. In Fig. 2.2 the 6-lb force **F** exerted by the hand *on the box* is directed 30° upward to the right, and the 6-lb force **R** exerted by the box *on the hand* is directed 30° downward to the left.

The two forces in any pair are sometimes termed the *action* and the *reaction* forces. This is somewhat misleading terminology, since it implies that one of the forces comes into existence in response to the other. This is not the case. Both forces come into existence simultaneously, and it is arbitrary which force is called the action and which the reaction. Usually the force acting on the object of primary interest is called the action, but this is only an arbitrary convention. In Fig. 2.2, if the box were the object of main interest, **F** would be the action and **R** the reaction; but if the hand were the focus of attention, **R** might be called the action and **F** the reaction.

It is of *extreme importance* to realize that *the action and reaction forces always act on two different objects.* Property 1 states that a force is always applied by one object to another. Newton's third law (Property 3) goes further and says that each object exerts a force on the other.

The principles of mechanics were first formalized by Isaac Newton (1642-1727) in his great book "Principia Mathematica," published in 1687. In this book he stated the principles in terms of three laws of motion. The first law is given below (Property 5), and the second law is discussed in Chap. 4 (Property 7). The third law is Property 3. It is one of the basic laws of nature, though it is by no means obvious. Only after studying the consequences of this property in some detail will it become clear why it must be true for the world to be as it is.

**Property 4**   *If two (or more) forces act simultaneously on the same object, their effect is the same as that of a single force equal to the* vector sum *of the individual forces.*

Figure 2.3 shows an object subjected to two forces **F**₁ and **F**₂. The vector sum **S** of these forces is found in the following way (Fig. 2.4). Draw an arrow to represent **F**₁, and from the head of **F**₁ draw a second arrow to represent **F**₂. Connect the *tail* of **F**₁ to the *head* of **F**₂ by a third arrow. This third arrow represents the vector sum **S** of **F**₁ and **F**₂. In other words, the meaning of the equation

$$\mathbf{S} = \mathbf{F}_1 + \mathbf{F}_2$$

is that the three vectors **S**, **F**₁, and **F**₂ form a closed triangle when drawn end to end as in Fig. 2.4.

**REMARK**   In going from Fig. 2.3 to Fig. 2.4 the vector **F**₂ was moved so that its tail touched the head of **F**₁. For the purpose of adding vectors it is permissible to move a vector from one position to another as long as its length and direction are not changed.

Property 4 allows us to find the combined effect of several forces acting on an object. Of course we have not yet said what the effect

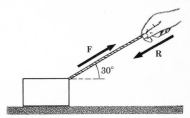

FIGURE 2.2
The force **R** applied by the box to the hand is the reaction to the force **F** applied by the hand to the box.

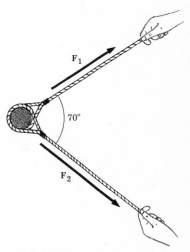

FIGURE 2.3
An object subjected to the two forces **F**₁ and **F**₂.

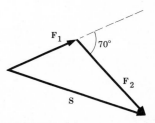

FIGURE 2.4
The force **S** is the vector sum of **F**₁ and **F**₂.

of a force is, i.e., what a force does. This information is given by Newton's first and second laws (Properties 5 to 7). Properties 6 and 7 are discussed in Chaps. 3 and 4, respectively, and only Property 5 will be discussed now. However it is enough to enable us to solve a number of interesting problems.

**Property 5 (Newton's first law of motion)** *For an object to remain at rest, i.e., to be in equilibrium, it is necessary that the vector sum of all the forces acting on the object be zero.* This is only a necessary condition for an object to be at rest. That is, if an object is to remain at rest, the net force on it must be zero, and if the net force is not zero, the object cannot remain at rest. However, it is possible for an object to move even though the net force on it is zero because there is another condition (Property 6) that must also be satisfied for an object to remain at rest. This is discussed further in Sec. 3.1.

A complete understanding of these five properties of force is possible only after studying how they are applied to specific problems. The remainder of this chapter is devoted to utilizing these properties in a variety of situations. While studying these applications you should constantly refer back to the statements of the five properties in this section to make sure you understand their relevancy to the problem at hand.

## 2.2  SOME SPECIFIC FORCES

The last section dealt with very general properties common to all forces. In this section we discuss some common everyday forces and the special features peculiar to them.

### Force of gravity

The name *gravity* is given to the force with which the earth attracts all objects. For a given material object, say a block of lead, this force has nearly the same magnitude everywhere on the surface of the earth. It might vary by as much as 0.5 percent between the pole and the equator, but this small difference can be neglected for our purposes.

The direction of the force is toward the center of the earth, and so the direction of the force of gravity on the lead block at the pole is different from the direction at the equator (see Fig. 2.5). The lead block is said to be attracted toward the center of the earth by the force of gravity. Then by the third law (Property 3), the block must attract the earth with a force of equal magnitude and opposite direction to the force exerted by the earth on the block.

This last point is more than a mere intellectual whim, for it implies that the block also has the ability to attract things gravitationally. Thus gravitational attraction cannot be a special property of the earth but must be present in all objects. We shall return to this interesting conclusion in Sec. 5.4.

The force of gravity can be used to establish a unit of force. First, a well-machined block of metal is produced and arbitrarily designated as the standard. That is, the force of gravity on this block is taken

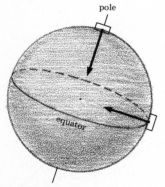

**FIGURE 2.5**
The force of gravity on a lead block at the pole and at the equator.

*by definition* to be 1 unit of force. [In the English system of units this unit is called the *pound* (lb).] Second, a number of secondary standards are produced from the primary one. To do this a balance (Fig. 2.6) is used. The primary standard is put on one side, and powdered metal is added to the other side until the balance is level. The powdered metal is then melted into a single block. Once secondary standards are produced, the primary standard is put in a vault and used only occasionally to verify that the secondary standards have not changed. (It is not desirable to handle the primary standard too much because dirt, scratches, and dents over a period of years would change the force of gravity on it.)

Third, metal blocks of $\frac{1}{4}$, $\frac{1}{2}$, 5, 10, etc., lb are produced from the secondary standards, and a complete set of comparison weights is obtained. The force of gravity on any unknown object is found by placing it on one side of a balance and adding these calibrated weights to the other side until the scale is level. To measure forces other than gravity, indirect methods are required. We next discuss one such indirect method.

**REMARK** Since the force of gravity varies slightly from point to point on the surface of the earth, this method of defining the unit of force is not entirely satisfactory. It does illustrate what is involved in establishing a unit, however, and is good enough for the present. Another method of establishing a unit of force is discussed in Sec. 4.4.

### Spring force

Figure 2.7*a* shows a spring suspended from a bar and hanging vertically alongside a ruler. When a 1-lb weight is attached to the lower end of the spring, the spring stretches a certain distance, say 0.75 cm (Fig. 2.7*b*). When a 2-lb weight is attached to the spring, the spring stretches 1.5 cm, or twice as much as with 1 lb. In general, it is found that the amount that the spring is stretched (up to a limit) is proportional to the force of gravity on the weight. The magnitude $F_g$ of this force is thus related to the distance $x$ that the spring is stretched by

$$F_g = kx$$

where $k$ is a constant characteristic of the spring. In the present case

$$k = \frac{1 \text{ lb}}{0.75 \text{ cm}} = \frac{2 \text{ lb}}{1.5 \text{ cm}} = 1.33 \text{ lb/cm}$$

The dimension of $k$ is [force/length], or $[f/l]$.

Since a block hanging from the spring is clearly in equilibrium, the first law (Property 5) says that the total force on it must be zero. But the earth is exerting a force of gravity $\mathbf{F}_g$ on it, directed downward. Therefore there must be another force $\mathbf{F}_k$ on the block to cancel $\mathbf{F}_g$. This force is exerted by the spring and arises from the molecular rearrangements produced in the metal when the spring is stretched.

The spring force is found as follows. The first law says that the vector sum of $\mathbf{F}_g$ and $\mathbf{F}_k$ is zero, since the block is in equilibrium. This can be written

$$\mathbf{F}_g + \mathbf{F}_k = 0 \qquad 2.1$$

FIGURE 2.6
A balance used to produce a secondary standard.

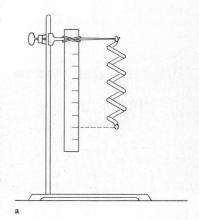

a

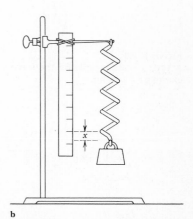

b

FIGURE 2.7
A spring stretched a distance $x$ by a weight suspended from it.

To add these forces, Property 4 says that $\mathbf{F}_k$ must first be drawn with its tail on the head of $\mathbf{F}_g$. The sum of these vectors is then the vector drawn from the tail of $\mathbf{F}_g$ to the head of $\mathbf{F}_k$. Since this sum is to be zero, the tail of $\mathbf{F}_g$ must coincide with the head of $\mathbf{F}_k$. Figure 2.8 shows this: $\mathbf{F}_k$ must have the same magnitude as $\mathbf{F}_g$ and be oppositely directed.

The normal rules of algebra can be used to solve Eq. 2.1 for $\mathbf{F}_k$. The obvious result is

$$\mathbf{F}_k = -\mathbf{F}_g$$

What is not so obvious is the meaning of the minus sign in front of a vector. In this case we know that $\mathbf{F}_k$ is equal in magnitude and opposite in direction to $\mathbf{F}_g$. This suggests the general rule: *For any force* $\mathbf{F}$*, the force* $-\mathbf{F}$ *has the same magnitude as* $\mathbf{F}$ *and is oppositely directed.*

Springs are used to make crude force-measuring devices. These are suitable for simple lecture demonstrations but are not accurate enough for commercial purposes. However, with a spring scale forces other than the force of gravity can be measured. Figure 2.9 shows a spring scale measuring the force exerted by the hand.

### Contact force

A block resting on a table (or a student sitting in his chair) is clearly in equilibrium and so again, by the first law, there must be another force acting on it in addition to the force of gravity. The surface of the table is slightly deformed by the block and, as a consequence, exerts an upward force $\mathbf{F}_c$ on the block. This force, which is directed perpendicular to the surface that produces it, is called the *contact force*. Since the sum of $\mathbf{F}_g$, the force of gravity on the block, and $\mathbf{F}_c$, the contact force on the block, is zero by the first law, it follows that $\mathbf{F}_c$ must be directed opposite to $\mathbf{F}_g$, or

$$\mathbf{F}_c = -\mathbf{F}_g \qquad\qquad 2.2$$

Contact forces are exerted by solid bodies on other objects in contact with them. They are real forces and are accompanied by small distortions in the surfaces of the bodies that produce them. In fact, a contact force differs from a spring force only in the degree to which distortion occurs. A spring is designed to produce a noticeable change of shape when it applies a force, whereas only minute changes occur in the surface of a solid body exerting a contact force. Thus in a sense a solid body acts like a very stiff spring. Consequently a solid body can exert different contact forces in different circumstances without any noticeable change in its appearance. However, this should not mislead you into thinking that these forces are any less real.

Figure 2.10 shows the two forces $\mathbf{F}_g$ and $\mathbf{F}_c$ that act on the block on the table. Because these forces have equal magnitude and opposite direction, students (and, alas, many textbooks) sometimes call $\mathbf{F}_c$ the reaction to $\mathbf{F}_g$. *This is a horrendous error.* Equation 2.2 is a consequence of the first law and the fact that the block is in equilibrium. Furthermore, the two forces involved act on the same object whereas action and reaction forces always act on different objects.

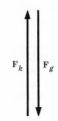

**FIGURE 2.8**
The sum of the forces $\mathbf{F}_g$ and $\mathbf{F}_k$ is zero.

**FIGURE 2.9**
A spring scale measuring the force exerted by a hand.

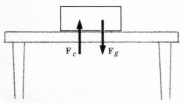

**FIGURE 2.10**
The force of gravity $\mathbf{F}_g$ and the contact force $\mathbf{F}_c$ acting on a block at rest on a table.

The reaction to $\mathbf{F}_g$ is the gravitational force $\mathbf{R}_g$ that the block exerts on the earth: it acts on the earth (at its center). The contact force $\mathbf{F}_c$ also has a reaction force $\mathbf{R}_c$, which is a contact force and acts on whatever is producing $\mathbf{F}_c$. Since the table produces $\mathbf{F}_c$, the reaction to $\mathbf{F}_c$ is the force $\mathbf{R}_c$ the block exerts on the table. Figure 2.11 shows all these forces. Remember, in applying the first law, only the forces acting on the object in question must sum to zero.

### Friction force

A friction force is present whenever one object slides over another. It generally acts to oppose any externally applied force. Thus again consider a block on a table and imagine that a small force $\mathbf{F}_a$ is applied to it, parallel to the surface of the table (Fig. 2.12). If $\mathbf{F}_a$ is small enough, the block will not move, so that there must be some other force $\mathbf{F}_f$ acting on the block. As long as the block remains at rest, the first law requires that $\mathbf{F}_f = -\mathbf{F}_a$.

The force of friction $\mathbf{F}_f$ is always directed parallel to the surface that produces it. This is what distinguishes it from the contact force, which is the force a surface exerts perpendicular to itself. The sum of the contact and friction forces, $\mathbf{F}_c + \mathbf{F}_f$, is the total *surface force* $\mathbf{F}_s$ that the surface exerts on an object (Fig. 2.13). The magnitude of $\mathbf{F}_f$ can change to equal $F_a$, just as the magnitude of $\mathbf{F}_c$ can change to equal $F_g$. However, there is a maximum value to the magnitude of $\mathbf{F}_f$, and if $F_a$ exceeds this value, the force of friction no longer balances it and the block will move. ($\mathbf{F}_c$ also has a maximum value, namely, the value at which the surface breaks. This value is not usually attained in everyday life, though it can be if a heavy person sits in a weak chair.)

The maximum force of friction depends on the nature of the two surfaces involved. It is greater for a wood block on a wood table than for a waxed ski on ice. It usually also depends on the magnitude of the contact force $\mathbf{F}_c$. For simple problems we assume that the maximum frictional force is just proportional to $F_c$:

$$\text{Maximum force of friction} = \mu_s F_c \qquad 2.3$$

As long as the body remains at rest, the actual force of friction on it will be less than (or equal to) this, so

$$F_f \leq \mu_s F_c \qquad 2.4$$

The symbol $\leq$ means "less than or equal to," and Eq. 2.4 says that the force of friction is less than or equal to a certain fraction of the contact force.

The constant† $\mu_s$ is called the *coefficient of static friction*, and Table 2.1 gives its value for some common materials. The term static refers to the fact that Eq. 2.4 holds only as long as the body is at rest.

† $\mu$ is the Greek letter mu.

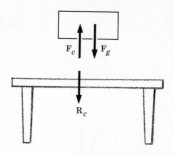

FIGURE 2.11
The reaction to $\mathbf{F}_g$ is a force $\mathbf{R}_g$ that acts on the earth, and the reaction to $\mathbf{F}_c$ is a force $\mathbf{R}_c$ that acts on the table.

TABLE 2.1 Coefficient of static friction of some common materials

| Material | Coefficient of static friction $\mu_s$ |
|---|---|
| Ice on ice | 0.05–0.15 |
| Leather, on wood | 0.3–0.4 |
| On metal | 0.6 |
| Ski wax on dry snow | 0.04 |
| Steel on steel, dry | 0.6 |
| Lubricated | 0.10 |
| Wood, on wood, dry | 0.25–0.50 |
| On metal, dry | 0.2–0.6 |

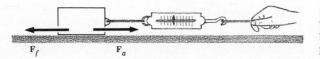

FIGURE 2.12
The force of friction $\mathbf{F}_f$ on the block opposes a force $\mathbf{F}_a$ applied parallel to the surface.

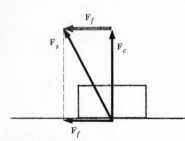

**FIGURE 2.13**
The surface force $\mathbf{F}_s$ is the sum of the contact force $\mathbf{F}_c$ and the force of friction $\mathbf{F}_f$.

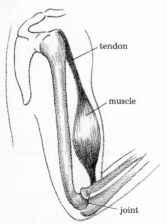

**FIGURE 2.14**
A muscle attaches to two bones across a joint.

Once $F_a$ exceeds $\mu_s F_c$, the body is no longer in equilibrium and so it will start to slide. The frictional force now changes in character and depends on the speed of the object. For simple problems this dependence is neglected, and the moving friction is taken to be a constant, also proportional to $F_c$. For moving bodies we have

$$F_f = \mu_k F_c \qquad\qquad 2.5$$

where $\mu_k$, the *coefficient* of moving, or *kinetic, friction*, is usually less than $\mu_s$.

**Muscle force**

Animal posture and motion are controlled by forces produced by muscles. A muscle consists of a large number of *fibers*, the cells of which are capable of contraction when stimulated by impulses coming to them from nerves. A muscle is usually attached to two different bones at either end by *tendons* (Fig. 2.14). The two bones are linked by a flexible connection called a *joint*. The contraction of the muscle produces two pairs of forces that act on the two bones and the muscles at the point where the tendons attach. These are the action-reaction forces between each bone and the muscle.

The maximum force that a muscle can exert depends on its cross-sectional area, and in man it is about 40 to 50 lb/in.². That is, to produce a muscle force $F_m$ of 200 lb requires a muscle with a cross-sectional area of 4 or 5 in.². This idea was used in Sec. 1.4 when we said that strength scaled as $L^2$. An animal whose (linear) size is $L$ times as large as a smaller animal of similar shape has muscles with $L^2$ times the cross-sectional area of the smaller animal, and so the larger animal has $L^2$ times the strength of the smaller one.

Of the two bones connected by a muscle, one is often more movable than the other. For example, Fig. 2.14 shows how the bicep connects at one end to the radius in the forearm and at the other end to the shoulder. When the muscle is contracted, the more movable bone is usually moved toward the less movable one.

The point of attachment to the less movable member is called the *origin* of the muscle, and the point of attachment to the more movable member is called the *insertion*. These terms are often arbitrary. For example, in lifting a stone the bicep moves the forearm to the shoulder, whereas in chinning oneself the shoulder (and the entire body) moves toward the stationary forearm. The important point to remember is that as a consequence of the third law, the muscle must exert forces of equal magnitude at the origin and the insertion.

The study of how muscle forces act to produce human motion and balance is called *kinesiology* or *biomechanics*. It is of special importance to athletes and physical therapists, who need to understand what forces are required to produce specific body movements. Some aspects of this subject will be studied in the examples and problems in this book.

**Compression and tension**

A solid block that has two opposing forces $\mathbf{F}_1$ and $\mathbf{F}_2 = -\mathbf{F}_1$ pressing it on either side (Fig. 2.15) will be in equilibrium. Nevertheless it clearly differs in some sense from a block that has no forces acting on it. When the opposing forces act, the block is said to be *compressed* or in a state of *compression;* the magnitude $C$ of the compression is equal to the magnitude of either force acting on it. That is, $C = F_1 = F_2$. (Remember we use italic symbols to indicate only the magnitude of the force.)

Likewise a block in equilibrium could have two opposing forces pulling on it, as in Fig. 2.16. In this case the block is said to be in a state of *tension,* and the magnitude $T$ of the tension is again equal to the magnitude of either force acting on it ($T = F_1 = F_2$).

### Flexible cord

A flexible cord, such as a string, rope, or tendon, has a number of special properties:

1 It can be in a state of tension but not compression.
2 It can transmit a force only along its length. (This is in contrast to a solid rod, e.g., a golf club or baseball bat, which can exert a force either along its length or perpendicular to it.)
3 In the absence of friction, the tension is the same at all points along a cord.

To understand these properties consider Fig. 2.17, which shows two weights, $A$ and $B$, connected by a cord passing over a frictionless pulley. Assume that the whole system is in equilibrium, and let $\mathbf{F}_g$ be the force of gravity on weight $A$. What is the force of gravity $\mathbf{F}_g'$ on weight $B$?

By the first law the cord must exert a force $\mathbf{F}_c = -\mathbf{F}_g$ upward on $A$, and by the third law the weight then exerts the reaction force $\mathbf{R}_c = -\mathbf{F}_c = -(-\mathbf{F}_g) = \mathbf{F}_g$ downward on the cord (Fig. 2.18). Although $\mathbf{R}_c$ is equal to the force of gravity on the weight, it is not itself a gravitational force.

With the force $\mathbf{R}_c$ applied to one end, the cord has a tension $T = R_c = F_g$. This is the same at all points along the cord (since the pulley is frictionless), and so the force $\mathbf{R}_c'$ that weight $B$ exerts downward on the cord is also equal in magnitude to $T = F_g$. The reaction to $\mathbf{R}_c'$ is $\mathbf{F}_c' = -\mathbf{R}_c'$, which is an upward force on weight $B$ (Fig. 2.18). By the first law again, the total force on $B$ must be zero, so the force of gravity $\mathbf{F}_g'$ on it is equal to $-\mathbf{F}_c' = \mathbf{R}_c'$ and has a magnitude equal to $T = F_g$. Thus for equilibrium, the two weights must be equal.

This lengthy discussion shows that a cord can be used to change the direction of a force without changing its magnitude. This is very important in biomechanics, where tendons are used to change the direction of the force of a muscle. These tendons pass over bones instead of pulleys. Lubricating fluids in the body reduce the friction between the tendon and the bone almost to zero.

## 2.3 EXAMPLES INVOLVING FORCES IN A LINE

Two forces are said to lie in a line if they have the same or opposite directions. The forces $\mathbf{F}_1$ and $\mathbf{F}_2$ in Fig. 2.19 have the same direction, and Fig. 2.20 shows that the magnitude of their sum $\mathbf{S} = \mathbf{F}_1 + \mathbf{F}_2$ is

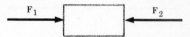

FIGURE 2.15
A block compressed by two opposing forces pressing on it.

FIGURE 2.16
A block in tension from two opposing forces pulling on it.

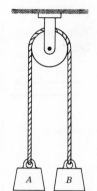

FIGURE 2.17
Two weights connected by a cord passing over a frictionless pulley.

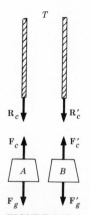

FIGURE 2.18
The forces acting on the two weights and the cord shown in Fig. 2.17.

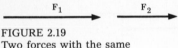

FIGURE 2.19
Two forces with the same
direction.

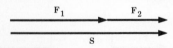

FIGURE 2.20
The sum of the two forces in
Fig. 2.19.

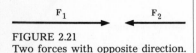

FIGURE 2.21
Two forces with opposite direction.

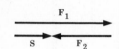

FIGURE 2.22
The sum of the two forces in
Fig. 2.21.

equal to the sum of the magnitudes of $F_1$ and $F_2$. That is, if the magnitude of $F_1$ is 8 lb and the magnitude of $F_2$ is 5 lb, the magnitude of $S$ is 13 lb.

Figure 2.21 shows two forces $F_1$ and $F_2$ that have opposite directions. In this case the magnitude of their sum $S = F_1 + F_2$ is equal to the *difference* of the magnitudes of $F_1$ and $F_2$ (Fig. 2.22). That is, if the magnitude of $F_1$ is 8 lb and the magnitude of $F_2$ is 5 lb, the magnitude of $S$ is 3 lb.

If two forces do not lie in a line, their sum is not equal to either the sum or the difference of the magnitudes of the individual forces, as can be seen in Fig. 2.4. We postpone discussion of this more complicated situation to Sec. 2.5 and consider now some examples of problems in which the forces all lie in a line.

For example, consider Fig. 2.23, which shows the head of a patient in neck traction. A force is applied to the head by means of the chin strap in order to put tension on the cervical structures (the seven vertebrae in the neck). The chin strap is pulled by a cord that passes over the pulley and attaches to a 12-lb weight. (The head rests on a movable platform in order to eliminate friction between the head and the table.) With this arrangement, what is the tension $T_n$ in the neck vertebrae?

The tension $T_c$ in the cord is 12 lb. This follows from arguments identical to those used in Sec. 2.2 in discussing Figs. 2.17 and 2.18. Since the 12-lb weight is in equilibrium, the cord must exert an upward force of 12 lb on it to balance the force of gravity. The reaction to the upward force exerted by the cord on the weight is a downward force of 12 lb exerted by the weight on the cord. By definition, the tension $T_c$ in the cord is equal to the magnitude of the force applied to either end of the cord, and so it is 12 lb.

By the nature of a flexible cord (Sec. 2.2), this tension is the same at either end, and so the head must exert a 12-lb force to the right on the cord. The reaction to this is a 12-lb force $F_a$ to the left, exerted by the cord on the head. The whole purpose of the weight, pulley, cord, and chin strap is to apply this force $F_a$ to the head. It may be intuitively obvious to the reader that this is what the traction device is doing, but remember that the 12-lb force of gravity acts only on the weight. We must make repeated use of the first and third laws (Properties 5 and 3) to deduce that the force applied to the head by the cord (which is not a gravitational force) is also 12 lb.

Having got this far, we can analyze the forces on the head. Since the head is in equilibrium, the total force on it must be zero. Therefore there must be another force on the head equal to $-F_a$, as shown

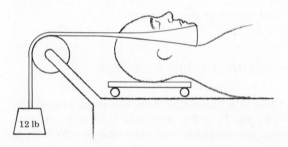

FIGURE 2.23
A patient in neck traction.

in Fig. 2.24. The only object that can apply such a force is the neck, and so we conclude that there is a contact force $\mathbf{F}_c = -\mathbf{F}_a$ exerted on the head by the neck. The reaction to $\mathbf{F}_c$ is the contact force $\mathbf{R}_c = -\mathbf{F}_c = \mathbf{F}_a$ that the head applies to the neck. So again by use of the first and third laws we conclude that there is a force of magnitude 12 lb pulling to the left on the neck.

Since the neck is also in equilibrium, there is a 12-lb force exerted to the right on the neck by the rest of the spinal column. The neck is thus in tension, and the magnitude of this tension $T_n$ is 12 lb.

This result is not very surprising. In fact you may have known the answer from the beginning and wonder why such involved reasoning was needed to obtain it. The point is that the first and third laws, when properly used, lead to results in agreement with our expectations. Such results help to establish confidence in the validity of these laws.

As another example, let us analyze what happens when two weights $A$ and $B$ are put on a scale. Weight $A$ weighs 5 lb, and weight $B$ weighs 10 lb. This means that if $A$ is put on the scale alone, the scale will read 5 lb, and if $B$ is put on the scale alone, the scale will read 10 lb. What happens if $A$ is put on the scale and $B$ is put on top of $A$? Everyone knows that the scale will read 15 lb. The same principle applies when you stand on a scale. The scale reads the sum of the gravitational forces on all the parts of your body, even though only your feet are in contact with the scale.

The more one thinks about these common everyday experiences, the more mysterious they seem. There must be some fundamental law of nature which accounts for them. This is Newton's third law of motion (Property 3). As mentioned in Sec. 2.1, the verification of this law is its ability to explain some of the most commonplace experiences of everyday life.

To analyze the two weights on the scale we start by considering $B$, the 10-lb weight (Fig. 2.25). Since there is the 10-lb force of gravity $\mathbf{F}_g$ acting downward on it, by the first law there must be an upward force of 10 lb as well. The only possible source for this second force is the contact force $\mathbf{F}_c$ that $A$ exerts on $B$. Therefore $\mathbf{F}_c$ is a 10-lb force directed upward.

Weight $A$ is acted upon by the 5-lb force of gravity $\mathbf{F}_g'$ and by the upward contact force $\mathbf{F}_c'$ exerted by the scale (Fig. 2.26). However, in this case the magnitude of $\mathbf{F}_c'$ does not equal 5 lb, because $\mathbf{F}_g'$ and $\mathbf{F}_c'$ are not the only forces acting on $A$. The third law says that there is a reaction force $\mathbf{R}_c$ to $\mathbf{F}_c$: $\mathbf{R}_c$ is the 10-lb contact force that $B$ exerts against $A$. The first law requires that the sum $\mathbf{F}_g' + \mathbf{F}_c' + \mathbf{R}_c$ be zero. The sum $\mathbf{F}_g' + \mathbf{R}_c$ is a 15-lb force directed downward, so $\mathbf{F}_c'$ is a 15-lb force directed upward.

Finally, by the third law again, the reaction $\mathbf{R}_c'$ to $\mathbf{F}_c'$ is a 15-lb force exerted by $A$ downward on the scale. It is this force on the scale that the scale registers. It is the third law that guarantees that the force on the scale (which is a contact force and not a gravitational force) is equal in magnitude to the total gravitational force on all the objects sitting on the scale.

It is particularly important to note that in this last example the upward contact force $\mathbf{F}_c'$ on the lower weight $A$ is greater than the upward contact force $\mathbf{F}_c$ on the upper weight $B$. It is obviously true

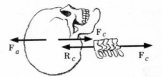

FIGURE 2.24
The forces on the head of a patient in neck traction, without friction.

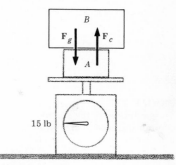

FIGURE 2.25
The forces acting on $B$, the 10-lb weight.

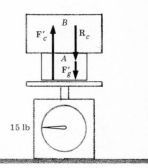

FIGURE 2.26
The forces acting on $A$, the 5-lb weight.

that in any vertical structure the contact force on a part near the bottom of the structure is greater than the contact force on a part near the top because each part supports the entire weight of all the parts above it. In many structures, both natural and man-made, the lower parts are larger than the higher parts in order to be strong enough to bear the larger force. In the spinal column of man, for instance, the vertebrae increase steadily in size from top to bottom for this reason (Fig. 2.27).

As a final example, let us consider again the patient in neck traction. We complicate the problem a bit by removing the movable platform and letting the head rest on the bench. Now there is a force of friction $\mathbf{F}_f$ on the head in addition to the forces $\mathbf{F}_a$ and $\mathbf{F}_c$ exerted by the cord and the neck. Equation 2.4 says that the magnitude of $\mathbf{F}_f$ is less than or equal to the coefficient of friction times the contact force. One may be tempted to jump to conclusions and say that the contact force to use in Eq. 2.4 is $\mathbf{F}_c$, the 12-lb force exerted on the head by the neck. This is wrong. Reread the discussion of friction in Sec. 2.2. The contact force required in Eq. 2.4 is the one exerted by the surface which is producing the frictional force.

Figure 2.28 shows the two vertical forces on the head: the downward gravitational force $\mathbf{F}_g$ and the upward contact force $\mathbf{F}'_c$. Suppose the magnitude of $\mathbf{F}_g$ is 10 lb. Then the magnitude of $\mathbf{F}'_c$, the contact force exerted by the bench on the head, is also 10 lb. This is the contact force to use in Eq. 2.4. If the coefficient of friction $\mu_s$ between the head and the bench is 0.15, then

$$F_f \leq \mu_s F'_c = 0.15 \times 10 \text{ lb}$$

or
$$F_f \leq 1.5 \text{ lb}$$

**REMARK** The condition that the vector sum of all the forces on a body at rest be zero is satisfied if the vertical and horizontal forces separately sum to zero. This is explained more fully in Sec. 2.5.

Before $\mathbf{F}_a$ is applied, both $\mathbf{F}_f$ and $\mathbf{F}_c$ are zero. Then if $\mathbf{F}_a$ is applied gradually, the force of friction will increase in magnitude to oppose $\mathbf{F}_a$ until $F_f$ reaches its maximum value. Thereafter the contact force will increase to maintain equilibrium. Thus, when $F_a$ has reached its full value of 12 lb, $\mathbf{F}_f$ is a 1.5-lb force directed to the right. By the first law, the sum of the three horizontal forces $\mathbf{F}_a$, $\mathbf{F}_c$, and $\mathbf{F}_f$ is zero. Figure 2.29 shows that this implies that the magnitude of $\mathbf{F}_c$ is only 10.5 lb, i.e., it is 1.5 lb less than it was when the frictional force was absent. With the same reasoning as for Fig. 2.24, the tension in the neck is found to be 10.5 lb. The frictional force on the head has decreased the tension applied to the neck.

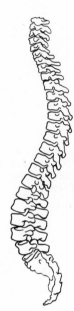

**FIGURE 2.27**
The vertebrae in the spinal column increase in size from top to bottom.

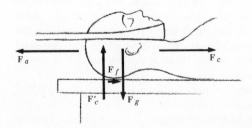

**FIGURE 2.28**
The forces on the head of a patient in neck traction, with friction.

**REMARK**  The magnitude of $\mathbf{F}_f$ depends on just how the forces are applied. For instance, if the patient lifts his head slightly, the force of friction is reduced to zero and the magnitude of $\mathbf{F}_c$ immediately equals $F_a$. The force of friction remains zero when the patient puts his head down again, because now $\mathbf{F}_a + \mathbf{F}_c$ is zero. Thus the tension in the patient's neck can vary as he shifts his head about. To avoid this complication, a patient's head is often placed on a movable platform, as in Fig. 2.23.

The ability of frictional forces to decrease the tension applied to a structure is very important. It makes it possible, for instance, to apply tension to the neck without putting the entire spinal column in tension. This is because the frictional force on the shoulders and back can be so large that no tension is transmitted to the rest of the spine. To apply tension to the whole spine, the entire trunk of the body must be placed on a movable platform. In this case the frictional forces on the hips and legs prevent the tension from being transmitted beyond the spine.

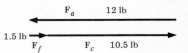

FIGURE 2.29
The vector sum of the horizontal forces on the head of a patient in neck traction.

## 2.4  COMPONENTS OF FORCE

Sometimes it is convenient to replace a force $\mathbf{F}$ in a problem by two *perpendicular forces* $\mathbf{F}_x$ and $\mathbf{F}_y$ whose vector sum equals $\mathbf{F}$:

$$\mathbf{F} = \mathbf{F}_x + \mathbf{F}_y$$

The forces $\mathbf{F}_x$ and $\mathbf{F}_y$ are called the *components* of $\mathbf{F}$, and the operation of finding $\mathbf{F}_x$ and $\mathbf{F}_y$ is called *resolving* $\mathbf{F}$ into its components.

Figure 2.30 shows the force $\mathbf{F}$ exerted by the deltoid muscle on the humerus when the arm is held in a horizontal position. From x-ray studies it is found that the muscle exerts this force at approximately 15° to the humerus. It will be shown in Chap. 3 how the magnitude of this force can be calculated, but now we just assume it is 60 lb. This force is performing two distinct functions: (1) *supporting* the arm against the force of gravity and (2) *stabilizing* the joint by pulling the humerus against the scapula. The magnitude of the force involved in each of these functions is found by resolving $\mathbf{F}$ into its components parallel and perpendicular to the humerus.

This can be done by graphical construction, similar to the method used to find the distance across the river in Sec. 1.2:

1 Draw a horizontal line $OO'$ to represent the direction of the humerus and, using a protractor, draw a dotted line at an angle of 15° to $OO'$ (Fig. 2.31a). This dotted line is drawn from an arbitrary point $A$ on $OO'$ and represents the direction of $\mathbf{F}$.

2 Using some convenient scale (say 5 lb = 1 cm), mark off a vector

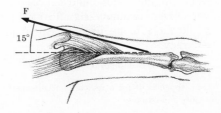

FIGURE 2.30
The force exerted on the humerus by the deltoid muscle.

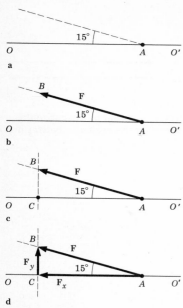

**FIGURE 2.31**
Graphical procedure for finding the components of a given vector parallel and perpendicular to a given line.

of appropriate length (12 cm) on the dotted line to represent **F** (Fig. 2.31*b*).

*3* From the head of **F** (point *B*) draw a line perpendicular to *OO'* (Fig. 2.31*c*). Let *C* denote the point of intersection of this line with *OO'*.

*4* Draw a vector **F**$_x$ from *A* to *C* and a vector **F**$_y$ from *C* to *B* (Fig. 2.31*d*). These are the desired components of **F** since they are parallel and perpendicular to the humerus (*OO'*) and their vector sum is **F**.

The magnitudes of **F**$_x$ and **F**$_y$ are found by measuring the lengths *AC* and *CB* with a ruler. In this case, with the scale 5 lb = 1 cm, the length of *AC* should be about 11.5 cm, and the length of *CB* should be about 3.0 cm. Therefore we find

$$F_x = 11.5 \text{ cm} \times 5 \text{ lb/cm} = 57.5 \text{ lb}$$

and
$$F_y = 3.0 \text{ cm} \times 5 \text{ lb/cm} = 15.0 \text{ lb}$$

**REMARK**  The sum of the magnitudes of $F_x$ and $F_y$ is greater than the magnitude of $F$ itself. This is because the two forces **F**$_x$ and **F**$_y$ do not lie in a line. Remember, in general vector addition involves both magnitude and direction and not just magnitude alone.

This graphical method of finding the components of a vector is direct and easy to learn. With reasonable care it will give results that are accurate enough for most of our purposes. However, it is rather tedious and of limited accuracy, especially if the angle is close to 0 or 90°. It is quicker and more accurate to use trigonometry to find the magnitude of the components.

In the trigonometric method, you need only draw a rough sketch of **F** and its components, being careful to make **F** the hypotenuse of a right triangle. The sketch will look like Fig. 2.31*d*, from which it is evident that the sine (opposite over hypotenuse†) of 15° is $F_y/F$,

$$\sin 15° = \frac{F_y}{F}$$

so

$$F_y = F \times \sin 15° = 60 \text{ lb} \times 0.259 = 15.5 \text{ lb}$$

Likewise the cosine (adjacent over hypotenuse) of 15° is $F_x/F$,

$$\cos 15° = \frac{F_x}{F}$$

so

$$F_x = F \times \cos 15° = 60 \text{ lb} \times 0.966 = 58.0 \text{ lb}$$

These results appear to be more accurate than those obtained using the graphical method. However, in this case $F$ is really known only to two significant figures (60 lb), so that the difference between the graphical and trigonometrical results is meaningless. Of course, if $F$ were given to many significant figures, the trigonometric method would have to be used to calculate the components to the same accuracy.

The graphical method should be used in solving most problems

† See Appendix II for a review of trigonometry and the inside back cover for a table of sines and cosines.

involving forces, at least at first. The trigonometric method should be used only when you see a real advantage in it.

## 2.5  EXAMPLES INVOLVING FORCES IN A PLANE

A typical problem in mechanics is to find some of the forces acting on a body at rest, given all the other forces. The problem is solved using the first law (Property 5) and the rule for the vector addition of forces (Property 4). If the forces lie in a plane, and not just in a line, two unknown quantities can be found this way. These can be the magnitude and direction of a single force, the magnitudes of two forces, or (less commonly) the directions of two forces.

A classic example is the problem of finding the magnitudes of the contact and frictional forces that act on a block resting on an inclined plane. Figure 2.32 shows an 8-lb block sitting on a board that is inclined 25° to the horizontal. The force of gravity $\mathbf{F}_g$ on the block is an 8-lb force directed vertically downward. Therefore the surface force $\mathbf{F}_s$ on the block must be an 8-lb force directed vertically upward. By definition, the contact and frictional forces, $\mathbf{F}_c$ and $\mathbf{F}_f$, are the components of $\mathbf{F}_s$ perpendicular and parallel to the incline (Fig. 2.33).

The magnitudes of $\mathbf{F}_c$ and $\mathbf{F}_f$ can easily be found by the methods of Sec. 2.4 once the angle $\theta$ between $\mathbf{F}_s$ and $\mathbf{F}_c$ is known. A theorem in geometry says that if the corresponding sides of two angles are perpendicular, the angles are equal (Theorem 3, Appendix II). Thus, since $\mathbf{F}_s$ is perpendicular to the horizontal and $\mathbf{F}_c$ is perpendicular to the inclined surface, the angle $\theta$ between them is equal to the angle of 25° between the incline and the horizontal. Then from Fig. 2.33 we see that

$$F_c = F_s \times \cos\theta = 8\ \text{lb} \times \cos 25° = 7.25\ \text{lb}$$
$$F_f = F_s \times \sin\theta = 8\ \text{lb} \times \sin 25° = 3.38\ \text{lb}$$

You should check this by finding $F_c$ and $F_f$ graphically.

With this result the inequality (Eq. 2.4)

$$F_f \leq \mu_s F_c$$

becomes

$$3.38\ \text{lb} \leq \mu_s \times 7.25\ \text{lb}$$

Dividing both sides of this inequality by 7.25 lb, we get

$$\mu_s \geq \frac{3.38\ \text{lb}}{7.25\ \text{lb}} = 0.466$$

This means that if the coefficient of friction were less than 0.466, the block could not rest on a board inclined at 25°.

As a slightly more complex example, consider again the outstretched arm in Fig. 2.30. Figure 2.34 shows the three forces acting on it: the muscle force $\mathbf{F}_m$ exerted by the deltoid muscle, the gravitational force $\mathbf{F}_g$, and the surface force $\mathbf{F}_s$ applied to the humerus at the joint. Assuming that the arm weighs 7 lb and that the magnitude of $\mathbf{F}_m$ is 60 lb, what is the magnitude and direction of $\mathbf{F}_s$? That is, what must $\mathbf{F}_s$ be so that the sum of all three forces is zero?

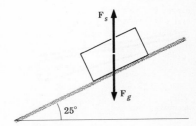

**FIGURE 2.32**
A block sitting on a board inclined 25° to the horizontal.

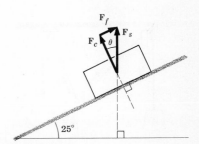

**FIGURE 2.33**
The contact force $\mathbf{F}_c$ and the frictional force $\mathbf{F}_f$ are the components of the surface force $\mathbf{F}_s$ perpendicular and parallel to the inclined surface.

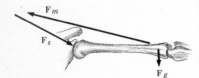

**FIGURE 2.34**
The forces acting on an outstretched arm.

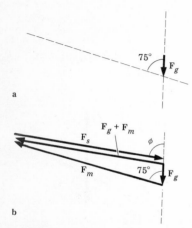

a

b

**FIGURE 2.35**
Graphical procedure for finding the force $\mathbf{F}_s$ acting on the arm in Fig. 2.34.

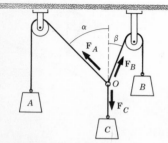

**FIGURE 2.36**
A device for applying three forces to a common point $O$.

To answer this, you first must find the sum of the known forces, $\mathbf{F}_m$ and $\mathbf{F}_g$. Choose a convenient scale (say 5 lb = 1 cm) and draw a line of appropriate length (1.4 cm) to represent $\mathbf{F}_g$ (Fig. 2.35$a$). With a protractor draw a line through the head of $\mathbf{F}_g$ at an angle to $\mathbf{F}_g$ of 75°, the angle between $\mathbf{F}_m$ and $\mathbf{F}_g$ in Fig. 2.34. From the head of $\mathbf{F}_g$ draw a vector of appropriate length (12 cm) along this line to represent $\mathbf{F}_m$ (Fig. 2.35$b$). Then the vector drawn from the tail of $\mathbf{F}_g$ to the head of $\mathbf{F}_m$ is the sum of $\mathbf{F}_g$ and $\mathbf{F}_m$. When added to $\mathbf{F}_g + \mathbf{F}_m$, the surface force must result in a zero total force. This is possible only if $\mathbf{F}_s$ equals $-(\mathbf{F}_g + \mathbf{F}_m)$, that is, if it is directed opposite to the sum of $\mathbf{F}_g$ and $\mathbf{F}_m$. This is shown in Fig. 2.35$b$: the sum of the three vectors $\mathbf{F}_g$, $\mathbf{F}_m$, and $\mathbf{F}_s$ is zero.

The magnitude and direction of $\mathbf{F}_s$ are found from the scale drawing by measuring its length and the angle it makes with $\mathbf{F}_g$. You should make such a drawing. You will find that the magnitude of $F_s$ is about 58.6 lb and that the angle $\phi$ it makes with the vertical is about 81°.

**REMARK** The force at the joint has been called a surface force because, from the information given, there is no way to determine whether it is a pure contact force or not. To be a pure contact force it must be perpendicular to the surface of the humerus at the point where the humerus touches the joint. If it is not exactly perpendicular, then only its component perpendicular to the surface of the humerus is a contact force and its component parallel to this surface is a frictional force. It is known from physiology, however, that the coefficient of friction between bones in a joint is exceedingly small, so that there is very little friction. Consequently $\mathbf{F}_s$ is almost perpendicular to the surface of the humerus, and it is almost a pure contact force.

We consider, finally, the device shown in Fig. 2.36, which consists of three cords attached together at the common point $O$. The other ends of the cords are attached to three different weights $A$, $B$, and $C$. Each weight puts a tension on its cord equal to the force of gravity on the weight, and each cord exerts a force on point $O$ equal in magnitude to its tension. Thus each cord exerts a force on $O$ equal in magnitude to the weight attached to it. The two pulleys allow the directions of forces $\mathbf{F}_A$ and $\mathbf{F}_B$ to change until $O$ is in equilibrium.

This device is often used to demonstrate Newton's first law and the vector addition of forces (Properties 4 and 5). The magnitude and direction of each of the three forces applied to $O$ are found by measuring the weights and the angles between the cords. It can then be verified that the vector sum of these forces is zero. One way to do this is to measure the weight $C$ and the angles $\alpha$ and $\beta$ the cords make with the vertical and use Properties 4 and 5 to calculate the magnitudes of $\mathbf{F}_A$ and $\mathbf{F}_B$. The result can then be compared to the magnitudes found by measuring weights $A$ and $B$ directly.

Thus suppose that $C$ weighs 2 lb, $\alpha = 45°$, and $\beta = 20°$, so that $\mathbf{F}_C$ is a force of magnitude 2 lb directed vertically downward and $\mathbf{F}_A$ and $\mathbf{F}_B$ are forces of known direction and unknown magnitude. It is important to recognize that this situation is very similar to the problem of the block on the incline. Here, as there, the sum of the unknown forces is known:

$$\mathbf{F}_A + \mathbf{F}_B = -\mathbf{F}_C$$

To find the magnitudes of $\mathbf{F}_A$ and $\mathbf{F}_B$, draw a vector to represent $-\mathbf{F}_C$. This is a force of 2 lb directed vertically upward, and with a scale of 1 lb = 10 cm it would be drawn 20 cm long (Fig. 2.37a). From the tail of $-\mathbf{F}_C$ draw a dotted line at an angle $\alpha = 45°$ to $-\mathbf{F}_C$ and from the head of $-\mathbf{F}_C$ draw a dotted line at an angle $\beta = 20°$ to $-\mathbf{F}_C$. These lines are in the directions of $\mathbf{F}_A$ and $\mathbf{F}_B$, and their point of intersection determines the magnitudes of $\mathbf{F}_A$ and $\mathbf{F}_B$ (Fig. 2.37b). You should perform this construction and find that $F_A = 0.75$ lb and $F_B = 1.6$ lb.

This last example differs from the problem of the block on the incline only in that the angle $\alpha + \beta$ between the unknown forces is not 90°. Consequently $\mathbf{F}_A$ and $\mathbf{F}_B$ are not the components of $-\mathbf{F}_C$, and so, for instance, $F_A$ does not equal 2 lb $\times$ cos $\alpha$. However, components can still be used to solve this problem.

Figure 2.38 shows the three vectors $\mathbf{F}_A$, $\mathbf{F}_B$, and $\mathbf{F}_C$ added together and the components of $\mathbf{F}_B$ and $\mathbf{F}_C$ parallel and perpendicular to $\mathbf{F}_A$. It is clear from this figure that the sums of the parallel and perpendicular forces are separately equal to zero. (This fact was used in Sec. 2.3.) Thus we have

$$\mathbf{F}_{By} + \mathbf{F}_{Cy} = 0 \qquad 2.6$$

$$\mathbf{F}_A + \mathbf{F}_{Bx} + \mathbf{F}_{Cx} = 0 \qquad 2.7$$

The components of $\mathbf{F}_C$ are

$$F_{Cx} = F_C \cos \alpha = 2 \text{ lb} \times \cos 45° = 1.41 \text{ lb}$$

$$F_{Cy} = F_C \sin \alpha = 2 \text{ lb} \times \sin 45° = 1.41 \text{ lb}$$

and the components of $\mathbf{F}_B$ are

$$F_{Bx} = F_B \sin \phi$$

$$F_{By} = F_B \cos \phi$$

Since $RST$ is a right triangle, the sum of the angles $\alpha$ and $\phi + \beta$ is 90°, so

$$\phi = 90° - \alpha - \beta = 25°$$

From Eq. 2.6 and Fig. 2.38 we have

$$F_{By} = F_{Cy}$$

or

$$F_B \cos 25° = 1.41 \text{ lb}$$

which gives $F_B = 1.56$ lb. From Eq. 2.7 and Fig. 2.38 we have

$$F_{Cx} = F_A + F_{Bx}$$

or

$$F_A = 1.41 \text{ lb} - 1.56 \text{ lb} \times \sin 25° = 0.75 \text{ lb}$$

This last example shows how trigonometry can be used to solve complicated force problems. But although it gives more accurate results than the graphical method, the trigonometric method is more difficult and requires some algebra. It is best to use the graphical method for most problems, except perhaps for the very simple problem of finding the components of a force.

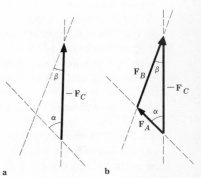

a          b

FIGURE 2.37
Graphical procedure for finding the magnitudes of the forces $\mathbf{F}_A$ and $\mathbf{F}_B$ in Fig. 2.36.

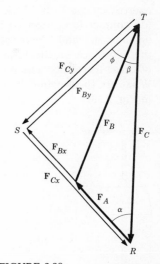

FIGURE 2.38
The vectors $\mathbf{F}_A$, $\mathbf{F}_B$, and $\mathbf{F}_C$ added together and the components of $\mathbf{F}_B$ and $\mathbf{F}_C$ parallel and perpendicular to $\mathbf{F}_A$.

## PROBLEMS

1 A 7-lb block hangs from a cord attached to a hook in the ceiling. The hook weighs 0.1 lb, and the weight of the cord can be neglected. Give the magnitude and direction of the following forces: (a) force of gravity on the block, (b) contact force exerted by the cord on the block, (c) contact force exerted by the cord on the hook, (d) force of gravity on the hook, (e) contact force exerted by the hook on the cord, and (f) contact force exerted by the block on the hook. (g) Which of these forces are action-reaction pairs? (h) What is the tension in the cord?
Ans. (a) 7 lb down; (b) 7 lb up; (c) 7 lb down; (d) 0.1 lb down; (e) 7 lb up; (f) 0 lb; (g) those in parts (c) and (e); (h) 7 lb

2 A 12-lb block sits on top of a 4-lb block that sits on a scale. Give the magnitude and directions of the following forces: (a) force of gravity on the 4-lb block, (b) contact force exerted by the scale on the 4-lb block, (c) contact force exerted by the 12-lb block on the 4-lb block, (d) contact force exerted by the 12-lb block on the scale, and (e) contact force exerted by the 4-lb block on the 12-lb block. (f) Which of these forces are action-reaction pairs?

3 What are the tensions $T_1$ and $T_2$ in the cords in Fig. 2.39? Ans. 3 and 11 lb

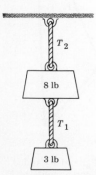

FIGURE 2.39   Problem 3.

4 What is the tension in the cord in Fig. 2.40?
5 (a) Find the tensions $T_1$, $T_2$, and $T_3$ in the three cords in Fig. 2.41. (b) What force must be

FIGURE 2.40   Problem 4.

applied to the cord by the hand to support the 50-lb weight?
Ans. (a) 75, 50, and 25 lb; (b) 25 lb

6 Find the tensions $T_1$, $T_2$, and $T_3$ in the three cords in Fig. 2.42.

7 Figure 2.43 shows a 140-lb man standing with the weights of various parts of his body indicated. (a) What is the magnitude of the contact force supporting the head and neck? (This is exerted mainly by the seventh cervical vertebra.) (b) What is the force supporting one arm? (This force is exerted by the muscles and ligaments that span the shoulder joint.) (c) What is the total force that supports the trunk at the two hip joints? (If the man is standing straight, about half of this force is exerted at each hip joint.) (d) What is the

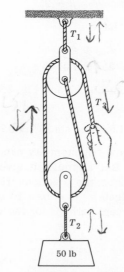

FIGURE 2.41   Problem 5.

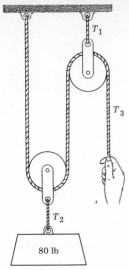

FIGURE 2.42   Problem 6.

total contact force at the knee joints? (*e*) If the man stands on one foot, what is the contact force in the knee joint on which he is standing? (*f*) What is the force in the knee joint that supports the leg that is off the floor? *Ans.* (*a*) 10 lb; (*b*) 7 lb; (*c*) 98 lb; (*d*) 124 lb; (*e*) 132 lb; (*f*) 8 lb

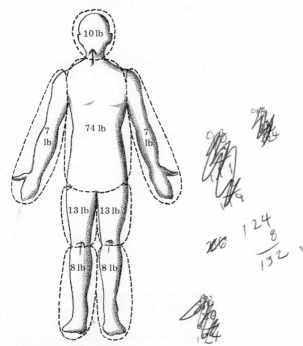

FIGURE 2.43   Problem 7.

8   What is the magnitude of the horizontal force required to push a 120-lb crate across the floor if the coefficient of static friction between the crate and the floor is 0.45?

9   A 2-lb block of wood on a wooden table just starts to slide when a horizontal force of 0.8 lb is applied to it. (*a*) What is the coefficient of friction between the block and the table? (*b*) A 5-lb weight is placed on top of the block. What is the magnitude of the horizontal force required to move the block now?
*Ans.* (*a*) 0.40; (*b*) 2.8 lb

10   A 110-lb skier requires a 6-lb push to start moving on a horizontal snow-covered surface. What is the magnitude of the push required to start a 180-lb skier moving?

11   Two 12-lb blocks are connected by a cord, as shown in Fig. 2.44. The coefficient of static friction is 0.4 between block *A* and the table and 0.3 between block *B* and the table. (*a*) What minimum force $F_a$ must be applied to block *B* to move the entire assembly? (*b*) What will be the tension *T* in the connecting cord when the assembly just starts to move?   *Ans.* (*a*) 8.4 lb; (*b*) 4.8 lb

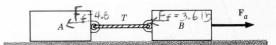

FIGURE 2.44   Problem 11.

12   A 20-lb block rests on top of a 10-lb block that is sitting on a table. The coefficient of static friction is 0.50 between the two blocks and 0.30 between the 10-lb block and the table. (*a*) What minimum force must be applied to the 20-lb block to slide the two blocks together on the table? (*b*) What is the maximum force that can be applied to the 20-lb block without its sliding off the 10-lb block?

13   The length of a spring increases 2.7 in. when a 3-lb weight is suspended from it. (*a*) What is the spring constant? (*b*) The spring is stretched 4.5 in. when another object is suspended from it. What is the weight of the object?   *Ans.* (*a*) 1.11 lb/in.; (*b*) 5.0 lb

14   The effective spring constant of a block of wood is $10^7$ lb/in. (*a*) How much is the block

compressed when a 10-lb weight is placed on it? (*b*) How much is the block compressed when a 1000-lb weight is placed on it?

**REMARK** This last problem shows that the deformation of a solid object, like that of a spring, varies with the force applied to it. However, the deformation is so small that it is normally unnoticed.

15 (*a*) Find the sum of $\mathbf{F}_1$ and $\mathbf{F}_2$, shown in Fig. 2.45. (*b*) Find the sum of $\mathbf{F}_2$ and $\mathbf{F}_3$. (*c*) Find the sum of $\mathbf{F}_1$, $\mathbf{F}_2$, and $\mathbf{F}_3$. (*d*) Find the components of $\mathbf{F}_1$ parallel and perpendicular to $\mathbf{F}_3$. *Ans.* (*a*) 16.4 lb; (*b*) 20.5 lb; (*c*) 11 lb; (*d*) −5.7 and 8.2 lb

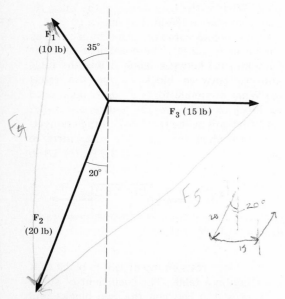

FIGURE 2.45 Problem 15.

16 A 30-lb block rests on a plane inclined 28° to the horizontal. (*a*) Find the magnitudes of the contact force and the force of friction on the block. (*b*) Given that the block remains at rest, what is the minimum value of the coefficient of friction between the block and the plane?

17 The biceps tendon in Fig. 2.46 exerts a 15-lb force $\mathbf{F}_m$ on the forearm. The arm is flexed so that this force makes an angle of 40° to the forearm. Find the components of $\mathbf{F}_m$ (*a*) parallel to the forearm (the stabilizing

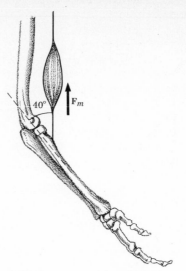

FIGURE 2.46 Problem 17.

force) and (*b*) perpendicular to the forearm (the supporting force).

*Ans.* (*a*) 11.5 lb; (*b*) 9.6 lb

18 A pencil with a rubber eraser is held at an angle of 25° to the surface of a table (Fig. 2.47). A 2-lb force is exerted downward along the length of the pencil. Neglect the weight of the pencil itself. (*a*) What are the vertical and horizontal components of this applied force? (*b*) If the coefficient of static friction between the pencil and the table is 0.40, what is the maximum force of friction the table can exert against the pencil? (*c*) Will the pencil move? (*d*) Repeat parts (*a*) and (*b*) with an angle of 70°. You should find that now the pencil will not move. What force must be applied along the pencil in order to get it to move? Experiment with a real pencil.

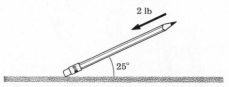

FIGURE 2.47 Problem 18.

19 A method for determining the coefficient of friction $\mu_s$ between a block and a surface is to tilt the surface until the block just starts to slide. Show that the angle $\theta$ between the inclined surface and the horizontal when the

block starts to slide is related to $\mu_s$ by $\mu_s = \tan \theta$.

20 Figure 2.48 shows an elastic cord attached to two back teeth and stretched across a front tooth. The purpose of this arrangement is to apply a force **F** to the front tooth. (The figure has been simplified by running the cord straight from the front tooth to the back teeth.) If the tension in the cord is 0.25 lb, what· is the magnitude and direction of the force **F** applied to the front tooth?

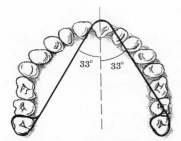

FIGURE 2.48    Problem 20.

21 Figure 2.49 shows the shape of the quadriceps tendon as it passes across the patella (knee-cap). If the tension $T$ in the tendon is 300 lb, what is (a) the magnitude and (b) direction of the contact force $\mathbf{F}_c$ exerted on the patella by the femur?          *Ans.* (a) 314 lb; (b) 21.5°

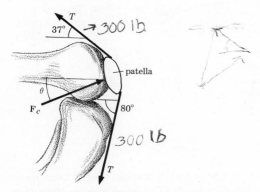

FIGURE 2.49    Problem 21.
[*After M. Williams and H. R. Lissner, "Biomechanics of Human Motion," W. B. Saunders Company, Philadelphia, 1962; used by permission.*]

22 The hip abduction muscle, which connects the hip to the femur, consists of three independent muscles acting at different angles. Figure 2.50 shows the results of measurements of the

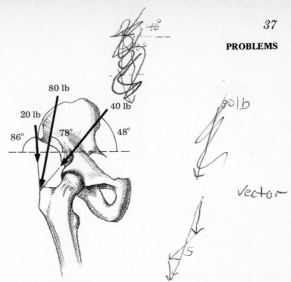

FIGURE 2.50    Problem 22.
[*After M. Williams and H. R. Lissner, "Biomechanics of Human Motion," W. B. Saunders Company, Philadelphia, 1962; used by permission.*]

force exerted separately by each muscle. Find the total force exerted by the three muscles together.

23 Find the total force applied to the patient's head by the traction device shown in Fig. 2.51.
          *Ans.* 5.6 lb

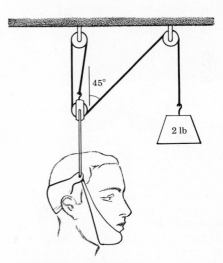

FIGURE 2.51    Problem 23.
[*After M. Williams and H. R. Lissner, "Biomechanics of Human Motion," W. B. Saunders Company, Philadelphia, 1962; used by permission.*]

24 Find the angle $\theta$ and the tension $T$ of the cord supporting the pulley in Fig. 2.52.

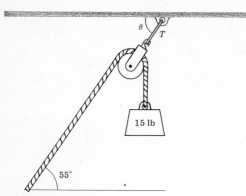

FIGURE 2.52    Problem 24.

25  Find the force that the traction device in Fig.
2.53 exerts on the foot.           *Ans.* 9.2 lb

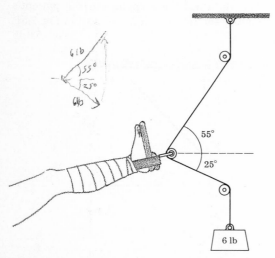

FIGURE 2.53    Problem 25.

26  A 12-lb weight is suspended by two spring
scales as shown in Fig. 2.54.  One scale registers 10 lb and is inclined 35° to the vertical.
Find the reading on the other scale and the
angle $\theta$ it makes with the vertical.

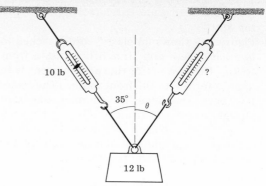

FIGURE 2.54    Problem 26.

27  Figure 2.55 shows a Russell traction apparatus
for femoral fixation.  (*a*) Find the total force
$\mathbf{F}_a$ applied to the leg by this apparatus when
an 8-lb weight $W$ is suspended from it.  (*b*) If
the leg weighs 8 lb, what is the force $\mathbf{F}_a + \mathbf{F}_g$
on the leg?  (*c*) What is the force $\mathbf{R}_c$ exerted
on the femur by the lower leg.
*Ans.* (*a*) 16 lb; (*b*) 13.9 lb; (*c*) 13.9 lb; note that
$\mathbf{R}_c$ is horizontal

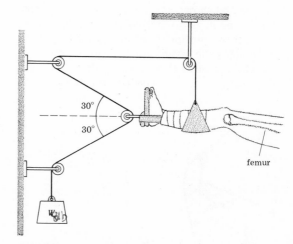

FIGURE 2.55    Problem 27.

28  Figure 2.56 shows a student's head bent over
his physics book.  The head weighs 10 lb and
is supported by the muscle force $\mathbf{F}_m$ exerted
by the neck extensor muscles and by the contact force $\mathbf{F}_c$ exerted at the atlantooccipetal
joint.  Given that the magnitude of $\mathbf{F}_m$ is 12 lb
and that it is directed 35° below the horizontal, find (*a*) the magnitude and (*b*) the direction of $\mathbf{F}_c$.

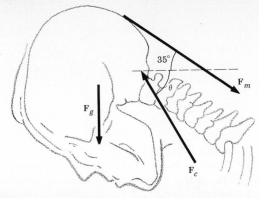

FIGURE 2.56   Problem 28.
[*After M. Williams and H. R. Lissner, "Biomechanics of Human Motion," W. B. Saunders Company, Philadelphia, 1962; used by permission.*]

## BIBLIOGRAPHY

STEINDLER, ARTHUR: "Kinesiology of the Human Body," Charles C Thomas, Publisher, Springfield, Ill., 1970. A detailed and authoritative treatment of the physics and physiology of human motion.

WELLS, KATHARINE, and JANET WESSEL: "Kinesiology," 5th ed., W. B. Saunders Company, Philadelphia, 1971. A textbook for physical therapists.

WILLIAMS, MARIAN, and HERBERT R. LISSNER: "Biomechanics of Human Motion," W. B. Saunders Company, Philadelphia, 1962. The principles of mechanics applied to a variety of problems in functional anatomy and treatment. The book is primarily for students of physical therapy and physical education.

25)

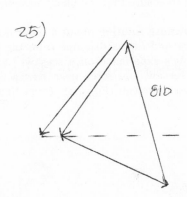

810

# CHAPTER 3 torque

Newton's first law of motion (Property 5, Sec. 2.2) is a necessary condition for an object to be in equilibrium. In the last chapter it was used to calculate some of the forces in a physical situation. In this chapter a second necessary condition for equilibrium (Property 6) is introduced and used to determine more of the forces. Properties 5 and 6 together are the necessary and sufficient conditions for an object to be in equilibrium.

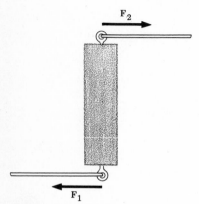

**FIGURE 3.1**
Two equal and opposite forces acting on a block.

## 3.1 PROPERTY 6 OF FORCE

It was emphasized in Sec. 2.1 that Property 5 is only a necessary condition for an object to be in equilibrium. For example, consider the two forces $F_1$ and $F_2$ acting on the block in Fig. 3.1. Even if $F_2 = -F_1$, so that the total force on the block is zero, the block will move. In fact it will rotate. The condition $F_1 + F_2 = 0$ only ensures that one point of the block (its center of gravity) remains at rest. A second condition is required to ensure that the block does not start to rotate about this point.

The tendency of a force to cause rotation about a point depends on the magnitude of the force and on its distance from the point. This fact is in accord with one's experience on a seesaw. When a child sits on each end of a seesaw, the force each exerts on the board tends to rotate it in opposite senses (Fig. 3.2). From Newton's

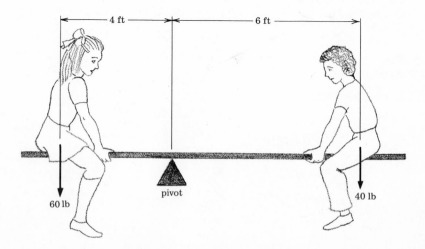

**FIGURE 3.2**
A 40-lb and a 60-lb girl balanced on a seesaw.

first and third laws (Properties 5 and 3) and the discussion in Sec. 2.3, we know that if the 40-lb boy on the right is at rest, he exerts a downward force of 40 lb on the board. This force is the reaction to the contact force that the board exerts on the boy, and it tends to rotate the board clockwise. Likewise, the 60-lb force the 60-lb girl on the left exerts on the board tends to rotate it counterclockwise. In spite of the inequality of these forces, the seesaw can be balanced (in equilibrium) if the boy sits farther from the pivot than the girl.

The rule is that the seesaw balances if the force the boy exerts times his distance from the pivot is equal to the force the girl exerts times her distance from the pivot. Thus if the boy sits 6 ft from the pivot, he can balance the girl if she sits 4 ft from the pivot, since

$$40 \text{ lb} \times 6 \text{ ft} = 240 \text{ lb-ft} = 60 \text{ lb} \times 4 \text{ ft}$$

In order to generalize this rule to more complex situations, the concept of *torque* is introduced. The torque† $\tau$ exerted by a force **F** about a point $O$ is equal to the magnitude of **F** times its perpendicular distance $d$ from $O$:

$$\tau = Fd \qquad \qquad 3.1$$

The sign of $\tau$ is taken to be positive if **F** tends to produce counterclockwise rotation about $O$ and negative if **F** tends to produce clockwise rotation. Torque is a quantitative measure of the tendency of a force to produce rotation about a point. Its unit is the pound-foot (lb-ft).

There are two very important characteristics of torque: (1) the magnitude and sign of the torque produced by a given force depend on the point $O$ about which it is calculated, and (2) the distance $d$ in Eq. 3.1 is the perpendicular distance from the point $O$ to the *line of action* of the force. The line of action is the straight line in the direction of the force passing through the point where the force is applied.

These characteristics can be illustrated by considering the torques exerted by a 5-lb weight held by a man's outstretched arm (Fig. 3.3). The weight exerts a 5-lb contact force $F_c$ downward on the hand, and so the line of action of the force is a vertical line passing through the hand. The perpendicular distance from the wrist (point $O''$) to this line is 3 in., so the torque exerted about the wrist is

$$\tau_c'' = -5 \text{ lb} \times 3 \text{ in.} = -5 \text{ lb} \times 0.25 \text{ ft} = -1.25 \text{ lb-ft}$$

The sign is negative because $F_c$ tends to rotate the hand clockwise about the wrist.

---

† $\tau$ is the Greek letter tau.

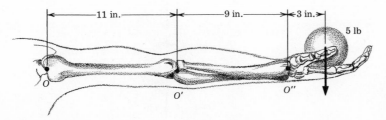

FIGURE 3.3
A 5-lb weight held by an outstretched arm.

The torque exerted by this same force about the elbow (point $O'$) is

$$\tau'_c = -5 \text{ lb} \times 12 \text{ in.} = -5 \text{ lb} \times 1 \text{ ft} = -5 \text{ lb-ft}$$

because the perpendicular distance from $O'$ to the line of action of $\mathbf{F}_c$ is 12 in. Again the sign is negative because $\mathbf{F}_c$ tends to rotate the forearm clockwise about the elbow. Similarly, the torque about the shoulder (point $O$) is $-9.6$ lb-ft. The value of the torque thus depends on the point about which it is calculated. The physical significance of this is that the tendency of a force to produce rotation about a point increases with the perpendicular distance from the point to the force.

When the arm is held at 30° to the body (Fig. 3.4), the torques about the wrist, elbow, and shoulder are different from the torques when the arm is horizontal. This is because the perpendicular distances from these points to the line of action of the force are no longer the distances measured along the arm. To find the torque about the elbow, for instance, the perpendicular distance $d'$ from $O'$ to the force must be found. This can be done either by making a scale drawing and measuring $d'$ or by using a little trigonometry. Figure 3.5 shows the important geometrical relations without the nonessential anatomical detail. The solid vertical line represents the body, and the parallel dotted line represents the line of action of the force. The arm is represented by the line $OH$, which is inclined 30° to the body and the line of action.

The line $O'H$ from the elbow to the hand is 12 in., and it is the hypotenuse of the right triangle $HP'O'$. The distance $d'$ is the side of this triangle opposite the 30° angle, so

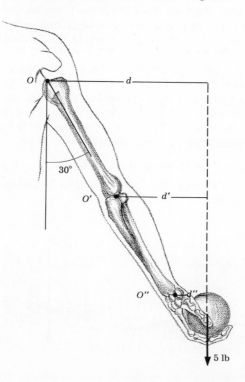

FIGURE 3.4
A 5-lb weight held by an arm at 30° to the vertical.

$$\sin 30° = \frac{d'}{O'H} = \frac{d'}{12 \text{ in.}}$$

Therefore the distance $d'$ is

$$d' = 12 \text{ in.} \times \sin 30° = 6 \text{ in.}$$

and the torque produced by $\mathbf{F}_c$ about $O'$ is

$$\tau'_c = -5 \text{ lb} \times 6 \text{ in.} = -2.5 \text{ lb-ft}$$

The torque about $O'$ is less now than when the arm was horizontal because the perpendicular distance is less. The torques about $O$ and $O''$ can be calculated similarly.

The importance of torque in mechanics is stated in Property 6 of force.

**Property 6**  *For an object to remain at rest, i.e., to be in equilibrium, it is necessary that the sum of the torques produced by all the forces acting on the object be zero.* Furthermore, if the vector sum of the forces acting on an object is zero and the sum of their torques is zero, the object will be in equilibrium. That is, the necessary and sufficient conditions for an object to be in equilibrium are that the sum of all the forces acting on it be zero (Property 5) and the sum of all the torques acting on it be zero (Property 6).

In applying Property 6 all the torques must be calculated about the same point. However, if the object is in equilibrium, it does not matter where this point is located.

For example, consider again the children on the seesaw, and assume, as before, that the board is in equilibrium (Fig. 3.2). To apply Properties 5 and 6 one must first determine what the object is and what the forces on it are. In this case the object is the board, and the forces are the contact forces exerted on it by the children and the pivot (Fig. 3.6). The torque about the pivot (point $O$) produced by the force $\mathbf{F}_1$ (exerted by the boy on the board) is

$$\tau_1 = -F_1 \times 6 \text{ ft} = -40 \text{ lb} \times 6 \text{ ft} = -240 \text{ lb-ft}$$

The torque about $O$ produced by the force $\mathbf{F}_2$ (exerted by the girl on the board) is

$$\tau_2 = F_2 \times 4 \text{ ft} = 60 \text{ lb} \times 4 \text{ ft} = 240 \text{ lb-ft}$$

The torque about $O$ produced by the force $\mathbf{F}_c$ (exerted on the board by the pivot) is zero, since the line of action of this force passes through $O$. That is, the perpendicular distance from $O$ to the line of action of $\mathbf{F}_c$ is zero, so

$$\tau_c = F_c \times 0 = 0$$

The sum of these three torques is clearly zero, as required for the board to be in equilibrium.

The only virtue in calculating the torques about $O$ is that knowledge of $\mathbf{F}_c$ is not needed. However, from the first law (Property 5) we know that $\mathbf{F}_c = -(\mathbf{F}_1 + \mathbf{F}_2)$, so that $\mathbf{F}_c$ is a 100-lb force directed upward. Therefore it is just as easy to calculate the torques about any other point. For instance, about the point $O'$, where the girl sits, the torques are

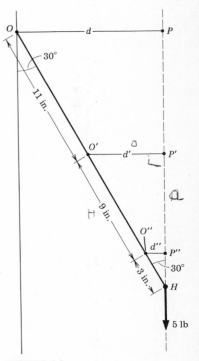

FIGURE 3.5
The geometrical relations of the arm in Fig. 3.4.

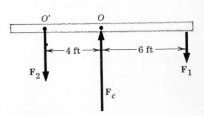

FIGURE 3.6
The forces on the seesaw in Fig. 3.2.

$$\tau_1' = -40 \text{ lb} \times 10 \text{ ft} = -400 \text{ lb-ft}$$
$$\tau_2' = 60 \text{ lb} \times 0 = 0$$
$$\tau_c' = 100 \text{ lb} \times 4 \text{ ft} = 400 \text{ lb-ft}$$

and again their sum is zero. Thus, while the individual torques change when the point about which they are calculated changes, the sum of the torques is zero regardless of the point chosen.

**REMARK** Be sure you understand what has just been said. To apply Property 6 in an equilibrium situation all the torques must be calculated about the same point, but any point can be used. This remarkable fact is proved in Appendix III.

## 3.2 CENTER OF GRAVITY

The problem of calculating the torque $\tau_g$ produced by the force of gravity on an extended object requires special consideration because gravity acts on every point in the object. Thus in the case of the outstretched arm in Fig. 3.7, there are gravitational forces on the hand, the bones of the wrist, the lower arm, and, in fact, on every cell and atom in the arm. Each of these forces has its own line of action and produces its own torque. The sum of all these forces is the total force of gravity $\mathbf{F}_g$ on the arm, and the sum of these torques is the total torque $\tau_g$ due to gravity.

In every such extended object there is a point, called the *center of gravity*, where, for the purpose of calculating $\tau_g$, the total force of gravity $\mathbf{F}_g$ is considered to act. For instance, the center of gravity (cg) of the outstretched arm in Fig. 3.7 is located near the elbow, 11 in. from the shoulder joint (point $O$). Thus, the distance $d$ from $O$ to the line of action of $\mathbf{F}_g$ is 11 in., and so if the arm weighs 7 lb, the torque about $O$ produced by the force of gravity on the arm is

$$\tau_g = -F_g d = -7 \text{ lb} \times 11 \text{ in.} = -6.4 \text{ lb-ft}$$

**REMARK** The torques calculated in Sec. 3.1 were all produced by contact forces acting on an object. Here the torque produced by the force of gravity on the object itself has been calculated.

For objects of very simple shape the center of gravity can be easily calculated. Figure 3.8 shows a 2-ft bar (of negligible weight) with a 5-lb weight attached to one end and a 10-lb weight attached to the other. The torque about the point $O$ produced by the force of gravity

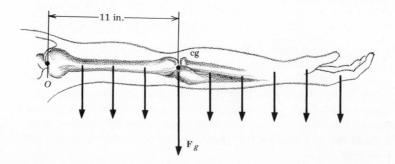

FIGURE 3.7
The force of gravity on an outstretched arm.

on the 5-lb weight is 2.5 lb-ft, and the torque about $O$ produced by the force of gravity on the 10-lb weight is 25 lb-ft. Thus the total torque about $O$ is 27.5 lb-ft, and it is equal, by definition, to the total force of gravity (15 lb) times the perpendicular distance $d$ from $O$ to the vertical line that passes through the center of gravity:

$$\tau_g = 27.5 \text{ lb-ft} = 15 \text{ lb} \times d$$

Therefore the center of gravity of the two weights lies somewhere on the vertical line $AA'$ located 27.5 lb-ft/15 lb = 1.83 ft to the left of $O$ (1.33 ft to the left of the 5-lb weight).

A second calculation is required to locate the center of gravity on the line $AA'$. For instance, we can repeat the above calculation with the bar and weights inclined 30° to the vertical, as shown in Fig. 3.9. The torque about $O$ produced by the force of gravity on the 10-lb weight is still 25 lb-ft, but the torque produced by the 5-lb weight is now 5 lb × 1.5 ft = 7.5 lb-ft. The center of gravity thus lies somewhere on the vertical line $BB'$ located 32.5 lb-ft/15 lb = 2.16 ft to the left of $O$, and so it is located at the intersection of lines $AA'$ and $BB'$. If the bar and weights are inclined at any other angle, the line on which the center of gravity lies will be found to intersect the bar at this same point. The center of gravity is a point fixed with respect to an object. As the object moves, the center of gravity moves with it.

For objects of more complicated shape, e.g., an arm, an automobile, or a human body, the center of gravity is found experimentally. Methods for doing this are explained in Sec. 3.5. These methods are important in biomechanics because knowledge of the position of the center of gravity of each body segment is required to understand the mechanics of body motion.

A number of features of the center of gravity follow directly from its definition and the examples given. These are summarized for convenience.

1 *The force of gravity on an object produces zero torque about the center of gravity of the object.* This is obvious since the line of action of the force of gravity passes through the center of gravity, by definition, so the distance from the center of gravity to this line is zero.

2 *The center of gravity of a rigid object is the balance point.* If a single support is placed directly under the center of gravity of an object (Fig. 3.10), the contact force $\mathbf{F}_c$ it exerts on the object equals $-\mathbf{F}_g$, so the total force on the object is zero. Furthermore, both $\mathbf{F}_g$ and $\mathbf{F}_c$ produce zero torques about the center of gravity since their lines of action pass through it. Consequently the total torque about the center of gravity is zero, and so the object is balanced (in equilibrium). (Note that since the object is in equilibrium, the total torque about any other point is also zero.)

3 *If an object is divided into two pieces by cutting it through the center of gravity, the two pieces will not necessarily weigh the same.* This is clear from the example of the bar and weights in Fig. 3.10, for which the two pieces weigh 5 and 10 lb, respectively. The small weight far from the center of gravity balances the large weight close to the center of gravity. For an object of nearly uniform density, on the other hand, the center of gravity is near the geometrical center.

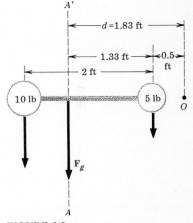

FIGURE 3.8
The center of gravity of the two weights lies somewhere on the line $AA'$.

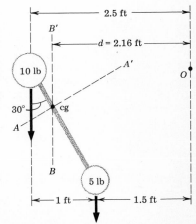

FIGURE 3.9
The center of gravity of the two weights is located at the intersection of lines $AA'$ and $BB'$.

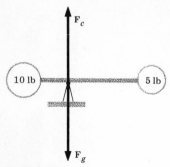

**FIGURE 3.10**
An object balanced on a single support placed directly under its center of gravity.

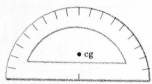

**FIGURE 3.11**
The center of gravity of a protractor lies in the hole.

*4 For a rigid object the center of gravity is a fixed point relative to the object, though it does not necessarily lie in the object itself.* The center of gravity of the bar and weights is a fixed point on the bar, and it does not change its position on the bar when the bar moves. The center of gravity of a protractor (Fig. 3.11) lies in the hole, as will be demonstrated in Sec. 3.5.

*5 For a flexible object, such as the human body, the position of the center of gravity relative to the object changes as the object changes shape.* The center of gravity of a man standing erect is located at the level of the second sacral vertebra on a vertical line touching the floor about 3 cm in front of the ankle joint (Fig. 3.12). If the man raises his arms up over his head, the center of gravity will rise several centimeters. During a jackknife dive (Fig. 3.13) the center of gravity lies outside the body altogether. The ability to change the position of the center of gravity of the body by moving body segments is of critical importance in maintaining balance while walking and in the successful performance of many athletic feats.

### 3.3   BALANCE

For an object to be in equilibrium, both the sum of the forces and the sum of the torques acting on the object must separately be zero. If the total torque is not zero, the object is *unbalanced* and will rotate in the sense of the total nonzero torque acting on it. An object in contact with a solid surface becomes unbalanced when the line of action of the total contact force exerted on it by the surface no longer passes through its center of gravity.

For example, consider the book on the table in Fig. 3.14. The contact force $\mathbf{F}_c$ the table exerts on the book is distributed over the entire area of contact between the book and the table, but like the force of gravity, the total contact force can be considered to act at a single point somewhere inside this area. As long as the center of gravity of the book is over the area of contact, the point of application of $\mathbf{F}_c$ will lie directly under it (Fig. 3.14). In this circumstance the total torque and the total force on the book are both zero, so the book is in equilibrium.

As the book is moved farther off the table, the point of application of $\mathbf{F}_c$ moves toward the edge of the table in order to remain under the center of gravity. However, since $\mathbf{F}_c$ is exerted by the table itself, its point of application cannot move beyond the edge of the table. When the center of gravity is moved past the edge of the table (Fig. 3.15), the contact force remains at the edge and the total torque on the book is no longer zero. In fact, the total torque about point $O$ is $-F_g d$. This is a clockwise torque, and it rotates the book clockwise off the table.

This example illustrates the *principle of balance: If $\mathbf{F}_c$ and $\mathbf{F}_g$ are the only forces acting on an object, the object will be balanced if and only if its center of gravity is located over the area of support.* This principle follows directly from Properties 5 and 6 of force and the nature of contact forces. It is emphasized here because of its great importance for the proper understanding of body balance.

The principle of balance requires that the center of gravity of the standing body lie on a vertical line that passes somewhere inside the area of support defined by the position of the feet (Fig. 3.16). When a person bends over to touch his toes without bending his knees, his center of gravity tends to move forward, beyond this area of contact. To prevent this, his legs and buttocks move backward, so that the body remains balanced over the feet (Fig. 3.17). The exercise is impossible without this backward motion of the lower extremities. You can demonstrate this by trying to touch your toes from a position with your heels and back against a wall. The wall prevents the body from keeping the center of gravity over the area of contact so that balance cannot be maintained.

When a person stands on his toes, his center of gravity must be brought forward over the narrow area of support. Because the area is so small, it is difficult to maintain balance in this position. Furthermore, if you stand with your feet on either side of an open door and your nose touching the edge of the door, you cannot rise on your toes. The door prevents your center of gravity from moving forward over the area of contact between your toes.

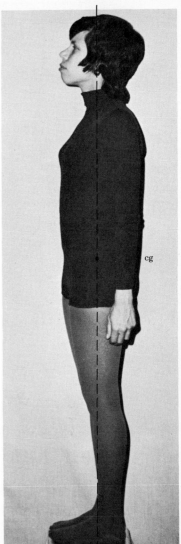

FIGURE 3.12
The center of gravity of a person standing erect lies on a vertical line that touches the floor 3 cm in front of the ankle joint.

FIGURE 3.13
During a jackknife dive, the center of gravity lies outside the body. [*United Press International photo.*]

48

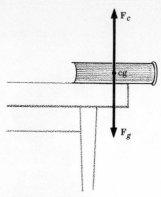

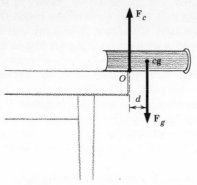

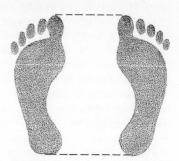

FIGURE 3.14
The forces on a book resting on a table. As long as the center of gravity of the book is above the table, the line of action of the total contact force passes through it.

FIGURE 3.15
The line of action of the total contact force on the book remains at the edge of the table when the center of gravity of the book is moved past the table edge.

FIGURE 3.16
Area of support defined by the position of the feet.

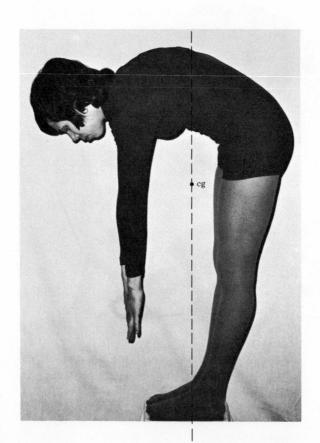

FIGURE 3.17
When a person bends over, the legs and buttocks move backward to keep the center of gravity over the area of support.

In an erect posture the center of gravity of the body normally lies on a line that passes about 3 cm in front of the ankle joint. From the principle of balance this implies that $\mathbf{F}_g$ and the contact force $\mathbf{F}_c$ at the ankle joint are not the only forces on the body above the ankle. A third force is required to maintain balance and prevent the body from rotating forward about the ankle joint. This force is applied by the tendo Achillis muscle in each leg, which attaches at the ankle. The centers of gravity of most segments of the body do not lie above their supporting joints, so that muscle forces are required for equilibrium. This is discussed more fully in the next section.

The problem of maintaining balance while walking is even greater than while standing erect. As one foot is lifted off the ground, the center of gravity of the body must move over the supporting foot. This requires the whole body to move sideways. While walking, the body continually rocks from side to side to keep the center of gravity over a constantly shifting area of support. This is easily observed by having someone walk toward you and noting the side-to-side motion of his body relative to a line on the wall in back of him.

In practice the principle of balance is not sufficient to guarantee equilibrium. For instance, it might be possible to balance a ruler upright on one end momentarily (Fig. 3.18). However, because its center of gravity is so high above a very small area of support, any small vibration of the table causes the center of gravity to move outside the support area. As soon as this happens, the torque on the ruler causes it to fall over (Fig. 3.18). Such an equilibrium, which can be permanently destroyed by a small disturbance, is said to be *unstable*. On the other hand, a box sitting on a table is *stable* because if it is tipped slightly so that it is no longer in equilibrium, the total torque on it rotates it back to its original equilibrium position (Fig. 3.19).

Good stability is obtained by having the center of gravity of an object situated low over a large area of support. For a four-legged animal (quadruped) the area of support is the large area between the four legs, which makes the animal very stable. A man standing erect has a relatively small support area and is mechanically not very stable. A football guard assumes a position (Fig. 3.20) with a low center of gravity and large support area to increase his stability against sudden upset.

Great stability is achieved if the center of gravity is actually below the area of support. The wooden figure in Fig. 3.21 is a common toy that marvelously rights itself no matter how far over it is tipped. The figure carries heavy weights that reach down below its support and cause the center of gravity of the whole figure to be below the support. When the figure is tipped forward, the total torque on it always acts to rotate it back to its equilibrium position.

Mammals in general stand on four legs (quadrupedalism). The two-legged posture of man (bipedalism) has evolved only in the last million years or so from quadrupedal origins. This has required a number of makeshift changes in the human anatomy to take care of the many physical difficulties associated with this unmammalian posture. In addition to the tapered spinal column already mentioned (Sec. 2.3), the extensor muscles of the back, legs, and hips are enlarged to hold the trunk upright, and the knee has the unique feature (shared

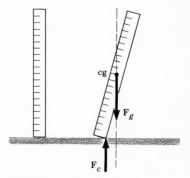

FIGURE 3.18
An example of unstable equilibrium. A ruler, because of its small area of support, topples over when it is slightly disturbed.

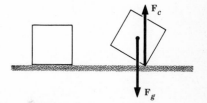

FIGURE 3.19
An example of stable equilibrium. The box, because of its large area of support, returns to its equilibrium position when it is slightly disturbed.

FIGURE 3.20
The position of a football guard is very stable.

only with the elephant) of locking the leg when it is fully extended. These changes are often inadequate for the purposes of bipedalism, and the result is an animal that suffers numerous ailments, e.g., back trouble, because of its peculiar posture.

Birds have been bipedal for over a hundred million years and are in some ways much better adapted for this mode of support than man. For instance, in man the center of gravity of the head, arms, and trunk is located in the chest, high above the hips that support it, whereas in birds the center of gravity of the head, wings, and body is below the hips (Fig. 3.22). Thus a bird hangs from its hips in a very stable configuration whereas a man balances precariously over his.

Because of the poor stability of the erect human posture, a complex control system is required to maintain balance. While standing, an automatic neuromuscular mechanism is constantly repositioning the center of gravity. Minute shifts in the position of the center of gravity are detected by kinesthetic receptors, and the necessary adjustments in the muscles of the body are made to bring the center of gravity back over the center of the support area. This constant realigning of the center of gravity causes it to move in an approximately figure-eight pattern about a vertical line through the center of the area of support.

## 3.4 EXAMPLES INVOLVING TORQUE

Property 6 is an additional condition that the forces on an object in equilibrium must satisfy. Together with Property 5, it increases the number of unknown forces that can be determined in a problem. In this section these properties are applied to some problems in bio-mechanics. However, the techniques used here are quite general and can be used just as well to solve equilibrium problems in engineering and mechanics. Examples of such problems will be found in the problems at the end of the chapter.

As a first example we investigate the forces on a 180-lb man who is standing erect. Let $\mathbf{F}_R$ and $\mathbf{F}_L$ be the upward contact forces on his right and left foot, respectively, and assume that he is standing

**FIGURE 3.21**
The "bird" always returns to the upright position because its center of gravity is below its area of support.

straight so that his center of gravity lies on a line that passes midway between his feet (Fig. 3.23). From the first law (Property 5) the sum of the forces on the man is zero,

$$\mathbf{F}_R + \mathbf{F}_L + \mathbf{F}_g = 0$$

and so, since these forces are parallel, their magnitudes satisfy the relation

$$F_R + F_L = 180 \text{ lb} \qquad 3.2$$

To find the individual magnitudes $F_R$ and $F_L$ we use Property 6. The torques may be taken about any point, but in this case it is convenient to take them about the point $O$, where $F_L$ acts on the left foot. If the man's feet are 1 ft apart, the torques about this point are

$$\tau_L = F_L \times 0 = 0$$
$$\tau_R = -F_R \times 1 \text{ ft}$$
$$\tau_g = F_g \times \tfrac{1}{2} \text{ ft} = 90 \text{ lb-ft}$$

Since, by Property 6, their sum is zero, we have

$$-F_R \times 1 \text{ ft} + 90 \text{ lb-ft} = 0$$

so

$$F_R = 90 \text{ lb}$$

Putting this result into Eq. 3.2 gives

$$90 \text{ lb} + F_L = 180 \text{ lb}$$

so

$$F_L = 90 \text{ lb}$$

**REMARK** The point $O$ is convenient in this case because one of the two unknowns passes through it. This unknown is thus eliminated from the first part of the problem since its torque about $O$ is zero. A little bit more algebra is needed if the torques are taken about the point $O'$, which lies on the line of action of $F_g$. The torques about $O'$ are

$$\tau'_L = F_L \times \tfrac{1}{2} \text{ ft}$$
$$\tau'_R = -F_R \times \tfrac{1}{2} \text{ ft}$$
$$\tau'_g = F_g \times 0 = 0$$

and now both unknown forces produce nonzero torques. The sum of all these torques is still zero, so

$$\tfrac{1}{2}F_L - \tfrac{1}{2}F_R = 0 \qquad 3.3$$

Equations 3.2 and 3.3 must be solved simultaneously to find the two unknown forces. This is not very difficult, and the result is $F_L = F_R = 90$ lb, as before. The final values are the same using either $O$ or $O'$, but the algebra is a little simpler with $O$.

What happens if one of the man's legs is injured so that it cannot take its full share of the weight? For instance, suppose the man's left ankle is injured so he cannot have a contact force on it greater than 45 lb without great pain. The total force on him is still zero,

FIGURE 3.22
The center of gravity of a bird's body is below its hips.

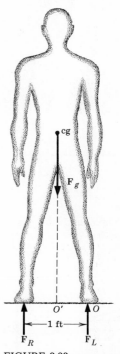

FIGURE 3.23
The forces on a man standing erect, with his center of gravity midway between his feet.

so with $F_L = 45$ lb, Eq. 3.2 requires that $F_R = 135$ lb. If the man stood straight, the torques about $O$ would now be

$$\tau_L = 45 \text{ lb} \times 0 = 0$$
$$\tau_R = -135 \text{ lb} \times 1 \text{ ft} = -135 \text{ lb-ft}$$
$$\tau_g = 180 \text{ lb} \times \tfrac{1}{2} \text{ ft} = 90 \text{ lb-ft}$$

The sum of these torques is $-45$ lb-ft, and so, by Property 6, he would not be in equilibrium. This is impossible. In reality the man would not stand straight but would lean to shift his center of gravity toward the right foot (Fig. 3.24).

To find where the center of gravity is located, the calculation is repeated, starting with the knowledge that $F_L = 45$ lb; the distance $d$ of the center of gravity to the left foot is unknown. From Eq. 3.2 we again find $F_R = 135$ lb, so the torques about $O$ are

$$\tau_L = 45 \text{ lb} \times 0 = 0$$
$$\tau_R = -135 \text{ lb} \times 1 \text{ ft} = -135 \text{ lb-ft}$$
$$\tau_g = 180 \text{ lb} \times d$$

and their sum is

$$-135 \text{ lb-ft} + 180 \text{ lb} \times d = 0$$

Thus the position of the center of gravity is

$$d = \frac{135 \text{ lb-ft}}{180 \text{ lb}} = 0.75 \text{ ft}$$

and it is shifted away from the injured foot toward the good foot. This is accomplished by bending the body toward the right and assuming a typical limping posture.

Compare this last example with the example of the seesaw in Sec. 3.1. While the circumstances are different, the basic physics is the same: in both problems an object in equilibrium is being acted upon by three parallel forces with different lines of action.

As another example we analyze once again the forces on the outstretched arm in Figs. 2.30, 2.37, 3.7, and 3.25. There are three forces acting on it: the force of gravity $\mathbf{F}_g$, the muscle force $\mathbf{F}_m$ applied by the deltoid muscle, and the contact force $\mathbf{F}_c$ exerted at the shoulder joint. The arm weighs 7 lb; both the magnitude and direction of $\mathbf{F}_g$ are known. The direction of $\mathbf{F}_m$ is found from x-ray studies of the muscle. In Sec. 2.5 we assumed, furthermore, a value for the magnitude of $\mathbf{F}_m$ and showed how the first law enables us to find the magnitude and direction of $\mathbf{F}_c$. Now we show how the magnitude of $\mathbf{F}_m$ can be calculated using Property 6.

Torques are taken about the shoulder joint (point $O$) because the unknown force $\mathbf{F}_c$ exerts zero torque about this point, i.e.,

$$\tau_c = F_c \times 0 = 0$$

The center of gravity of the arm can be determined by methods discussed in the next section. For now it is assumed to be at the elbow, 11 in. from $O$, so the torque produced by $\mathbf{F}_g$ about $O$ is

$$\tau_g = -7 \text{ lb} \times 11 \text{ in.} = -77 \text{ lb-in.}$$

(For convenience the pound-inch is used here as the unit of torque

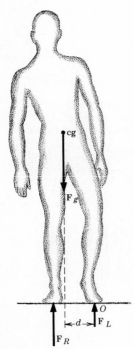

**FIGURE 3.24**
A man with an injured left ankle shifts his center of gravity toward his right foot.

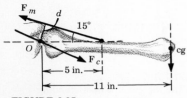

**FIGURE 3.25**
The forces on an outstretched arm.

instead of the pound-foot.) The torque exerted by $\mathbf{F}_m$ depends on its line of action. This can be estimated from x-ray studies of living subjects and from anatomical examinations. In Fig. 3.25 it is shown at an angle of 15° to the humerus and intersecting the humerus 5 in. from $O$.

The torque produced by $\mathbf{F}_m$ about $O$ is

$$\tau_m = F_m d$$

where $d$ is the perpendicular distance from $O$ to the line of action of $F_m$. Figure 3.26 shows more clearly the geometrical construction used to calculate $d$. This distance can be found either by drawing Fig. 3.26 to scale or by using the trigonometric relation

$$\sin 15° = \frac{d}{5 \text{ in.}}$$

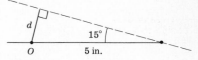

FIGURE 3.26
Geometrical construction, based on
Fig. 3.25, used to calculate $d$.

so

$$d = 5 \text{ in.} \times \sin 15° = 1.3 \text{ in.}$$

Either way one gets

$$\tau_m = F_m \times 1.3 \text{ in.}$$

Property 6 says the sum of these torques is zero, so

$$-77 \text{ lb-in.} + F_m \times 1.3 \text{ in.} = 0$$

Solving for $F_m$, we get

$$F_m = \frac{77 \text{ lb-in.}}{1.3 \text{ in.}} = 59 \text{ lb}$$

The forces $\mathbf{F}_g$ and $\mathbf{F}_m$ are now known completely, so Property 5 can be used to determine $\mathbf{F}_c$. This has already been done in Sec. 2.5, where $F_m$ was assumed to be 60 lb for simplicity. Thus Properties 5 and 6 together enable us to find three unknowns: the magnitudes of $\mathbf{F}_m$ and $\mathbf{F}_c$ and the direction of $\mathbf{F}_c$.

Suppose now a 5-lb weight is held in the hand of the outstretched arm at a distance of 23 in. from the shoulder joint (Fig. 3.3). This weight exerts a 5-lb contact force on the arm, and its torque about $O$ is $-5 \text{ lb} \times 23 \text{ in.} = -115 \text{ lb-in.}$ The total torque about $O$ now is

$$-77 \text{ lb-in.} + F_m \times 1.3 \text{ in.} - 115 \text{ lb-in.} = 0$$

so

$$F_m = \frac{192 \text{ lb-in.}}{1.3 \text{ in.}} = 148 \text{ lb}$$

Because the distance from the joint to the hand is much larger than the distance from the joint to the muscle, a force of only 5 lb exerted on the hand requires the force in the muscle to increase by 89 lb. It is impossible for an average person to hold a weight of more than 10 or 15 lb in this outstretched position.

Small changes in body posture can produce surprisingly large changes in the forces acting inside the body. As an example, we shall examine the forces involved in standing on one foot. First, consider a 200-lb man standing straight so that his center of gravity lies on a line passing midway between his feet. Each of his legs weighs 30 lb

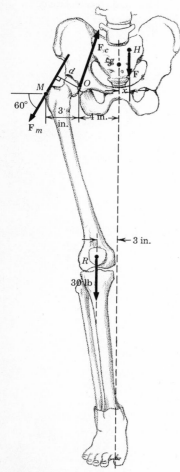

and his head, arms, and trunk (called HAT for short) together weigh 140 lb. This weight is supported by contact forces of 70 lb exerted by the femurs at each hip joint.

The man now lifts his left foot off the ground. In order to maintain balance the center of gravity must shift over to the right foot. Figure 3.27 shows it on a line that passes 3 in. from the center of gravity of the right leg (point $R$). The head, arms, trunk, and left leg (called HATL for short) can be considered as a single object with three forces acting on it: the 170-lb force of gravity $\mathbf{F}_g$, the force $\mathbf{F}_m$ exerted by the hip abductor muscles, and the contact force $\mathbf{F}_c$ exerted by the femur on the hip. These forces can be calculated in the same way as the forces on the arm.

There is one complication. The center of gravity of the HATL (point $H$) is not on a vertical line with the center of gravity of the entire body (point cg), but lies on a line a distance $x$ to the left. The total force of gravity on the body has been split into the 30-lb force on the right leg and the 170-lb force on the HATL. The sum of the torques of these two forces about cg must be zero since these forces are equivalent to the total force of gravity acting at cg. That is, $x$ is determined by the condition

$$30 \text{ lb} \times 3 \text{ in.} - 170 \text{ lb} \times x = 0$$

so

$$x = \frac{90 \text{ lb-in.}}{170 \text{ lb}} = 0.53 \text{ in.}$$

The line of action of $\mathbf{F}_m$ is known from anatomical and x-ray studies of the hip abductor group. It makes an angle of 60° to the horizontal and intersects the femur at point $M$, as shown in Fig. 3.27. The point of application of $\mathbf{F}_c$ is not known exactly, but the point must lie somewhere in the joint. For the purposes of this calculation it is taken to be point $O$, which is 3 in. to the left of $M$ and 4 in. to the right of the vertical line passing through cg. Torques are taken about $O$ because $\mathbf{F}_c$ produces zero torque about it. The perpendicular distance $d$ from $O$ to $\mathbf{F}_m$ is found by studying Fig. 3.27. It is

$$d = 3 \text{ in.} \times \sin 60° = 2.60 \text{ in.}$$

so the total torque about $O$ is

$$F_m \times 2.60 \text{ in.} - 170 \text{ lb} \times 4.53 \text{ in.} = 0$$

The force in the muscle then is

$$F_m = \frac{770 \text{ lb-in.}}{2.60 \text{ in.}} = 296 \text{ lb}$$

**FIGURE 3.27**
The forces on the HATL (head, arms, trunk, and left leg) of a 200-lb man standing on one foot. The right leg (shown) supports the HATL. The left leg is not shown. [*After M. Williams and H. R. Lissner, "Biomechanics of Human Motion," W. B. Saunders Company, Philadelphia, 1962, used by permission.*]

The contact force can now be found by using Property 5. The forces $\mathbf{F}_m$ and $\mathbf{F}_g$ are drawn head to tail (Fig. 3.28), and $\mathbf{F}_c$ is the force that, added to these, completes the triangle. The details have been discussed in Sec. 2.5. It is found that the magnitude of $\mathbf{F}_c$ is about 450 lb, or more than 6 times the contact force exerted on the hip when standing on both feet. Thus as a person walks, the contact force at the hip changes enormously as the body is alternately supported by one foot and the other, subjecting the joint to a considerable pounding. Of course the joint is designed to absorb these shocks without ill effect, but if the joint malfunctions, this pounding may produce trauma.

**REMARK**   The details of this calculation should not obscure the basic physical reason for the large value of $F_c$. Because the center of gravity of the HATL is 4.53 in. from $O$, the downward force of gravity tends to rotate it clockwise about $O$. To prevent this, the hip muscle pulls down on the other side of $O$, and since the line of action of $\mathbf{F}_m$ is only 2.55 in. from $O$, its magnitude must be larger than $F_g$. A still larger upward contact force is then required to counteract these two downward forces.

## 3.5   LOCATING THE CENTER OF GRAVITY

In order to analyze the forces on an object the position of its center of gravity must be known. There are various methods for locating it in both animate and inanimate objects. These methods make use of the special nature of the center of gravity, in particular the fact that the total force of gravity on an object produces zero torque about its center of gravity and so the center of gravity is the balance point of a rigid object (Sec. 3.2).

### Suspension

When a rigid object is suspended from a point in the object, the object hangs with its center of gravity on the vertical line that passes through the point of suspension. This is because the contact force at the point of suspension, being the only force other than gravity on the object, must also produce zero torque about the center of gravity. The center of gravity of a rigid object can thus be found by suspending it from a point and marking the vertical line through the point of suspension. Figure 3.29 shows a protractor suspended from the point $A$, together with the vertical line through this point determined with a plumb line. The object is then suspended from another point and a second vertical line is marked on it. Figure 3.30 shows the protractor suspended from point $B$. These two lines intersect at the center of gravity of the object, which in this case is located inside the hole in the protractor. The centers of gravity of various parts of the human body have been located by suspending frozen specimens taken from cadavers.

### Calculation

The center of gravity of an object composed of several parts can be calculated if the weight and the position of the center of gravity of each part are known. Consider first an object composed of two parts whose centers of gravity, $A$ and $B$, are separated by a distance $d$ (Fig. 3.31). Since the common center of gravity cg of the whole object is its balance point, it must lie on the straight line connecting $A$ and $B$. Then if $x$ is the distance from cg to $A$, the condition that cg be the balance point is

$$W_1 x - W_2 (d - x) = 0$$

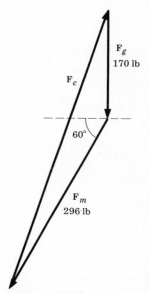

FIGURE 3.28
The vector diagram used to find $\mathbf{F}_c$.

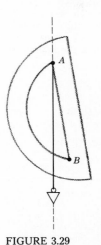

FIGURE 3.29
A protractor suspended from point $A$ hangs with its center of gravity somewhere along the vertical line defined by the plumb line.

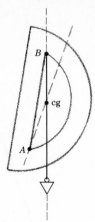

FIGURE 3.30
The protractor is suspended from a second point (B) to determine a second line on which the center of gravity must lie. The center of gravity is located at the intersection of these two lines.

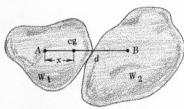

FIGURE 3.31
Locating the common center of gravity (cg) of two objects, given the positions A and B of the centers of gravity of the individual objects.

where $W_1$ and $W_2$ are the weights of the two parts. Solving this equation for $x$ gives

$$x = \frac{W_2}{W_1 + W_2} d \qquad 3.4$$

which is the general formula for finding the center of gravity of an object composed of two parts. This is a slight generalization of the method used to find the center of gravity of the bar and weights in Sec. 3.2.

As an example we shall locate the center of gravity of the body (minus the arms) of the athlete in the position shown in Fig. 3.32. The positions of the centers of gravity of the head (point $H$), trunk (point $T$), and legs (point $L$) are assumed to be known from studies on cadavers. For a 6-ft-tall athlete weighing 200 lb, the normal values for the weights of the head, trunk, and legs are $W_H = 12$ lb, $W_T = 98$ lb, and $W_L = 60$ lb.

The distance $y$ from $T$ to the common center of gravity (point $A$) of the head (12 lb) and trunk (98 lb) is first found. The distance $d$ between $H$ and $T$ is 18 in., so Eq. 3.4 determines $y$ to be

$$y = \frac{12 \text{ lb}}{110 \text{ lb}} \times 18 \text{ in.} = 2 \text{ in.}$$

Then the common center of gravity (cg) of the head and trunk (110 lb) and legs (60 lb) lies on the line between $A$ and $L$. The length $l$ of this line is determined from the distances and angles given in the figure either by making a scale drawing or noting, in this case, that $AOL$ is an equilateral triangle, so $l = 17$ in. The distance $x$ from $A$ to cg is obtained using Eq. 3.4 with the head and trunk as one part and the legs as the other. The result is

$$x = \frac{60 \text{ lb}}{170 \text{ lb}} \times 17 \text{ in.} = 6 \text{ in.}$$

so cg is located outside the body, as shown.

There is a gymnastic stunt called the half lever (or L position) in

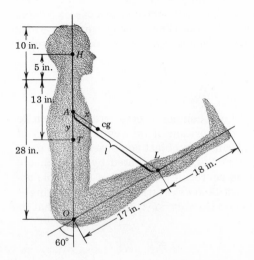

FIGURE 3.32
Locating the center of gravity of an athlete in a bent posture.

which the performer sits on the floor and raises his body on his arms, as shown in Fig. 3.33. Since the center of gravity of the head, trunk, and legs is in front of the chest, the body swings slightly backward so that the center of gravity lies directly under the shoulders. This shows how the position of the center of gravity of the body depends on posture and how it influences body movement.

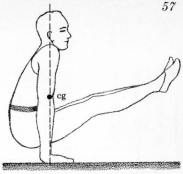

**FIGURE 3.33**
When a gymnast does a half lever, the center of gravity of his head, trunk, and legs lies on a vertical line passing through his shoulders.

## Weighing

The method of suspension is not very practical for locating the center of gravity of a living person. Instead what is done is to weigh him on a board supported on two scales (Fig. 3.34). The board rests on wedges so that the exact position of the contact forces on it are known. The system is adjusted so that when a subject lies on the board, his head and feet are even with the wedges. The position of the subject's center of gravity is determined from the readings $W_1$ and $W_2$ of the two scales, since these equal the contact forces on the scale. This has already been done in Sec. 3.4, where the center of gravity of the limping man was found from the contact forces on his feet. It is easy to show that if the distance between the wedges is $d$, the distance $x$ from the center of gravity to the scale reading $W_1$ is given by Eq. 3.4.

The front-to-back position of the center of gravity is found similarly by having the subject stand with the balls of his feet on one scale and his heels on another.

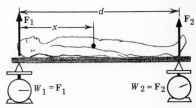

**FIGURE 3.34**
Locating the center of gravity of a person by weighing on two scales.

## PROBLEMS

1 The forearm in Fig. 3.35 is held at 90° to the upper arm, and a 15-lb weight is held in the hand. Neglect the weight of the forearm itself. (a) What is the torque about the elbow joint (point $O$) produced by the 15-lb weight? (b) What is the torque about $O$ produced by the force $\mathbf{F}_m$ exerted on the forearm by the biceps (use Property 6)? (c) What is the magnitude of $\mathbf{F}_m$?
*Ans.* (a) −195 lb-in.; (b) +195 lb-in.; (c) 130 lb

2 Repeat Prob. 1 assuming the forearm and hand together weigh 3 lb and their center of gravity is 6 in. from $O$.

3 With the forearm held horizontally, as shown in Fig. 3.36, the hand exerts an 18-lb force on the scale. Find the magnitudes of the forces $\mathbf{F}_m$ and $\mathbf{F}_c$ that the triceps muscle and the humerus exert on the forearm. (Neglect the weight of the forearm.)
*Ans.* 270 and 288 lb

4 Repeat Prob. 3 assuming that the forearm and

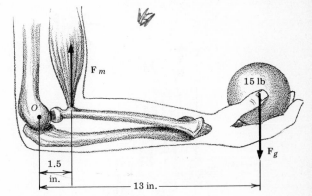

**FIGURE 3.35** Problems 1 and 2.

hand together weigh 5 lb and their center of gravity is 7 in. from $O$.

5 Young adult men can exert a maximum force of 84 lb to the test strap shown in Fig. 3.37. If the test strap is 11 in. from the elbow and the biceps muscle attaches 2 in. from the el-

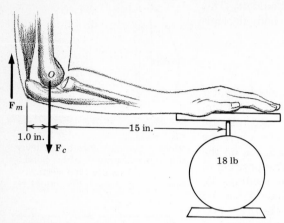

FIGURE 3.36   Problems 3 and 4.

bow, what are the magnitudes of the forces exerted by (*a*) the biceps and (*b*) the humerus?

*Ans.* (*a*) 462 lb; (*b*) 378 lb

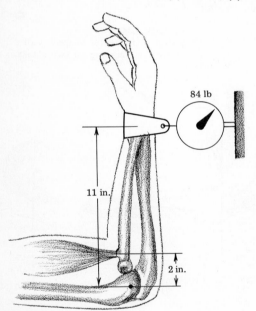

FIGURE 3.37   Problem 5.

6 Joe and Sam are carrying a 120-lb weight which rests on a 10-ft board, as shown in Fig. 3.38. The board alone weighs 25 lb, and its center of gravity is at its midpoint. The 120-lb weight is 3 ft from Joe's end of the board. What are the magnitudes of the forces Joe and Sam must exert to support this load?

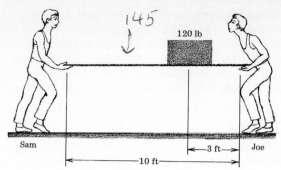

FIGURE 3.38   Problem 6.

7 Figure 3.39 shows an athlete about to do a push-up. He weighs 180 lb, and his center of gravity is located above a point $P$ on the ground 3 ft from his toes and 2 ft from his shoulders. What are the forces exerted by the floor on the athlete's hands and feet?

*Ans.* 72 and 108 lb

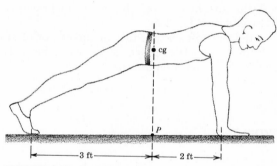

FIGURE 3.39   Problem 7.

8 A 200-lb man stands 5 ft from one end of a 20-ft scaffold (Fig. 3.40). The scaffold weighs 150 lb. What are the tensions $T_1$ and $T_2$ in the ropes supporting the scaffold?

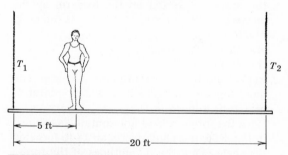

FIGURE 3.40   Problem 8.

9 A man carries an 8-ft board with one hand pushing on it with a force $\mathbf{F}_1$ at one end and the other hand lifting it with a force $\mathbf{F}_2$ applied 1 ft from this end (Fig. 3.41). The board

weighs 25 lb, and its center of gravity is at its center. Find $F_1$ and $F_2$.

Ans. 75 and 100 lb

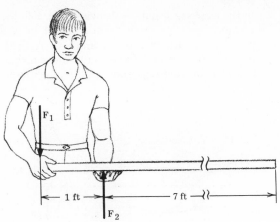

FIGURE 3.41    Problem 9.

10 Figure 3.42 shows a mobile consisting of four ornaments attached by threads to light rods. The distances (in inches) between the ornaments and the supporting threads, and the weight (in pounds) of one ornament are indicated. If the mobile is to hang balanced, as shown, what are the weights of the other ornaments? Neglect the weight of the rods and thread. (*Hint*: Find the weight of $A$ first.)

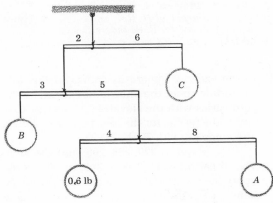

FIGURE 3.42    Problem 10.

11 The man in Fig. 3.43 is in the process of putting his canoe up on his shoulders. The canoe is 18 ft long and weighs 84 lb; its center of gravity is at its midpoint. What is the magnitude of the force $F_1$ that the man applies to the canoe while in the position shown? His wife, ignorant of physics, tries to help him by lifting the canoe at point $A$. Explain why this does not help the man. Where should she lift to be of help? (Knowledge of the angle $\theta$ is not required for this problem, but if it will help, you may assume it to be 30°.)

Ans. 63 lb

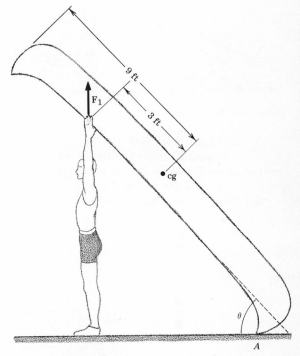

FIGURE 3.43    Problem 11.

12 Figure 3.44 shows a 9-ft board hinged at one end (point $O$) and maintained at an angle of 30° to the horizontal by a 50-lb weight suspended from a cord attached to the other end of the board. The center of gravity of the board is 4 ft from the hinged end. (*a*) What is the magnitude of the force $F_t$ exerted on the board by the cord? (*b*) What is the torque exerted by $F_t$ about $O$? (*c*) What is the weight of the board? (*d*) What is the magnitude of the contact force exerted by the hinge on the board?

13 A 12-ft ladder leans against a wall at an angle of 34° as shown in Fig. 3.45. A 200-lb painter is standing on the ladder, 3 ft from the top. (*a*) What is the torque about point $O$ exerted

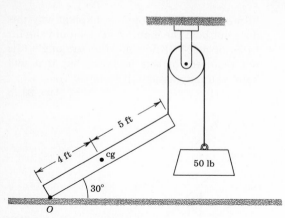

FIGURE 3.44   Problem 12.

by the painter's weight? (*b*) Assume the wall is frictionless so that the force $F_1$ exerted on the ladder by the wall is perpendicular to the wall. What is the magnitude of $F_1$? (Neglect the weight of the ladder.) Find (*c*) the magni-

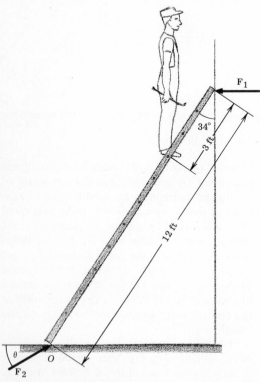

FIGURE 3.45   Problem 13.

tude and (*d*) direction of the force $F_2$ exerted by the floor on the ladder at point $O$.
*Ans.* (*a*) 1006 lb-ft; (*b*) 101 lb; (*c*) 224 lb; (*d*) 63°

14  In the exercise situation shown in Fig. 3.46 the torque about the knee joint exerted by the 20-lb weight attached to the ankle varies with the elevation of the leg. Calculate the torque for the four positions shown.

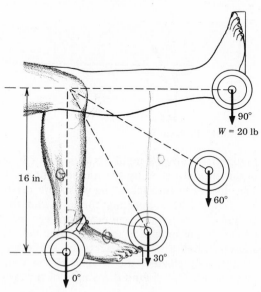

FIGURE 3.46   Problem 14.
[*After M. Williams and H. R. Lissner, "Biomechanics of Human Motion," W. B. Saunders Company, Philadelphia, 1962, used by permission.*]

15  In the exercise apparatus shown in Fig. 3.47 the torque about the knee joint exerted by the cord varies with the elevation of the leg. Calculate the torque for the four positions shown, and compare with Prob. 14.
            *Ans.* 320, 288, 206, and 104 lb-in.

16  The forearm in Fig. 3.48 is held at 50° to the upper arm, and a 15-lb weight is held in the hand. (*a*) What is the magnitude of the force exerted on the forearm by the biceps muscle? (Neglect the weight of the forearm.) (*b*) Find the magnitude of the force exerted by the elbow on the forearm.

17  Repeat Prob. 16 assuming that the forearm and hand together weigh 6 lb and their center of gravity is 8 in. from the elbow.
            *Ans.* (*a*) 144 lb; (*b*) 123 lb

18  In an erect posture the center of gravity of

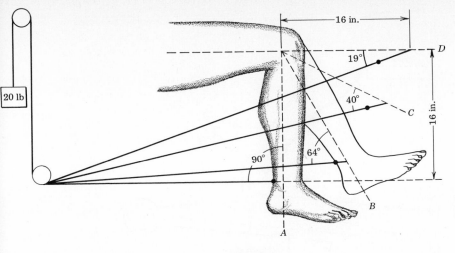

FIGURE 3.47   Problem 15.
[*After M. Williams and H. R.
Lissner, "Biomechanics of Human
Motion," W. B. Saunders
Company, Philadelphia, 1962, used
by permission.*]

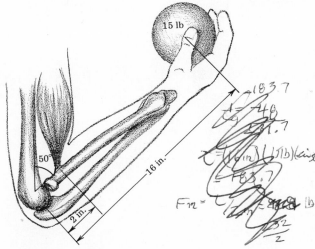

FIGURE 3.48   Problems 16 and 17.

the tension in the cord. For this problem
assume the weight $W$ is 4 lb and that the
board weighs 10 lb and is 4 ft long. (*a*) Calculate individually the torques about the hinge
(point $O$) produced by the weight of the board
and by the 4-lb weight when the board is
horizontal ($\theta = 0$). (The center of gravity of
the board can be assumed to be in the middle
of the board.) (*b*) Calculate individually the
torques about $O$ produced by the weight of the
board and by the 4-lb weight when the board
is raised so that $\theta = 26°$. (*c*) When the board
is held at 26°, what is the torque about $O$
produced by the force $\mathbf{F}_t$ exerted by the cord?

the body lies on a line which falls 1.25 in. in
front of the ankle joint (Fig. 3.49). The calf
muscle (the tendo Achillis muscle group) attaches at the ankle 1.75 in. in back of the joint
and passes up at an angle of 83°. (*a*) Find the
force $\mathbf{F}_m$ in this muscle for a 150-lb man standing erect. (Remember, each leg supports half
the man's weight.) (*b*) What is the contact
force $\mathbf{F}_c$ exerted at the ankle joint?

19  Figure 3.50 shows Storms' exercise apparatus
for quadriceps strengthening. It consists of a
board hinged at one end (point $O$) and a cable
to which a pulley is attached. The patient sits
on the edge of a table and lifts the board by
a harness attached to one foot and connected
by a cord to the pulley. A weight $W$ at the
far end of the board can be varied to change

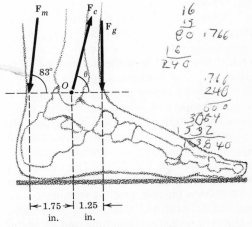

Figure 3.49   Problem 18.
[*After M. Williams and H. R. Lissner, "Biomechanics
of Human Motion," W. B. Saunders Company,
Philadelphia, 1962, used by permission.*]

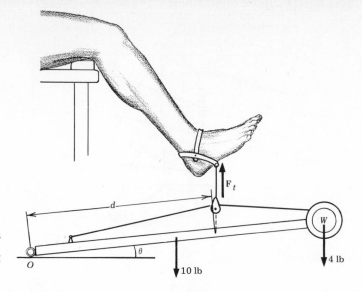

FIGURE 3.50  Problem 19.
[*After M. Williams and H. B. Lissner, "Biomechanics of Human Motion," W. B. Saunders Company, Philadelphia, 1962, used by permission.*]

Is this more or less than the torque produced by $\mathbf{F}_t$ when $\theta = 0$? (*d*) When $\theta = 26°$, the distance $d$ from the pulley to $O$ is 3 ft, measured along the board. What is $F'_t$? (*e*) What should the weight $W$ at the end of the board be for a 14-lb force to be applied to the foot when $\theta = 26°$?

*Ans.* (*a*) $-20$ and $-16$ ft-lb; (*b*) $-18$ and $-14.4$ ft-lb; (*c*) 32.4 ft-lb; (*d*) 12 lb; (*e*) 5.5 lb

**REMARK**  As the board is raised, the torque that $\mathbf{F}_t$ must produce decreases. In addition, as the leg lifts, the pulley moves away from $O$ so that $\mathbf{F}_t$ is applied farther from $O$. These two factors decrease the force that the leg must exert as it reaches its extended position, where the muscle's ability to exert a force is smallest.

20  In mountain climbing the standard method of descending a vertical wall is by *rappelling*. The climber attaches one end of a rope to a secure rock at the top of the cliff and the other end to a special buckle strapped to his waist, 0.5 ft above his center of gravity. (The buckle holds the rope and allows the climber to increase the length of the rope by feeding the free end of the rope through the buckle. In this way he is able to walk down the wall with ease, as shown in Fig. 3.51.) The climber in the figure weighs 180 lb, his center of gravity is 3.0 ft from his feet, and his rope makes an angle of 25° with the wall. Find the tension in the rope and the magnitude and direction of the surface force exerted by the wall on his feet.

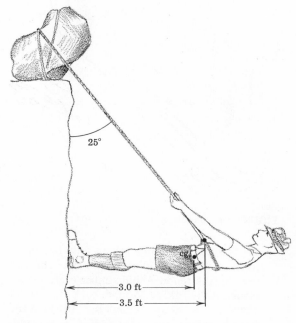

FIGURE 3.51  Problem 20.

21  Figure 3.52 shows the forces on the foot of a 200-lb man in a squatting position. Determine (*a*) the magnitude of the force $F_m$ exerted by the Achilles tendon and (*b*) the magnitude and (*c*) the direction of the contact force $\mathbf{F}_c$ exerted at the ankle joint.

*Ans.* (*a*) 127 lb; (*b*) 205 lb; (*c*) 61°

**REMARK**  Figure 3.52 shows the forces on the foot,

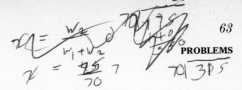

PROBLEMS

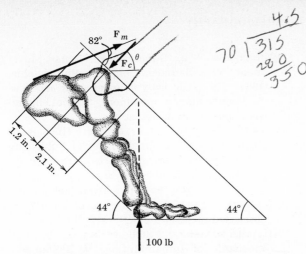

FIGURE 3.52  Problem 21.
[*After M. Williams and H. B. Lissner, "Biomechanics of Human Motion," W. B. Saunders Company, Philadelphia, 1962, used by permission.*]

whereas Fig. 3.49 shows the forces on the body above the foot.

22  The 30-ft crane in Fig. 3.53 weighs 5000 lb and is lifting a 3000-lb weight. Find the tension in the supporting cable and the force $\mathbf{F}_s$ at the pivot.

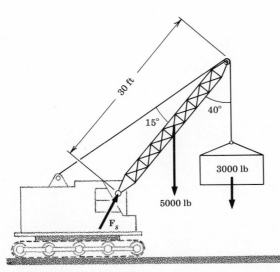

FIGURE 3.53  Problem 22.

23  An object rests on two scales 7 ft apart (Fig. 3.54). The left-hand scale reads 45 lb, and the right-hand scale reads 25 lb. (*a*) What is the weight of the object? (*b*) What is the perpendicular distance $d$ from $O$ to the vertical line passing through the center of gravity (cg) of the object?          *Ans.* (*a*) 70 lb; (*b*) 2.5 ft

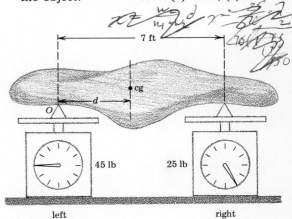

FIGURE 3.54  Problem 23.

24  Locate the center of gravity of the uniformly thick L-shaped plate in Fig. 3.55.

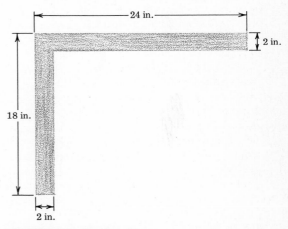

FIGURE 3.55  Problem 24.

25  Locate the position of the center of gravity of the three objects shown in Fig. 3.56.

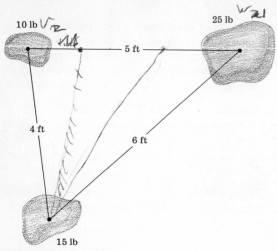

FIGURE 3.56   Problem 25.

**BIBLIOGRAPHY**

BRODER, MARION R.: "Efficiency of Human Movement," W. B. Saunders Company, Philadelphia, 1960. The basic laws of mechanics are used to analyze human movement. Special emphasis is placed on athletics.

TRICKER, R. A. R., and B. J. K. TRICKER: "The Science of Mechanics," American Elsevier Publishing Company, Inc., New York, 1967. The general principles of mechanics are illustrated by many interesting examples from biology and athletics. The biological aspects of balance are discussed in several chapters.

WILLIAMS, MARIAN, and HERBERT R. LISSNER: "Biomechanics of Human Motion," W. B. Saunders Company, Philadelphia, 1962. Contains many worked examples of statics problems based on realistic medical data taken from the scientific literature.

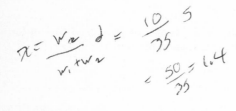

# dynamics CHAPTER 4

The last two chapters dealt with statics, the physics of objects in equilibrium. There equilibrium was taken to mean "at rest," but in this chapter it will be shown to also mean "in uniform motion." Dynamics is the physics of objects that are not in equilibrium, i.e., are neither at rest nor in uniform motion. Nonuniform motion occurs whenever the total force on an object is not zero.

## 4.1 FRAMES OF REFERENCE

The last two chapters discussed fully the necessary and sufficient conditions for an object to remain at rest. The phrase "at rest," however, is very ambiguous, because an object can be at rest in one system while moving in another. For example, a man sitting in an airplane is at rest relative to the airplane, but he is moving 600 mi/h relative to the ground. The airplane and the earth are two different reference systems, or *frames of reference*, with respect to which the man's motion can be referred. Because these two frames of reference are moving with respect to each other, the man can be at rest in one frame while moving in the other.

The phrase "at rest" is clearly meaningful only in a particular frame of reference. When the equilibrium conditions (Properties 5 and 6) were applied in Chaps. 2 and 3, this frame was taken to be the earth. This seems reasonable, and in fact, up to the time of Galileo it was thought that the earth was at rest at the center of the universe. However, since the discovery that the earth is in motion around the sun, it has been realized that there is nothing special about the earth's frame of reference. If Properties 5 and 6 are true in the frame of the earth, they are probably true in other frames as well.

Let us concentrate on Newton's first law of motion (Property 5, Sec. 2.1). Briefly, it states that the total force on an object at rest is zero. This is certainly true in frames of reference other than the earth, such as the airplane going 600 mi/h. A physics experiment, like the one in Fig. 2.36, gives the same result in the airplane as it does on earth. Even just sitting in the airplane is an experiment of sorts, because the muscle forces and contact forces on each part of the body must keep that part at rest. Since a person sitting in the airplane experiences no sensation of motion, these forces must be the same for him as they are for a person sitting on earth. Thus if the first law is true on earth, it is also true in a frame of reference moving 600 mi/h relative to the earth.

On the other hand, the first law is not true in all frames. For example, it is not true in the frame of reference of a roller coaster going through a dip. That is, the total force on a person at rest with respect to the roller coaster is not zero. This is evident from the additional forces a rider feels on his body as the roller coaster goes through the dip.

Therefore it is meaningful to ask in which frames of reference Newton's first law is true. These frames, which are of special importance in physics, are called *inertial frames*. Both the earth and the airplane are inertial frames, but the dipping roller coaster is not.

Galileo investigated this question for many years. He wanted to convince people that the earth could be moving even without any apparent sensation of motion. The result of his investigations was the discovery of the fundamental principle that *any frame that is moving uniformly with respect to an inertial frame is itself an inertial frame*. The earth can be an inertial frame provided its motion with respect to the sun is uniform. In this case a person at rest on the earth experiences no sensation of motion because the total force on him is zero. Furthermore, if the earth is an inertial frame, then the airplane moving uniformly with respect to the earth is an inertial frame also, and consequently a person at rest in the airplane has no sensation of motion.

**REMARK**  Uniform motion is motion with constant speed in a straight line. This is discussed more fully in the next section. The roller coaster is not an inertial frame because it moves in a curved path as it goes through a dip. The earth is not truly an inertial frame either, because objects at rest on its surface move through a large circle every day as the earth rotates on its axis. This motion has little effect, however, and for the purposes of this book the earth is considered to be an inertial frame of reference.

An immediate consequence of Galileo's principle is that the total force on an object is zero not only if the object is at rest but also if it is moving uniformly (with respect to an inertial frame). This follows because an object moving uniformly with respect to the earth (or any other inertial frame) is itself an inertial frame. Since the object is clearly at rest in its own frame of reference, the total force on it is zero. The first law, as stated in Property 5, includes the case of uniform motion if *equilibrium* is understood to mean rest or *uniform motion*.

The first law of motion is true in all inertial frames, by definition. But what about the other laws of physics, such as Property 6 (Sec. 3.1); are they also true in all inertial frames? Nineteenth-century physicists thought that all the laws of mechanics are true in all inertial frames but that other laws, especially laws concerning light and electromagnetism, are exactly true only in one special inertial frame, called the *ether*. The ether idea was abandoned after Einstein showed that the laws of light and electromagnetism, when properly understood, are exactly true in all inertial frames. His results are summarized in the principle of (special) relativity: *All the laws of physics are true in all inertial frames.*

Remember, an inertial frame is defined only as a frame of reference in which the first law is true. Galileo's principle asserts further that any frame moving uniformly with respect to an inertial frame is itself an inertial frame. Then, according to the principle of relativity, all the laws of physics are true in all inertial frames. The principle of relativity thus establishes inertial frames as the proper frames of reference in which to describe the laws of physics.

**REMARK** We are not yet in a position to appreciate fully the importance of the principle of relativity or the genius required for its discovery. It is only in the study of laws other than those of mechanics, especially the laws of light and electromagnetism, that relativity is seen to be a truly revolutionary concept in violent contradiction to our common sense concepts of space and time. This is discussed further in Sec. 14.6.

## 4.2  VELOCITY AND ACCELERATION

The concept of uniform motion must now be made more precise. All motion must be described with respect to some frame of reference, which we shall always take to be an inertial frame. However, it is not necessary to state which particular inertial frame we are using because the principle of relativity assures us that all inertial frames are equivalent for the purposes of physics.

An object moves with constant speed $v$ if the distance $d$ that it moves in time $t$ is given by

$$d = vt \qquad\qquad 4.1$$

for all values of $t$. The constant $v$ is the *speed* of the object. The dimension of speed is $[l/t]$, and common units of speed are feet per second (ft/s), miles per hour (mi/h), and meters per second (m/s). Figure 4.1 is a plot of $d$ against $t$ for the case in which $v = 5$ ft/s.

The *velocity* **v** of a moving object is a vector quantity with magnitude $v$ (the speed of the object) and direction (the direction of motion of the object). *Uniform motion* is just motion with constant velocity. Since velocity is a vector, constant velocity implies two things: (1) the speed $v$ is not changing (constant speed), and (2) the direction of motion is not changing (motion in a straight line). Thus uniform motion (constant velocity) is motion in a straight line with constant speed. An object at rest is a special case of uniform motion in which the velocity is zero.

An object that is not moving with constant velocity is said to be *accelerating*. Clearly an accelerating object is not moving in a straight line, or not moving with constant speed, or both. Figure 4.2 shows a car moving with constant speed around a circular racetrack. The arrows, which represent the car's velocity at different positions on the track, have the same length but are pointing in different directions. This car is accelerating because its direction of motion is continuously changing. Figure 4.3 shows a car that is increasing its speed steadily while traveling in a straight line. Here the arrows, which represent the car's velocity at different positions, have different lengths but are pointing in the same direction. This car is accelerating because its speed is changing. Situations in which an object is simultaneously changing its speed and direction are also common, but we shall not have to consider them.

In fact, for the most part, we shall have to consider only the simplest type of acceleration, that of *constant linear acceleration*.

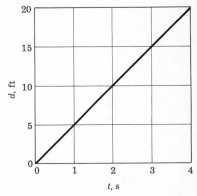

**FIGURE 4.1**
Plot of distance $d$ against time $t$ for an object moving with a constant speed $v$ of 5 ft/s.

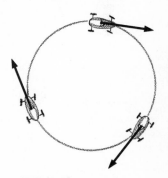

**FIGURE 4.2**
A car accelerating because its direction is changing.

**FIGURE 4.3**
A car accelerating because its speed is changing.

This is acceleration in which an object moves in a straight line with a changing speed $v$ that is given at any time $t$ by

$$v = v_0 + at \qquad 4.2$$

where $v_0$ is the object's speed at time $t = 0$. Here $a$ is a constant, called the *acceleration*. It can be either positive or negative, depending on whether the speed is increasing or decreasing. The dimension of $a$ is $[l/t^2]$ since when multiplied by $t$ in Eq. 4.2, the product $at$ must have the dimension of speed $[l/t]$. Common units of $a$ are feet per second per second (ft/s²), miles per hour per second (mi/h-s), and meters per second per second (m/s²).

It is possible to assign a direction to acceleration. For constant linear acceleration, the direction is parallel to the velocity **v** when the speed is increasing and antiparallel to **v** when the speed is decreasing. Acceleration is a vector quantity, like force and velocity, and so it should be written **a** (Appendix IV). The magnitude of **a** is $|a|$, the positive value of $a$. For instance, an object that is slowing down might have $a = -10$ ft/s². In this case, the vector **a** points opposite to the direction of motion of the object, and its magnitude is 10 ft/s². (The magnitude of a vector is always a positive number.)

The force of gravity is the total force on an unsupported object. The object therefore cannot be in equilibrium, and so it falls. From his study of the motion of falling objects, Galileo found that all objects fall with constant linear acceleration and that (in vacuum) the magnitude of the acceleration is the same for all objects. This is not obvious: a feather certainly does not fall as fast as a rock. But this is because the air exerts a relatively large upward force on the falling feather (air resistance) which partially supports it. In a vacuum, a feather and a rock do fall with the same acceleration.

The symbol $g$ is used for the acceleration due to gravity. The value of $g$ is the same for all objects at a given site, but it varies slightly with geographical location, as shown in Table 4.1. This geographical variation depends on the latitude and to a lesser extent on the altitude and geology of the site. In doing problems it is sufficient to use the value of 32 ft/s² or 9.8 m/s² for $g$.

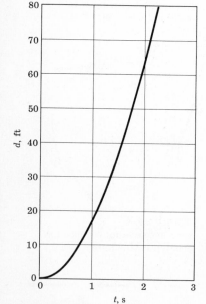

FIGURE 4.4
Plot of distance $d$ against time $t$ for an object moving with a constant acceleration $a$ of 32 ft/s².

TABLE 4.1 **Acceleration due to gravity at various locations on earth**
The values shown for specific places are measured, whereas the values shown for 0, 70, 80 and 90° are calculated from a formula of the U.S. Coast and Geodetic Survey. For a given latitude, the elevation and geology of a site can produce variation in the last two decimal places.

| Location | Latitude (north or south) | $g$ ft/s² | m/s² |
|---|---|---|---|
| Equator | 0.0° | 32.0878 | 9.78039 |
| Canal Zone, Panama | 8.9°N | 32.0950 | 9.78243 |
| Khartoum, Sudan | 15.6°N | 32.0966 | 9.78308 |
| Buenos Aires | 34.6°S | 32.1413 | 9.79669 |
| New York | 40.8°N | 32.1609 | 9.80267 |
| Greenwich, England | 51.5°N | 32.1911 | 9.81188 |
| Helsinki, Finland | 60.2°N | 32.2149 | 9.81912 |
| Theoretical | 70.0° | 32.2377 | 9.82608 |
| | 80.0° | 32.2525 | 9.83059 |
| Poles | 90.0° | 32.2577 | 9.83217 |

With $a = g = 32$ ft/s$^2$ in Eq. 4.2, we see that an object falling from rest ($v_0 = 0$) has a speed of 16 ft/s at $t = 0.5$ s, a speed of 32 ft/s at $t = 1.0$ s, and a speed of 64 ft/s at $t = 2.0$ s. How far does it fall in 1 s? At first you might be tempted to use Eq. 4.1 with the speed of 32 ft/s that the object has after 1 s. The result would be

$$d = vt = 32 \text{ ft/s} \times 1 \text{ s} = 32 \text{ ft}$$

This is wrong, however, because the speed of 32 ft/s is obtained only at the end of the first second. Since the speed of the object was zero at the beginning of the first second, the average speed $\bar{v}$ during the first 1-s interval is

$$\bar{v} = \tfrac{1}{2}(0 \text{ ft/s} + 32 \text{ ft/s}) = 16 \text{ ft/s}$$

Using the average speed in Eq. 4.1, we get the correct distance traveled:

$$d = \bar{v}t = 16 \text{ ft/s} \times 1 \text{ s} = 16 \text{ ft}$$

The general formula for the distance $d$ traveled in time $t$ by an object that starts from rest and accelerates with constant acceleration $a$ is

$$\boxed{d = \tfrac{1}{2}at^2} \qquad 4.3$$

Figure 4.4 is a plot of $d$ against $t$ for the case in which $a = 32$ ft/s$^2$. (Compare this with Fig. 4.1.) If the object starts with an initial speed $v_0$, the distance traveled in time $t$ is given by

$$\boxed{d = v_0 t + \tfrac{1}{2}at^2} \qquad 4.4$$

Figure 4.5 is a multiflash photograph of a ball dropped from rest. The time intervals between adjacent images are equal, so that the distance between adjacent images is large when the speed is large and small when the speed is small. This photograph clearly shows that the speed of a falling object starts out small and increases with time, in accordance with Eq. 4.2. In addition, the photograph shows that the distance increases with the square of the time (in accordance with Eq. 4.3), since (for instance) image 8 is 4 times farther from the starting point than image 4†.

As an example of how to use Eqs. 4.2 to 4.4, let us calculate the height reached by a ball thrown straight up in the air with an initial speed of 40 ft/s. We are given $v_0$ and asked to find $d$. Neither Eq. 4.2 nor Eq. 4.4 seems to be very useful, since both involve $t$, which is unknown. Clearly, therefore, we must first find $t$. To do this we must realize that as the ball travels upward, its speed is steadily decreasing, until it reaches its highest point, where its speed is momentarily zero. This is seen in Fig. 4.6, which is a multiflash photograph of a ball thrown up in the air. As its highest point, the adjacent images of the ball are very close together, showing that the ball is moving very slowly there.

While the ball is moving upward, its acceleration $a$ is $-32$ ft/s$^2$, so Eq. 4.2 can be written

$$v = 40 \text{ ft/s} - 32 \text{ ft/s}^2 \times t$$

† The time interval between successive images in Fig. 4.5 is $\frac{1}{30}$ s. However, the first image corresponds to both $t = 0$ and $t = \frac{1}{30}$ s, because the ball did not move enough to produce separate images.

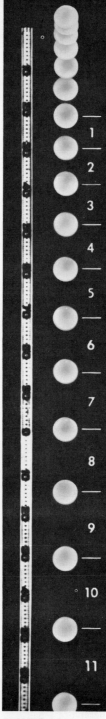

FIGURE 4.5
Multiflash photograph of a ball dropped from rest. [*From "PSSC Physics," 3d ed., copyright © 1971 by Education Development Center, Inc. Reprinted by permission of D. C. Heath and Co.*]

The time $t$ at which the ball reaches its highest point is found by setting $v$ to zero in this equation and solving for $t$. The result is

$$0 = 40 \text{ ft/s} - 32 \text{ ft/s}^2 \times t$$

or
$$t = \frac{40 \text{ ft/s}}{32 \text{ ft/s}^2} = 1.25 \text{ s}$$

This time is then substituted into Eq. 4.4 to find $d$:

$$d = 40 \text{ ft/s} \times 1.25 \text{ s} - \tfrac{1}{2} \times 32 \text{ ft/s}^2 \times 1.25^2$$
$$= 50 \text{ ft} - 25 \text{ ft} = 25 \text{ ft}$$

The ball reaches its maximum height of 25 ft in 1.25 s.

After the ball reaches 25 ft, it starts to fall back down. The time it takes to fall back to the ground is found from Eq. 4.3, or from Eq. 4.4 with $v_0 = 0$, since its speed is zero at its highest point. During this phase of the motion its speed is increasing, and so $a = +32 \text{ ft/s}^2$.

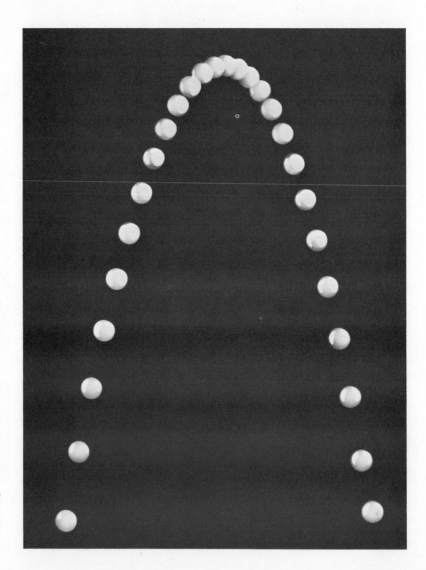

FIGURE 4.6
Multiflash photograph of a ball thrown into the air. [*From "PSSC Physics," 3d ed., copyright © 1971 by Education Development Center, Inc. Reprinted by permission of D. C. Heath and Co.*]

Solving Eq. 4.3 for $t^2$, we get

$$t^2 = \frac{2d}{a}$$

$$= \frac{2 \times 25 \text{ ft}}{32 \text{ ft/s}^2} = 1.56 \text{ s}^2$$

so that

$$t = \sqrt{1.56 \text{ s}^2} = 1.25 \text{ s}$$

The ball falls back to the ground in the same time that it took to go up. Its speed when it reaches the ground is given by Eq. 4.2 with $v_0 = 0$. The result is

$$v = 0 + 32 \text{ ft/s}^2 \times 1.25 \text{ s}$$
$$= 40 \text{ ft/s}$$

It returns to the ground with the same speed with which it was thrown. Its velocity is not the same, however, because initially its velocity had a magnitude of 40 ft/s and was directed upward, whereas at the end its velocity has a magnitude of 40 ft/s and is directed downward. Figure 4.6 shows that at any given distance below the highest point, the speed of the ball is the same for both upward and downward motion.

_Uniform circular motion_ is motion in a circle of radius $R$ with constant speed $v$. It is another special case of accelerated motion that requires mention. The acceleration in this case is due to the changing direction of the motion rather than to a change of speed. Such acceleration is called _centripetal_ acceleration because at any instant it is directed toward the center of the circle. Thus as the object moves, the direction of the acceleration changes, as shown in Fig. 4.7. The magnitude of this acceleration is related to the speed $v$ and the radius $R$ by†

$$a = \frac{v^2}{R} \qquad \qquad 4.5$$

Notice that $v^2/R$ has the correct dimension of acceleration:

$$\left[ \frac{v^2}{R} \right] = \frac{[l/t]^2}{[l]} = \left[ \frac{l}{t^2} \right]$$

The distance traveled (measured around the circle) is given by Eq. 4.1, since the speed is constant.

**REMARK** An object at rest on the earth has a small centripetal acceleration because of the earth's rotation about its axis. It is this acceleration, which varies with latitude, that causes most of the variation in $g$ shown in Table 4.1.

## 4.3 NEWTON'S SECOND LAW: PROPERTY 7 OF FORCE

The basic problem of dynamics is to describe the motion of an object acted upon by a total force **F**. Newton's first law of motion (Property 5, Sec. 2.1) states that when **F** is zero, the object remains at rest or

† The properties of centripetal acceleration are proved in Appendix IV.

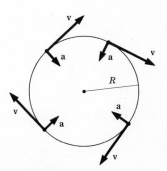

FIGURE 4.7
Centripetal acceleration **a** and velocity **v** of an object in uniform circular motion.

moves with constant velocity. Newton's second law of motion describes what happens when **F** is not zero. It should be clear that in this case the object accelerates; i.e., the object does not move with constant velocity. The second law goes further, however, and states that the acceleration **a** of the object is proportional to **F**. This means that **a** has the same direction as **F** and that the magnitude of **a** is proportional to the magnitude of **F**.

To be more precise, suppose an object is acted upon by a constant force $\mathbf{F}_1$. Then the object has a constant acceleration $\mathbf{a}_1$ in the direction of $\mathbf{F}_1$, with the magnitude

$$a_1 = kF_1$$

where $k$ is a constant of proportionality that is characteristic of the object. If the same object is acted upon by a different force $\mathbf{F}_2$, it has a different acceleration $\mathbf{a}_2$ whose magnitude is given by

$$a_2 = kF_2$$

with the same constant $k$. This relation holds generally, even if the force is not constant. Of course, in this case the acceleration is not constant either. But at every instant of time the acceleration **a** is related to the total force **F** by

$$\mathbf{a} = k\mathbf{F} \qquad\qquad 4.6$$

This is written as a vector relation to emphasize that **a** has the same direction as **F**.

It is customary to work with the constant

$$m = \frac{1}{k}$$

rather than with $k$. Noting that

$$k = \frac{1}{m}$$

we see that Eq. 4.6 can be written as

$$\mathbf{a} = \frac{1}{m}\,\mathbf{F} = \frac{\mathbf{F}}{m}$$

or $$\mathbf{F} = m\mathbf{a} \qquad\qquad 4.7$$

The constant $m$ is the *mass*, an intrinsic property of the object. Equation 4.7 is Newton's second law of motion, which can be stated formally as a property of force.

**Property 7 (Newton's second law of motion)**  *An object acted upon by a total force* **F** *has an acceleration* **a** *in the direction of* **F**. *The magnitude of* **a** *is $F/m$, where $F$ is the magnitude of the force and $m$ is the mass of the object.*

The total force on an unsupported object is the force of gravity $\mathbf{F}_g$. Consequently the object falls with an acceleration given by Eq. 4.7:

$$a = \frac{F_g}{m}$$

On the other hand, we know that all objects fall with the same acceleration $g$. Therefore we have

$$g = \frac{F_g}{m}$$

or
$$F_g = mg \qquad 4.8$$

which shows that the force of gravity on an object is proportional to its mass. In a spaceship, where the value of $g$ is much smaller than on earth, the force of gravity on an object is proportionally smaller. However, the mass of an object, being an intrinsic property, remains the same regardless of where the object is located.

Equation 4.7 completely determines the motion of an object in terms of the forces acting on it. This is clear when the force is constant because then Eq. 4.7 determines the constant acceleration to be used in Eqs. 4.2 and 4.4 for the speed and position of the object. When **F** is not constant, the acceleration determined by Eq. 4.7 is not constant, in which case Eqs. 4.2 and 4.4 do not apply. However, using more advanced mathematics, we can still calculate the subsequent speed and position of the object from its acceleration. In this sense Eq. 4.7 provides a complete description of the motion.

**REMARK** The motion of objects acted upon by torques is analogous to the motion of objects acted upon by forces. An object, such as a wheel, rotates at constant speed if the total torque on it is zero. The wheel will rotate with increasing (or decreasing) speed when the total torque on it is not zero. We shall not go further into this except to note that these properties can be derived from Newton's second law, so that no new principles are required.

## 4.4  SYSTEMS OF UNITS

The dimensions of mass $[m]$, length $[l]$, and time $[t]$ are considered to be fundamental because the dimensions of all other quantities in mechanics can be written in terms of them.† For example, the dimension of force is found from Eq. 4.7 to be

$$
\begin{aligned}
[f] &= [m][a] \\
&= [m]\left[\frac{l}{t^2}\right] \\
&= \left[\frac{ml}{t^2}\right]
\end{aligned}
$$

It is useful to have a coherent system of units in which standard units are adopted for length, mass, and time. These units are then used consistently in all calculations. For instance, if the unit of length is the foot, all lengths are measured in feet (rather than inches, miles, meters, etc.), all areas are measured in square feet (rather than acres, square inches, etc.), and all volumes are measured in cubic feet (rather than gallons, liters, hogsheads, etc.). In a coherent system the unit of force is given naturally in terms of the fundamental units. There are, unfortunately, several different systems with which you must be familiar. Three of the most important systems are described here.

---

† Two other fundamental dimensions are introduced later in this book in connection with temperature and electric charge.

## Mks system

This is part of the *International System* (SI) *of Units* recommended by the International Conference on Weights and Measures in 1960. In this system the unit of length is the *meter*, the unit of mass is the *kilogram*, and the unit of time is the *second*. The abbreviations for these units are m, kg, and s, respectively; hence the name mks. A kilogram is a unit of mass equal to the mass of a particular cylinder of platinum-iridium alloy kept in a vault in Sèvres, France.

In the mks system the unit of speed is meters per second, the unit of acceleration is meters per second per second, and the unit of force is kilogram-meters per second per second (kg-m/s$^2$). This unit of force is called a *newton* (N). From Eq. 4.7 it is seen that a force of one newton causes a mass of one kilogram to have an acceleration of one meter per second per second.

The acceleration due to gravity in mks units is approximately

$$g = 32 \text{ ft/s}^2 = 32 \text{ ft/s}^2 \times 0.305 \text{ m/ft}$$
$$= 9.8 \text{ m/s}^2$$

Table 4.1 gives the exact value of $g$ in mks units for various locations on earth. For most exercises it is sufficient to use the value of 9.8 m/s$^2$. For example, the force of gravity on 1 kg is found from Eq. 4.8 to be

$$F_g = mg = 1 \text{ kg} \times 9.8 \text{ m/s}^2$$
$$= 9.8 \text{ kg-m/s}^2 = 9.8 \text{ N}$$

The mks system is the primary system of units used in the rest of the book. It is hoped that the mks system will eventually replace all others, so that there will be a single international system of units. At present, however, many specialized fields still use their own preferred units. In this book every new quantity is first expressed in mks units, and then specialized units commonly used in the life-science literature are introduced.

## Cgs system

In this system the unit of length is the *centimeter*, the unit of mass is the *gram*, and the unit of time is the *second*. The abbreviations for these units are cm, g, and s, respectively; hence the name cgs. This system, like the mks system, is based on the metric system. The centimeter and gram are related to the meter and kilogram by

$$1 \text{ cm} = 10^{-2} \text{ m} \qquad \text{and} \qquad 1 \text{ g} = 10^{-3} \text{ kg}$$

In the cgs system the unit of speed is centimeters per second, the unit of acceleration is centimeters per second per second, and the unit of force is gram-centimeters per second per second (g-cm/s$^2$). This unit of force is called a *dyne* (dyn). From Eq. 4.7 it is seen that a force of one dyne causes a mass of one gram to have an acceleration of one centimeter per second per second.

The conversion between mks and cgs units is relatively easy. For example, using the basic relations

$$1 \text{ kg} = 10^3 \text{ g} \qquad \text{and} \qquad 1 \text{ m} = 10^2 \text{ cm}$$

we find

$$1 \text{ N} = 1 \text{ kg-m/s}^2$$
$$= 10^3 \text{ g} \times 10^2 \text{ cm/s}^2$$
$$= 10^5 \text{ g-cm/s}^2 = 10^5 \text{ dyn}$$

The acceleration due to gravity in cgs units is approximately

$$g = 9.8 \text{ m/s}^2 = 9.8 \times 100 \text{ cm/s}^2$$
$$= 980 \text{ cm/s}^2$$

The exact value of $g$ in cgs units is found by multiplying the mks value of $g$ in Table 4.1 by 100. The force of gravity on 1 g is found from Eq. 4.8 to be

$$F_g = mg = 1 \text{ g} \times 980 \text{ cm/s}^2$$
$$= 980 \text{ g-cm/s}^2 = 980 \text{ dyn}$$
$$= 980 \times 10^{-5} \text{ N} = 0.00980 \text{ N}$$

**REMARKS** (*1*) The symbols $m$ for mass and $g$ for acceleration due to gravity are easily distinguished in print from the abbreviations m for meter and g for gram. In order not to confuse these symbols in your handwritten notes, you should get in the habit of underlining the symbols for mass and acceleration due to gravity: $\underline{m}$ and $\underline{g}$. (2) The units of volume in the mks and cgs systems are cubic meters and cubic centimeters, respectively. A closely related unit in common use is the *liter* (l), which is 1000.027 cm³. However, for our purposes it is sufficient to take 1 l to be $10^3$ cm³ $= 10^{-3}$ m³ and to take 1 milliliter (1 ml $= 10^{-3}$ l) to be 1 cm³.

### English system

In this system force rather than mass is taken to be a fundamental unit. The unit of force is the *pound* (lb), the unit of length is the *foot* (ft), and the unit of time is the *second* (s).

In the English system the unit of speed is feet per second, and the unit of acceleration is feet per second per second. The unit of mass is found from Eq. 4.7 to be lb-s²/ft, called a *slug*. A force of one pound causes a mass of one slug to have an acceleration of one foot per second per second.

The force of gravity on 1 slug is

$$F_g = mg = 1 \text{ slug} \times 32 \text{ ft/s}^2$$
$$= 32 \text{ slug-ft/s}^2$$

These units can be simplified by using the result that 1 slug = 1 lb-s²/ft, so that

$$F_g = 32 \, \frac{\text{lb-s}^2}{\text{ft}} \, \frac{\text{ft}}{\text{s}^2} = 32 \text{ lb}$$

Thus an object that weighs about 32 lb has a mass of 1 slug.

The conversion between slugs and kilograms is

$$1 \text{ slug} = 14.594 \text{ kg}$$

and the conversion between pounds and newtons is

$$1 \text{ lb} = 4.4485 \text{ N}$$

**REMARK**  The kilogram and the gram are sometimes used as units of force in laboratory work. A kilogram force ($kg_f$) is the force of gravity on a mass of one kilogram. Since the magnitude of this force depends on the value of $g$, the standard value of $9.80665$ m/s$^2$ is used when great accuracy is required. Thus the conversion from kilogram force to newtons is

$$1 \text{ kg}_f = 1 \text{ kg} \times 9.80665 \text{ m/s}^2$$
$$= 98.0665 \text{ N}$$

## 4.5  EXAMPLES INVOLVING NEWTON'S SECOND LAW OF MOTION

The variety of problems that can be solved using Newton's second law is surprisingly large. In fact, all problems in dynamics can be solved using it, though often mathematical techniques are needed that are beyond the scope of this book. Examples of problems are given here to show the variety of dynamical problems that can be solved with simple algebra.

First, consider the 60-kg block resting on the horizontal surface in Fig. 4.8. A constant horizontal force $\mathbf{F}_a$ is applied to the block, causing the block to accelerate. If the magnitude of $\mathbf{F}_a$ is 150 N, how far has the block moved after 3 s and what is its speed?

The force of gravity $\mathbf{F}_g$ and the contact force $\mathbf{F}_c$ also act on the block, so the total force $\mathbf{F}$ on it is

$$\mathbf{F} = \mathbf{F}_g + \mathbf{F}_c + \mathbf{F}_a$$

But $\mathbf{F}_g + \mathbf{F}_c = 0$, since the block does not move in the vertical direction. Therefore $\mathbf{F}_a$ is the total force $\mathbf{F}$ on the block, so that the block accelerates in the direction of $\mathbf{F}_a$. Since the magnitude of $\mathbf{F}$ is 150 N, the acceleration of the block is found from Eq. 4.7 to be

$$a = \frac{F}{m} = \frac{150 \text{ N}}{60 \text{ kg}} = 2.5 \text{ m/s}^2$$

We know that $a$ will be in the mks unit of acceleration because the force and mass are in the proper mks units. This illustrates the advantage of using a coherent system of units.

The speed and position of the block are calculated using Eqs. 4.2 and 4.4. Initially, i.e., at time $t = 0$, the block was at rest, so $v_0 = 0$. Then at time $t = 3$ s its speed is

$$v = v_0 + at$$
$$= 0 + 2.5 \text{ m/s}^2 \times 3 \text{ s} = 7.5 \text{ m/s}$$

and the distance it has moved is

$$d = v_0 t + \tfrac{1}{2} a t^2$$
$$= 0 + 0.5 \times 2.5 \text{ m/s}^2 \times (3 \text{ s})^2$$
$$= 11.25 \text{ m}$$

An interesting variation of this last problem is to calculate the force on an object given its initial and final speed. For example, a water skier (Fig. 4.9) starting from rest in the water is accelerated until she reaches a speed of about 10 m/s (22.4 mi/h). Thereafter she moves with approximately constant velocity, so that the total force on her is zero. (The force $\mathbf{F}_a$ exerted by the towline is then equal

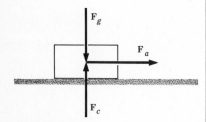

FIGURE 4.8
A constant horizontal force $\mathbf{F}_a$ applied to a block on a horizontal surface.

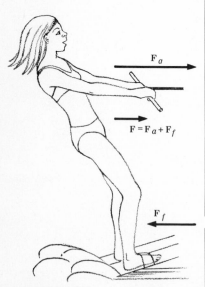

FIGURE 4.9
The forces on a water skier.

in magnitude but opposite in direction to the frictional force $\mathbf{F}_f$ exerted by the water.) While she is accelerating, however, the total force $\mathbf{F} = \mathbf{F}_a + \mathbf{F}_f$ is not zero. It can be found by doing the calculation of the last problem in reverse order.

Let us assume that the mass of the skier is 45 kg and that it takes 5 s for her to be accelerated from rest to a speed of 10 m/s. Thus we have $v_0 = 0$, $v = 10$ m/s, and $t = 5$ s. Putting these into Eq. 4.2, we find that

$$10 \text{ m/s} = 0 + a \times 5 \text{ s}$$

or

$$a = \frac{10 \text{ m/s}}{5 \text{ s}} = 2 \text{ m/s}^2$$

The magnitude of the total force $\mathbf{F}$ on the skier is then found from Eq. 4.7 to be

$$F = ma = 45 \text{ kg} \times 2 \text{ m/s}^2 = 90 \text{ N}$$

Again we know that $F$ is in newtons because both $m$ and $a$ are in mks units.

This force can be converted to pounds by using the conversion equation in Sec. 2.4:

$$90 \text{ N} = 90 \text{ N} \times \frac{1 \text{ lb}}{4.45 \text{ N}} = 20 \text{ lb}$$

Since the force of friction $\mathbf{F}_f$ is in the opposite direction to $\mathbf{F}_a$, the magnitude of the total force $\mathbf{F}$ is equal to the difference of the magnitudes of $\mathbf{F}_a$ and $\mathbf{F}_f$ (Sec. 2.3):

$$F = F_a - F_f$$

For instance, if $F_f$ is 25 lb, $F_a$ must be 45 lb. This is the force on the skier's arms. Because it is so large, beginners have difficulty holding onto the towline while they are being accelerated.

As another example, consider a 5-kg block on a frictionless surface inclined 35° to the horizontal (Fig. 4.10). We want to find the acceleration of the block as it slides down the incline. The magnitude of the force of gravity $\mathbf{F}_g$ on the block is

$$F_g = mg = 5 \text{ kg} \times 9.8 \text{ m/s}^2 = 49 \text{ N}$$

and $\mathbf{F}_g$ is directed vertically downward. The angle $\theta$ between $\mathbf{F}_g$ and a line perpendicular to the inclined surface is equal to 35°, the angle of the incline, because the corresponding sides of these two angles are perpendicular to each other (Appendix II). Therefore the magnitudes of the components of $\mathbf{F}_g$ parallel and perpendicular to the inclined surface are

$$F_\perp = F_g \cos 35° = 49 \text{ N} \times 0.82 = 40 \text{ N}$$
$$F_\parallel = F_g \sin 35° = 49 \text{ N} \times 0.57 = 28 \text{ N}$$

Since the block does not move perpendicularly to the inclined surface, the sum of $\mathbf{F}_\perp$ and the contact force $\mathbf{F}_c$ is zero, so that $\mathbf{F}_\parallel$ is the total force on the block. The acceleration then is

$$a = \frac{F_\parallel}{m} = \frac{28 \text{ N}}{5 \text{ kg}} = 5.6 \text{ m/s}^2$$

This is less than $g$ by the factor $\sin 35° = F_\parallel / F_g$.

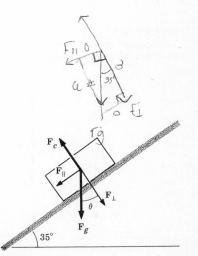

FIGURE 4.10
A block on an inclined surface.

**REMARK**  Galileo was unable to study the acceleration of freely falling objects in detail because their acceleration is too large. Instead, he studied the motion of objects rolled down an inclined plane. By making the angle of inclination very small, he reduced the acceleration enough to be able to study it with the crude means available to him.

If the incline is not frictionless, there will be a force of friction $\mathbf{F}_f$ parallel to the incline, directed opposite to $\mathbf{F}_\parallel$. For a moving object the magnitude of $\mathbf{F}_f$ is related to the coefficient of kinetic friction $\mu_k$ (Sec. 2.2) by

$$F_f = \mu_k F_c$$

Now $F_c = F_\perp$, so if $\mu_k = 0.3$, the force of friction is

$$F_f = 0.3 \times 40 \text{ N} = 12 \text{ N}$$

The total force on the block then is

$$F = F_\parallel - F_f = 28 \text{ N} - 12 \text{ N} = 16 \text{ N}$$

so now the acceleration of the block is only

$$a = \frac{F}{m} = \frac{16 \text{ N}}{5 \text{ kg}} = 3.2 \text{ m/s}^2$$

When the angle of inclination is so small that $F_\parallel$ is less than the maximum force of static friction, the block will remain at rest on the incline.

A rather different type of problem is shown in Fig. 4.11. Here two masses, $m_1$ and $m_2$, are connected by a flexible cord that passes over a pulley. If $m_1$ is greater than $m_2$, it will accelerate downward while $m_2$ accelerates upward. Since the masses are connected by the cord, they must move together, so that the magnitudes of their accelerations are equal. We want to find this acceleration. The force on each mass consists of the downward force of gravity and the upward force exerted by the cord. The difficulty of this problem is that both the acceleration and the tension of the cord are unknown.

The magnitude of the force exerted by the cord on each mass is equal to the tension $T$, which is the same throughout the cord (Sec. 2.2). We know that $T$ is less than $m_1 g$, the force of gravity on $m_1$, because $m_1$ accelerates downward (Fig. 4.12). Thus the total (downward) force on $m_1$ is $m_1 g - T$, and it is related to the (downward) acceleration $a$ of $m_1$ by Eq. 4.7:

$$m_1 g - T = m_1 a \qquad\qquad 4.9$$

Likewise $T$ is greater than $m_2 g$ because $m_2$ accelerates upward. The total (upward) force $T - m_2 g$ on $m_2$ is related to the (upward) acceleration $a$ of $m_2$ by

$$T - m_2 g = m_2 a \qquad\qquad 4.10$$

Here we have used the fact that the accelerations of $m_1$ and $m_2$ are equal.

Equations 4.9 and 4.10 are two linear equations in the two unknowns $T$ and $a$. To solve them for $a$ we must first eliminate $T$

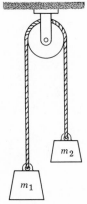

**FIGURE 4.11**
Two masses connected by a flexible cord passing over a pulley.

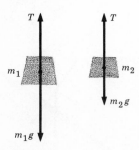

**FIGURE 4.12**
The forces on the two masses in Fig. 4.11.

between them. This is easily done by just adding the equations together. The result is

$$(m_1g - T) + (T - m_2g) = m_1a + m_2a$$

After some simple algebra this becomes

$$m_1g - m_2g = (m_1 + m_2)a$$

so that the acceleration is given by

$$a = \frac{m_1 - m_2}{m_1 + m_2}g$$

For instance, if $m_1 = 7.5$ kg and $m_2 = 4.5$ kg, the acceleration is

$$a = \frac{7.5 \text{ kg} - 4.5 \text{ kg}}{7.5 \text{ kg} + 4.5 \text{ kg}} \times 9.8 \text{ m/s}^2$$

$$= \frac{3 \text{ kg}}{12 \text{ kg}} \times 9.8 \text{ m/s}^2 = 2.45 \text{ m/s}^2$$

Finally, we give an example involving centripetal acceleration. Consider an artificial satellite of mass $m$ moving around the earth in a circular orbit of radius $R$ at constant speed $v$ (Fig. 4.13). The only force on it is the force of gravity $\mathbf{F}_g$, which is directed toward the center of the earth. The centripetal acceleration of the satellite is also directed toward the center of the earth and, from Eq. 4.5, its magnitude is $v^2/R$. Equation 4.7 applies to centripetal acceleration as well as to linear acceleration, so that these quantities are related by

$$F_g = \frac{mv^2}{R} \qquad\qquad 4.11$$

which determines the speed of the satellite if $F_g$ is known.

The magnitude of $F_g$ decreases with distance from the earth (Sec. 5.4), but for distances of only a few hundred miles this decrease is small. Thus the force of gravity on a low-orbiting satellite is approximately $mg$, the same as on earth. A satellite orbiting at a height $h$ above the earth moves in a circle of radius

$$R = R_e + h$$

about the center of the earth (Fig. 4.13), where $R_e$ is the radius of the earth. Again, for a low-orbiting satellite $h$ is small compared to $R_e$, so that $R$ is approximately $R_e$. Thus for low-orbiting satellites we have

$$F_g = mg \qquad \text{and} \qquad R = R_e$$

Substituting these into Eq. 4.11, we get

$$mg = \frac{mv^2}{R_e}$$

or

$$v^2 = gR_e \qquad\qquad 4.12$$

Notice that the mass of the satellite has canceled on both sides of the equation. This is an important result because it demonstrates that

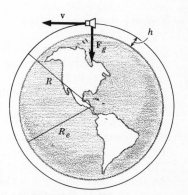

FIGURE 4.13
An artificial satellite in circular orbit about the earth.

all low-orbiting satellites have the same speed, regardless of their mass.

With the values $R_e = 6.4 \times 10^6$ m and $g = 9.8$ m/s$^2$, Eq. 4.12 gives the result

$$v^2 = (6.4 \times 10^6 \text{ m})(9.8 \text{ m/s}^2)$$
$$= 62.6 \times 10^6 \text{ m}^2/\text{s}^2$$

or
$$v = \sqrt{62.6 \times 10^6 \text{ m}^2/\text{s}^2} = 7.9 \times 10^3 \text{ m/s}$$
$$= 7.9 \text{ km/s}$$

for the speed of a low-orbiting satellite. From the relation

$$1 \text{ km} = 0.615 \text{ mi}$$

the speed in miles per second is found to be

$$v = 7.9 \text{ km/s} \times 0.615 \text{ mi/km} = 4.9 \text{ mi/s}$$

The time $t$ it takes the satellite to complete one revolution about the earth is called the *period*. The distance $d$ traveled in one revolution is equal to the circumference of the earth

$$d = 2\pi R_e = 6.28 \times 6.4 \times 10^6 \text{ m}$$
$$= 40 \times 10^6 \text{ m}$$

The period then is found from Eq. 4.1 to be

$$t = \frac{d}{v} = \frac{40 \times 10^6 \text{ m}}{7.9 \times 10^3 \text{ m/s}} = 5.1 \times 10^3 \text{ s}$$

or about 85 min. The period of a low-orbiting satellite is actually about 90 min, a number familiar to everyone since the beginning of the space age. Our calculation gives a slightly smaller value because it neglected the height of the satellite above the surface of the earth.

## PROBLEMS

1 An object moving at constant speed travels 20 m in 4 s. (*a*) What is the speed of the object? (*b*) How far does the object travel in 9 s? (*c*) How long does it take the object to travel 75 m?

    *Ans.* (*a*) 5 m/s; (*b*) 45 m; (*c*) 15 s

2 An earth satellite moving at constant speed travels 810 km in 3 min. (*a*) What is the speed of the satellite? (*b*) How far does the satellite travel in 1 h? (*c*) How long does it take the satellite to complete one orbit about the earth if the circumference of the orbit is $1.26 \times 10^5$ km?

3 The nearest star is about $4 \times 10^{16}$ m away. At what speed would a rocket ship have to travel to get to this star in 10 years?

    *Ans.* $1.27 \times 10^5$ km/s

4 A bee leaves an experimental feeding station at 3:15 P.M. and returns at 3:22 P.M. The hive is known to be 875 m from the station. What is the minimum flying speed of the bee?

5 Figure 4.14 is a graph of distance against time for an object moving with constant speed. (*a*) How far did the object travel between $t = 2$ s and $t = 5$ s? (*b*) What is the speed of the object?     *Ans.* (*a*) 22.5 m; (*b*) 7.5 m/s

6 Draw a graph of distance against time for an object moving at a constant speed of 12 m/s. Let $d = 0$ when $t = 0$.

7 A car accelerates from rest to 60 mi/h in 11 s. (*a*) What is its acceleration during this period? (*b*) How far does it travel during the acceleration period?

    *Ans.* (*a*) 8.0 ft/s$^2$; (*b*) 484 ft

8 (*a*) How long does it take an object starting from rest to fall 40 ft? (*b*) How far would the object fall in twice that time?

9 Figure 4.15 is a graph of distance against time for an object moving with constant acceleration. (*a*) What is the average speed of the

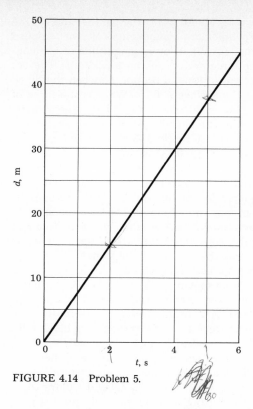

FIGURE 4.14  Problem 5.

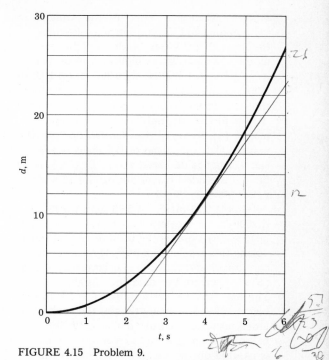

FIGURE 4.15  Problem 9.

object in the interval between $t = 0$ and $t = 4$ s? (*b*) What is the average speed of the object in the interval between $t = 4$ s and $t = 6$ s? (*c*) What is the acceleration of the object, assuming it starts from rest at $t = 0$? (*d*) What is the speed of the object at $t = 4$ s? *Ans.* (*a*) 3 m/s; (*b*) 7.5 m/s; (*c*) 1.5 m/s²; (*d*) 6 m/s

10 Plot the distance that the ball in Fig. 4.5 has fallen against time. The first image occurs at $t = 0$, and each successive image occurs $\frac{1}{30}$ s later. The bottom of the first image ($t = 0$) is at the zero mark of the scale, and the bottom of the third image ($t = \frac{2}{30}$ s) is about 2 cm below the zero mark. The second image cannot be distinguished from the first image in this picture. From your graph determine the acceleration of the object. What do you expect the acceleration to be?

11 Draw a graph of distance against time for an object moving with a constant acceleration of 4 m/s². Take $v_0$ to be zero.

12 In a head-on collision, a car going 60 mi/h is brought to rest in 0.1 s. What is the acceleration during the collision?

13 A ball is thrown straight up in the air with an initial speed of 12 m/s. (*a*) How long does it take the ball to reach its highest point? (*b*) How high does the ball go? (*c*) What is the time interval between the instant the ball leaves the hand until it is caught again?
*Ans.* (*a*) 1.22 s; (*b*) 7.29 m; (*c*) 2.44 s

14 A high jumper jumps 1.2 m straight up in the air. With what speed did he leave the ground? (*Hint*: Remember that an object will hit the ground with the same speed that it left the ground.)

15 When making a vertical jump, a grasshopper extends its legs 2.5 cm in 0.025 s. (*a*) What is the acceleration of the grasshopper while extending its legs? (*b*) What is the speed of the grasshopper as it leaves the ground, i.e., at the instant its legs are fully extended? (*c*) How high does the grasshopper rise from the ground?
*Ans.* (*a*) 80 m/s²; (*b*) 2 m/s; (*c*) 0.20 m

16 (*a*) What is the weight in pounds of 1 kg? (*b*) What is the mass in kilograms of a 1-lb weight?

17 The acceleration due to gravity on the moon

is $g = 1.63$ m/s². What is the weight in (*a*) newtons and (*b*) pounds of a 1-kg mass on the moon? (*c*) What is the weight on the moon of a man who weighs 180 lb on earth?

*Ans.* (*a*) 1.63 N; (*b*) 0.367 lb; (*c*) 29.9 lb

18 What is the mass in kilograms of an object that weighs 65 lb on the moon (see Prob. 17)?

19 A 15-N force is the only force acting on a 4-kg mass. (*a*) What is the acceleration of the mass? (*b*) If the speed of the mass is zero at $t = 0$, what is its speed at $t = 10$ s? (*c*) How far did the mass travel during the 10-s interval?

*Ans.* (*a*) 3.75 m/s²; (*b*) 3.75 m/s; (*c*) 187.5 m

20 A 2000-lb car accelerates from rest to 60 mi/h in 12 s. (*a*) What is the total force on the car during this acceleration? (*b*) How far does the car travel during the acceleration period? (*Hint:* Convert the data into English units.)

21 The Saturn V rocket weighs $6.1 \times 10^6$ lb just before takeoff. During takeoff the rocket engines produce a thrust of $7.5 \times 10^6$ lb. This means that an upward force of this magnitude is exerted on the rocket. (*a*) What is the acceleration of the rocket during takeoff? (Remember, the thrust is not the only force on the rocket.) (*b*) If this acceleration remains constant, what is the speed of the rocket after 60 s?     *Ans.* (*a*) 7.37 ft/s²; (*b*) 301 mi/h

22 The brakes are suddenly applied to a 1000-kg car going 25 m/s. The wheels lock, and the car skids to a stop in 5 s. (*a*) What is the force of friction on the car while it skids to a stop? (*b*) What is the coefficient of kinetic friction $\mu_k$ between the tires and the road? (*c*) How far does the car travel while skidding to a stop?

23 A 3-kg block on a horizontal surface is accelerated along the surface by a 10-N force applied at 40° to the horizontal (Fig. 4.16). Find

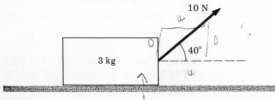

FIGURE 4.16   Problem 23.

the magnitudes of (*a*) the parallel and (*b*) the perpendicular components of the applied force. (*c*) What is the magnitude of the contact force acting on the block? (*d*) The coefficient of kinetic friction between the block and the surface is 0.25. What is the force of friction on the block? (*e*) What is the acceleration of the block?

*Ans.* (*a*) 7.66 N; (*b*) 6.43 N; (*c*) 23.0 N; (*d*) 5.75 N; (*e*) 0.637 m/s²

24 A skier slides down a slope inclined 9° to the horizontal. (*a*) Neglecting friction, what is the acceleration of the skier? (*b*) If the slope is 50 m long, how long does it take the skier to reach the bottom? Assume he starts from rest at the top. (*c*) What is his speed when he reaches the bottom of the slope?

25 A 5.1-kg block is attached by a cord to a 7.4-kg block, as shown in Fig. 4.17. The 5.1-kg block hangs freely, while the 7.4-kg block is supported on a frictionless table. Initially both blocks are at rest, but as the 5.1-kg block falls, it pulls the 7.4-kg block along with it. Since the blocks are attached, they have the same speed and acceleration at any instant. Write equations similar to Eqs. 4.9 and 4.10 for these masses. (*a*) Eliminate the tension between these equations to find the acceleration of the masses. (*b*) What is the tension in the cord?

*Ans.* (*a*) 4 m/s²; (*b*) 29.6 N

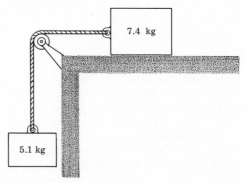

FIGURE 4.17   Problem 25.

26 If the masses $m_1$ and $m_2$ in Fig. 4.11 are 5 and 3 kg, respectively, what is (*a*) the acceleration of the masses and (*b*) the tension in the cord?

27 Find (*a*) the speed and (*b*) the period of a spaceship orbiting around the moon. The moon's radius is $1.74 \times 10^6$ m, and the acceleration due to gravity on the moon is

1.63 m/s². (Assume the spaceship is orbiting just above the moon's surface.)

*Ans.* (*a*) 1.68 km/s; (*b*) 106 min

28 The moon travels in an approximately circular orbit of radius $3.8 \times 10^8$ m about the earth, completing one revolution every 27.3 days. What is (*a*) the orbital speed and (*b*) the centripetal acceleration of the moon?

**REMARK** The moon's acceleration is produced by the earth's gravitational attraction of the moon. The calculation in Prob. 28 was first done by Newton to find how the earth's gravity decreases with distance from the earth (Sec. 5.4).

## BIBLIOGRAPHY

BUTTERFIELD, HERBERT: "The Origins of Modern Science, 1300–1800," rev. ed., Free Press, New York, 1965. A provocative account of the intellectual struggles that led to the scientific revolution. The critical issues that confronted men like Copernicus and Galileo are placed in their historical context.

GALILEI, GALILEO: "Dialogue concerning the Two Chief World Systems," trans. by Stillman Drake, rev. ed., University of California Press, Berkeley, 1967. Galileo's masterful account of the physical implications of Ptolemy's geocentric and Copernicus' heliocentric world systems. It is written in the form of a dialogue among three intelligent Venetian gentlemen, who meet to discuss the merits of these two systems. In the course of their witty and far-ranging discussion, most of Galileo's research on motion is expounded.

SHAMOS, MORRIS H.: "Great Experiments in Physics," Holt, Rinehart and Winston, New York, 1959. A compilation of excerpts from the works of great physicists. Chapter 2 contains passages concerning accelerated motion from Galileo's "Dialogue." Chapter 4 contains passages concerning the laws of motion from Newton's "Principia."

TRICKER, R. A. R., and B. J. K. TRICKER: "The Science of Mechanics," American Elsevier Publishing Company, Inc., New York, 1967. The principles of linear and rotational dynamics are applied to specific biological and athletic events.

# CHAPTER 5 energy

A complete description of the motion of an object can be obtained from Newton's second law (Property 7, Sec. 4.3), which relates the total force on an object to its acceleration. However, this usually requires complex mathematics, whereas for many purposes such a complete description of the motion is not desired. Often relevant information can be found more easily by using the concepts of work and energy introduced in this chapter. The concepts of kinetic and potential energy are especially useful in situations where there is no friction, such as the motion of astronomical bodies in outer space and the motion of atoms inside molecules. Kinetic and potential energy are only two of the many forms of energy, all of which are related by the law of conservation of energy.

## 5.1 WORK

Forces differ in the amount of effort required to maintain them. For now, effort is loosely taken to be the changes in the environment that accompany certain forces. For example, no effort is required to maintain the contact force $F_c$ exerted by the table on the block in Fig. 5.1 because nothing in the block's environment is changing. On the other hand, continuous effort is necessary to maintain the applied force $F_a$ that is moving the block horizontally across the table in Fig. 5.2. Maintaining this force requires continuous change in the block's environment, which in this case is the change in the position of the weight.

The difference between $F_c$ and $F_a$ is not just that $F_a$ is applied to a moving block whereas $F_c$ is applied to a block at rest. After all, the contact force $F_c$ continues to act while the block in Fig. 5.2 is moving, but still no effort is needed to maintain it. What is significant is that $F_a$ is applied parallel to the direction of motion whereas $F_c$ is perpendicular to the motion. It only requires effort to maintain a force parallel to the direction of motion.

The amount of effort expended in maintaining a force is related to the *work* done by the force. The work done by a constant force $F$ in moving an object a distance $d$ is

$$W = F_x d \qquad\qquad 5.1$$

where $F_x$ is the magnitude of the component of $F$ parallel to the direction of motion (Fig. 5.3). The sign of $W$ is taken to be positive when $F_x$ points in the direction in which the object is moving and negative when $F_x$ points in the opposite direction. Positive work, which is work done on the object by the environment, requires the environment to expend effort, whereas negative work, which is work done by the object on the environment, requires the object to expend effort.

For example, $\mathbf{F}_a$ is parallel to the direction of motion of the block in Fig. 5.2, so the work $W_a$ done by $\mathbf{F}_a$ in moving the block a distance $d$ to the right is just $F_a d$ and is positive. Thus if $F_a = 5$ N and $d = 3$ m, this work is

$$W_a = F_a d = 5 \text{ N} \times 3 \text{ m} = 15 \text{ N-m}$$

Since $\mathbf{F}_c$ is perpendicular to the direction of motion, it has no component parallel to the motion. Consequently the work $W_c$ done by $\mathbf{F}_c$ is zero. Likewise the work $W_g$ done by the force of gravity $\mathbf{F}_g$ is also zero in this case. The force of friction $\mathbf{F}_f$ on the block is opposite to the direction of motion, so if $F_f = 3$ N, the work $W_f$ done by $\mathbf{F}_f$ is

$$W_f = -F_f d = -3 \text{ N} \times 3 \text{ m} = -9 \text{ N-m}$$

This is work done by the block on the table. The total work done by all the forces acting on the block is

$$\begin{aligned} W &= W_a + W_c + W_g + W_f \\ &= 15 \text{ N-m} + 0 + 0 - 9 \text{ N-m} = 6 \text{ N-m} \end{aligned}$$

The effort required to move the block is related to the work $W_a$ done by $\mathbf{F}_a$. The difference between $W$, the total work done on the block, and $W_a$ is the frictional work the block does on the table.

Figure 5.3 shows a 5-N force $\mathbf{F}_a$ being applied to a block at an angle of 30° to the horizontal. The magnitude $F_{ax}$ of its component parallel to the direction of motion is

$$F_{ax} = F_a \cos 30° = 5 \text{ N} \times 0.86 = 4.4 \text{ N}$$

so the work done in moving the block 3 m is

$$W_a = F_{ax} d = 4.4 \text{ N} \times 3 \text{ m} = 13.2 \text{ N-m}$$

Only the component of a force parallel to the direction of motion enters into the calculation of work.

In the mks system, the unit of work is the newton-meter (N-m). Because it is such an important unit, it is given its own name, the *joule* (J):

$$1 \text{ J} = 1 \text{ N-m}$$

In the cgs system, the unit of work is the dyne-centimeter (dyn-cm), or *erg*:

$$1 \text{ erg} = 1 \text{ dyn-cm}$$

Since $1 \text{ m} = 10^2$ cm and $1 \text{ N} = 10^5$ dyn, the conversion from joules to ergs is

$$\begin{aligned} 1 \text{ J} &= 1 \text{ N-m} = 10^5 \text{ dyn} \times 10^2 \text{ cm} \\ &= 10^7 \text{ dyn-cm} = 10^7 \text{ ergs} \end{aligned}$$

In the English system, the unit of work is the foot-pound (ft-lb), which equals 1.35 J.

**REMARK** Notice that torque and work have the same dimension [force ×

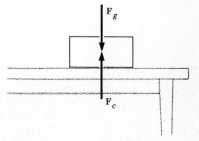

FIGURE 5.1
Forces on a block resting on a table.

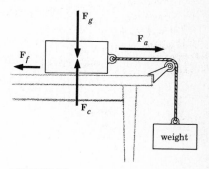

FIGURE 5.2
Forces on a block moving horizontally across a table.

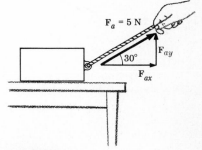

FIGURE 5.3
A 5-N force $\mathbf{F}_a$ applied to a block at an angle of 30° to the horizontal.

length] and the same units. To distinguish between them, the unit of torque in the English system is usually written lb-ft, whereas the unit of work is written ft-lb. In the mks and cgs systems no such convention is needed since the units of work have their own name. In spite of their common dimension, torque and work are entirely different concepts.

Although effort is always expended in doing work, the reverse is not necessarily true. That is, effort can be expended without any work being done. For instance, effort is required to hold a weight in a person's hand, even though no work is done as long as the weight is not moved. The effort expended in this case is evident by the chemical changes that occur in the muscles used to hold the weight.

Effort is not a technical word. It has been used loosely here to refer to the changes in the environment that accompany certain mechanical processes. Quantitatively these changes are measured by *energy*, which is equal to the maximum work the change could produce. Conversely, the work done by a force is equal to the minimum amount of energy that must be expended in maintaining the force. That is, work always requires the expenditure of energy, but energy can be expended without doing work. For instance, since zero work is done in supporting a weight, the weight can be supported without expending energy, e.g., by putting it on a table, although energy will be expended if the weight is held in a person's hand. On the other hand, at least 15 J of energy must be expended by whatever mechanism applies the 5-N force $\mathbf{F}_a$ to move the block 3 m in Fig. 5.2. If the force is applied by a hand pulling the cord, more than 15 J of energy will actually be expended by the muscles.

## 5.2 KINETIC ENERGY

There are many forms of energy, each of which is recognized by its capacity to do work. The most familiar form of energy is the energy of motion, or *kinetic energy*. The fundamental relation between kinetic energy and work will be derived by considering a mass $m$ acted upon by a constant force $\mathbf{F}$. Because $\mathbf{F}$ is constant, the work $W$ done by $\mathbf{F}$ in moving the mass a distance $d$ is

$$W = Fd \qquad 5.2$$

and from Eq. 4.7 the constant acceleration $a$ of the mass is

$$a = \frac{F}{m} \qquad 5.3$$

Suppose that the mass, starting from rest, moves the distance $d$ in the time $t$. Then from Eqs. 4.2 and 4.3 we have that the distance moved is

$$d = \tfrac{1}{2}at^2 \qquad 5.4$$

and that the speed attained is

$$v = at \qquad 5.5$$

These two equations relate distance to time and speed to time, respectively. From them a relation between distance and speed can be derived by eliminating the time.

First Eq. 5.5 is solved for $t$,

$$t = \frac{v}{a}$$

and this is substituted into Eq. 5.4 to give

$$d = \tfrac{1}{2}a\left(\frac{v}{a}\right)^2 = \tfrac{1}{2}a\frac{v^2}{a^2} = \frac{1}{2}\frac{v^2}{a}$$

Second, this is multiplied on both sides by $a$

$$da = \tfrac{1}{2}v^2$$

and Eq. 5.3 is substituted for $a$:

$$\frac{dF}{m} = \tfrac{1}{2}v^2$$

Finally this is multiplied on both sides by $m$ to give

$$Fd = \tfrac{1}{2}mv^2 \qquad\qquad 5.6$$

The quantity $\tfrac{1}{2}mv^2$ is the *kinetic energy* $K$ of the mass,

$$K = \tfrac{1}{2}mv^2 \qquad\qquad 5.7$$

The left-hand side of Eq. 5.6 is the work done on the mass by $\mathbf{F}$ (Eq. 5.2), so Eq. 5.6 says that the work done is equal to the kinetic energy attained.

If the force $\mathbf{F}$ is applied to a mass that is moving initially with a speed $v_0$, the more general result is

$$W = \tfrac{1}{2}mv^2 - \tfrac{1}{2}mv_0{}^2 \qquad\qquad 5.8$$

That is, *the work W done on an object is equal to the change in its kinetic energy*. This is a statement of the *work-energy theorem*. For a constant force $\mathbf{F}$, Eq. 5.8 can be derived from Eqs. 4.2 and 4.4 in the same way that Eq. 5.6 was derived from Eqs. 4.2 and 4.3. However, Eq. 5.8 is true in general, even if the force changes in magnitude and direction while it acts on the object.

The work-energy theorem proves that $\tfrac{1}{2}mv^2$ is a form of energy because it shows that $\tfrac{1}{2}mv^2$ is the amount of work that can be obtained from a moving object. For example, consider the block in Fig. 5.4, which initially is moving to the left with speed $v_0$. As it moves, it slows down, looses kinetic energy, and does work against the force of friction. When it comes to rest, its kinetic energy is zero, so from Eq. 5.8 the work done on the block in changing its kinetic energy from $\tfrac{1}{2}mv_0{}^2$ to zero is

$$W = 0 - \tfrac{1}{2}mv_0{}^2 = -\tfrac{1}{2}mv_0{}^2$$

The minus sign means that the block has done (positive) work on the environment by moving against the frictional force. The magnitude of this work is the change in the quantity $\tfrac{1}{2}mv^2$, which means that $\tfrac{1}{2}mv^2$ is a form of energy.

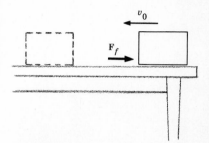

FIGURE 5.4
A block initially moving to the left with speed $v_0$.

**REMARK** The units of kinetic energy must be the same as the units of work. In the mks system, kinetic energy has the units

$$kg \times (m/s)^2 = kg\text{-}m^2/s^2$$

whereas work has the units newton-meters. These are the same, however, since

$$1 \text{ N} = 1 \text{ kg-m/s}^2$$

In applying the work-energy theorem, the total work done on the object must be used. If several forces are involved, it is the sum of the work done by each, with proper regard for sign, that is equal to the change in kinetic energy. For example, suppose a man pushes a crate at a constant speed a distance $d$ by applying a force $\mathbf{F}_a$. Since the crate's speed is constant, its kinetic energy does not change, so the total work done is zero. The work done by the man is $F_a d$. The total force on the crate is zero, since it is not accelerating, so there must be a frictional force $\mathbf{F}_f = -\mathbf{F}_a$ on the crate. The work $W_f$ done by this force is $W_f = -F_a d$, so the total work, $W_a + W_f$, is zero. In this case the work the man does on the crate is converted into the frictional work that the crate does on the floor, rather than into kinetic energy.

**Speed of animals**

An interesting application of the work-energy theorem is provided by the analysis of the running speed in animals. Consider an animal running at constant speed $v$. Once it is in motion, no work is required to keep its body moving (except for a small amount needed because of air resistance). However, as one foot strikes the ground, it is momentarily brought to rest while the rest of the animal's body continues to move forward. When the foot is picked up, it is accelerated forward by one set of muscles in order to catch up with the body. As the foot moves past the body, it is slowed down by a set of antagonistic muscles that bring it to rest on the ground again. One of these sets of muscles does work in changing the kinetic energy of the leg from 0 to $\frac{1}{2}m_l v^2$ while the other does work in changing it back from $\frac{1}{2}m_l v^2$ to 0. Here $m_l$ is the mass of the leg.

**REMARK** The main effort in running is in the work required to continually accelerate and decelerate the legs. The enormous advantage of the wheel is that it permits locomotion without the need for this continual starting and stopping (Sec. 6.3).

The work $W = \frac{1}{2}m_l v^2$ is done by a set of muscles which exert a force $F_m$ while contracting a distance $d$. From the work-energy theorem we have

$$F_m d = \tfrac{1}{2}m_l v^2$$

or

$$v^2 = \frac{2F_m d}{m_l} \qquad\qquad 5.9$$

This gives the animal's speed in terms of the force exerted by its muscles, the distance these muscles contract, and the mass of its leg.

From Eq. 5.9 we can determine how the speed of an animal depends on its size, i.e., how speed scales. The argument is similar to that given in Sec. 1.4. We consider two animals of different size but similar shape. If the quantities in Eq. 5.9 refer to the smaller animal, the speed of the larger animal is given by a similar equation,

$$v'^2 = \frac{2F'_m d'}{m'_l} \qquad 5.10$$

where the primed quantities all refer to the larger animal. If the larger animal has a scale factor $L$ relative to the smaller animal, all distances in the larger animal are $L$ times the corresponding distances in the smaller animal, so

$$d' = Ld$$

The force $F'_m$ and mass $m'$ are related to $F_m$ and $m$ by the scaling laws discussed in Sec. 1.4:

$$F'_m = L^2 F_m \qquad \text{and} \qquad m'_l = L^3 m_l$$

Substituting these expressions in Eq. 5.10, we get

$$v'^2 = \frac{2(L^2 F_m)(Ld)}{L^3 m_l}$$

$$= \frac{2F_m d}{m_l} = v^2$$

Thus the speed of the larger animal is the same as the speed of the smaller animal. Although this surprising result applies rigorously only to animals of similar shape, it is approximately true for animals ranging in size from rabbits to horses.

Equation 5.9 shows that an especially fast animal must have extra light legs and extra large leg muscles. This can be accomplished by placing the leg muscles inside the body, so that they do not move with the legs. Thus most animals designed for great speed, such as thoroughbreds, deer, and greyhounds, have extremely thin legs.

**REMARK** Most of the great cats have rather massive legs because they have adapted more for great initial acceleration than sustained speed. They rely on surprise and the sudden leap to catch their prey since they are not fast enough to run it down.

## 5.3 POTENTIAL ENERGY

As the object of mass $m$ in Fig. 5.5 is lowered vertically from point $A$ to point $C$, the work done on it by the force of gravity is

$$W_{AC} = F_g h = mgh$$

The sign is positive because $\mathbf{F}_g$ is directed from $A$ to $C$. The work done by gravity in moving the object horizontally from $C$ to $B$ is zero since $\mathbf{F}_g$ is perpendicular to the motion in this case. Thus

$$W_{CB} = 0$$

so the total work done on the object by gravity as it goes from $A$ to $C$, and then from $C$ to $B$, is

$$W_{ACB} = W_{AC} + W_{CB} = mgh + 0 = mgh$$

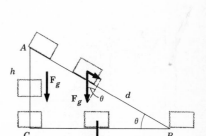

**FIGURE 5.5**
Comparison of the work done by the force of gravity on a mass moving from $A$ to $B$ along the incline with the work done on the mass moving from $A$ to $B$ along the path $ACB$.

The object can also go from $A$ to $B$ by moving diagonally along the incline. The work $W_{AB}$ done by gravity in going along this path is equal to the component of $\mathbf{F}_g$ parallel to the incline times $d$. From Fig. 5.5 this is seen to be

$$W_{AB} = mg \sin \theta \, d$$

But from the right triangle $ACB$ we have

$$\sin \theta = \frac{h}{d}$$

so

$$W_{AB} = mg \frac{h}{d} d = mgh = W_{ACB}$$

Thus the work done by gravity on the object in going from $A$ to $B$ is the same whether it moves along the path $ACB$ or along the diagonal path $AB$. In fact it can be shown that the work done by gravity on the object in going from $A$ to $B$ is the same no matter what path is taken. It is always equal to $mgh$, where $h$ is the separation in the vertical heights of $A$ and $B$.

The fact that the work done is independent of the path taken means that it depends only on the positions of the points $A$ and $B$ themselves. Let $h_A$ be the height of point $A$ above some reference surface, and let $h_B$ be the height of point $B$ from this same surface (Fig. 5.6). Then the work done by gravity on an object going from $A$ to $B$ along any path can be written

$$W_{AB} = mgh_A - mgh_B = mg(h_A - h_B) = mgh \qquad 5.11$$

This expresses the work as the difference of two quantities ($mgh_A$ and $mgh_B$) that depend only on the positions of the points $A$ and $B$, respectively.

When the object is at point $A$, it is said to have the *potential energy* $U_A$, where

$$U_A = mgh_A \qquad 5.12$$

Potential energy is the energy of position, just as kinetic energy is the energy of motion. It is a little more abstract, perhaps, because an object sitting at $A$ does not look any different from an object sitting at $B$, even though its potential energy $U_A = mgh_A$ at $A$ is greater than its potential energy $U_B = mgh_B$ at $B$. That $U$ really is a form of energy follows at once from Eq. 5.11, which shows that the work done in going from $A$ to $B$ is

$$W_{AB} = U_A - U_B \qquad 5.13$$

so that a change in potential energy can produce work.

**REMARK** Since the reference surface is arbitrary, the absolute value of the potential energy at a point is arbitrary. If the reference surface is changed, the potential energy at all points will change by the same amount. This is not important, however, since in practice it is only the difference between potential energies at two points that is of interest. This difference is independent of the position of the reference surface.

The work-energy theorem gives the change in the kinetic energy of an object in terms of the work done on it. If $K_A$ is the kinetic

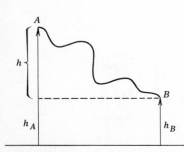

**FIGURE 5.6**
The work done by the force of gravity on a mass moving from point $A$ to point $B$ depends only on the difference $h$ between the heights of the points and not on the path taken.

energy of an object at $A$ and $K_B$ is its kinetic energy at $B$, the work-energy theorem says that

$$K_B - K_A = \text{total work done on object in going from } A \text{ to } B$$

Equation 5.13 gives the work done by gravity on the object in terms of the difference between its potential energies at $A$ and $B$. If gravity is the only force doing work on the object, the total work is given by Eq. 5.13, so

$$K_B - K_A = W_{AB} = U_A - U_B$$

or, rearranging terms,

$$K_A + U_A = K_B + U_B \qquad 5.14$$

The sum $K + U$ of the kinetic and potential energies of an object is called the *mechanical energy* $E_m$. Equation 5.14 says that in the absence of forces other than gravity, the mechanical energy of an object does not change as the object moves from $A$ to $B$. Since $A$ and $B$ are any two points, this means that $E_m$ remains constant as the object moves. This can be expressed as

$$K + U = E_m = \text{const}$$

or, using Eqs. 5.7 and 5.12,

$$\tfrac{1}{2}mv^2 + mgh = E_m = \text{const} \qquad 5.15$$

As an example, suppose a girl drops a 0.3-kg ball from a bridge 12 m above the water. What is the speed $v$ of the ball when it hits the water? Gravity is the only force on the ball (if air resistance is neglected), so Eqs. 5.14 and 5.15 apply. Let $A$ be the position of the ball as it leaves the girl's hand, and let $B$ be its position as it hits the water (Fig. 5.7). If the surface of the water is taken as the reference surface, the potential energy of the ball at $A$ is

$$\begin{aligned} U_A = mgh_A &= 0.3 \text{ kg} \times 9.8 \text{ m/s}^2 \times 12 \text{ m} \\ &= 35.3 \text{ kg-m}^2/\text{s}^2 = 35.3 \text{ N-m} \\ &= 35.3 \text{ J} \end{aligned}$$

and its potential energy at $B$ is

$$U_B = mgh_B = mg \times 0 = 0$$

The ball is at rest at the instant it is released from the girl's hand, so its kinetic energy at $A$ is zero:

$$K_A = \tfrac{1}{2}m \times 0^2 = 0$$

The mechanical energy at $A$ then is

$$E_m = K_A + U_A = 0 + 35.3 \text{ J} = 35.3 \text{ J}$$

and is equal to the mechanical energy at $B$,

$$35.3 \text{ J} = K_B + U_B = \tfrac{1}{2}mv^2$$

Solving this for $v$ gives

$$v = \sqrt{\frac{70.6 \text{ J}}{m}} = \sqrt{\frac{70.6 \text{ kg-m}^2/\text{s}^2}{0.3 \text{ kg}}}$$

$$= \sqrt{235 \text{ m}^2/\text{s}^2} = 15.3 \text{ m/s}$$

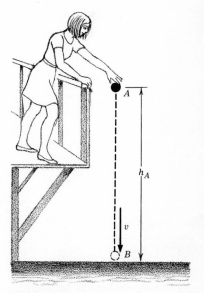

FIGURE 5.7
A girl dropping a ball from a bridge.

**REMARK** The speed $v$ is independent of the mass of the ball since the mass appears as a common factor in both the kinetic and potential energies. This peculiarity of the gravitational force has already been discussed in Chap. 4 in connection with the fact that all objects fall to earth with the same acceleration.

Next, suppose a boy passing under the bridge in a boat tries to return the ball by throwing it straight up in the air. If the speed of the ball as it leaves the boy's hand is 14 m/s, how high will it go? For simplicity assume the boy's hand is at the surface of the water, point $B$, when the ball leaves it. Then the kinetic and potential energies of the ball at point $B$ are

$$K_B = \tfrac{1}{2} \times 0.3 \text{ kg} \times (14 \text{ m/s})^2 = 29.4 \text{ J}$$

and $\qquad U_B = 0$

At point $C$, where the ball reaches its maximum height $h_C$ above the water, its potential and kinetic energies are

$$U_C = mgh_C \qquad \text{and} \qquad K_C = 0$$

The kinetic energy is zero because the ball is instantaneously at rest at its highest point (Sec. 4.2). Then from

$$K_B + U_B = K_C + U_C$$

we have

$$29.4 \text{ J} + 0 = 0 + mgh_C$$

Solving this for $h_C$ gives

$$h_C = \frac{29.4 \text{ J}}{mg} = \frac{29.4 \text{ J}}{0.3 \text{ kg} \times 9.8 \text{ m/s}^2} = 10 \text{ m}$$

Since the girl was 12 m above the water, the boy failed to get the ball back to her.

FIGURE 5.8
By rotating his body as he jumps, a high jumper has to lift his center of gravity only about half the height of the bar.

It is easy to apply these ideas to more complex objects because the potential energy of an extended object just depends on the position of its center of gravity, so that changes in the potential energy of the object depend only on changes in the position of its center of gravity. This is of great importance for the proper understanding of many athletic events. For instance, in high jumping the athlete's leg muscles first do work that gives him his initial kinetic energy. During the jump this initial kinetic energy is converted into the potential energy (height) needed to clear the bar. In order to raise his center of gravity by a height $h$, the athlete must leave the ground with a kinetic energy at least equal to $mgh$. However, the important point is that $h$ is not the height of the bar above the ground. Since the jumper starts his jump from an upright position, his center of gravity is already about 1 m above the ground. As he goes over the bar, the jumper rotates his body so that it is horizontal. In this way the height $h$ that his center of gravity rises is kept as small as possible (Fig. 5.8) and is about half the height of the bar above the ground.

A pendulum is a device that continually converts its energy back and forth between potential and kinetic energy. It consists of a heavy mass $m$ attached to one end of a light rod. The other end of the rod is pivoted freely from a hook (Fig. 5.9). At one end of its swing, point $A$ in Fig. 5.9, its kinetic energy is zero and its potential energy is $mgh$, where $h$ is the vertical height of the mass above its lowest point, point $B$. As the pendulum swings toward point $B$, it loses potential energy and gains kinetic energy until, at point $B$ itself, its potential energy is zero and its kinetic energy equals $mgh$. Then, as it continues to point $C$, it loses kinetic energy and gains potential energy. At point $C$ the kinetic energy is again zero, and all the energy is back in the form of potential energy. If there are no frictional forces on the pendulum, the height it reaches at $C$ is equal to its height at $A$, so that a frictionless pendulum can oscillate indefinitely between $A$ and $C$.

In reality there is always some friction in the pivot and with the air (air resistance). These frictional forces do negative work because they are directed opposite to the motion. Consequently the mechanical energy of the pendulum steadily decreases, as it does work against these forces. Since the pendulum arrives at $C$ with less potential energy than it had at $A$, its height at $C$ is less than $h$. Furthermore, when it swings back to $A$, it does not rise as high as it did at $C$. Thus as it swings back and forth the pendulum continually loses mechanical energy, until it is brought to rest.

The effect of friction can be dramatically demonstrated by allowing a pendulum to fall freely from the inverted position shown in Fig. 5.10. If there were no friction, all the potential energy the mass had in the inverted position would be converted into kinetic energy when it reached the lowest position. This would be just enough kinetic energy to get the mass back up to the inverted position again, and so the rod and mass would revolve indefinitely about the pivot. Of course this does not happen. Because of friction, the mass arrives at point $B$ with less energy than it had at $A$, so it cannot rise all the way back to $A$ again. It gets only as far as point $C$ before it starts to fall back toward $B$. It then continues to oscillate about $B$ with decreasing amplitude.

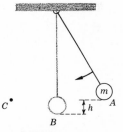

FIGURE 5.9
The mass of a pendulum falls through the vertical distance $h$ as it swings from $A$ to $B$.

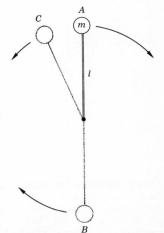

FIGURE 5.10
A pendulum released from rest at $A$ gets only as far as $C$ before it starts to fall back toward $B$.

In gymnastics there is a beautiful stunt done on the high horizontal bar called the *giant swing*. The performer, starting in the inverted position labeled $A$ in Fig. 5.11, falls freely with his body fully extended. He is able, however, to complete a full circle and rise back to the inverted position, so that he can keep rotating in a giant circle about the bar in apparent defiance of the energy loss due to friction. This stunt is worth analyzing in detail because it illustrates several important physical principles.

When the performer starts in position $A$, his center of gravity is a distance of (say) 1.25 m above the bar. If the bar itself is 2.50 m above the ground, his initial potential energy (relative to the ground) is

$$U_A = mg \times 3.75 \text{ m}$$

Since his kinetic energy at $A$ is zero, his mechanical energy at $A$ is

$$E_m(A) = U_A + K_A = mg \times 3.75 \text{ m}$$

For instance, the mechanical energy of a 90-kg gymnast at $A$ is

$$E_m(A) = 90 \text{ kg} \times 9.8 \text{ m/s}^2 \times 3.75 \text{ m} = 3308 \text{ J}$$

If there were no energy loss due to friction, this would also be his mechanical energy at points $B$, $C$, and $D$. There is always some energy loss, however, so that his mechanical energy at $B$ is slightly less than at $A$. Suppose that 25 J is lost on his way down from $A$ to $B$ and that another 25 J is lost on his way back up to position $D$. Then in position $D$ his total mechanical energy is

$$E_m(D) = 3308 \text{ J} - 50 \text{ J} = 3258 \text{ J}$$

and this is all potential energy. The height of his center of gravity at $D$ then is given by

$$mgh_D = 3258 \text{ J}$$

so

$$h_D = \frac{3258 \text{ J}}{90 \text{ kg} \times 9.8 \text{ m/s}^2} = 3.694 \text{ m}$$

This is 0.056 m less than the height of 3.75 m from which he started in position $A$.

For a pendulum this would mean that it could not reach position $A$ but instead would fall back toward $B$. The gymnast can reach $A$, however, because as he passes through the horizontal position $C$ he bends his body slightly at the waist and pulls in toward the bar with his arms. This moves his center of gravity the necessary 0.056 m closer to the bar so that he will rise up over the bar in position $D$ with his body bent. While in this position he straightens his body, raising his center of gravity 0.056 m. This is accomplished by the forces exerted by his muscles. The work done by his muscles equals the work done in raising 90 kg a distance of 0.056 m or

$$W = 90 \text{ kg} \times 9.8 \text{ m/s}^2 \times 0.056 \text{ m} = 50 \text{ J}$$

which is just equal to the energy lost in friction. Thus the gymnast is able to keep returning to his original position by supplying with

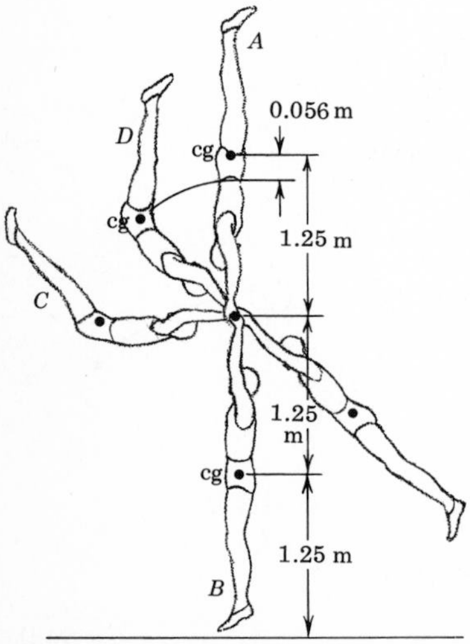

FIGURE 5.11
A gymnast starting from rest at $A$ is able to swing completely around the bar because at $C$ he pulls himself in toward the bar. As he returns to $A$, his center of gravity is lower than when he started, but by straightening his body at this point he lifts his center of gravity back to its original height.

his arm muscles the energy lost in friction. A similar principle is
involved in "pumping" a swing or trapeze.

## 5.4  OTHER POTENTIALS

The fact that the sum of the kinetic and potential energies is a constant in the absence of friction may appear to be of limited use
because there always seems to be some friction present. Actually,
although friction is common in everyday life, it plays no role in
astronomical situations (such as the motion of the moon about the
earth or the earth about the sun) or in the workings of atoms and
molecules. In these domains of basic physics, the mechanical energy
is absolutely conserved. However, the potential energies in these
situations are different in form from the potential used in the last
section.

The expression $mgh$ for the gravitational potential energy is valid
only near the surface of the earth, where the force of gravity can
be considered to be approximately constant. For problems involving
objects far from the earth's surface, such as an artificial satellite or
the moon itself, account must be taken of the variation of the force
of gravity with distance. The variation was discovered by Newton,
who showed that the force acting on the moon to hold it in orbit
about the earth is the same gravitational force (diminished by distance) acting on objects at the surface of the earth. Furthermore
he showed that gravity is not unique to the earth but is a force that
exists between any two objects. Newton made these discoveries by
deducing the force required to explain the known facts about planetary motion. We cannot go into his method here but only state the
results.

**Universal law of gravitation**  *Between any two objects of mass $m_1$ and
$m_2$ there exists an attractive force proportional to the product of the
masses and inversely proportional to the square of the distance between
them.* In symbols this force is written

$$F = G\frac{m_1 m_2}{r^2} \qquad\qquad 5.16$$

where $r$ is the distance between the objects and $G$ is a universal
constant of nature. In mks units

$$G = (6.673 \pm 0.003) \times 10^{-11} \text{ N-m}^2/\text{kg}^2$$

**REMARKS**  (1) Forces always come in pairs, of course, and Eq. 5.16 gives the
magnitude of either member of the pair. Since the force is attractive, the force
$\mathbf{F}_1$ on $m_1$ is directed toward $m_2$ and the (reaction) force $\mathbf{F}_2$ on $m_2$ is directed
toward $m_1$ (Fig. 5.12). (2) Equation 5.16 holds exactly only for point objects
(particles). It holds for extended objects only if $r$ is very much larger than
the objects themselves. The one very important exception to this occurs when
the object is a sphere. For spherical objects it is a remarkable fact, first proved
by Newton, that Eq. 5.16 holds if the distance $r$ is measured from the center
of the sphere.

Since the earth is nearly spherical, Eq. 5.16 can be used for objects

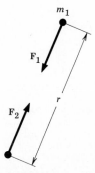

FIGURE 5.12
The forces on two masses
attracted to each other by gravity.

on its surface. The force of gravity on an object of mass $m$ near the earth's surface then is

$$F_g = G \frac{m_e m}{R_e^2}$$

where $m_e$ is the mass of the earth and $R_e = 6.37 \times 10^6$ m is the radius of the earth. The acceleration $a$ of the mass is given in terms of $F_g$ by Eq. 4.7: $F_g = ma$. But since all objects have the same acceleration $g$ near the earth's surface, we also have $F_g = mg$, so

$$G \frac{m_e m}{R_e^2} = mg$$

After canceling the mass $m$ on both sides of this equation, we get a very useful relation between the acceleration of gravity $g$ and the mass and the radius of the earth:

$$g = \frac{Gm_e}{R_e^2} \qquad 5.17$$

For instance, the mass of the earth can be determined from this relation because all the other quantities are known. Solving Eq. 5.17 for $m_e$, we get

$$m_e = \frac{gR_e^2}{G} = \frac{(9.8 \text{ m/s}^2)(6.37 \times 10^6 \text{ m})^2}{6.67 \times 10^{-11} \text{ N-m}^2/\text{kg}^2}$$

$$= 6.0 \times 10^{24} \text{ m-kg}^2/\text{N-s}^2$$

$$= 6.0 \times 10^{24} \text{ kg}$$

where we have used the fact that 1 N = 1 kg-m/s$^2$ to properly reduce the units to kilograms.

Equation 5.16 must be used when dealing with objects that change their distance from the center of the earth. The work done by this force in moving the mass $m_2$ from one point to another (Fig. 5.13) does not depend on the path taken, which means that the work can be expressed in terms of the potential energy at the two points. With the integral calculus it can be shown that the work done by this force in moving $m_2$ from point $A$ to point $B$ is

$$W_{AB} = U_A - U_B$$

where the potential energy $U$ of $m_2$ at a point a distance $r$ from $m_1$ is

$$U = \frac{-Gm_1 m_2}{r} \qquad 5.18$$

Equation 5.18 is the proper form of the gravitational potential energy to use for objects that are a large distance from the earth. It is also valid for objects close to the earth, of course. Thus the potential energy of a mass $m$ at point $B$ on the surface of the earth (Fig. 5.14) is

$$U_B = \frac{-Gm_e m}{R_e}$$

At point $A$, a height $h$ above the surface of the earth, its potential energy is

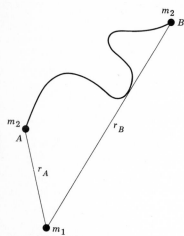

FIGURE 5.13
The work done by gravity on a mass moving from $A$ to $B$ does not depend on the path taken.

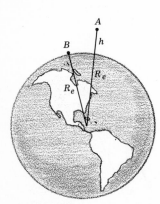

FIGURE 5.14
The potential energy of a mass at point $A$ is greater than at point $B$.

$$U_A = \frac{-Gm_em}{R_e + h}$$

The difference in potential energy between these two positions is

$$U_A - U_B = \frac{-Gm_em}{R_e + h} - \frac{-Gm_em}{R_e}$$

This looks very different from the result $mgh$ that was given for this potential difference in Eq. 5.11. Actually, however, this result is not as different as it looks. Using Eq. 5.17, we can rewrite it

$$U_A - U_B = gR_e{}^2m \left( \frac{1}{R_e} - \frac{1}{R_e + h} \right)$$

$$= gR_e{}^2m \frac{(R_e + h) - R_e}{R_e(R_e + h)}$$

$$= mgR_e{}^2 \frac{h}{R_e(R_e + h)}$$

$$= mgh \frac{R_e}{R_e + h}$$

If $h$ is very small compared to $R_e$, then $R_e + h$ nearly equals $R_e$, in which case this last equation becomes

$$U_A - U_B = mgh$$

Thus, although the values for the gravitational potential energy at points $A$ and $B$ given by Eq. 5.18 are completely different from the values given by Eq. 5.12, both equations give the same value for the *difference* between the potential energies at these points when $h$ is very much less than the radius of the earth. Equation 5.18 is the correct form to use when $h$ is not small compared to $R_e$.

**REMARK** In Eq. 5.12 the potential energy is arbitrarily taken to be zero at some conveniently located reference surface near the surface of the earth, whereas in Eq. 5.18 the potential energy is taken to be zero at an infinite distance from the earth. Since the potential energy increases with distance from the earth, the potential energy given by Eq. 5.18 must be negative for all finite distances from the earth. Equation 5.12 cannot be used to calculate the potential energy at very large distances from the earth because it increases without limit as the distance from the earth becomes large.

As an example of the use of Eq. 5.18, imagine a bit of interplanetary debris of mass $m$ floating about in space very far from the earth. The potential energy of this mass, according to Eq. 5.18, is almost zero since the distance $r$ is very large. If the mass is moving very slowly as well, so that its kinetic energy is also nearly zero, its mechanical energy is (approximately) zero. However the mass may start to drift toward the earth. Once it starts being attracted to the earth, it will pick up speed as it falls to earth. Since the mechanical energy of the mass is a constant, its potential energy must become more negative as its kinetic energy increases. When the mass (which may be a meteor) reaches the earth's atmosphere, its potential and kinetic energies are

$$U = \frac{-Gm_em}{R_e} \quad \text{and} \quad K = \tfrac{1}{2}mv^2$$

Since the mechanical energy is still zero, we have

$$0 = \tfrac{1}{2}mv^2 - \frac{Gm_em}{R_e}$$

(The distance from the earth's surface to the top of the earth's atmosphere is negligible compared to the radius of the earth.) Solving this last equation for $v$, we get

$$v = \sqrt{\frac{2Gm_e}{R_e}} = \sqrt{2gR_e} = 11.2 \text{ km/s}$$

which is the speed with which a meteor enters the earth's atmosphere. At this high speed, the frictional forces on the meteor due to air resistance are so enormous that it usually burns up. A spacecraft reentering the earth's atmosphere must have a specially designed nose cone to protect it from the tremendous heat generated.

**REMARK**  Compare this last example with the example of the ball dropped by the girl in Sec. 5.3. Although the form of the potential energy used is different in these two examples, the reasoning, based on the fact that the mechanical energy is a constant, is identical.

As another example, we shall calculate the speed with which a rocket of mass $m$ must be shot from the earth in order to completely escape from the earth's gravity. At the surface of the earth its potential energy is

$$U_e = -\frac{Gm_em}{R_e}$$

and at infinite distance from the earth its potential energy is $U_\infty = 0$. We want the minimum speed the rocket must have to escape, so we can assume it has zero kinetic energy when it finally gets to infinity, $K_\infty = 0$. The mechanical energy of the rocket is a constant, so

$$U_e + K_e = U_\infty + K_\infty = 0$$

and
$$K_e = -U_e = \frac{Gm_em}{R_e}$$

or
$$v^2 = \frac{2Gm_e}{R_e} = 2gR_e$$

This gives $v = 11.2$ km/s, which is the same as the speed with which a meteor enters the earth's atmosphere. Since this is also the minimum speed needed to escape completely from the earth, it is called the *escape velocity*. It is approximately the speed a rocket ship has as it leaves the earth on a journey to the moon.

It is sometimes useful to plot a graph of the potential energy of an object. The potential energy of a 1-kg mass at the surface of the earth is

$$U = -\frac{Gm_e}{R_e} \times 1 \text{ kg}$$
$$= -gR_e \times 1 \text{ kg} = -6.3 \times 10^7 \text{ J}$$

and it increases, i.e., becomes less negative, as the mass moves away from the earth. Figure 5.15 is a plot of the potential energy $U$ of the mass against its distance $r$ from the center of the earth. If the mass were projected upward from the surface of the earth with a kinetic energy of $5.0 \times 10^7$ J, its total energy would be $-1.3 \times 10^7$ J. This energy is represented by the horizontal line $ABC$ on the potential-energy diagram (Fig. 5.15). The mass starts at point $A$, where its potential energy is $-6.3 \times 10^7$ J. Its kinetic energy at this point ($5.0 \times 10^7$ J) is equal to the difference between the horizontal line and the potential-energy curve. As the mass moves away from the earth, its potential energy increases and its kinetic energy decreases. At point $B$, for instance, a distance $2R_e$ from the center of the earth, its potential energy is $-3.15 \times 10^7$ J, and its kinetic energy is only $1.85 \times 10^7$ J. The object continues to move out to point $C$, about $4.5R_e$ from the center of the earth, where its potential energy is equal to its total mechanical energy. At this point the kinetic energy is zero. The mass is momentarily at rest at $C$ before it starts to fall back to earth.

It is clear from this diagram that a 1-kg mass on the surface of the earth must be given a kinetic energy of at least $6.3 \times 10^7$ J in order to escape from the earth completely. In the language of chemistry, we can say that the mass at rest on the earth is bound to the earth with a binding energy of $6.3 \times 10^7$ J. This is the minimum energy required to dissociate this earth-mass "molecule." If this much energy is supplied, the mechanical energy of the mass would be represented by a horizontal line along the zero-energy axis. The mass could then move to infinity before stopping, thus dissociating the "molecule."

The binding of atoms in a molecule is very similar to the binding of the mass to the earth. The atoms in a molecule are attracted to each other by electrostatic forces that are similar in some respects

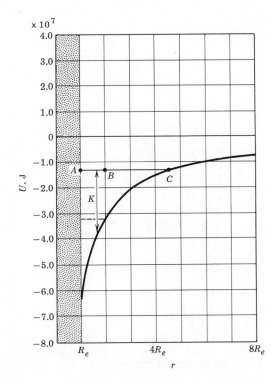

FIGURE 5.15
Plot of the potential energy $U$ of a 1-kg mass against its distance $r$ from the center of the earth.

to the gravitational force. However, because of the complex quantum-mechanical behavior of the electrons in the atoms, the characteristics of the force and the corresponding potential energy are greatly modified (Sec. 19.5). Although no simple formula can be given for the potential energy of an atom in a molecule, a plot of the potential similar to Fig. 5.15 can be drawn.

For example, Fig. 5.16 plots the potential energy $U$ of a hydrogen atom against its distance $r$ from a fluorine atom. Note that as the hydrogen atom gets close to the fluorine atom, the potential energy increases sharply. This rapid increase in potential energy represents, on an atomic scale, the surface of the fluorine atom. When the hydrogen atom gets very close to the fluorine atom, it experiences a repulsive force which is the atomic equivalent of a contact force.

A hydrogen atom with $-4 \times 10^{-19}$ J of mechanical energy is represented by the horizontal line $ABC$ in Fig. 5.16. This atom is momentarily at rest at point $C$ ($1.6 \times 10^{-10}$ m from the center of the fluorine atom), where its kinetic energy is zero. However, it is attracted to the fluorine atom, so it starts to move. At point $B$ ($1.0 \times 10^{-10}$ m from the fluorine atom) it has a kinetic energy of $5.6 \times 10^{-19}$ J. As the hydrogen atom moves still closer to the fluorine atom, it starts to touch the outer surface of the fluorine atom. This surface exerts a repulsive force on the atom that slows it down. At point $A$ the kinetic energy of the hydrogen atom is again momentarily zero, but since it is now being repelled by the fluorine atom, it starts moving back toward $C$. The hydrogen atom thus vibrates in and out, much like the back-and-forth motion of a pendulum. This is the way a hydrogen atom is bound to a fluorine atom in a hydrogen fluoride molecule.

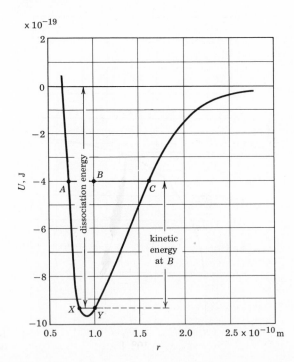

**FIGURE 5.16**
Plot of the potential energy $U$ of a hydrogen atom against its distance $r$ from a fluorine atom.

The hydrogen atom can have a number of different negative energies and still be bound to the fluorine. However, in its normal state, the hydrogen atom is almost at rest. It has an energy of about $-9.5 \times 10^{-19}$ J and vibrates very slightly between points $X$ and $Y$ in Fig. 5.16. To dissociate a molecule in this state, the hydrogen atom must get enough energy to enable it to separate from the fluorine atom completely. From the potential-energy diagram it is clear that the hydrogen atom must have at least zero energy to separate, so that at least $9.5 \times 10^{-19}$ J must be supplied to dissociate the molecule. Likewise, if a free hydrogen atom gets captured by a fluorine atom, this much energy would be released. Chemists usually give the energy of formation of a mole ($6.02 \times 10^{23}$ molecules). In this case the energy of formation of a mole of hydrogen fluoride is

$$(6.02 \times 10^{23})(9.5 \times 10^{-19}) \text{ J} = 5.7 \times 10^5 \text{ J}$$

This is also the "strength" of the chemical bond between hydrogen and fluorine.

## 5.5  HEAT AND THE CONSERVATION OF ENERGY

The results of this chapter up to now can be summarized as follows:
1 Work-energy theorem: The total work done on an object is equal to the change in its kinetic energy.
2 In the absence of frictional forces the sum of the potential and kinetic energies of an object, i.e., its mechanical energy, is a constant.

In the presence of frictional forces, the mechanical energy is not a constant but steadily decreases. To be perfectly clear about this, let us consider a 3-kg block that slides down the incline shown in Fig. 5.17. The work done by gravity on the block is

$$W_g = U_A - U_B = mgh = 3 \text{ kg} \times 9.8 \text{ m/s}^2 \times 6 \text{ m}$$
$$= 176 \text{ J}$$

The contact force $\mathbf{F}_c$ is perpendicular to the incline, so it does no work, but the frictional force $\mathbf{F}_f$ is directed opposite to the motion of the block, so it does negative work. If the frictional force is 12 N, the work it does is

$$W_f = -12 \text{ N} \times 9 \text{ m} = -108 \text{ J}$$

and so the total work done on the block is

$$W = W_g + W_f + W_c = 176 \text{ J} - 108 \text{ J} + 0 = 68 \text{ J}$$

From the work-energy theorem, this is equal to the change in the kinetic energy of the block in going from $A$ to $B$:

$$W = W_g + W_f = U_A - U_B + W_f = K_B - K_A = 68 \text{ J}$$

By rearranging terms this equation can be written

$$K_B + U_B = K_A + U_A + W_f$$

or

$$E_m(B) - E_m(A) = W_f = -108 \text{ J}$$

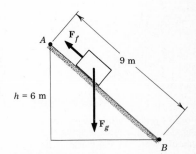

FIGURE 5.17
A 3-kg block sliding down an incline.

That is, the mechanical energy of the block at $B$ is less than its mechanical energy at $A$ by just the amount of work done by friction.

The question naturally arises: What happened to this energy? Is it lost, or has it been converted into still another form? It was over 150 years after the publication of Newton's "Principia" before these questions found satisfactory answers. During the 1840s James Prescott Joule (1818-1889) did a series of experiments showing that the loss of mechanical energy is accompanied by the creation of a definite amount of *heat*. These experiments demonstrated that heat is another form of energy, an idea that had been suggested as early as 1798 by the American expatriate Benjamin Thompson (1753-1814).

**REMARK** Thompson is one of the first great American scientists (along with Benjamin Franklin), and certainly the most colorful. He was born in Woburn, Massachusetts, in 1753, but after supporting the wrong side in the War of Independence, he moved to Europe, where he spent the rest of his life. He served as the war minister for Bavaria, where he received the title Count Rumford.

According to Joule, then, the decrease in the mechanical energy of the block as it goes from $A$ to $B$ is equal to the heat created by friction. Of course it is well known that friction can produce heat. Primitive people have various devices for rubbing wood together to create fire, and we rub our hands together to make them warm. The important physical fact that Joule discovered is that a definite amount of heat is produced whenever a definite amount of mechanical energy is lost producing it.

Heat is measured by the rise in temperature it causes. One unit of heat is the *calorie* (cal), defined as the amount of heat required to raise the temperature of one gram of water one Celsius degree.† Since careful measurements show that the amount of heat required depends on the temperature of the water, the temperature is stated in defining the calorie. By agreement the calorie used today is the 15° calorie, i.e., the amount of heat required to raise the temperature of water from 14.5 to 15.5°C. In terms of this calorie, it requires 1.004 cal to raise the temperature of one gram of water from 4 to 5°C and 0.998 cal to raise its temperature from 30 to 31°C. This small variation with temperature can be neglected for most purposes.

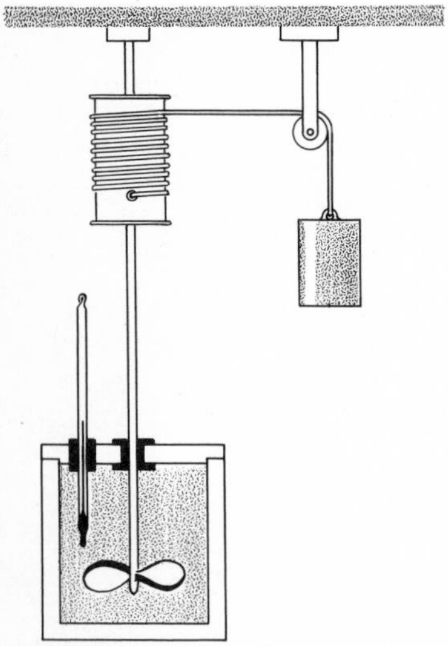

In a series of careful measurements, Joule showed that the loss of a definite amount of mechanical energy results in a definite rise in temperature and so is equivalent to a definite amount of heat. His experiments were simple in concept but very difficult to execute properly. His most famous device consisted of a weight connected by a cord to a paddle wheel in such a way that as the weight fell, its potential energy was converted into the kinetic energy of the paddle wheel (Fig. 5.18). The paddle wheel was immersed in an insulated pail filled with water, so that its kinetic energy could be converted into heat. A thermometer measured the rise in the temperature of the water produced by the fall of a known mass through a known distance.

The experiment is difficult to do properly because a large amount of mechanical energy must be expended to produce even a small rise in temperature. Great care must be taken to prevent the water from cooling while the experiment is in progress. To get a temperature

FIGURE 5.18
Joule's paddle-wheel experiment. As the weight falls, it spins the paddle wheel in the water. The temperature rise of the water measures the amount of heat produced from the expenditure of a known amount of mechanical energy.

† The Celsius temperature scale was formerly called the centigrade scale.

rise of 0.30°C in a pail containing 6.0 kg of water, Joule had to repeat-edly lower a 25-kg mass a distance of 1.5 m.† The potential energy of the mass before it is lowered is

$$25 \text{ kg} \times 9.8 \text{ m/s}^2 \times 1.5 \text{ m} = 367 \text{ J}$$

Each time the mass is lowered this much energy is converted into kinetic energy in the paddle wheel and then dissipated as heat in the water. Joule lowered the mass 20 times to get a 0.30°C temperature rise, so that a total of $20 \times 367 \text{ J} = 7340 \text{ J}$ was converted into heat.

The 0.30°C rise in the temperature of 6000 g of water corresponds to $0.30 \times 6000 = 1800$ cal. This is equivalent to 7340 J of mechanical energy, so

$$1 \text{ J} = \frac{1800 \text{ cal}}{7340} = 0.245 \text{ cal}$$

Joule repeated the experiment many times under various conditions and always found that the loss of 1 J of mechanical energy was accompanied by the appearance of about 0.24 cal of heat. This con-version factor is called the *mechanical equivalent of heat.* Modern measurements give its value as 0.23895 cal/J or 4.185 J/cal.

The calorie used in evaluating the nutrition of food is a kilocalorie (kcal), which is the amount of heat required to raise the temperature of one kilogram of water from 14.5 to 15.5°C. Its relation to the calorie and the joule is

$$1 \text{ kcal} = 10^3 \text{ cal} = 4185 \text{ J}$$

Thus a daily diet containing 2500 kcal is an energy intake of

$$2500 \times 4185 \approx 10^7 \text{ J}$$

If a person on this diet is neither gaining nor losing weight, all this energy is expended during each 24-h period. This rate of energy expenditure is about the same as that of a 100-watt light bulb.

There are other forms of energy besides kinetic energy, potential energy, and heat. Light, sound, electric energy, chemical energy, and nuclear energy are forms of energy that will be discussed later in this book. Each form of energy is characterized by its ability to transform, under suitable conditions, into another form of energy and by its ability to do work. After the experiments of Joule it gradually became evident that although energy may change from one form to another, the total amount of energy does not change. This fact has been demonstrated in thousands of experiments performed over the last hundred years and is now believed to be a fundamental law of physics. It can be stated formally as follows.

**Conservation of energy** *Energy can neither be created nor destroyed but only transformed from one form into another.* Consequently in a closed system, in which energy can neither enter nor leave, *the total energy is a constant.*

As an example, consider a hydroelectric power station converting the potential energy of water into electric energy. The potential energy of the water at the top of the dam is transformed into the kinetic energy of a turbine. The turbine connects to a generator that

---

† These are not the units Joule originally used. His own account of his experiment can be found in Shamos (1959).

converts this into an equal amount of electric energy. Finally, this electric energy reaches the consumer, who converts it into heat and light energy in a light bulb. For each joule of energy that comes out one end of the system, a joule of energy has to be put in at the other end.

This observation leads to an alternate statement of the conservation of energy: *it is impossible to build a perpetual motion machine.* A perpetual motion machine is any device that without any other change produces work indefinitely. Such a device, contrary to the hopes of many uninformed inventors, is impossible because it would be creating energy from nothing.

With the exception of nuclear energy, all the major sources of energy used on earth are derived from the sun. The sun supplies energy in the form of light, which plants are able to convert into chemical energy by photosynthesis. The entire food chain is based on this process, since all animals derive their energy by eating plants or other animals that eat plants. The potential energy of the water in the hydroelectric station comes from the energy of the sun (in the form of heat), which lifts the water from the oceans (by evaporation) and deposits it on the land (rain). Even our fossil fuels, coal and oil, derive their energy from the sun, which grew the prehistoric forests from which these fuels were produced.

The energy of the sun comes from nuclear reactions deep in its interior. The sun is not a perpetual motion machine but is gradually changing its composition as it converts its hydrogen into helium (Sec. 20.3). In time it will exhaust its energy supply and burn out, as will every other star in the universe. Fortunately, the sun still has a few billion years left.

There is one puzzling aspect. If energy is never destroyed and the earth keeps receiving more energy from the sun, why is the world in continuous search for more energy sources? Why don't we just reuse the energy we have? What becomes of all the energy we do use?

If you think about this a while, you will realize that the energy in the light bulb, the energy of an automobile engine, the energy in your food is all eventually converted into heat. The light from the bulb is absorbed by the walls of the room and converted to heat; the work done by the car's engine is converted into heat produced by friction; the chemical energy in your food is ultimately converted into body heat.

This heat is radiated back into the universe at night. It is absolutely essential that this happen because of another fundamental law of physics, the *second law of thermodynamics.* This law (discussed more fully in Chap. 11) says that heat can be converted into usable work only if it can flow from a high-temperature region to a low-temperature region. Without night, the heat generated on earth would accumulate until the earth became uniformly hot. Without any cold region, all this heat would be worthless, because it could not be used to do work.

Thus, although the total energy of the universe is a constant, it is all gradually being converted into heat. When all the energy of the universe is finally in this form, the universe will be a formless void at the end of time.

$KE = \frac{1}{2}uv^2$

$v^2 = \frac{2KE}{m}$

*1* A horizontal force of 24 N pulls a 4-kg mass a distance of 3 m. (*a*) What is the work done by this force? (*b*) If the mass starts from rest, what is its kinetic energy after moving 3 m? (*c*) What is its speed after moving 3 m?

Ans. (*a*) 72 J; (*b*) 72 J; (*c*) 6 m/s

*2* A horizontal force of 60 N acts on a 3-kg mass. If the initial speed of the mass is 5 m/s, what is its speed after moving 5 m?

*3* The brakes are suddenly applied to a 1000-kg car going 25 m/s. The wheels lock, and the car skids 62 m before stopping. (*a*) What is the force of friction on the car? (*b*) What is the coefficient of kinetic friction $\mu_k$ between the tires and the road? (*c*) Show that in general the stopping distance is given by $v^2/2\mu_k g$, independent of mass.

$F_f = U_k F_c$

$F_g < U_k$

$F_c$

Ans. (*a*) 5040 N; (*b*) 0.514

*4* The nearest star is about $4 \times 10^{16}$ m (4 light-years) from earth. (*a*) What speed would a rocket need to get to this star in 8 years? (*b*) How much energy is required to accelerate a $3 \times 10^5$ kg rocket ship to this speed?

**REMARK** The energy required to send a rocket ship to a star in a reasonable time is many times the total annual energy requirement of the United States.

*5* A 10-N force at an angle of 35° to the horizontal is applied to the 5-kg block in Fig. 5.19. In addition, there is a 7-N frictional force acting on the block as shown. (*a*) Find the work

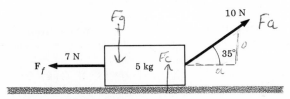

FIGURE 5.19   Problem 5.

done by each of the four forces acting on the block in moving the block 12 m. (*b*) What is the total work done on the block? (*c*) If the block starts from rest, what is its speed after moving the 12 m?

Ans. (*b*) 14.3 J; (*c*) 2.39 m/s

*6* In shuffleboard, a long stick is used to give a shove to a 0.3-kg wooden puck and start it

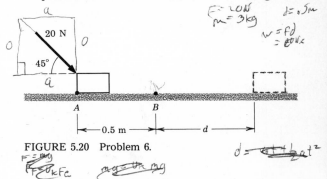

moving across the floor (Fig. 5.20). A player pushes down with a force of 20 N along the

FIGURE 5.20   Problem 6.

$F = mg$

$F_F = U_k F_c$      $mg = U_k mg$

stick, which is inclined 45° to the ground. (*a*) How much work is done on the puck by the stick as the stick pushes the puck from point *A* to point *B*, a distance of 0.5 m? (*b*) If the puck starts from rest at *A*, what is its speed at *B* (neglect friction)? (*c*) From point *B*, where the stick stops pushing, the puck slides a distance *d* before stopping. If the force of friction is 1 N, what is *d*?    $F = 1N$   $\frac{d}{W = Fd}$   $d = \frac{v}{F}$

*7* A bullet of mass $5 \times 10^{-3}$ kg is moving horizontally with a speed of 400 m/s. (Neglect gravity in this problem.) (*a*) What is the kinetic energy of the bullet? (*b*) The bullet penetrates a 5-cm-thick piece of wood, emerging with a speed of 200 m/s. What is the work done on the bullet by the wood? (*c*) What is the force exerted on the bullet by the wood?

Ans. (*a*) 400 J; (*b*) −300 J; (*c*) 6000 N

*8* A 0.2-kg hockey puck is given an initial speed of 12 m/s. If the coefficient of friction between the puck and the ice is 0.1, what is the puck's speed after traveling 15 m?

*9* How does the maximum work done by animals of similar shape depend on size?

Ans. $L^3$

*10* A 1000-kg car traveling 25 m/s collides with a brick wall. The car moves 0.5 m before stopping. What is the force exerted on the car by the wall during the collision?

*11* A 6-kg block starting from rest slides 4 m down the incline in Fig. 5.21. (*a*) What is the potential energy of the block (relative to the bottom of the incline) when it is at the top?

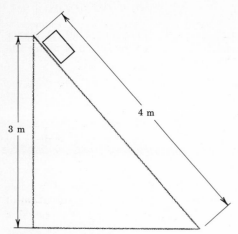

FIGURE 5.21   Problem 11.

(*b*) If the incline is frictionless, what is the speed of the block when it reaches the bottom of the incline? (*c*) If there is a constant 8-N force of friction on the block while it slides down the incline, what is its speed at the bottom?

*Ans.* (*a*) 176 J; (*b*) 7.66 m/s; (*c*) 6.93 m/s

12 A 100-kg skier, skiing along level ground at a speed of 7 m/s, comes to a hill (Fig. 5.22).

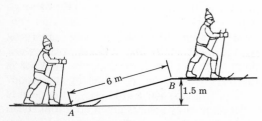

FIGURE 5.22   Problem 12.

(*a*) If the skier coasts up the hill, what is his speed at point *B*, assuming the slope exerts no friction on the skis? (*b*) What is his speed at point *B*, assuming the slope exerts a constant 75-N force of friction on the skis?

13 A man takes a running jump off the high-dive board at the local pool. His speed as he leaves the board is 3 m/s, and the board is 5 m above the water. What is his speed as he hits the water? *Ans.* 10.3 m/s

14 Show that the vertical height $h$ that similar animals can jump is independent of size. (*Hint*: The height depends on the work done by the muscles in the animal's legs, which in turn depends on the force exerted by the muscles and the distance they contract.)

15 A 7-g bullet shot straight up in the air with an initial speed of 200 m/s reaches a vertical height of 900 m. What is the average frictional force on the bullet? *Ans.* 0.087 N

16 Using the principles of scaling, show that the period of a pendulum is proportional to the square root of its length $l$ and is independent of its mass $m$. (The period is the time required to complete one oscillation, and so it scales as $l/v$, where $v$ is the maximum speed of the pendulum.)

17 A girl throws a 0.2-kg ball a distance of 6 m straight up in the air. (*a*) What is the kinetic energy of the ball as it leaves the girl's hand? (*b*) How much work does the girl do in throwing the ball? (*c*) If the girl's arm muscle contracted a distance of 0.05 m while throwing the ball, what was the average force exerted by the muscle?

*Ans.* (*a*) 11.8 J; (*b*) 11.8 J; (*c*) 236 N

18 What is the force of gravity on an 80-kg astronaut when he is (*a*) on the surface of the earth, (*b*) in orbit 200 km above the earth's surface, (*c*) halfway to the moon, a distance of $1.9 \times 10^5$ km from the center of the earth?

19 The mass of the moon is $7.40 \times 10^{22}$ kg, and its radius is $1.74 \times 10^6$ m. (*a*) What is the acceleration of gravity $g$ on the moon? (*b*) What is the force of gravity on an 80-kg astronaut standing on the moon? (*c*) With what speed must a rocket be launched from the moon's surface in order to completely escape from the moon?

*Ans.* (*a*) 1.63 m/s²; (*b*) 130 N; (*c*) 2.4 km/s

20 With what speed must a rocket be launched from the surface of the earth in order to have a speed of 5 km/s when it is infinitely far from earth?

21 The mass of the sun is $2 \times 10^{30}$ kg, and it is $1.5 \times 10^{11}$ m from earth. (*a*) Find the gravitational potential energy of the earth due to the sun's gravitational force. (*b*) What speed would the earth need to escape completely from the sun?

*Ans.* (*a*) $-5.3 \times 10^{33}$ J; (*b*) 42 km/s

22 Calculate the earth's orbital speed about the sun. (*Hint:* Combine Eqs. 4.11 and 5.16, and use the data in Prob. 21.)

23 A rocket of mass 100 kg is shot straight up from the earth with an initial speed of 9000 m/s. (*a*) Using Eq. 5.12, calculate the height above the earth that the rocket reaches. (Equation 5.12 is not valid for the potential energy in this case because the rocket rises too high above the earth for the force of gravity to be considered constant.) (*b*) Using Eq. 5.18, calculate the true height above the earth that the rocket reaches. (*Hint:* First find *r*, the distance the rocket goes from the center of the earth, and then subtract the radius of the earth.)
      *Ans.* (*a*) $4.13 \times 10^6$ m; (*b*) $11.4 \times 10^6$ m

24 A hydrogen atom with a kinetic energy of $4.0 \times 10^{-19}$ J is $1.5 \times 10^{-10}$ m from a fluorine atom. (*a*) What is the potential energy of the hydrogen atom? (*b*) What is the maximum distance that the hydrogen atom can be from the fluorine atom?

25 How much heat is required to raise the temperature of 650 g of water from 22 to 85°C?
      *Ans.* 40.9 kcal

26 How much does 25 kcal raise the temperature of 1.25 kg of water?

27 A 0.4-kg rock falls 1200 m into a bucket containing 2.5 kg of water. How much does the temperature of the water rise?  *Ans.* 0.45°C

28 (*a*) How much work is done by a 50-kg woman in climbing a mountain 1500 m high? (Just consider the work done in lifting the woman the vertical distance.) (*b*) If the woman consumes 4 kcal of food energy for every 1 kcal of work performed, how much food energy is consumed in climbing the mountain?

29 In a paper published in 1845, Joule mentioned that the temperature of the water at the bottom of Niagara Falls is greater than at the top because of the conversion of mechanical energy to heat. Assuming the falls are 50 m high, how much is the temperature rise?
      *Ans.* 0.12°C

**REMARK**  Joule took a very large thermometer with him on his honeymoon in Germany in order to measure the temperature of the water at the top and bottom of waterfalls.

## BIBLIOGRAPHY

FEYNMAN, RICHARD P.: "The Character of Physical Law," M.I.T. Press, Cambridge, Mass., 1965. Chapter 1 gives a very readable account of the origin and development of the law of gravitation. Chapter 3 discusses the conservation of energy and other conservation laws in physics.

————, ROBERT B. LEIGHTON, and MATTHEW SANDS: "Lectures on Physics," Addison-Wesley Publishing Company, Inc., Reading, Mass., 1963. Chapter 4 gives an amusing and insightful account of the law of energy conservation.

RYABOV, Y.: "An Elementary Survey of Celestial Mechanics," Dover Publications, Inc., New York, 1961. A detailed description of the role of gravity in planetary motion. There is an excellent account of the history of the law of gravitation, with some of Newton's original proofs included in an appendix. These proofs can be understood with a knowledge of plane geometry.

SHAMOS, MORRIS H.: "Great Experiments in Physics," Holt, Rinehart and Winston, New York, 1959. Chapter 12 contains Joule's account of his famous paddle-wheel experiment measuring the mechanical equivalent of heat. Also of interest is chap. 6, which contains Cavendish's account of his measurement of the gravitational constant $G$.

SMITH, J. MAYNARD: "Mathematical Ideas in Biology," Cambridge University Press, Cambridge, 1968. The physics of running is treated in chap. 1.

# CHAPTER 6 machines

The last four chapters have considered the effects of the forces that act on an object. In this chapter attention is focused on the thing that applies the force. Of particular interest is the class of devices, called *engines*, that are capable of transforming energy of various forms into work. Engines are necessary for the motion of both machines and animals. The muscles of an animal are its engines, and they serve the same function as the engine in an automobile. Thus there is a deep similarity between animals and engine-driven machines that has intrigued men ever since such machines came into general use.

## 6.1 POWER AND EFFICIENCY

Figure 6.1 shows a moving block acted upon by some familiar forces. The force of gravity $\mathbf{F}_g$ is perpendicular to the horizontal surface, the contact force $\mathbf{F}_c$ is perpendicular to the inclined surface, and the frictional force $\mathbf{F}_f$ and the applied force $\mathbf{F}_a$ are both parallel to the inclined surface. In accordance with Property 1 (Sec. 2.1) each of these forces is exerted by some other object. These forces are all different in character, and the nature of the work that each one does on the block is also different.

The work done by the force of gravity $\mathbf{F}_g$ is equal to the change in the gravitational potential energy of the block. Forces with this property are called *conservative*. If $\mathbf{F}_g$ were the only force acting on the block, the total mechanical energy of the block (the sum of its potential and kinetic energies) would be constant (Sec. 5.3).

No work is done by the contact force because it is always perpendicular to the surface and thus has no component in the direction of motion of the block.

The work done by the force of kinetic friction is always negative. Forces with this property are called *dissipative*. A dissipative force constantly converts kinetic energy into heat, dissipating the mechanical energy of an object. An object that has only the forces $\mathbf{F}_g$, $\mathbf{F}_c$, and $\mathbf{F}_f$ acting on it eventually comes to rest, because the force of friction converts all the object's mechanical energy into heat.

**REMARK** Kinetic friction is the parallel force between two surfaces moving relative to each other. Static friction, which is the parallel force between two stationary surfaces, is not dissipative. For example, the force between a rolling wheel and the ground is static friction, not kinetic, because the wheel is not moving relative to the ground at the point of contact. In such a situation static friction can do either positive or negative work.

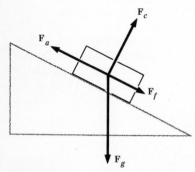

**FIGURE 6.1**
Forces on a block being pushed up an incline.

Only the applied force can do positive work on an object and at the same time increase its mechanical energy. This force is applied by a device, called an *engine*, that converts nonmechanical energy into work. The details of the mechanism used by an engine to do this are not of concern here, but the general nature of the process is characterized by the equation

$$E_{\text{input}} = W + E_{\text{output}} \qquad 6.1$$

This says that for a given amount of energy $E_{\text{input}}$ consumed by the engine, a definite amount of work $W$ is produced; the difference is the amount of energy $E_{\text{output}}$ expelled from the system, usually in the form of heat. Equation 6.1 is a statement of the conservation of energy, since work can always be transformed into an equal amount of energy.

The *efficiency e* of an engine is the ratio of the work it produces to the energy it consumes:

$$e = \frac{W}{E_{\text{input}}} \qquad 6.2$$

It is sometimes quoted as a percentage. For engines, such as gasoline engines and animal muscles, that operate on chemical energy, the efficiency is only about 25 percent; i.e., for every 100 J of energy consumed, only 25 J is converted into usable work. The remaining 75 J is expelled in the form of heat. The adequate removal of this heat is a problem for any engine, whether it is an electric power station or a human athlete.

**REMARK**  Power stations expel enormous quantities of heat into the environment, creating a problem of thermal pollution. It may be wondered why this energy is not recycled. That is, why isn't the energy $E_{\text{output}}$ fed back to the engine as input energy? The answer is that this heat is at too low a temperature to be converted into work. As a consequence of the second law of thermodynamics (Sec. 11.3), there is a fundamental limitation to the efficiency of any engine, which for most practical engines seldom exceeds 25 percent or so.

The *power P* of an engine is the rate at which it produces work. Thus if an engine produces an amount of work $W$ in the time $t$, its power is

$$P = \frac{W}{t} \qquad 6.3$$

For example, the power of an engine that produces 75 J of work in 3 s is

$$P = \frac{75 \text{ J}}{3 \text{ s}} = 25 \text{ J/s}$$

The unit of power is joules per second or *watts* (W):

$$1 \text{ W} = 1 \text{ J/s}$$

A *kilowatt* (kW) is equal to $10^3$ W.

In the English system of units, the unit of work is foot-pounds, and the unit of power is foot-pounds per second. A more common unit is the *horsepower* (hp), which is equal to 550 ft-lb/s. The conversions between these units and the mks units are

$$1 \text{ hp} = 550 \text{ ft-lb/s} = 746 \text{ W} = 0.746 \text{ kW}$$

and
$$1 \text{ W} = 0.737 \text{ ft-lb/s} = 10^{-3} \text{ kW}$$

For example, suppose the power of an automobile engine is 85 hp. This is equal to

$$P = 85 \text{ hp} \times 746 \text{ W/hp} = 63,400 \text{ W} = 63.4 \text{ kW}$$

The rate of energy consumption of the engine is much greater than this, however, because the efficiency is much less than 1. If the efficiency is 0.25, Eq. 6.2 shows that for every joule of work done the energy input is

$$E_{\text{input}} = \frac{W}{e} = \frac{1 \text{ J}}{0.25} = 4 \text{ J}$$

The rate of energy consumption is thus $1/e$ times the power:

$$\text{Rate of energy consumption} = \frac{P}{e} \qquad 6.4$$

In this example this is

$$\frac{63.4 \text{ kW}}{0.25} = 254 \text{ kW}$$

There is another useful equation for power that can be derived from Eq. 6.3 by remembering that work is just force times distance (Sec. 5.1). Thus substituting the expression

$$W = Fd$$

into Eq. 6.3, we get

$$P = \frac{Fd}{t} = Fv \qquad 6.5$$

where $v = d/t$ is the speed of the object on which the work is being done.

For instance, if the automobile in the previous example is moving at a constant speed of 27 m/s (about 60 mi/h), the applied force on it is

$$F_a = \frac{P}{v} = \frac{63,400 \text{ W}}{27 \text{ m/s}}$$
$$= 2350 \text{ J/m} = 2350 \text{ N} = 528 \text{ lb}$$

This force is the force of static friction the ground exerts on the wheels (Fig. 6.2). Since it is directed in the direction of motion of the car, this force does positive work on the car. Of course, this work is generated by the car's engine (Sec. 6.3). The total force on the car must be zero because the car is not accelerating, so there is another 528-lb force $\mathbf{F}_f$ directed against the forward motion of the car. This is the frictional force (wind resistance) exerted by the wind on the car. Since it is directed against the motion of the car, this force does negative work on the car.

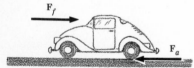

FIGURE 6.2
Forces on an automobile moving at constant speed.

A simple machine is a device that changes the magnitude or direction of an applied force. Most complex machines consist of an engine that produces a force and a series of simple machines that transform this force into a form suited to the required task. In an automobile the force of the engine is transformed by a series of gears and wheels into a force applied to the ground. In the human body the muscles are the engines, and the forces they apply are transformed by levers, composed of bones and joints, to the hands and feet (the details have already been discussed in Chap. 3).

The basic simple machines are the wheel, the lever, the inclined plane, and the screw. They all operate in the same way. A force $F_1$ is applied to the machine, doing work $W_1$ on the machine, while at the same time the machine applies a force $R_2$ to an external object, doing work $W_2$ on the object. Conservation of energy requires that $W_2$ equal $W_1$ minus the work that gets converted into heat by the frictional forces inside the machine:

$$W_1 = W_2 + E_{output} \qquad 6.6$$

where $E_{output}$ is the heat generated by friction.

**REMARK** Different letters are used for these forces to emphasize that they act on different objects. Different subscripts are used to emphasize that these forces are not an action-reaction pair.

The efficiency of a simple machine is the ratio of the output work $W_2$ to the input work $W_1$:

$$e = \frac{W_2}{W_1} \qquad 6.7$$

Equation 6.6 shows that $W_1 > W_2$, so that $e$ is always less than 1. However, the efficiency can be made very close to 1 by using lubricants to decrease internal friction. To simplify the following discussion we shall consider ideal (frictionless) simple machines, which have an efficiency of 1.

All simple machines operate according to the same principle. The force $F_1$ applied to the machines moves through a distance $d_1$, doing the work $W_1 = F_1 d_1$ on the machine. At the same time the machine applies the force $R_2$ through the distance $d_2$, doing the work $W_2 = R_2 d_2$ on an external object. If no energy is lost through friction, $W_1$ equals $W_2$, so that

$$F_1 d_1 = R_2 d_2$$

or
$$\frac{R_2}{F_1} = \frac{d_1}{d_2} \qquad 6.8$$

The ratio of the magnitude of the force $R_2$ applied by the machine to the magnitude of the force $F_1$ applied to the machine is called the *mechanical advantage M* of the machine. From Eq. 6.8 we see that

$$M = \frac{R_2}{F_1} = \frac{d_1}{d_2} \qquad 6.9$$

Figure 6.3 shows a pulley that is used to transform the downward force $F_1$ into the upward force $R_2$. The work $W_1 = F_1 d_1$ done by $F_1$

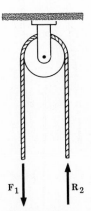

$F_1$     $R_2$

FIGURE 6.3
The pulley transforms the downward force $F_1$ (applied to the cord) into an upward force $R_2$ (applied by the cord).

in moving the cord down a distance $d_1$ is equal to the work $W_2 = R_2 d_2$ done by $\mathbf{R}_2$ in moving up the distance $d_2$. Since the distances $d_1$ and $d_2$ are equal, the forces $F_1$ and $R_2$ are also equal. The mechanical advantage of this machine is 1.0.

Figures 6.4 and 6.5 show arrangements in which wheels are used to change the magnitude of the applied force. In Fig. 6.4 the force $\mathbf{F}_1$ is applied to the inner circumference of the wheel by means of a drive belt. The force $\mathbf{R}_2$ is applied by the outer circumference to a second drive belt. The force $\mathbf{F}_1$ moves through the distance $d_1 = 2\pi r_1$, equal to the circumference of the inner circle, every time the wheel makes one revolution. At the same time the force $\mathbf{R}_2$ moves through the distance $d_2 = 2\pi r_2$, equal to the circumference of the outer wheel. Therefore the mechanical advantage of the machine is

$$M = \frac{d_1}{d_2} = \frac{r_1}{r_2}$$

Since $M$ is less than 1 in this case, the force $R_2$ is less than $F_1$.

Figure 6.5 shows a block-and-tackle arrangement with a mechanical advantage of 2. The distance $d_1$ that the force $\mathbf{F}_1$ moves is twice the distance $d_2$ that $\mathbf{R}_2$ moves. For instance, if $\mathbf{F}_1$ moves section $a$ of the cord down 2 ft, sections $b$ and $c$ will each be shortened 1 ft, and so $\mathbf{R}_2$ will be moved 1 ft. A force $F_1$ of 50 lb applied to point $O$ results in a force $R_2$ of 100 lb being applied to point $P$.

An inclined plane is not in itself a machine. A mechanical advantage can be defined for it, however, by noting in Fig. 6.6 that a force $\mathbf{F}_1$ applied parallel to the incline for the distance $d_1$ raises the force $\mathbf{F}_2$ a vertical distance $d_2$. Consequently the mechanical advantage of the incline is $d_1/d_2$. The wedge shown in Fig. 6.7 is a machine in the form of an inclined plane. By applying a force of magnitude $F_1$ to the wedge, a force of magnitude $R_2 = (d_1/d_2)F_1$ is applied by the wedge to the object. A screw is a machine in the form of a spiral inclined plane. The jackscrew in Fig. 6.8 has a very large mechanical advantage because the screw advances only a very short distance $d_2$ when the handle is moved a large distance $d_1$.

While the wheel is the simple machine most commonly used in mechanical devices, the lever is the simple machine used in animals. The force $\mathbf{F}_1$ applied by the muscle to the arm in Fig. 6.9 moves a distance $d_1$ while the hand moves the distance $d_2$. Since the mechanical advantage $d_1/d_2$ of this lever is less than 1, the force $R_2$ that the hand exerts on the ball is less than $F_1$. Similar results were obtained in Sec. 3.4 using torques and the conditions for equilibrium. The same method can be applied to a moving arm provided it moves slowly enough to be nearly in equilibrium.

The torques about point $O$ exerted by the forces acting on the arm are $F_1 s_1$ and $-F_2 s_2$. Here $\mathbf{F}_2$ is the force applied to the hand by the ball. (It is the reaction to the force $\mathbf{R}_2$ applied by the hand to the ball.) Since the arm is nearly in equilibrium, the sum of these torques is approximately zero, so

$$F_1 s_1 - F_2 s_2 = 0$$

or

$$\frac{F_2}{F_1} = \frac{s_1}{s_2}$$

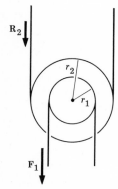

**FIGURE 6.4**
When the force $\mathbf{F}_1$ is applied to the inner circumference of the wheel, the outer circumference applies the force $\mathbf{R}_2$ to an external object.

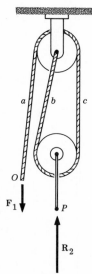

**FIGURE 6.5**
When the force $\mathbf{F}_1$ is applied to the block and tackle, the block and tackle applies a force $\mathbf{R}_2$ to an external object.

But the magnitude of $\mathbf{F}_2$ is equal to the magnitude of $\mathbf{R}_2$. Therefore the mechanical advantage of the lever is

$$M = \frac{R_2}{F_1} = \frac{F_2}{F_1} = \frac{s_1}{s_2} = \frac{d_1}{d_2}$$

The ratio $s_1/s_2$ is equal to $d_1/d_2$ because the triangles $BOC$ and $SOT$ are similar (Fig. 6.9). Thus the mechanical advantage of a lever can be calculated either by considering the distances through which the forces move or by using the method of torques discussed in Chap. 3.

## 6.3 LOCOMOTION

Some machines, such as automobiles and airplanes, are, like animals, capable of self-motion. The term *automaton* will be used to refer to any self-moving object, whether it is animate or inanimate. All automatons operate on the same principles: they all have parts that function as engines and parts that function as simple machines. Furthermore, during motion, the engines do work on the automaton itself.

To see how a machine can do work on itself, consider the machine in Fig. 6.10, which consists of an engine that can move a force $\mathbf{R}$ through a distance, i.e., do work. If the force $\mathbf{R}$ is applied by a cord to an immovable object, the machine can pull itself along the rope. In this case the force on the machine is the reaction $\mathbf{F} = -\mathbf{R}$ that the rope applies to the machine. Furthermore it is the reaction $\mathbf{F}$ that does the work in this case, since it is $\mathbf{F}$ that acts on a moving object (the machine). The force $\mathbf{R}$ applied by the machine does no work, since it acts on an immovable object.

Thus, in a machine, either the force $\mathbf{R}$ applied by the machine to an external object or the reaction $\mathbf{F}$ applied to the machine by the external object does the work $W$. In all automatons the reaction force does work on the automaton itself. This work is generated by the automaton's engines. Of course, only spiders and caterpillars use the technique of pulling on a cord. Most animals move by applying the force directly to a relatively immovable object, such as the ground.

To move parallel to the ground the force $\mathbf{R}$ must be applied parallel to the ground. This means that $\mathbf{R}$ is a frictional force and that $\mathbf{F}$ is the force of friction applied by the ground to the automaton. Friction, which is usually thought to oppose motion, is the force essential for locomotion on earth.

Consider walking, for example. As the foot strikes the ground, it is prevented from slipping parallel to the ground by the force of friction $\mathbf{F}$ applied to it by the ground. Since the foot is stationary while it is on the ground, $\mathbf{F}$ is a force of static friction. Nevertheless $\mathbf{F}$ does work, because it acts on the whole body, which is moving. This work is supplied by the muscles in the legs.

The wheel of an automobile has the same function as the leg of an animal. Like the foot, the wheel is at rest relative to the ground at the point of contact between it and the ground. Again, it is the frictional force $\mathbf{F}$ on the wheel at this point that moves the car forward. When this frictional force is removed, as when the wheel is

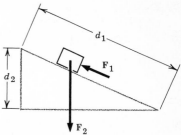

FIGURE 6.6
A force $\mathbf{F}_1$ applied parallel to an inclined plane lifts a weight $\mathbf{F}_2$ that is greater than $\mathbf{F}_1$.

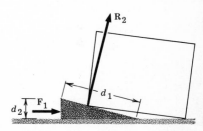

FIGURE 6.7
When a small force $\mathbf{F}_1$ is applied to the head of a wedge, the face of the wedge applies a large force $\mathbf{R}_2$ to an external object.

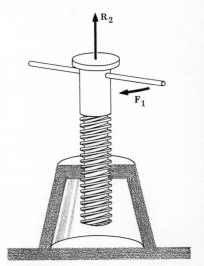

FIGURE 6.8
A jackscrew is an inclined plane wound into a spiral.

on ice, the wheel spins relative to the ground and no forward motion is possible.

A wheel is like a succession of legs all moving in the same direction (Fig. 6.11). One of the great advantages of the wheel is that it eliminates the wasteful back-and-forth motion involved in moving a leg. However, the advantages of a wheel can be realized only on a smooth surface. For moving over rugged terrain nothing can beat a good pair of legs.

To summarize: every self-moving object contains an engine and a series of simple machines that transform the force exerted by the engine into a force exerted by the automaton on an immovable object. The reaction of this force pushes the automaton forward, doing work on it. In this way the work generated by the engine is applied to the automaton itself.

The energy an animal gets from its food is expended in the form of work and heat. An animal that is neither gaining nor losing weight must, on the average, expend energy at the same rate at which it is supplied. Actually, even the energy that initially goes into work eventually gets converted into heat. For instance, in man the heart does work by contracting and pumping blood into the aorta at the rate of about 1 W (Sec. 7.5). At first this work is converted into the mechanical energy of the blood. But as the blood moves through the circulatory system, this energy is gradually converted into heat by the frictional forces exerted on the blood by the walls of the blood vessels.

A 70-kg (154-lb) man normally expends about 2500 kcal every day, the exact amount depending on his physical activity, i.e., on how much work (in the technical sense) he does. Since 1 kcal is equivalent to 4185 J, his daily energy expenditure is

$$2500 \text{ kcal} \times 4185 \text{ J/kcal} = 1.05 \times 10^7 \text{ J}$$

This is, on average, a rate of energy expenditure, or *metabolic rate*, of 121 W. The metabolic rate is only about 75 W during sleep but about 230 W while walking slowly.

The metabolic rate of a person engaged in a particular activity is measured by collecting all the air he exhales in about 5 min. The oxygen content of this expired air is then analyzed to determine the amount of oxygen consumed per minute. The consumed oxygen reacts with carbohydrates, fats, and protein in the body, releasing an average of about 4.8 kcal $= 2.0 \times 10^4$ J of energy for each liter

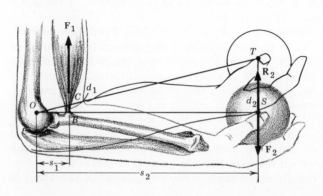

**FIGURE 6.9**
The lever action of the forearm. When a force $F_1$ is applied (by the muscle) to the arm, the hand exerts a force $R_2$ on an external object.

of oxygen consumed. Suppose, for example, that a man consumes 1.45 l of oxygen per minute during rapid bicycle riding. His metabolic rate then is

$$(1.45 \text{ l/min})(2.0 \times 10^4 \text{ J/l}) = 2.90 \times 10^4 \text{ J/min}$$
$$= 483 \text{ J/s} = 483 \text{ W}$$

**REMARK**  The energy for muscular contraction, and hence for work, comes from the conversion of adenosine triphosphate (ATP) into adenosine diphosphate (ADP) and adenosine monophosphate (AMP). The oxidation of carbohydrates, fats, and proteins supplies the energy to convert ADP and AMP back into ATP. Thus, although food is the ultimate energy source, ATP is the immediate energy source for motor activity.

It is interesting to try to calculate the energy consumed in a particular activity, such as running. In Sec. 5.2 it was shown that the major work done in running is the acceleration and deceleration of the legs with each stride. As a leg is lifted off the ground, it is brought from rest up to a speed $v$ that is approximately equal to the speed of the body. In the process, the muscles of the leg do work equal to the change of kinetic energy of the leg, i.e., equal to $\frac{1}{2}mv^2$, where $m$ is the mass of the leg. Then, as the leg is brought back to rest, the antagonistic muscles also do an amount of work equal to $\frac{1}{2}mv^2$. With each stride, therefore, the leg muscles do work approximately equal to $mv^2$.

Consider a 70-kg man running 3 m/s (approximately a 9-min mile). Each of his legs has a mass of about 10 kg, so the work done on a leg on each stride is

$$W = mv^2 = 10 \text{ kg} \times (3 \text{ m/s})^2 = 90 \text{ J}$$

Suppose the length of his stride—the distance between two successive footfalls of the same foot—is 2 m. Then the man takes 1.5 strides per second with each leg, so that the power input to his two legs is

$$P = 2(90 \text{ J/stride} \times 1.5 \text{ strides/s}) = 270 \text{ W}$$

Since the efficiency of muscle is only about 0.25, the rate of energy expenditure is

$$\frac{P}{e} = \frac{270 \text{ W}}{0.25} = 1080 \text{ W}$$

This is very close to the measured value of 1000 W.

Actually our estimate should have come out somewhat smaller than the measured value because our estimate does not take account of the energy consumed in all the other body functions, such as breathing and pumping blood. Nevertheless, this calculation shows that the energy involved in motor activity can be understood in terms of basic physical principles.

## 6.4  SCALING LAWS INVOLVING POWER

A person's metabolic rate depends on his size and physical activity. For medical purposes it is necessary to have a measure of metabolic

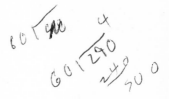

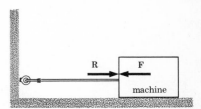

FIGURE 6.10
Locomotion. The machine pulls itself by applying a force **R** to an immovable object. The reaction force **F** does work on the machine.

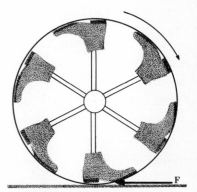

FIGURE 6.11
A wheel is like a succession of legs all moving in the same direction.

rate that controls these factors. To control the effect of physical activity the metabolic rate is measured under standardized conditions, called *basal conditions*, in which the person is awake but otherwise completely at rest. It is found that under these conditions the normal metabolic rate is proportional to the total surface area of the person's body. The effect of body size is controlled, therefore, by dividing the measured rate by the area of the body.

For example, while completely at rest a person consumes 15 l of oxygen per hour, so his metabolic rate is

$$(15 \text{ l/h})(2.0 \times 10^4 \text{ J/l}) = 3.0 \times 10^5 \text{ J/h}$$
$$= 83.3 \text{ W}$$

If the total surface area of his body is 1.7 m$^2$, his basal metabolic rate (BMR) is

$$\text{BMR} = \frac{83.3 \text{ W}}{1.7 \text{ m}^2} = 49 \text{ W/m}^2$$

The proportionality of BMR to body area is a scaling law. It is found that the maximum metabolic rate of similar animals is also proportional to their total surface area, so that metabolic rate scales as $L^2$. Thus an animal $L$ times larger than another animal can expend energy at $L^2$ times the rate of the smaller animal. Consequently the maximum power achieved by the larger animal is $L^2$ times that of the smaller since the power is just a multiple (the efficiency $e$) of the metabolic rate. This is a rather unexpected result. In Sec. 5.2 it was found that the work done by a muscle scales as $L^3$. Why does power scale as $L^2$?

One reason is that although the work done by a muscle scales as $L^3$, the speed $v$ with which a limb moves is independent of $L$ (Sec. 5.2). The time it takes the limb to move a distance $d$ is $d/v$, and hence time scales as $L$. The power achieved scales as work divided by time, or as $L^2$. Another reason is that all the energy used by the body is ultimately converted into heat, which must be removed from the body. Since this heat must escape through the surface of the body, the maximum rate of heat loss scales as $L^2$. The maximum metabolic rate cannot be greater than the maximum rate of heat loss, so it must also scale as $L^2$.

There are a number of interesting consequences of this scaling law.

**Heart rate**

Since the oxygen required for metabolism is supplied by the blood, the metabolic rate is proportional to the volume of blood pumped per second by the heart. This is proportional to the volume $V$ of the heart times the heart rate $r$ (the number of heart beats per second). We have then that

$$P \propto Vr$$

so that $r$ scales as $P/V$. Since $P$ scales as $L^2$ and $V$ scales as $L^3$, $r$ scales as $L^2/L^3 = L^{-1}$. Thus the heart of a large animal beats slower than the heart of a small animal. For example, the scale factor $L$ of man relative to the rhesus monkey is about 2.5. Therefore the human heart rate should be about $1/2.5 = 0.4$ times the monkey heart rate. This is close to the measured value of 0.5. The discrepancy is to be expected, since a scaling law is exactly true only for animals that are similar in all respects except size.

### Diving duration

The total energy expended by a sea mammal during an underwater dive is equal to its metabolic rate times the duration $t$ of the dive. This energy is proportional to the amount of oxygen stored in the lungs and blood of the mammal at the beginning of the dive:

Volume of oxygen stored $\propto$ energy expended = metabolic rate $\times$ $t$

Since the volume of oxygen scales as $L^3$ and the metabolic rate scales as $L^2$, the duration $t$ scales as $L$. A large mammal can stay underwater longer than a small mammal.

### Hovering

Some small birds, such as the hummingbird, can remain stationary in midair (hover) by beating their wings rapidly enough to produce a downward force on the air equal in magnitude to their weight $mg$. The reaction to this downward force on the air is an upward force of magnitude $mg$ on the bird, which maintains the bird in equilibrium. The power required to sustain this force is $mgv$, where $v$ is the speed with which the wings move. The maximum value of $v$ is independent of size (Sec. 5.2), so the power requirement scales as $L^3$. Since the maximum power available scales as $L^2$, the ratio of power required to power available scales as $L$. Because this ratio increases with size, only small birds, for which the ratio is less than 1, can hover. A similar argument shows that there is a maximum size for which horizontal flight is possible. The heaviest flying birds, the great bustards of Europe and Australia, weigh only 32 lb.

### Rowing

As a somewhat different example of scaling, we shall consider how the speed of a racing shell depends on the number of oarsmen.† Let $n$ be the number of oarsmen, and let $P$ be the power of each oarsman, so that the total power available is $nP$. The power required to move the boat at speed $v$ is equal to $F_f v$, where $F_f$ is the frictional force exerted on the shell by the water. This force is found experimentally to be proportional to the area of the boat submerged in the water times $v^2$. Therefore $F_f$ scales as $L^2 v^2$, where $L$ is the scale factor of the shell, and so the required power scales as $L^2 v^3$. Equating the power available to the power required, we get

$$nP \propto L^2 v^3$$

or
$$v^3 \propto \frac{n}{L^2} \qquad \qquad 6.10$$

The factor $P$ has been dropped because the oarsmen in different sized

† This discussion is based on McMahon (1971), to which the reader is referred for more details.

shells are considered to achieve the same power. The volume of the shell scales as $L^3$ and is proportional to the number of oarsmen, so that

$$L^3 \propto n$$

If each side of this proportion is raised to the $\frac{2}{3}$ power, it becomes

$$L^2 \propto n^{2/3}$$

Substituting this into Eq. 6.10, we get

$$v^3 \propto \frac{n}{n^{2/3}} = n^{1/3}$$

which, after taking the cube root of both sides, becomes

$$v \propto (n^{1/3})^{1/3} = n^{1/9} \qquad 6.11$$

This means that the speed of a shell increases with the ninth root of the number of oarsmen.

For instance, in the 1968 Olympics the eight-man shell won the 2000-m race with an average speed of 5.65 m/s, whereas the two-man shell won with a speed of 4.82 m/s. The ratio of these speeds is $5.65/4.82 = 1.17$, and according to Eq. 6.11, it should be equal to the ninth root of the ratio of the number of oarsmen in the two shells. The ninth root of $8/2 = 4$ is 1.167, which is very good agreement.

## 6.5 FEEDBACK AND CONTROL

The last four sections have presented some of the physical principles of machines and have tried to show the similarity of animals to mechanical machines. No one doubts that for the purpose of studying the power and energy requirements of an animal it is useful to think of the animal as a machine. However, many people balk at the notion that an animal is a machine. After all, they say, a machine lacks the volition and purposefulness characteristic of an animal.

It is certainly true that an animal is much more than an engine connected to some simple machines. This does not mean, however, that a very sophisticated machine cannot be designed that will closely resemble an animal. It does mean that additional concepts of machine design must be incorporated into such a machine.

The study of the animallike—even human—capabilities of machines is called *cybernetics*. This is an interdisciplinary field involving engineering, mathematics, biology, and psychology. In cybernetics problems of animal behavior (and even human thinking) are approached from an engineering viewpoint. A cyberneticist tries to understand how an animal performs a certain function by designing a machine that performs the same function. For example, to understand how a person is able to reach out and grasp an object in front of him, scientist have built a computer-controlled mechanical arm. The object of their study is to determine the principles of detection and control involved in the simple task of picking up an object.

Cyberneticists have found that one of the important differences between animals and most machines is that animals are controlled by *negative-feedback* circuits. Negative feedback is an engineering concept denoting an arrangement in which information from the external world is first sensed and then used to control the operation of a machine. Automation has resulted from the introduction of negative feedback into industry.

The best-known example of negative feedback is the heating system of a house. The operation of the furnace producing the heat is controlled by a thermostat that senses the temperature of the house and returns a signal to the furnace to turn on or turn off. For instance, when the thermostat is set at 70°F, it will turn the furnace on when the house temperature falls below a predetermined value (say 68°F), and it will turn the furnace off when the temperature rises above another predetermined value (say 72°F). Thus the furnace is on as long as the house temperature is below 72°F. The heat produced by the furnace raises the temperature of the house until it reaches 72°F, at which point the thermostat signals the furnace to turn off. The house then cools until it reaches 68°F, at which point the thermostat signals the furnace to turn on. In this way the operation of the furnace is controlled so that it maintains the temperature of the house between 68 and 72°F.

This type of feedback is called negative because the operation is self-limiting. Thus it is the heat produced by the furnace itself that eventually signals the furnace to turn off. One can get positive feedback by just switching the wires. That is, the thermostat could be made to turn on the furnace when the room temperature is above 72°F, and to turn it off when the temperature is below 68°F. Under such conditions the furnace would either stay on indefinitely or remain off indefinitely, depending on the initial temperature of the room. If the temperature is above 72°F, for instance, it will continue to increase. Such an arrangement lacks any element of control and is often catastrophic. The population growth is an example of catastrophic positive feedback. The more people there are, the more people reproduce. Positive feedback is similar to what is commonly called a vicious cycle.

Negative feedback is essential for any self-regulating system. Some form of negative feedback is involved in the control of most of the functions inside an animal's body, as well as in the control of interactions among animals in a social group. For example, the temperature of the human body is controlled by a thermostat in the preoptic area of the hypothalamus. The body has both heating and cooling mechanisms that are called into operation as the temperature of the thermostat rises and falls. On the other hand, honeybees maintain a constant temperature inside their hive by a socially controlled feedback mechanism, described by Lindauer (1967).

Positive feedback can occur in an organism under abnormal conditions. For example, with moderate blood loss, the negative-feedback mechanisms of the body are able to maintain blood pressure and blood flow by increasing the heart rate and constricting the blood vessels. However, positive feedback occurs when the blood loss is so severe that there is no longer sufficient blood to meet the energy requirement of the heart. The heart then slows down, further decreasing its own supply of blood. As a consequence, the heart beats still slower, decreasing its own blood supply even more. Since such positive feedback in the body is clearly inconsistent with life, death will rapidly follow unless emergency measures intervene.

A machine with even a few negative-feedback circuits begins to imitate an animal. In 1950 W. G. Walter built a little battery-operated machine that moved around the floor [Walter (1950)]. Called a *mechanical tortoise*, it backed away and changed direction whenever it hit an obstacle, so that its motion appeared spontaneous. A photocell was connected to the machine's steering mechanism in such a way that the machine would steer toward light of moderate intensity. However, the circuit was designed so that when its battery was fully charged, the machine would move away from a strong light (negative feedback).

A light was placed over the machine's "feeding box," which contained an electric outlet. Normally the machine would approach the light from a distance but then turn away as it got close. But when it got "hungry," i.e., when its battery began to run down, the machine was no longer repelled by the light. It then entered its feeding box and connected to the outlet, which recharged its battery. When it had "eaten," the machine was again repelled by the light, so that it moved out of the box and continued its exploratory behavior. Walter's mechanical tortoise thus demonstrated that a machine with negative feedback exhibits much of the apparent purposefulness and volition characteristic of an animal.

## PROBLEMS

*1* Assuming that muscles have an efficiency of 22 percent for converting energy into work, how much energy is expended by an 80-kg man in climbing a vertical distance of 15 m?
*Ans.* 12.8 kcal

*2* A steam engine expels 81 kcal of heat for every 100 kcal of heat it consumes. (*a*) How much work does the engine generate, and (*b*) what is the engine's efficiency?

*3* A 150-lb man runs up a flight of stairs 15 ft high in 6 s. What is his power output in horsepower?
*Ans.* 0.68 hp

*4* Find the conversion from horsepower to kilocalories per second.

*5* The power output of a cyclist moving at a constant speed of 6.0 m/s on a level road is 120 W. (*a*) What is the frictional force exerted on him (and the bicycle) by the air? (*b*) By bending low over the handlebars, the cyclist reduces the wind resistance to 18 N. If he maintains his former power output, what will his speed be? *Ans.* (*a*) 20 N; (*b*) 6.67 m/s

*6* Assume for the sake of simplicity that the total frictional force $F_f$ opposing the motion of an automobile traveling at a constant speed $v$ is proportional to $v^2$. (*a*) If the power output of a car is 20 hp at a speed of 30 mi/h, what is the power at 60 mi/h? (*b*) If the car goes 20 mi on a gallon of gasoline at 60 mi/h, how far can it go on a gallon of gasoline at 30 mi/h? (*c*) How many miles can the car go on a gallon of gasoline at 70 mi/h?

**REMARK** The last problem is somewhat unrealistic because the frictional force does not behave in the simple way assumed. However, the problem correctly illustrates how rapidly the fuel consumption of a car increases with speed.

*7* (*a*) What is the mechanical advantage of the nut cracker shown in Fig. 6.12? (*b*) If a force

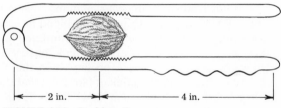

FIGURE 6.12   Problem 7.

of 10 lb is applied to the handle, what is the force applied to the nut?

Ans. (a) 3; (b) 30 lb

8 (a) What is the mechanical advantage of the wedge in Fig. 6.13? (b) When the top of the wedge is struck with a sledgehammer, a momentary force of 1000 lb is applied to the wedge. What is the force $R_2$ exerted by the face of the wedge?

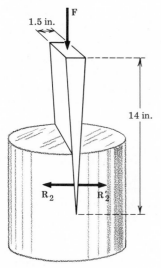

FIGURE 6.13    Problem 8.

9 Figure 6.14 shows the mechanical connections to the rear wheel of a bicycle. (a) What is the mechanical advantage $M$ of this arrangement; i.e., what is the ratio of the force $F_1$ applied by the foot to the pedal to the force $F_2$ applied by the road to the wheel? (b) The *speed ratio*

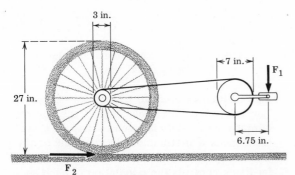

FIGURE 6.14    Problem 9.

is the ratio of the speed at which the bicycle moves to the speed at which the foot moves. Show that this ratio is equal to $1/M$. If the bicycle is traveling at a constant speed of 6.0 m/s against a frictional resistance of 20 N, (c) what is the force $F_1$ exerted by the cyclist on the pedal, and (d) what is the speed of his foot?

Ans. (a) 3:14; (c) 93.3 N; (d) 1.28 m/s

10 What force $F_1$ is required to pull a 250-lb block up the incline in Fig. 6.15? Neglect friction.

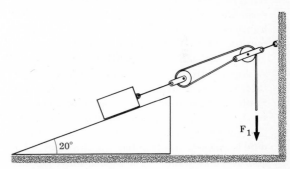

FIGURE 6.15    Problem 10.

11 A runner consumes oxygen at the rate of 4.1 l/min. What is his metabolic rate?

Ans. 1367 W

12 What is the rate of oxygen consumption during sleep assuming a metabolic rate of 75 W?

13 While resting, a man consumes 0.30 l of oxygen per minute. If the surface area of his body is 2.2 m², what is his basal metabolic rate?

Ans. 45.5 W/m²

14 Find how the speed with which animals of similar shape can run up a hill depends on size.

REMARK In Sec. 5.2 we showed that the maximum speed of similarly shaped animals is independent of size. That argument assumed the animals were running on level ground. The situation changes drastically on a hill, however, where a smaller animal can easily outrun a larger animal.

15 (a) Estimate the average speed of the winning four-man shell in a 2000-m race. (b) Estimate

the average speed of the winning one-man shell in a 2000-m race.

*Ans.* (*a*) 5.23 m/s; (*b*) 4.46 m/s

**REMARK** The estimates in Prob. 15 are in good agreement with the speeds of the winning shells in international championship competitions [McMahon (1971)].

*16* Show that the speed of a racing shell is proportional to $m^{2/9}$, where $m$ is the average mass of an oarsman. The average mass of a heavyweight oarsman is 86 kg, whereas the average mass of a lightweight oarsman is 73 kg. If the winning speed of an eight-man heavyweight team in a 2000-m race is 5.65 m/s, what is the estimated winning speed of an eight-man lightweight team? [*Hint:* Use the approximation $(1 + b)^a = 1 + ab$, which is valid when $b$ is much less than 1.]

*17* A dual-control electric blanket has separate thermostatic controls for each side of the bed. Suppose these controls are somehow reversed, so that the husband is unknowingly controlling the wife's side of the blanket while the wife is unknowingly controlling the husband's side of the blanket. Show that in this case the system has features of positive feedback.

**BIBLIOGRAPHY**

GRAY, JAMES: "Animal Locomotion," W. W. Norton & Company, Inc., New York, 1968. Comprehensive treatment of the motion of a wide variety of animals, including worms, insects, fish, birds, and mammals. Many interesting scaling laws are developed using only elementary physics.

GUYTON, ARTHUR C.: "Textbook of Medical Physiology," W. B. Saunders Company, Philadelphia, 1971. A standard physiology textbook for medical students. Chapter 1 gives the general theory of negative feedback in some detail and discusses its importance in the control of physiological processes.

LINDAUER, MARTIN: "Communication among Social Bees," rev. ed., Harvard University Press, Cambridge, Mass., 1971. An account of research on the behavior of bees. Chapter 1 describes the social mechanisms by which the temperature inside a hive is regulated.

McMAHON, THOMAS A.: Rowing: A Similarity Analysis, *Science*, **173**: 349 (1971). A study of how the speed of a rowing shell depends on the number of oarsmen. Besides obtaining the result given in Sec. 6.4, this paper gives suggestions, based on scaling laws, for increasing the speed of a racing shell.

———: Size and Shape in Biology, *Science*, **179**: 1201 (1973). Kleiber's law is a statement of the experimental fact that power in animals is proportional to $W^{0.75}$—where $W$ is the weight of the animal—rather than $L^2 \propto W^{2/3} = W^{0.67}$. The $L^2$ scaling law is only for animals of similar shape. This article uses criteria of mechanical stability to derive how the shape (ratio of width to length) of animals changes with size. Kleiber's law is then derived from a surface-area argument like the one used in Sec. 6.4 to derive the $L^2$ law for animals of similar shape.

MOREHOUSE, LAURENCE E., and AUGUSTUS T. MILLER: "Physiology of Exercise," The C. V. Mosby Company, St. Louis, 1971. The physiology of muscular activity for students of physical education. Chapter 19, which discusses the methods for measuring metabolic rate, has a table listing the metabolic rate for a variety of physical activities.

SCHMIDT-NIELSEN, KNUT: "How Animals Work," Cambridge University Press, London, 1972. Fascinating discussions of the physical mechanisms of various physiological functions. Chapter 6 contains an excellent treatment of scaling.

———: Locomotion: Energy Cost of Swimming, Flying, and Running, *Science*, **177**: 222 (1972). Comparisons of the energy expended by different animals with different methods of locomotion.

SMITH, JOHN MAYNARD: "Mathematical Ideas in Biology," Cambridge University Press, Cambridge, 1968. Chapter 1 gives a variety of scaling results based on the scaling law of power. The role of negative feedback in the control of muscular activity is given in chap. 7. Chapter 2 gives a mathematical treatment of the predator-prey system that shows how negative feedback controls the population of each species, an interesting example of negative feedback in ecology. Some knowledge of differential equations is needed to understand chaps. 2 and 7.

WALTER, W. GREY: An Imitation of Life, *Scientific American*, **183**: 42 (May 1950). The details of the mechanical tortoise are given in this article, together with a discussion of their deeper implication for biology.

WOOLDRIGE, DEANE.: "Mechanical Man: The Physical Basis of Intelligent Life," McGraw-Hill Book Company, New York, 1968. Animal and human behavior, from the simple reflex arch to free will, are shown to be understandable in terms of the machinelike operations of physical laws.

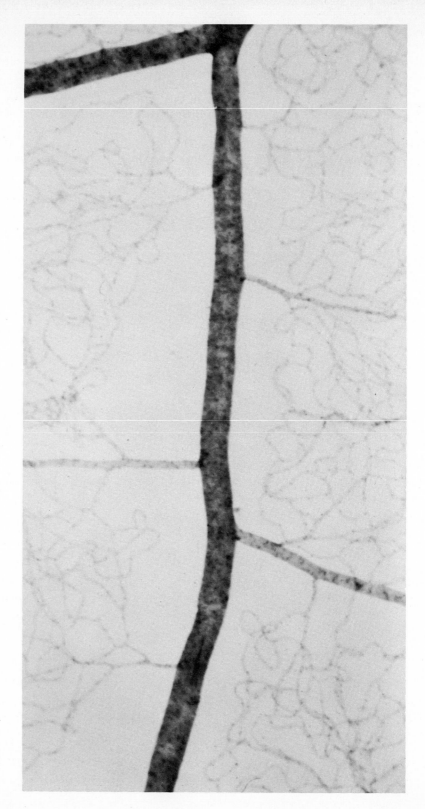

Photomicrograph of arteries and
capillaries in the retina of the eye.
The rate of flow of blood through
such vessels depends on their
diameter and on the viscosity and
pressure of the blood (Sec. 7.5).
[*Boston University Medical
School.*]

# II properties of matter

Part I dealt with general laws of physics that apply universally in all physical situations. In Part II, these general laws are used to study the more specialized laws that apply to matter in each of its three states: gas, liquid, and solid. Although not as general as Newton's laws of motion and the law of conservation of energy, these specialized laws are important because they describe the properties of the materials of which the world and all living things in the world are composed. As we shall see, knowledge of these properties provides the basis for understanding many basic physiological processes, such as respiration, circulation, and transpiration.

# CHAPTER 7 fluids

A solid is a rigid substance that maintains its shape against external distorting forces, whereas a fluid is a nonrigid substance (gas or liquid) that does not maintain its shape against such forces. Instead, a fluid always flows when acted upon by distorting forces. This chapter discusses the fundamental properties of fluids that are common to both gases and liquids. These properties apply as well to the flow of air through the bronchial tubes as to the flow of blood through blood vessels. The specific properties of gases and liquids are treated in Chaps. 8 and 9.

## 7.1 THE THREE STATES OF MATTER

In discussing the properties of matter, it is convenient to classify matter into three states: gas, liquid, and solid. Many substances can be transformed from one state to another by changes in temperature and pressure. The most familiar example is water. In fact, it is the only substance commonly found in all three states (steam, liquid water, and ice). You have undoubtedly also seen candle wax transformed from the solid to the liquid state when heated, and you may have seen a demonstration of liquid air, which is really nitrogen in its liquid state. Such transformations of substances from one state to another are common in many industrial and laboratory processes.

A *solid* is characterized by having a definite *volume* and *shape*. Its shape can be changed only by the application of considerable force, such as that required to bend a steel bar. This rigidity of form is the result of the strong forces between the molecules of the solid, which are packed tightly together in fixed positions. To bend a solid, this very stable molecular arrangement must be altered, which requires the application of a strong force.

A *liquid* is characterized by having a definite *volume* but not a definite shape. A liquid flows to conform to the shape of the container in which it is placed. It has a definite volume, however, which it maintains in spite of changes in shape. The molecules of a liquid are almost as close together as in a solid,† but they do not have fixed positions. It is because the molecules of a liquid are free to move relative to each other that a liquid has no rigidity. On the other hand, it is because of the strong attraction between its molecules that a given quantity of liquid occupies a definite volume, regardless of its shape.

---

† The molecules of $H_2O$ are actually closer together in water than in ice.

A *gas* is characterized by having neither a definite volume nor a definite shape. A gas will expand to fill any closed container in which it is placed, and if the container is opened, the gas will leak out the opening. (Only the gravitational attraction of the earth keeps the gaseous atmosphere from expanding into space. The moon, which exerts only one-sixth the gravitational force of the earth, cannot hold a gaseous atmosphere.) In a dilute gas, the molecules are so far apart that they exert forces on each other only when they collide. Consequently every molecule travels freely in a straight line until it hits another molecule or the walls of its container. It is this unconstrained molecular motion that gives a gas its inherent expandability. Furthermore, all very dilute gases tend to have the same properties because the frequency of molecular collisions is so small that the behavior of different gases is unaffected by differences in the nature of the forces between their molecules.

Gases have special properties that result from their expandability, whereas liquids have special properties that result from their having a surface. Nevertheless, gases and liquids have in common many properties that result from their nonrigidity. The word *fluid* is used to refer to gases and liquids when discussing those properties which are common to both. These common fluid properties are considered in this chapter, while the specific properties of gases and liquids are treated in Chaps. 8 and 9, respectively.

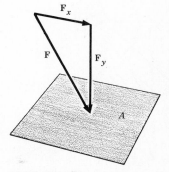

FIGURE 7.1
A force **F** acting on a surface of area $A$. $\mathbf{F}_x$ and $\mathbf{F}_y$ are the components of **F** parallel and perpendicular to the surface, respectively.

## 7.2 PRESSURE

The forces which a fluid exerts on its surroundings are characterized by a single quantity, the *pressure* in the fluid, which plays a role analogous to that of the tension in a flexible cord (Sec. 2.2). The pressure in a fluid can result from the application of an external force or from the weight of (the force of gravity on) the fluid itself. In order to consider these two origins of fluid pressure separately, the effects of gravity are neglected in this section.

Consider a force **F** acting on a surface of area $A$ (Fig. 7.1). The pressure $p$ on this surface is defined as the magnitude $F_y$ of the component of **F** perpendicular to the surface divided by $A$:

$$p = \frac{F_y}{A} \qquad 7.1$$

For example, the 5-kg block resting on the table in Fig. 7.2 exerts a force

$$F_y = 5 \text{ kg} \times 9.8 \text{ m/s}^2 = 49 \text{ N}$$

perpendicular to the table. If the area of the table is 1.4 m², the pressure on it is

$$p = \frac{F_y}{A} = \frac{49 \text{ N}}{1.4 \text{ m}^2} = 35 \text{ N/m}^2$$

**REMARK** It might be good to review the reasoning used to calculate $F_y$, for although $\mathbf{F}_y$ is equal in magnitude and direction to the force of gravity on the block, it is not the force of gravity on the block. In fact it is not a gravitational

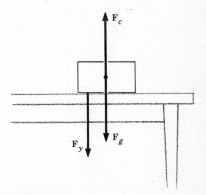

FIGURE 7.2
A 5-kg block resting on a table.

force at all, and it does not act on the block but on the table. The force of gravity $\mathbf{F}_g$ acts on the block. Since the block is in equilibrium, we know from Newton's first law (Property 5, Sec. 2.1) that the table exerts a contact force $\mathbf{F}_c$ on the block equal to $-\mathbf{F}_g$. From Newton's third law (Property 3, Sec. 2.1), the reaction to $\mathbf{F}_c$ is a contact force equal to $-\mathbf{F}_c = -(-\mathbf{F}_g) = \mathbf{F}_g$ that the block exerts on the table. It is this reaction force that we have now called $\mathbf{F}_y$.

As another example, consider an 80-kg skier going down a 20° slope (Fig. 7.3). The skis exert a vertical force of 80 kg $\times$ 9.8 m/s$^2$ = 784 N on the snow. The magnitude of the component of this force perpendicular to the slope is

$$F_y = 784 \text{ N} \times \cos 20° = 737 \text{ N}$$

If the area of the two skis together is 0.30 m$^2$, the pressure they exert on the slope is

$$p = \frac{737 \text{ N}}{0.30 \text{ m}^2} = 2460 \text{ N/m}^2$$

The concept of pressure is of limited utility in studying solids because, by definition, it involves only a part of the force that may be present. Furthermore, the value of the pressure depends on the area involved, which may be ambiguous. For instance, the pressure on the table in Fig. 7.2 is 35 N/m$^2$, but the pressure directly under the block is 490 N/m$^2$ if the area of the block is 0.1 m$^2$. Neither of these difficulties arises in fluids, however, because of two special fluid properties.

**Fluid Property 1** *A fluid at rest cannot exert a force parallel to a surface.* This remarkable fact is due to the nonrigidity of a fluid. If the fluid exerted a force parallel to a surface, the surface, of course, would exert a parallel force on the fluid. Figure 7.4 shows an object with forces $\mathbf{F}_1$ and $\mathbf{F}_2$ applied parallel to two sides and a force $\mathbf{F}_3 = -(\mathbf{F}_1 + \mathbf{F}_2)$ applied perpendicular to the base. The total force and total torque on the object are zero, so that the object is in equilibrium, provided it does not bend or break. A solid, which can resist bending (up to a point), will be in equilibrium under these conditions. A fluid, however, has no rigidity, so it will flow. A fluid cannot remain at rest if there are parallel forces applied to it, and so a fluid at rest cannot exert forces parallel to a surface.

Another way to say this is that *a fluid has no static coefficient of friction.* Imagine a wooden boat floating on water. If a force $\mathbf{F}$ is applied to the boat parallel to the water, the boat will not remain at rest no matter how small $\mathbf{F}$ is because the water cannot apply a parallel force to balance $\mathbf{F}$. Once the boat starts to move, however, the situation changes since the fluid is now moving relative to the boat. A moving fluid does exert a force parallel to a surface, the magnitude of which increases with speed (Sec. 7.5). Consequently the boat accelerates under the action of $\mathbf{F}$ until it reaches the speed at which the magnitude of the frictional force of the water on it equals the magnitude of $\mathbf{F}$. Similarly the air (which is a fluid) offers little resistance to slowly moving objects, but air resistance becomes large for high-speed objects.

**REMARK** A lubricant reduces the friction between two solid objects by introducing a thin layer of fluid, such as oil, between their surfaces. Since the fluid

**FIGURE 7.3**
A skier on a slope. $\mathbf{F}_y$ is the perpendicular (contact) force the skier exerts on the slope.

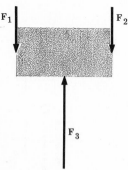

**FIGURE 7.4**
Three forces acting on an object. These forces tend to bend, rather than to move or rotate, the object.

itself can exert no static friction, the friction between the surfaces is greatly reduced. The motion of joints in the body is lubricated by the synovial fluid, which results in a static coefficient of friction of only 0.015. This is much smaller than can be obtained for mechanical surfaces. The small value is absolutely essential because of the large contact forces exerted on joints.

**Fluid Property 2 (Pascal's law)**  *In the absence of gravity,* i.e., neglecting the weight of the fluid itself, *the pressure in a fluid at rest is the same everywhere.*

This property is proved by showing that the pressure is the same at any two points $P$ and $Q$ in the fluid. Thus choose any two points $P$ and $Q$ in a fluid at rest and consider the fluid inside the cylindrical region shown in Fig. 7.5. Since the fluid is at rest everywhere, the total force on this cylinder of fluid, as on every other region of fluid, must be zero. Furthermore, from Fluid Property 1, the forces on this region are perpendicular to its surface. Therefore, if $p_P$ is the pressure at point $P$, and $p_Q$ is the pressure at point $Q$, there is a force of magnitude

$$F_P = p_P A$$

perpendicular to the cylinder at $P$ and a force of magnitude

$$F_Q = p_Q A$$

perpendicular to the cylinder at $Q$, where $A$ is the area of either end of the cylinder. Since these forces are parallel to the long axis of the cylinder whereas all the other forces are perpendicular to this axis, the forces $F_P$ and $F_Q$ must have the same magnitude if the total force parallel to the axis is to be zero. Thus we have

$$F_P = F_Q$$

and so $\qquad p_P A = p_Q A \qquad$ or $\qquad p_P = p_Q$

Since $P$ and $Q$ are any two points in the fluid, this proves that the pressure is the same everywhere in the fluid.

**REMARK**  Fluid Property 1 is essential in the proof of Fluid Property 2 because it assures us that the forces on the body of the cylinder have no components parallel to the axis of the cylinder.

To see how these fluid properties are applied in practice, consider a fluid contained in a cylinder of cross-sectional area $A$ (Fig. 7.6). If a downward force **F** is applied to the movable piston that closes the top of the cylinder, the fluid must apply the opposite force $-\mathbf{F}$ on the piston when the piston is at rest. Therefore, at equilibrium the pressure exerted by the fluid on the piston is $p = F/A$, which by Fluid Property 2 is also the pressure everywhere else in the fluid. (Remember, we are neglecting gravity in this section.)

Suppose now this cylinder is connected by a pipe to a smaller cylinder of cross-sectional area $A'$ (Fig. 7.7). What is the magnitude $F'$ of the force that must be applied to the smaller piston to maintain equilibrium? Since the pressure is the same everywhere in the fluid, the pressure exerted by the fluid on the small cylinder must also be $p = F/A$. On the other hand, the force exerted by the fluid on the smaller piston must have a magnitude of $F'$ to balance the applied force, so $p = F'/A'$. Equating these two evaluations of the pressure, we have

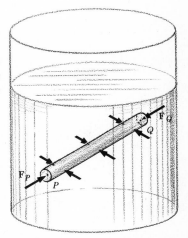

FIGURE 7.5
The forces on a cylindrical region in a fluid. The surface of the cylinder is an imaginary boundary defining the region of interest.

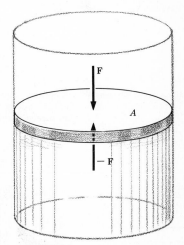

FIGURE 7.6
A fluid confined to a cylinder with a movable piston of area $A$.

$$p = \frac{F}{A} = \frac{F'}{A'}$$

or
$$F' = pA' = \frac{A'}{A}F$$

For instance, if $A = 0.1$ m$^2$ and $F = 900$ N, the pressure in the fluid is $p = 900$ N$/0.1$ m$^2 = 9000$ N/m$^2$. If $A' = 0.01$ m$^2$, the magnitude $F'$ of the force applied to the smaller piston is

$$F' = pA' = 9000 \text{ N/m}^2 \times 0.01 \text{ m}^2 = 90 \text{ N}$$

Thus a force of only 90 N on the small piston will support a force of 9000 N applied to the large piston.

The hydraulic lift, commonly used in garages to raise automobiles, uses this principle to lift large weights $F$ with a small applied force $F'$. The principle is similar to that of other simple machines (Sec. 6.2): the small piston moves through a large distance $d'$ in lifting the large piston a small distance $d$.

**REMARK** The two properties of a fluid just described are analogous to the properties of a flexible cord discussed in Sec. 2.2. A cord can exert tension (pull) but not pressure (push, compression), whereas a fluid can exert pressure (push) but not tension (pull).† The cord exerts a force only along its length, whereas a fluid exerts a force only perpendicular to a surface. Finally the tension is the same everywhere in a cord, just as the pressure is the same everywhere in a fluid. Thus a cord is like a one-dimensional fluid, except that it only pulls whereas a fluid only pushes. A two-dimensional system with analogous properties is discussed in Sec. 9.2.

The dimension of pressure is

$$[p] = \left[\frac{f}{l^2}\right] = \left[\frac{m-l}{s^2/l^2}\right] = \left[\frac{m}{s^2-l}\right]$$

The mks unit of pressure is the newton per square meter (N/m$^2$), as we have seen. Unfortunately this is not a very practical unit, so it is seldom used in practice. More common units are *atmospheres* (atm), pounds per square inch (lb/in.$^2$), millimeters of mercury‡ (mm

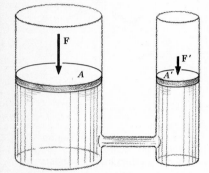

FIGURE 7.7
Two cylinders, each with a movable piston, connected by a tube.

† An exception to this is discussed in Sec. 9.5.
‡ Also called Torr, after the Italian mathematician and physicist Evangelista Torricelli (1608-1647).

TABLE 7.1 **Conversion factors between common units of pressure**
Each number gives the value of 1 unit of the pressure named at the left in terms of the unit named at the top. For example, 1 lb/in.$^2$ equals 51.7 mm Hg.

| | N/m$^2$ | dyn/cm$^2$ | lb/ft$^2$ | lb/in.$^2$ | atm | bar | mbar | mm Hg (Torr) at 0°C | cm H$_2$O at 4°C |
|---|---|---|---|---|---|---|---|---|---|
| N/m$^2$ | 1 | 10 | $2.09 \times 10^{-2}$ | $1.45 \times 10^{-4}$ | $9.87 \times 10^{-6}$ | $10^{-5}$ | $10^{-2}$ | $7.50 \times 10^{-3}$ | $1.02 \times 10^{-2}$ |
| dyn/cm$^2$ | $10^{-1}$ | 1 | $2.09 \times 10^{-3}$ | $1.45 \times 10^{-5}$ | $9.87 \times 10^{-7}$ | $10^{-6}$ | $10^{-3}$ | $7.50 \times 10^{-4}$ | $1.02 \times 10^{-3}$ |
| lb/ft$^2$ | 47.9 | 479 | 1 | $6.94 \times 10^{-3}$ | $4.73 \times 10^{-4}$ | $4.79 \times 10^{-4}$ | 0.479 | 0.359 | 0.488 |
| lb/in.$^2$ | $6.89 \times 10^3$ | $6.89 \times 10^4$ | 144 | 1 | $6.80 \times 10^{-2}$ | $6.89 \times 10^{-2}$ | 68.9 | 51.7 | 70.3 |
| atm | $1.01 \times 10^5$ | $1.01 \times 10^6$ | $2.12 \times 10^3$ | 14.7 | 1 | 1.01 | $1.01 \times 10^3$ | 760 | $1.03 \times 10^3$ |
| bar | $10^5$ | $10^6$ | $2.09 \times 10^3$ | 14.5 | 0.987 | 1 | $10^3$ | 750 | $1.02 \times 10^3$ |
| mbar | $10^2$ | $10^3$ | 2.09 | $1.45 \times 10^{-2}$ | $9.87 \times 10^{-4}$ | $10^{-3}$ | 1 | 0.750 | 1.02 |
| mm Hg (Torr) at 0°C | 133 | $1.33 \times 10^3$ | 2.78 | $1.93 \times 10^{-2}$ | $1.32 \times 10^{-3}$ | $1.33 \times 10^{-3}$ | 1.33 | 1 | 1.36 |
| cm H$_2$O at 4°C | 98.1 | 981 | 2.05 | $1.42 \times 10^{-2}$ | $9.68 \times 10^{-4}$ | $9.81 \times 10^{-4}$ | 0.981 | 0.736 | 1 |

Hg), and *millibars* (mbar). For instance, barometric pressure might be quoted as 989 mbar, whereas blood pressure is quoted as 120 mm Hg.

There are a number of other units that are frequently used in medicine and science, so you must be able to convert readily from one to another. Table 7.1 gives the conversion factors between the most common units. The origin and purpose of some of these units will be explained in the next section.

**REMARK** In Table 7.1 bar, atm, mm Hg, and cm $H_2O$ are abbreviations for the names of units. In a calculation the mm in mm Hg cannot be canceled the way the $m^2$ in $N/m^2$ might be. The four letters mm Hg are part of the abbreviation, just like the letters a-t-m. You cannot cancel part of an abbreviation.

## 7.3  THE EFFECT OF GRAVITY ON FLUIDS

Pascal's law (Fluid Property 2) is true only insofar as the force of gravity on a fluid can be neglected, in which case the pressure in the fluid can be considered to be produced entirely by external forces, e.g., the pistons in Figs. 7.6 and 7.7. The relative importance of the force of gravity on a fluid depends primarily on the density of the fluid.

### Density

The density† $\rho$ of a substance is the ratio of the mass $m$ of the substance to its volume $V$,

$$\rho = \frac{m}{V} \qquad 7.2$$

Density is a characteristic property of a substance, independent of its volume or mass. For example, the mass of 3 l (3000 $cm^3$) of ethanol (ethyl alcohol) is 2367 g. Therefore the density of ethanol is

$$\rho = \frac{m}{V} = \frac{2367 \text{ g}}{3000 \text{ cm}^3} = 0.79 \text{ g/cm}^3$$

The density of 3 $cm^3$ of ethanol is the same as that of 3000 $cm^3$, so the mass of 3 $cm^3$ of ethanol is

$$m = V\rho = 3 \text{ cm}^3 \times 0.79 \text{ g/cm}^3 = 2.37 \text{ g}$$

The mass of 1 $cm^3$ is clearly 0.79 g, which shows that the density of a substance is just its mass per unit volume.

The densities of some common solids, liquids, and gases are given in Table 7.2. These densities were measured at a pressure of 1 atm (760 mm Hg = $1.01 \times 10^5$ $N/m^2$) and at the temperatures indicated. The density of a solid or liquid varies only slightly with changes in temperature and pressure, whereas the density of a gas is strongly temperature- and pressure-dependent. This can be seen from Table 7.2 by comparing the temperature variation of the density of air with that of water.

† $\rho$ is the Greek letter rho.

TABLE 7.2 **Densities of some common substances at 1 atm (760 mm Hg)**

| Substance | Temperature, °C | Density | |
|---|---|---|---|
| | | g/cm³ | kg/m³ |
| *Solids* | | | |
| Aluminum | 20 | 2.7 | 2,700 |
| Bone | 20 | 1.6 | 1,600 |
| Copper | 20 | 8.5 | 8,500 |
| Glass | 20 | 2.6 | 2,600 |
| Granite | 20 | 2.7 | 2,700 |
| Iron | 20 | 7.7 | 7,700 |
| Lead | 20 | 11.3 | 11,300 |
| Steel | 20 | 7.7 | 7,700 |
| Water (ice) | 0 | 0.917 | 917 |
| Wood, maple | 20 | 0.7 | 700 |
| *Liquids* | | | |
| Air (liquid) | −183 | 1.14 | 1,140 |
| Blood plasma | 37 | 1.03 | 1,030 |
| Blood, whole | 37 | 1.05 | 1,050 |
| Ethanol (ethyl alcohol) | 20 | 0.791 | 791 |
| Glycerin | 0 | 1.26 | 1,260 |
| Hydrogen (liquid) | −253 | 0.07 | 70 |
| Mercury | 0 | 13.6 | 13,600 |
| Oxygen (liquid) | −183 | 1.14 | 1,140 |
| Trichloromethane (chloroform) | 20 | 1.483 | 1,483 |
| Water, pure | 4 | 1.00 | 1,000 |
| | 30 | 0.996 | 996 |
| | 100 | 0.958 | 958 |
| Sea | 15 | 1.025 | 1,025 |
| *Gases* | | | |
| Air | 0 | 0.00130 | 1.30 |
| | 10 | 0.00125 | 1.25 |
| | 20 | 0.00120 | 1.20 |
| | 30 | 0.00116 | 1.16 |
| Argon | 0 | 0.00178 | 1.78 |
| Carbon dioxide | 0 | 0.00198 | 1.98 |
| Helium | 0 | 0.000178 | 0.178 |
| Hydrogen | 0 | 0.0000899 | 0.0899 |
| Nitrogen | 0 | 0.00125 | 1.25 |
| Oxygen | 0 | 0.00143 | 1.43 |
| Water (steam) | 100 | 0.000596 | 0.596 |

**REMARK**  The density of water at 4°C is exactly 1.000 g/cm³ because the gram was originally defined as the mass of a cubic centimeter of pure water at this temperature.

Densities are most often given in grams per cubic centimeter, which is a cgs unit. The unit of density in the mks system is kilograms per cubic meter. It is easy to convert from one unit to another since $1 \text{ kg} = 10^3$ g and $1 \text{ m} = 10^2$ cm. Thus we have

$$1 \text{ kg/m}^3 = \frac{10^3 \text{ g}}{(10^2 \text{ cm})^3}$$

$$= \frac{10^3 \text{ g}}{10^6 \text{ cm}^3} = 10^{-3} \text{ g/cm}^3$$

and so  $1 \text{ g/cm}^3 = 10^3 \text{ kg/m}^3$

Thus to convert a density from grams per cubic centimeter to kilograms per cubic meter you need only multiply by $10^3$. Table 7.2 gives the densities in both units.

### Fluid Property 3

To study the effect of gravity on fluid pressure, we consider the fluid inside the cylinder in Fig. 7.8. A force $\mathbf{F}$ is applied perpendicular to the piston, which has a cross-sectional area $A$, so the pressure directly under the piston is

$$p_0 = \frac{F}{A}$$

The subscript 0 indicates that this is the pressure at the top of the fluid. From Pascal's law, the pressure $p_h$ at the bottom of the fluid would equal $p_0$ if gravity were neglected. However, because of gravity, the total downward force on the fluid is $\mathbf{F} + \mathbf{F}_g$, where $\mathbf{F}_g$ is the force of gravity on the fluid. Since the fluid is in equilibrium, there must be an upward contact force $\mathbf{F}_c = -(\mathbf{F} + \mathbf{F}_g)$ exerted on the fluid by the bottom of the cylinder. The reaction to $\mathbf{F}_c$ is the force $\mathbf{R}_c = -\mathbf{F}_c = \mathbf{F} + \mathbf{F}_g$ that the fluid exerts downward on the bottom of the cylinder. Thus the pressure $p_h$ at the bottom is

$$p_h = \frac{\mathbf{F} + \mathbf{F}_g}{A} = \frac{F}{A} + \frac{F_g}{A}$$
$$= p_0 + \frac{F_g}{A} \qquad 7.3$$

The pressure at the bottom of the fluid is greater than at the top because of the weight of the fluid itself.

This increase in pressure with depth is related to the density $\rho$ of the fluid. Since the volume of the fluid in the cylinder is $V = Ah$, where $h$ is the height of the fluid, the mass of the fluid is $m = \rho V = \rho Ah$. Therefore the weight of the fluid is

$$F_g = mg = \rho Ahg$$

and so Eq. 7.3 can be written

$$p_h = p_0 + \rho gh \qquad 7.4$$

or

$$p_h - p_0 = \rho gh \qquad 7.5$$

Equation 7.5 gives the correction to Pascal's law (Fluid Property 2) due to the weight of the fluid. If $\rho$ and $h$ are both small, the pressure difference due to gravity may be negligible. However, if either $\rho$ or $h$ is large, the pressure difference can be large.

Equation 7.4 is far more general than the simple derivation of it might indicate. For example, Fig. 7.9 shows a funnel-shaped container filled with water. The pressure $p_0$ at the top of the water is just the pressure of the atmosphere. The pressure $p_h$ at the bottom is given by Eq. 7.4, where $h$ is the vertical distance from the top to the bottom of the water. This somewhat surprising result follows from the fact that the force exerted by the bottom has to support only the column of water directly above it. The rest of the water is supported by the sides of the container.

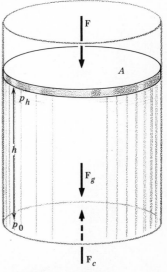

**FIGURE 7.8**
The forces on the fluid in a cylinder. $\mathbf{F}_g$ is the force of gravity (weight) of the fluid itself, whereas $\mathbf{F}$ and $\mathbf{F}_c$ are the forces exerted by the piston and the base of the cylinder on the fluid.

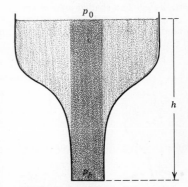

**FIGURE 7.9**
A funnel-shaped container filled with water. The base supports only the column of water directly above it, so $p_h$ is given by Eq. 7.4.

More surprising, perhaps, the pressure $p_h$ at the bottom of the container in Fig. 7.10 is also given by Eq. 7.4. The shaded region in this figure shows that there is a column of water of height $h$ over point $A$ on the bottom, so the pressure at $A$ is given by Eq. 7.4. But the pressure must be the same at all points on the bottom because if the pressure at $A$ were greater than the pressure at $B$, the fluid would flow from $A$ toward $B$.

The validity of Eq. 7.4 can be demonstrated by filling the containers in Figs. 7.9 and 7.10 with water and connecting them by a tube, as shown in Fig. 7.11. The water levels in the containers adjust themselves until, at equilibrium, the pressures at points $A$ and $B$ are equal. This means that if Eq. 7.4 is valid, the water will be at the same height in both containers. Experimentally this is found to be the case. The pressure at a point in a fluid depends on the distance of the point from the top of the fluid, but it does not depend on the shape of the container.

There is, of course, nothing special about the bottom of a container. The pressure at any point inside the fluid is also given by Eq. 7.4, where $h$ is the distance of the point from the surface. The pressure in a fluid thus increases steadily with depth below the surface.

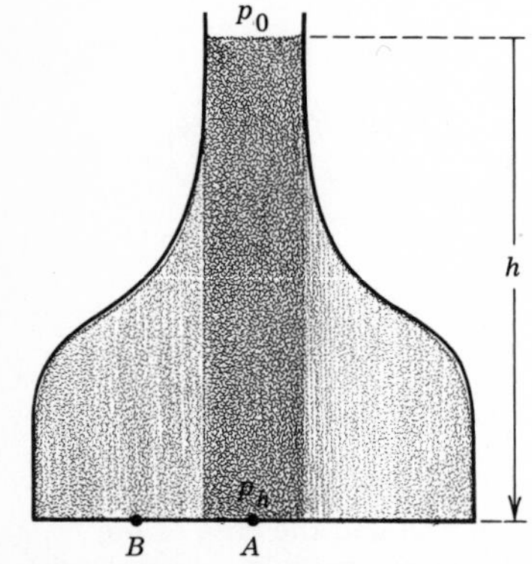

Consider two points $A$ and $B$ in a fluid (Fig. 7.12). Let $A$ be a distance $h_A$ below the surface, and let $B$ be a distance $h_B$ below the surface. According to Eq. 7.4, the pressures $p_A$ and $p_B$ at these points are

$$p_A = p_0 + \rho g h_A$$

and

$$p_B = p_0 + \rho g h_B$$

We subtract the second equation from the first to get

$$\begin{aligned} p_A - p_B &= (p_0 + \rho g h_A) - (p_0 + \rho g h_B) \\ &= \rho g h_A - \rho g h_B = \rho g (h_A - h_B) \\ &= \rho g h \end{aligned}$$

where $h$ is the vertical distance between points $A$ and $B$. This is the generalization of Pascal's law needed to include the effect of gravity. It can be stated formally as follows.

**Fluid Property 3** *The pressure in a fluid at rest is the same at all points at equal depth, and the difference in pressure between two points $A$ and $B$ at depths $h_A$ and $h_B$ is*

$$p_A - p_B = \rho g h_A - \rho g h_B = \rho g (h_A - h_B) \qquad 7.6$$

Here $h_A$ and $h_B$ are positive when measured downward from the surface of the fluid (Fig. 7.12).

**REMARKS** (1) Equation 7.6 looks very similar to Eq. 5.13 for the difference in potential energy of two points. In Eq. 7.6, $\rho$ replaces $m$, and $h_A$ and $h_B$ are positive when measured downward rather than upward. This shows that the pressure at a point is equal to the negative of the potential energy of a unit volume of the fluid at this point (plus the pressure $p_0$ at the top of the fluid). (2) Fluid Property 3 assumes that the density of the fluid is the same at all points between $A$ and $B$. The density of a liquid varies only slightly with changes in temperature and pressure, so that in most practical situations this assumption is valid. The density of a gas, on the other hand, does change

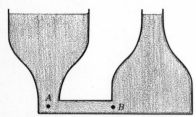

**FIGURE 7.10**
A funnel-shaped container filled with water. The shaded area of the base around point $A$ supports the column of water directly above it, so the pressure at $A$ is given by Eq. 7.4.

**FIGURE 7.11**
When the containers in Figs. 7.9 and 7.10 are connected, the water is the same height in both.

appreciably with changes in temperature and pressure, so that the density may not be the same at all points. In this case the average density between $A$ and $B$ should be used.

As an example of how Fluid Property 3 is used, let us calculate the pressure on a diver 10 m below the surface of the water. Let $p_0$ be the pressure at the top of the water (atmospheric pressure), and let $p_A$ be the pressure 10 m below the surface. With $h_0 = 0$, Eq. 7.6 can be written

$$p_A - p_0 = \rho g h_A - \rho g h_0 = \rho g h_A$$

The density of fresh water is $1.0 \times 10^3$ kg/m³ (Table 7.2), and so the pressure $p_A$ at a depth $h_A = 10$ m is

$$p_A = p_0 + (1.00 \times 10^3 \text{ kg/m}^3)(9.80 \text{ m/s}^2)(10 \text{ m})$$
$$= p_0 + 0.980 \times 10^5 \frac{\text{kg-m/s}^2}{\text{m}^2}$$
$$= p_0 + 0.980 \times 10^5 \text{ N/m}^2$$

Here we have used the fact that 1 N = 1 kg-m/s². The pressure $p_0$ at the surface is atmospheric pressure, which is about $1.0 \times 10^5$ N/m². The pressure 10 m below the surface is thus $p_A = 1.98 \times 10^5$ N/m², or nearly twice the pressure at the surface. If the diver were in seawater, the calculation would be the same except that the density of seawater should be used. Using the density of seawater given in Table 7.2, we have

$$p_A = p_0 + (1.025 \times 10^3 \text{ kg/m}^3)(9.80 \text{ m/s}^2)(10 \text{ m})$$
$$= p_0 + 1.04 \times 10^5 \text{ N/m}^2$$

**Atmospheric pressure**

We live at the bottom of a "sea" of air, the atmosphere, that exerts a pressure $p_0$ at sea level of about 14.7 lb/in.², or $1.01 \times 10^5$ N/m². The unit of pressure called the *atmosphere* is defined by the relation

$$1 \text{ atm} = 760 \text{ mm Hg} = 1.0133 \times 10^5 \text{ N/m}^2$$

This unit is equal to the average pressure of the atmosphere at sea level, though the actual pressure varies by about 5 percent depending on weather conditions. The average pressure at a site above sea level is less than at sea level because there is less air above it.

Equation 7.6 can be used to calculate the (average) air pressure $p_A$ at a site above sea level, but a little care is required. The distances in Eq. 7.6 are measured from the top of the fluid, whereas the elevation of a geographical site is measured from sea level, which is at the bottom of the fluid. However, since only the difference in the distances below the surface is used in Eq. 7.6, the distances can be measured downward from any convenient reference level.

For example, Mexico City is about 1500 m above sea level. To find the pressure $p_A$ at Mexico City it is convenient to measure distances from there. That is, we take $h_A$ to be zero, so that the depth at sea level is $h_0 = 1500$ m. (Remember, distances are measured positively downward.) Table 7.2 gives the density of air at sea level to be 1.2 kg/m³. At Mexico City the density is about 1.0 kg/m³, so the average density of 1.1 kg/m³ should be used. Then Eq. 7.6 gives

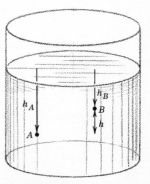

FIGURE 7.12
The pressure at two points $A$ and $B$ in a fluid.

$$p_A - p_0 = \rho g h_A - \rho g h_0 = -\rho g h_0$$
$$= -(1.1 \text{ kg/m}^3)(9.8 \text{ m/s}^2)(1.5 \times 10^3 \text{ m})$$
$$= -0.16 \times 10^5 \text{ N/m}^2$$

The pressure at Mexico City is less than the pressure at sea level, as expected. With $p_0 = 1.01 \times 10^5$ N/m$^2$ the pressure at Mexico City is

$$p_A = p_0 - 0.16 \times 10^5 \text{ N/m}^2$$
$$= 0.85 \times 10^5 \text{ N/m}^2$$

The density of air decreases with elevation above sea level because the pressure decreases. If the density were a constant, equal to its value at sea level, the height of the atmosphere could be readily calculated from Eq. 7.6. The pressure $p_A$ at the top of this hypothetical atmosphere is, of course, zero, and the distance $h_A$ of the top from the top is zero. If $h_0$ is the distance of sea level below the top, Eq. 7.6 gives

$$p_A - p_0 = \rho g h_A - \rho g h_0$$

or

$$0 - p_0 = 0 - \rho g h_0$$

so

$$h_0 = \frac{p_0}{\rho g} = \frac{1.0 \times 10^5 \text{ N/m}^2}{1.2 \text{ kg/m}^3 \times 9.8 \text{ m/s}^2}$$
$$= 8.5 \times 10^3 \text{ m} = 8.5 \text{ km}$$

Thus a constant-density atmosphere would be only 8.5 km (5.3 mi) high. The real atmosphere extends up much higher, but with decreasing density. The actual density at an altitude of 8.5 km is 0.5 kg/m$^3$, which is less than half the density at sea level. At an altitude of 18 km the density is only one-tenth the density at sea level.

Living organisms are not crushed by the pressure of the atmosphere because the fluids inside them are at about the same pressure. For instance, the pressure of the fluid inside a cell is the same as the pressure outside the cell, so that there is no net force on the cell wall. The blood pressure in the arteries is higher than atmospheric pressure, so that there is an outward force on the walls of the arteries. This is compensated by inward forces exerted by the tension in the walls.

### Gauge pressure

The pressure of a fluid at a point inside an animal is always quoted as the difference $\bar{p}$ between the *absolute pressure* $p$ of the fluid at that point and atmospheric pressure $p_0$. This difference is called *gauge pressure*:

$$\text{Gauge pressure} = \bar{p} = p - p_0 \qquad\qquad 7.7$$

For example, in man the average pressure of the blood, as it is pumped by the heart into the aorta, is about 2 lb/in.$^2$ (100 mm Hg). This is gauge pressure, the excess of the pressure in the blood over atmospheric pressure. It is the quantity of physiological interest, since it is the pressure that is actively maintained by the circulatory system. If atmospheric pressure is 14.7 lb/in.², Eq. 7.7 shows that the absolute pressure of the blood in the aorta is

$$p = p_0 + \overline{p} = 14.7 \text{ lb/in.}^2 + 2.0 \text{ lb/in.}^2$$
$$= 16.7 \text{ lb/in.}^2$$

**REMARK**  The actual gauge pressure in the aorta varies considerably during each cardiac cycle. The maximum (*systolic*) pressure, which is normally about 120 mm Hg, occurs when the heart contracts, and the minimum (*diastolic*) pressure, which is normally about 80 mm Hg, occurs when the heart relaxes. For purposes of discussion it is often only necessary to consider the average gauge pressure in the aorta, which is normally about 100 mm Hg.

The blood flows from the aorta into the major arteries of the body. These arteries, in turn, branch into smaller and smaller vessels, ultimately reaching the capillaries, which are the smallest vessels in the body. An artery with a diameter greater than 0.3 cm offers little resistance to the flow of blood, so that the pressure in it depends only on its vertical distance from the aorta, in accordance with Eq. 7.6. (The pressure in the smaller blood vessels is discussed in Sec. 7.5.)

This means, for example, that the pressure $p_B$ in an artery in the foot is greater than the pressure $p_A$ in the aorta by $\rho g h$, where $\rho = 1.05$ g/cm³ is the density of blood and $h$ is the vertical distance from the aorta to the foot. When one is standing erect, this distance is about 1.35 m, so the pressure difference is

$$p_B - p_A = \rho g h$$
$$= (1.05 \times 10^3 \text{ kg/m}^3)(9.8 \text{ m/s}^2)(1.35 \text{ m})$$
$$= 1.37 \times 10^4 \text{ N/m}^2$$
$$= (1.37 \times 10^4)(7.50 \times 10^{-3}) \text{ mm Hg}$$
$$= 103 \text{ mm Hg}$$

Here Table 7.1 has been used to convert from newtons per square meter to millimeters of mercury.

The difference $p_B - p_A$ of two absolute pressures is equal to the difference $\overline{p}_B - \overline{p}_A$ of the corresponding gauge pressures. This follows from Eq. 7.7 because

$$p_A = \overline{p}_A + p_0 \quad \text{and} \quad p_B = \overline{p}_B + p_0$$

so that

$$p_B - p_A = (\overline{p}_B + p_0) - (\overline{p}_A + p_0)$$
$$= \overline{p}_B - \overline{p}_A$$

Thus the arterial gauge pressure $\overline{p}_B$ in the feet is 103 mm Hg greater than the gauge pressure $\overline{p}_A$ in the aorta. If $\overline{p}_A$ is taken to be 100 mm Hg, this means that $\overline{p}_B$ is 203 mm Hg, or twice the pressure in the aorta. This large pressure sometimes causes swelling in the legs of people who must stand all day.

In an erect position, the top of the head is about 0.45 m above the aorta. By a calculation similar to the last one, the blood pressure in the head is found to be less than in the aorta by 35 mm Hg. These pressures are equal when the body is lying prone, since then the head and heart are at the same elevation. Therefore the blood pressure in the head drops from 100 to 65 mm Hg when a person rises from a prone to an erect position. To maintain a constant flow of blood to the brain, the arteries in the head expand to compensate for the pressure drop. Since this adjustment is not instantaneous, a momentary feeling of dizziness may result if a person sits up too rapidly.

Normally absolute pressure is positive,† whereas gauge pressure can be either positive or negative. A negative gauge pressure simply means that the absolute pressure is less than atmospheric pressure. For example, in breathing the lungs produce a pressure less than atmospheric, so that air at atmospheric pressure is forced into them. During quiet inhalation, the gauge pressure in the lungs is about $-0.1$ lb/in.². If atmospheric pressure is 14.7 lb/in.², Eq. 7.7 shows that the absolute pressure in the lungs is

$$
\begin{aligned}
p &= p_0 + \overline{p} \\
&= 14.7 \text{ lb/in.}^2 + (-0.1 \text{ lb/in.}^2) \\
&= 14.6 \text{ lb/in.}^2
\end{aligned}
$$

### Manometers and barometers

Gauge pressure is easily measured with a device called the *open-tube manometer*. It consists of a U-shaped tube partially filled with a liquid, usually either mercury or water. The tube is mounted in a vertical position with a measuring stick in back of it (Fig. 7.13). One end of the tube is connected to the vessel whose gauge pressure is to be measured, and the other end is open to the atmosphere. In Fig. 7.13 the manometer is measuring lung pressure $p$ during exhalation. The subject breathes into the left-hand side of the manometer, so that the pressure at point $B$ is $p$. Since the pressure at point $O$ is just atmospheric pressure $p_0$, the pressure $p_A$ at point $A$ is given by

$$
p_A - p_0 = \rho g(h_A - h_0) = \rho g h
$$

where $h_A$ and $h_0$ are the distances of points $A$ and $O$ from the top of the measuring rod. From Fluid Property 3 the pressure at $B$ is equal to the pressure at $A$ because these points are at the same elevation. Therefore the absolute pressure $p$ in the lungs is given by

$$
p = p_0 + \rho g h
$$

and the gauge pressure in the lungs is just $\rho g h$. The open-tube manometer thus measures gauge pressure directly in terms of the density of the liquid and the difference in the heights $h$ of the two columns of liquid.

For example, the liquid in a manometer used to measure lung pressure is water. A healthy subject can exert a pressure with his lungs sufficient to raise point $O$ about 65 cm above point $B$. Since 65 cm = 0.65 m and the density of water is $1.00 \times 10^3$ kg/m³, the gauge pressure in the lungs is

$$
\begin{aligned}
\overline{p} = \rho g h &= (1.00 \times 10^3 \text{ kg/m}^3)(9.80 \text{ m/s}^2)(0.65 \text{ m}) \\
&= 6.37 \times 10^3 \text{ N/m}^2
\end{aligned}
$$

This pressure can also be stated directly as 65 cm $H_2O$, meaning that it is a pressure sufficient to raise a column of water 65 cm. Thus 65 cm $H_2O$ is equal to $6.37 \times 10^3$ N/m², so

$$
1 \text{ cm } H_2O = \frac{6370 \text{ N/m}^2}{65} = 98 \text{ N/m}^2
$$

†Negative absolute pressure is discussed in Sec. 9.5.

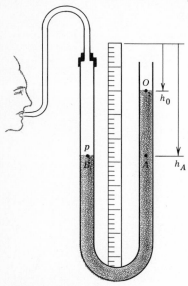

**FIGURE 7.13**
A manometer measuring lung pressure during exhalation.

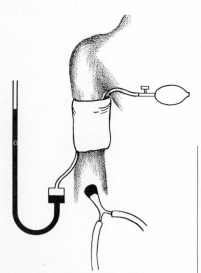

**FIGURE 7.14**
A mercury-filled manometer measuring blood pressure. The manometer is connected to a closed bag that can be pumped with air by squeezing the rubber bulb.

This is the conversion from centimeters of water to newtons per square meter given in Table 7.1.

Blood pressure is measured using a mercury-filled manometer. The instrument is attached to a closed bag that is wrapped around the upper arm (Fig. 7.14). The air pressure in the bag is first elevated well above the systolic blood pressure by pumping air into it. This collapses the brachial artery in the upper arm, cutting off the flow of blood into the arteries of the forearm. Next, the air is gradually released from the bag while a stethoscope is used to listen for the return of a pulse in the forearm. The first sound occurs when the pressure in the bag just equals the systolic blood pressure, because then some blood at peak pressure can squeeze through the collapsed artery. This restricted blood flow makes a characteristic tapping sound in the artery of the forearm, which is detected with the stethoscope. The difference in the heights (in millimeters) of the mercury columns at the first occurrence of this sound is thus equal to the systolic blood pressure in millimeters of mercury. Finally, more air is released from the bag to lower the pressure in it further. The tapping sound ceases when the pressure equals the diastolic blood pressure, because then blood at low pressure is able to pass through the artery in the arm. The difference in the heights of the two mercury columns when the sound first ceases is equal to the diastolic blood pressure in millimeters of mercury. To ensure that the pressures measured are equal to the pressures in the aorta, the bag must be attached to the arm at the elevation of the heart.

In principle a manometer can also be used to measure the pressure of the atmosphere. All that is needed is to reduce the pressure in one column to zero while the other column is left open to the atmosphere. This is done in Fig. 7.15 by attaching the left-hand column to a vacuum pump. The pressure at point $B$ is then just $\rho g h$ and is equal to the pressure of the atmosphere at $A$.

In practice the vacuum is obtained as follows. A glass tube, closed at one end, is filled with mercury and inverted into a dish of mercury (Fig. 7.16). If the tube is over 76 cm long, some of the mercury drops away from the closed end, leaving a vacuum there. The pressure (in millimeters of mercury) at point $A$ is equal to the height (in millimeters) of the mercury column, while the pressure at point $B$ is equal to atmospheric pressure. Since these points are at the same elevation, the height of the column measures the atmospheric pressure. This instrument is called a *barometer*.

The average atmospheric pressure at sea level will support a column of mercury 760 mm high (about 30 in., or 2.5 ft). It can support a column of water 34 ft high, the ratio of 34 to 2.5 ft being equal to the ratio of the density of mercury to the density of water. Water is pumped out of a well by lowering the air pressure at the upper end of a pipe. The higher pressure of the atmosphere at the lower end of the pipe forces the water up the pipe, as in Fig. 7.15. Since the lowest pressure the pump can produce is zero, the maximum height that water can be lifted in this way is 34 ft. (The mechanism which enables tall trees to lift water hundreds of feet from their roots to their upper branches is explained in Sec. 9.5.)

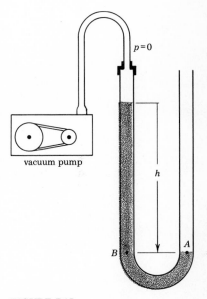

FIGURE 7.15
A manometer measuring air pressure. One side of the manometer is evacuated with a vacuum pump.

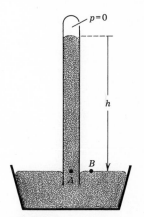

FIGURE 7.16
A mercury barometer.

The buoyant force $\mathbf{F}_b$ that a fluid exerts on an object immersed in the fluid can be calculated from the fluid properties discussed in Secs. 7.2 and 7.3. Thus consider the forces exerted on the block immersed in the fluid in Fig. 7.17. The force $\mathbf{F}_B$ exerted by the fluid on the top face is directed downward (Fluid Property 1), and its magnitude is $p_B A$, where $A$ is the area of this face and $p_B$ is the pressure in the fluid at this depth. Likewise the force $\mathbf{F}_A$ on the bottom face is directed upward, and its magnitude is $p_A A$. The sum of these forces has the magnitude

$$F_A - F_B = p_A A - p_B A$$

and is directed upward, since $F_A > F_B$. From Eq. 7.6 (Fluid Property 3) this magnitude can be written

$$\begin{aligned} F_A - F_B &= A\rho_F g(h_A - h_B) \\ &= A\rho_F g h \end{aligned} \qquad 7.8$$

where $\rho_F$ is the density of the fluid and $h$ is the height of the block.

The sum of the forces on the other faces of the block is zero because for any region on one vertical face, such as $S$ in Fig. 7.17, there is another region $S'$ on the opposite face with the same area and pressure. Since the forces on these two regions have equal magnitude but opposite direction, their sum is zero. All other regions on the vertical faces can be similarly paired, so that the total force on these faces is zero. Equation 7.8 thus gives the magnitude of the total

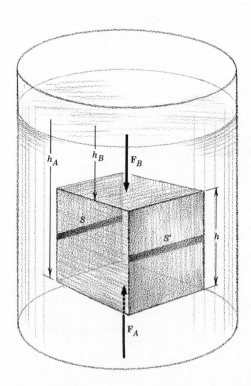

**FIGURE 7.17**
The forces on a block immersed in a fluid.

force exerted by the fluid on the block. This is the buoyant force $\mathbf{F}_b$.

The buoyant force can be written somewhat differently by noting that $Ah$ is the volume $V$ of the block, so that $\rho_F Ah = \rho_F V$ is the mass $m_F$ of an equal volume of fluid. The buoyant force on an object of volume $V$ then is

$$F_b = F_A - F_B = \rho_F V g$$
$$= m_F g = \text{weight of an equal volume of fluid}$$

This result, although derived for the special case of a rectangular block, is true in general. It can be stated formally as follows.

**Fluid Property 4 (Archimedes' principle)** *The buoyant force exerted by a fluid on an object is equal to the weight of the fluid displaced by the object.* Of course, the volume of fluid displaced by an object that is totally submerged in the fluid is equal to the volume of the object itself. Archimedes' principle is stated so as to include cases in which the object floats. In these cases only part of the object is in the fluid, and so the volume of fluid displaced is equal only to the volume of the object below the surface of the fluid.

**REMARK** It is not difficult to see why Archimedes' principle is true even for irregularly shaped objects. The thing to remember is that since the fluid is in contact only with the surface of the object, the force it exerts on the object depends only on the object's shape and not on the material of which the object is made. Thus the buoyant force on the irregularly shaped object in Fig. 7.18 is the same as the buoyant force on any other object of the same shape. In Fig. 7.19 the object has been removed, but a region of fluid of the same shape as the object has been outlined. The fluid inside this region is in equilibrium under the actions of the force of gravity and the buoyant force due to the rest of the fluid, so that the buoyant force on the region is equal to the weight of the fluid inside the region. This same buoyant force acts on any other object of the same shape.

The total force $\mathbf{F}$ acting on an object entirely immersed in a fluid is the sum of the buoyant force and the force of gravity:

$$\mathbf{F} = \mathbf{F}_b + \mathbf{F}_g$$

In general this force will not be zero, so that the object will not be in equilibrium. The force of gravity is directed downward, and its magnitude is

$$F_g = m_O g = \rho_O V g$$

where $m_O$ = mass of object
$\rho_O$ = density of object
$V$ = volume of object

The buoyant force is directed upward, and its magnitude is given by Eq. 7.8. The magnitude of the total force then is

$$F = F_b - F_g$$
$$= \rho_F g V - \rho_O g V$$
$$= (\rho_F - \rho_O) g V$$

If $F$ is positive, i.e., if the density $\rho_F$ of the fluid is greater than the density $\rho_O$ of the object, the total force is upward, and so the object rises to the top of the fluid. At equilibrium the object floats

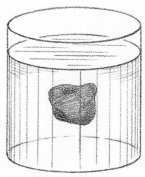

FIGURE 7.18
An irregularly shaped object immersed in a fluid.

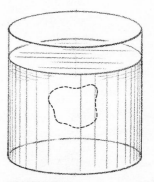

FIGURE 7.19
A region of fluid of the same shape as the object in Fig. 7.18.

with only part of its volume $V'$ submerged. The buoyant force $\rho_F g V'$ is then equal to the force of gravity $\rho_O g V$ on the object,

$$\rho_F g V' = \rho_O g V \qquad \text{or} \qquad \frac{V'}{V} = \frac{\rho_O}{\rho_F}$$

This says that the fraction of the object's volume that is submerged is equal to the ratio of the density of the object to the density of the fluid. For example, the density of a human being is about 0.98 g/cm³. A person floating motionless in fresh water will have the fraction

$$\frac{V'}{V} = \frac{0.98 \text{ g/cm}^3}{1.00 \text{ g/cm}^3} = 0.98$$

of his body submerged. Thus only 2 percent of his body is above water, which is not enough to include the nose or mouth.

If $F$ is negative, i.e., if the density $\rho_F$ of the fluid is less than the density $\rho_O$ of the object, the total force is downward and the object sinks to the bottom of the fluid. At equilibrium the bottom exerts an upward surface force of magnitude

$$\begin{aligned} F_s &= -F = -(\rho_F - \rho_O)gV \\ &= -(W_F - W_O) \\ &= W_O - W_F \end{aligned} \qquad 7.9$$

where $W_O$ = weight of object
$W_F = F_b$ = weight of equal volume of fluid

The surface force is less than it would be in the absence of the fluid by just the magnitude of the buoyancy.

An object can stay suspended in a fluid only if $\rho_O$ is exactly equal to $\rho_F$. Fish are able to satisfy this condition because they have a *swim bladder* under their spinal column (Fig. 7.20). This thin-walled sac is filled with a mixture of oxygen and nitrogen obtained from the blood. By varying the amount of gas in the sac, the volume of the entire fish can be varied without changing its mass, allowing the fish to adjust its density. To float suspended in the water, the density is adjusted to equal the density of the surrounding water.

When two immiscible fluids of different densities are mixed, the fluid with the smaller density floats on top of the fluid with the larger density. For example, oil floats on top of water because its density is less than the density of water. Even if the two fluids are miscible, the less dense fluid will float on top of the denser fluid if care is taken not to mix them. This fact is of fundamental importance for the circulation of water in a lake and the circulation of the air in the atmosphere. From Table 7.1 it is seen that both these fluids are denser at low temperatures than at high temperatures. In summer, as the surface water of a lake is warmed by the sun, it becomes less dense than the cooler water underneath it. This warm surface water thus floats on the cooler water and is prevented from mixing with the water in the lower levels of the lake by its own buoyancy. As a consequence the lower level remains stagnant as its oxygen content becomes depleted. In autumn, as the lake cools, the temperature of the water becomes uniform throughout, so that the wind is able to mix the upper and lower levels, replenishing the oxygen content in

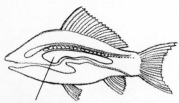
swim bladder
FIGURE 7.20
The swim bladder of a fish.

the lower level. In winter the surface water freezes. It is a remarkable fact of nature that water is one of the few substance that is less dense in its solid state than in its liquid state. Thus the ice floats on the surface. This prevents the lake from freezing to the bottom, which would, of course, kill all the fish. The water beneath the ice is all at a temperature just above freezing. In spring, as the ice breaks up, the lake is still at a uniform temperature, so that the wind again is able to mix the water in the upper and lower levels.

The circulation in the atmosphere is just the opposite. During the day, the sun warms the surface of the earth, which in turn warms the air in the lower atmosphere. This warmer air rises, while the cooler denser upper air falls. Thus normally there is a constant vertical mixing of the surface and upper-level air. This circulation is temporarily halted under certain weather conditions in which abnormally warm upper-atmosphere air moves in over cooler surface air. During such a *thermal inversion* the daily outpouring of air pollutants accumulates over a city, causing great misery and even death to people who suffer from lung disease. (The occurrence of thermal inversions dramatically demonstrates our dependence on the normal circulation of air to remove air pollution.)

## 7.5 FLUID FLOW

The last three sections discussed the physics of fluids at rest. Fluid dynamics, the physics of fluid motion, is much more complex, and a full treatment of it is beyond the scope of this book. It is of great importance, however, for understanding such diverse phenomena as the flight of airplanes, birds, and insects through the air, the flow of blood through the vessels of the circulatory system, and the circulation of air in the atmosphere. Although the underlying principles of fluid dynamics are just Newton's laws of motion, the mathematical equations that describe how these laws govern the behavior of a moving fluid are in general very complex. In this section we consider a few simple aspects of fluid motion related to the flow of fluids through pipes and apply them to the flow of blood through the vessels of the circulatory system.

### Viscosity

One of the main differences between a moving fluid and a fluid at rest is that a moving fluid exerts a force parallel to a surface whereas a fluid at rest does not (Fluid Property 1, Sec. 7.2). As a fluid flows past a surface, it exerts a force $\mathbf{F}_\parallel$ parallel to the surface in the direction of flow. The reaction $\mathbf{F}_v$ to $\mathbf{F}_\parallel$ is a force exerted by the surface on the fluid directed opposite to the direction of flow. This force, called the *viscous force*, plays a role in fluid flow similar to that of friction in the motion of one solid upon another. That is, the viscous force opposes the motion. In order to maintain a steady flow an external driving force must be applied to the fluid to balance the viscous force.

The simplest case is that of a fluid flowing along a flat surface

$S_1$, as shown in Fig. 7.21. To maintain the flow, a second surface $S_2$ placed on top of the fluid is moved in the direction of fluid flow with constant speed $v$. Since the fluid exerts the force $\mathbf{F}_\parallel$ parallel to $S_1$, an external force $\mathbf{F}_a = -\mathbf{F}_\parallel$ must be applied to $S_1$ to keep it at rest.† The viscous force $\mathbf{F}_v$ is the reaction to $\mathbf{F}_\parallel$, so that

$$\mathbf{F}_v = -\mathbf{F}_\parallel = -(-\mathbf{F}_a) = \mathbf{F}_a$$

Thus $\mathbf{F}_v$ is determined by measuring the external force $\mathbf{F}_a$ needed to keep $S_1$ from moving.

**REMARK** $\mathbf{F}_a$ and $\mathbf{F}_v$ are two different forces that happen to be equal. $\mathbf{F}_a$ is the force applied to $S_1$ by some outside agent, whereas $\mathbf{F}_v$ is the viscous force applied by $S_1$ to the fluid.

From studies of $F_v$ made using an arrangement of plates similar to Fig. 7.21, it is found that the magnitude of $F_v$ is directly proportional to the speed $v$ of $S_2$ and to the area $A$ of $S_1$ and inversely proportional to the distance $d$ between the surfaces. In symbols this is expressed as

$$F_v = \frac{\nu A v}{d} \qquad\qquad 7.10$$

where $\nu$‡ is a constant, called the *viscosity*, that is characteristic of the fluid.

To understand more clearly what is happening, it is necessary to examine the motion of the fluid in more detail. Studies have shown that the layer of fluid adjacent to a surface adheres to the surface and moves with it. That is, there is no motion of this layer of fluid relative to the surface. For example, the layer of fluid adjacent to $S_2$ does not slip along the surface: it moves with the same speed $v$ as $S_2$ (Fig. 7.22). The layer of fluid immediately below it is dragged along by this first layer, but with a slightly smaller speed because one fluid layer slips over the other. This second fluid layer in turn drags the next layer with a still smaller speed, and so on, until the layer of fluid adjacent to $S_1$ is reached. Since this layer adheres to the stationary surface $S_1$, it has zero speed. This is shown schematically in Fig. 7.22, where the speed in each layer is indicated by an arrow whose length is proportional to the speed. The fluid moves with different speeds at different distances from $S_1$, but all the layers of fluid move parallel to each other. This is *laminar flow*, the simplest form of fluid motion.

In reality, therefore, the viscous force does not have its origin in the slipping of the fluid along a surface (as with friction) but in the slipping of one layer of fluid along another. The viscous force is large when the force needed to produce this slippage is large. The viscosity $\nu$ is a measure of how much force is required to slip one layer of fluid over another. A large value of $\nu$ corresponds to a very viscous fluid like glycerin or oil, whereas a small value corresponds to a light fluid like water or ether. The viscosity is an intrinsic property of a fluid and does not depend on the nature of the surface along which the fluid is moving.

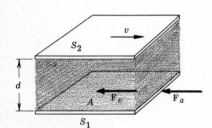

**FIGURE 7.21**
Fluid flowing along a flat surface $S_1$. The upper surface $S_2$ moves with constant speed $v$ to maintain the flow, while $S_1$ is held stationary.

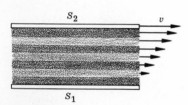

**FIGURE 7.22**
The speed of flow between the two surfaces in Fig. 7.21 varies continuously from $v$, for fluid in contact with $S_2$, to zero, for fluid in contact with $S_1$.

---

† The fluid may also apply a force perpendicular to $S_1$, but this is just the fluid pressure that has already been studied.
‡ $\nu$ is the Greek letter nu.

From Eq. 7.10 the dimensions of viscosity are found to be

$$[\nu] = \frac{[f][l]}{[l^2][v]} = \frac{[f]}{[l][l/t]}$$
$$= \frac{[f][t]}{[l^2]}$$

In the mks system, the units of viscosity are N-s/m², whereas in the cgs system the units are dyn-s/cm². The cgs unit is called a *poise* (P). The relation between these units is

$$1 \text{ N-s/m}^2 = \frac{10^5 \text{ dyn-s}}{(10^2 \text{ cm})^2}$$
$$= 10 \text{ dyn-s/cm}^2$$
$$= 10 \text{ P}$$

Table 7.3 gives the viscosity of some common liquids and gases in both mks and cgs units. Since the viscosity of a fluid varies rapidly with its temperature, the temperature at which the viscosity is measured must be given. Note that the viscosity of a gas is much less than the viscosity of a liquid.

**REMARK** Values of viscosity are usually quoted in poises or centipoises (1 cP = $10^{-2}$ P). Since these are cgs units, the viscosity must be converted into mks units before doing a calculation in mks units.

### Fluid flow through pipes

We now consider the problem of a fluid flowing through a pipe of radius $r$ and length $L$ (Fig. 7.23). The speed of the layer of fluid

TABLE 7.3 **Viscosity of some common liquids and gases**

| Fluid | Temperature, °C | Viscosity dyn-s/cm² (P) | N-s/m² |
|---|---|---|---|
| *Liquids* | | | |
| Acetone | 25 | 0.00316 | 0.000316 |
| Blood plasma | 37 | 0.015 | 0.0015 |
| Blood, whole | 37 | 0.04 | 0.004 |
| Ethanol | 20 | 0.0120 | 0.00120 |
| Ether | 20 | 0.00233 | 0.000233 |
| Glycerin | 20 | 14.9 | 1.49 |
| Mercury | 20 | 0.0155 | 0.00155 |
| Oil, light machine | 16 | 1.13 | 0.113 |
| | 38 | 0.34 | 0.034 |
| Water | 0 | 0.0179 | 0.00179 |
| | 20 | 0.0100 | 0.00100 |
| | 37 | 0.00691 | 0.000691 |
| | 100 | 0.00282 | 0.000282 |
| *Gases* | | | |
| Air | 0 | 0.000171 | 0.0000171 |
| | 18 | 0.000183 | 0.0000183 |
| | 40 | 0.000190 | 0.0000190 |
| Helium | 20 | 0.000194 | 0.0000194 |
| Water vapor | 100 | 0.000125 | 0.0000125 |

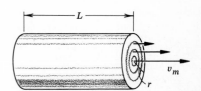

FIGURE 7.23
Fluid flowing through a pipe of radius $r$ and length $L$. The speed of the fluid varies from zero, for fluid in contact with the pipe, to $v_m$, for fluid in the center of the pipe.

adjacent to the wall of the pipe is zero, and the fluid moves with the maximum speed $v_m$ along the central axis of the pipe. The speed in each concentric layer of fluid varies steadily from $v_m$ to zero in going outward from the central axis. Therefore the average speed of the fluid is $\bar{v} = \frac{1}{2} v_m$.

The motion of the fluid through the pipe is opposed by the viscous force exerted on the fluid by the walls of the pipe. Equation 7.10 applies only to the case of two flat surfaces, but it can be used to estimate $F_v$ in the present case. For the area in Eq. 7.10 we take the area of the pipe wall, because this is the area in contact with the fluid. The area of the pipe wall is equal to the circumference of the pipe times its length or

$$A = 2\pi r L$$

For the distance $d$ in Eq. 7.10 we take the radius $r$ of the pipe because this is the distance over which the fluid speed varies from $v_m$ to zero. Finally, for $v$ we take the maximum fluid speed $v_m$. The result is

$$F_v = \nu \frac{2\pi r L v_m}{r} = 2\pi \nu L v_m$$

This estimate gives the correct dependence of $F_v$ on $\nu$, $L$, and $v_m$, but it is too small by a factor of 2. A more careful analysis shows that

$$F_v = 4\pi \nu L v_m \qquad \qquad 7.11$$

This viscous force opposes the fluid flow, so that to maintain constant flow there must be a driving force of magnitude $F_v$. If gravity is neglected, the only other force on the fluid is caused by the fluid pressure. Fluid entering the pipe on the left at pressure $p_1$ exerts a force $p_1 A$ to the right on the fluid inside the pipe, where now $A = \pi r^2$ is the cross-sectional area of the pipe. The fluid leaving the pipe on the right at pressure $p_2$ exerts a force $p_2 A$ to the left on the fluid inside the pipe. If $p_2$ is less than $p_1$, there will be a net driving force of

$$p_1 A - p_2 A = (p_1 - p_2)A$$
$$= (p_1 - p_2)\pi r^2$$

to the right on the fluid. The condition for steady flow is that this driving force just equal $F_v$,

$$(p_1 - p_2)\pi r^2 = 4\pi \nu L v_m$$

so that

$$v_m = \frac{(p_1 - p_2)r^2}{4\nu L} \qquad \qquad 7.12$$

This equation gives the speed in the center of a pipe in terms of the pressure difference between the ends of the pipe, the radius and length of the pipe, and the viscosity of the fluid. For a given pressure difference $v_m$ increases with $r$ and decreases with $\nu$ and $L$, which is reasonable. Solving Eq. 7.12 for $p_1 - p_2$, we have

$$p_1 - p_2 = \frac{4\nu L v_m}{r^2} \qquad \qquad 7.13$$

which gives the pressure difference produced in a fluid as it moves through a pipe.

An interesting example is the pressure drop in the blood as it passes through a capillary. A typical capillary is about 1 mm long ($L = 10^{-3}$ m). It has a radius of 2 $\mu$m ($r = 2 \times 10^{-6}$ m), and the speed of blood through its center is 0.66 mm/s ($v_m = 0.66 \times 10^{-3}$ m/s). The viscosity of whole blood is about $4 \times 10^{-3}$ N-s/m$^2$ (Table 7.3). With these values in Eq. 7.13, the pressure drop in passing through a capillary is found to be

$$p_1 - p_2 = \frac{4(4 \times 10^{-3}\ \text{N-s/m}^2)(10^{-3}\ \text{m})(0.66 \times 10^{-3}\ \text{m/s})}{(2 \times 10^{-6}\ \text{m})^2}$$

$$= 0.26 \times 10^4\ \text{N/m}^2$$
$$= (0.26 \times 10^4)(7.50 \times 10^{-3})\text{mm Hg}$$
$$= 19.5\ \text{mm Hg}$$

This is typical of the pressure drop across a capillary, although there is great variation from capillary to capillary.

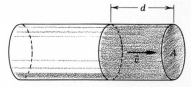

FIGURE 7.24
A section of fluid of length $d$ moving with average speed $\bar{v}$.

The quantity of primary interest is not the speed of flow but the volume of fluid that passes through a vessel per second. This is called the *fluid flow* $Q$, and it is related to the average speed of flow $\bar{v}$ and the cross-sectional area of the vessel. Figure 7.24 shows a section of fluid of length $d$ about to flow out of a pipe of radius $r$. If the section is moving with average speed $\bar{v}$, it will pass out of the pipe in the time $t = d/\bar{v}$. The volume $V$ of fluid in this section is equal to the cross-sectional area $A = \pi r^2$ of the pipe times the length $d$ of the section. Therefore the fluid flow is

$$Q = \frac{V}{t} = \frac{Ad}{d/\bar{v}} = A\bar{v}$$
$$= \pi r^2 \bar{v} \qquad\qquad 7.14$$

If the radius of the pipe changes, the speed of flow will change to keep the fluid flow constant.

For example, in a normal adult at rest, the average fluid speed through the aorta is $\bar{v} = 0.33$ m/s. The radius of the aorta is $r = 0.9 \times 10^{-2}$ m (0.9 cm), so its cross-sectional area is $A = \pi r^2 = 2.5 \times 10^{-4}$ m$^2$. Therefore the fluid flow is

$$Q = (2.5 \times 10^{-4}\ \text{m}^2)(0.33\ \text{m/s})$$
$$= 0.83 \times 10^{-4}\ \text{m}^3/\text{s}$$
$$= 83\ \text{cm}^3/\text{s}$$

From the aorta the blood flows into the major arteries, then into the smaller arteries (arterioles), and finally into the capillaries. At each stage these vessels divide into many smaller vessels. But, although the cross-sectional area of each artery is smaller than the area of the aorta, the total cross-sectional area of all the major arteries is $20 \times 10^{-4}$ m$^2$. This is shown schematically in Fig. 7.25. Since the total fluid flow through all these arteries is the same as through the aorta, Eq. 7.14 can be used to calculate the average speed of the blood in the arteries. The result is

$$\bar{v} = \frac{Q}{A} = \frac{0.83 \times 10^{-4}\ \text{m}^3/\text{s}}{20 \times 10^{-4}\ \text{m}^2}$$
$$= 0.041\ \text{m/s}$$

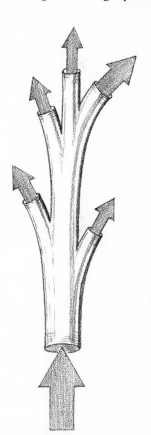

FIGURE 7.25
Blood flowing from the aorta into a number of smaller arteries. The combined cross-sectional areas of these arteries is greater than the area of the aorta, so the average speed through them is slower than through the aorta.

Thus the blood moves more slowly in the arteries than in the aorta, because the total cross-sectional area of the arteries is larger than the area of the aorta. The total cross-sectional area of all the capillaries is 0.25 m², so the average speed through them is only $0.33 \times 10^{-3}$ m/s.

Equations 7.12 and 7.14 can be combined to give an important equation relating the fluid flow to the pressure difference in a pipe. To do this we recall that the speed in the pipe varies from zero at the wall to $v_m$ at the center. Therefore the average speed $\overline{v}$ is just $\frac{1}{2}v_m$, and so Eq. 7.14 can be written

$$Q = \tfrac{1}{2}\pi r^2 v_m$$

Substituting the right-hand side of Eq. 7.12 for $v_m$ in this last expression, we get the equation

$$Q = \frac{\pi r^4 (p_1 - p_2)}{8 \nu L} \qquad 7.15$$

which is known as *Poiseuille's law*. This law says that *the amount of fluid that flows through a pipe is proportional to the pressure drop along the pipe and to the fourth power of the radius of the pipe.* That is, for same pressure difference, 16 times as much fluid will flow through a pipe of radius 2 mm as will flow through a pipe of radius 1 mm. Equation 7.15 is fundamental for a proper understanding of blood flow in the body.

**REMARK** Poiseuille's law is really only an approximation, valid when the flow speed $\overline{v}$ is sufficiently small. When the flow speed is large, the flow no longer consists of concentric layers of fluid all moving parallel to each other at different speeds (laminar flow). Instead turbulence develops, in which some of the fluid moves in small circles inside the pipe (eddies). The theory of turbulent flow is very complex, and the phenomenon of turbulence is still not completely understood.

**Blood flow**

The aorta is so large ($r = 9$ mm) that a pressure difference of only 3 mm Hg is required to maintain normal blood flow through it. Thus if the pressure of the blood is 100 mm Hg when the blood enters the aorta, it is reduced to 97 mm Hg when the blood enters the major arteries. Because these vessels have much smaller radii than the aorta, a pressure drop of 17 mm Hg is required to maintain the flow through them. Therefore the pressure is only 85 mm Hg when the blood enters the arterioles (small arteries). These vessels have still smaller radii, so that a pressure drop of 55 mm Hg is required to maintain flow through them. Finally, there is a further drop of 20 mm Hg when the blood passes through the capillaries. (The pressure drop in the capillaries is less than in the arterioles, even though the capillaries have much smaller radii, because the number of capillaries is so large that the blood flow through each one is very small.) Thus the blood pressure drops to only 10 mm Hg by the time it reaches the veins. Figure 7.26 shows schematically the pressure variations in the blood as it moves through the systemic circulatory system.

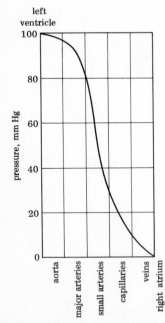

**FIGURE 7.26**
Schematic representation of the pressure variation in the blood as it moves through the systemic circulatory system.

It is convenient to write Eq. 7.15 in the form

$$Q = \frac{p_1 - p_2}{R} \qquad 7.16$$

where

$$R = \frac{8\nu L}{\pi r^4} \qquad 7.17$$

is the *resistance* of a single vessel. Equation 7.16 is also valid for a complex network of interconnected vessels, such as the blood vessels in the circulatory system, if $R$ is taken to be the total resistance of the network. This total resistance is calculated from the resistances of the individual vessels of the network by the same procedure used to calculate the total resistance of an electric circuit (Sec. 17.2). The importance of Eq. 7.16 is that it shows the relation between blood pressure and total resistance. For example, it has already been noted that in a normal adult $Q = 0.83 \times 10^{-4}$ m$^3$/s and that the total pressure drop from the aorta through the capillaries is

$$p_1 - p_2 = 90 \text{ mm Hg} = 1.2 \times 10^4 \text{ N/m}^2$$

Therefore the total resistance of all the arteries, arterioles, and capillaries in the body is

$$R = \frac{p_1 - p_2}{Q} = \frac{1.2 \times 10^4 \text{ N/m}^2}{0.83 \times 10^{-4} \text{ m}^3/\text{s}}$$

$$= 1.44 \times 10^8 \text{ N-s/m}^5$$

If the total body resistance becomes abnormally large, the blood pressure must increase to maintain normal blood flow. This is the condition of hypertension (high blood pressure), which is the cause of 12 percent of all deaths. On the other hand, if the resistance is lowered while the blood pressure is unchanged, the blood flow $Q$ is increased. During exercise there is both an increase in blood pressure and a decrease in total resistance, producing the greatly increased blood flow that is required. The decrease in resistance is caused by an increase in the radii of the blood vessels (vasodilation). Since $R$ is inversely proportional to the fourth power of $r$, a small increase in radius produces a large decrease in resistance.

The effect of high blood pressure is to cause the heart to work harder than normal. The power output $P$ of the heart is the work done by the heart per second in pumping the blood. It is equal to the average force $F$ exerted by the heart on the blood times the distance $d$ that the blood moves in 1 s.

$$P = Fd$$

The force is just the pressure $p$ exerted by the heart on the aorta times the cross-sectional area $A$ of the aorta,

$$F = pA$$

The blood flow $Q$ is the volume of blood that passes through the aorta in 1 s, so in 1 s the volume of blood moves the distance

$$d = \frac{Q}{A}$$

Therefore the output of the heart is

$$P = Fd$$

$$= pA \frac{Q}{A}$$

$$= pQ \qquad \qquad 7.18$$

The average blood pressure of a normal adult is $p = 100$ mm Hg $= 1.3 \times 10^4$ N/m², so

$$P = (1.3 \times 10^4 \text{ N/m}^2)(0.83 \times 10^{-4} \text{ m}^3/\text{s})$$
$$= 1.1 \text{ N-m/s} = 1.1 \text{ J/s} = 1.1 \text{ W}$$

Thus the normal power output of the heart is only 1 W or 1 percent of the total power dissipated by the body† (Sec. 6.3). Equation 7.18 shows that the heart of a person with abnormally high blood pressure does more work per second to maintain the same blood flow.

† This is only the mechanical energy dissipated by the heart. The heart also dissipates 3 to 4 W of heat for each watt of mechanical power produced.

## PROBLEMS

1 A 110-lb ballet dancer is standing on one toe. What is the pressure on the area of the floor under her toe if the area of her toe is 3.5 in.²?
*Ans.* 31.4 lb/in.²

2 The edge of a chisel has an area of 0.12 in.². When struck with a hammer, the chisel exerts a momentary force of 20 lb on a brick. What is the pressure exerted directly under the edge of the chisel?

3 An explosion creates a momentary increase in the pressure of the surrounding air (*overpressure*). Calculate the total force exerted on the wall of a building 20 ft high and 30 ft wide by an overpressure of 0.4 lb/in.².
*Ans.* 34,560 lb

4 A patient's systolic blood pressure is 220 mm Hg. Convert this pressure into (*a*) newtons per square meter, (*b*) pounds per square inch, and (*c*) centimeters of water.

5 The (gauge) pressure of the air delivered to a patient by a respirator is 20 cm H₂O. Convert this pressure into (*a*) newtons per square meter, (*b*) pounds per square inch, and (*c*) millimeters of mercury.
*Ans.* (*a*) 1962 N/m²; (*b*) 0.284 lb/in.²;
(*c*) 14.7 mm Hg

6 The diameters of the large and small pistons in a hydraulic lift (Fig. 7.7) are 6.0 and 1.5 in., respectively. (*a*) What force must be applied to the smaller piston to lift a 2000-lb car sup-

ported on the large piston? (*b*) If the small piston is depressed 5 in., how much is the large piston raised? (*c*) What is the mechanical advantage of the lift?

7 A force of 4 N is applied to the plunger of a hypodermic syringe that has a cross-sectional area of 2.5 cm² (Fig. 7.27). (*a*) What is the (gauge) pressure in the fluid inside the syringe? (*b*) The fluid passes through a hypo-

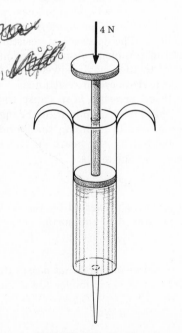

FIGURE 7.27   Problem 7.

dermic needle with a cross-sectional area of 0.008 cm². What force would have to be applied to the end of the needle to prevent the fluid from coming out? (*c*) What is the minimum force that must be applied to the plunger to inject fluid into a vein in which the blood pressure is 12 mm Hg?

*Ans.* (*a*) $1.6 \times 10^4$ N/m²; (*b*) 0.0128 N; (*c*) 0.40 N

8 The heart pumps blood into the aorta with an average pressure of 100 mm Hg. If the cross-sectional area of the aorta is 3 cm², what is the average force exerted by the heart on the blood entering the aorta?

9 What is the mass of 200 ml of trichloromethane? *Ans.* 297 g

10 Calculate the mass of the air in a room 6 m wide, 10 m long, and 4 m high.

11 (*a*) Calculate the mass of an aluminum cylinder 10 cm long and 4 cm in diameter. (*b*) The mass of a tungsten cylinder of the same size and shape is 1758 g. What is the density of tungsten? *Ans.* (*a*) 339 g; (*b*) 14 g/cm³

12 Inches of water (in. H₂O), a unit of pressure sometimes used in respiratory therapy, is the pressure exerted by a column of water 1 in. high. Find the conversion from inches of water to (*a*) centimeters of water and (*b*) millimeters of mercury.

13 Inches of mercury is a unit of pressure sometimes used in meteorology. Find the conversion from inches of mercury to (*a*) millimeters of mercury and (*b*) atmospheres.

*Ans.* (*a*) 1 in. Hg = 25.4 mm Hg; (*b*) 1 in. Hg = 0.0333 atm

14 A dike develops a leak at a point 4 m below the surface of the water. If the area of the hole is 1.5 cm², what force must a Dutch boy apply to the hole to prevent water from coming through?

15 (*a*) Plasma flows from a bag through a tube into a patient's vein. When the bag is held 1.5 m above the patient's arm, what is the pressure of the plasma as it enters the vein? (*b*) If the blood pressure in the vein is 12 mm Hg, what is the minimum height at which the bag must be held for the plasma to flow into the vein? (*c*) Suppose an astronaut needed a transfusion on the moon. What would the minimum height of the bag have to be in this case? On the moon *g* is 1.63 m/s².

*Ans.* (*a*) 114 mm Hg; (*b*) 0.16 m (because of the viscosity of the plasma, the bag must be

held much higher than this to get any appreciable flow); (*c*) 0.95 m

16 In an early experiment to demonstrate the existence of blood pressure, blood from an artery of a horse was allowed to flow into the bottom of a vertical tube. To what height did the blood rise in the tube? Assume that the blood pressure of the horse is 80 mm Hg and that the density of horse blood is the same as human blood.

17 Some people experience ear trouble going up in an elevator because of the pressure change. If the pressure in back of the eardrum does not change during the ascent, the decrease in the outside pressure causes a net outward force on the eardrum. (*a*) What is the change in air pressure in going up 100 m in an elevator? (*b*) What is the net force on an eardrum with an area of 0.6 cm²?

*Ans.* (*a*) 1176 N/m²; (*b*) 0.07 N

18 Around 1646 Pascal performed the experiment shown in Fig. 7.28. A very long tube, with cross-sectional area $A = 3 \times 10^{-5}$ m², was connected to a wine barrel which had a lid of area $A' = 0.12$ m². The barrel was first filled with water and then water was added to the tube until the barrel burst. This happened when the column of water was 12 m high. Just before the barrel burst, what was (*a*) the weight of the water in the tube, (*b*) the (gauge) pressure of the water on the lid of the barrel, (*c*) the net force exerted on the lid?

**REMARK** Note that the water in the tube, although it weighed less than 1 lb, was able to exert a force of thousands of pounds on the barrel lid.

19 With strong inspiratory effort, i.e., by sucking hard, the gauge pressure in the lungs can be reduced to −80 mm Hg. (*a*) What is the greatest height that water can be sucked up a straw? (*b*) Gin has a density of 920 kg/m³. What is the greatest height that gin can be sucked up a straw?

*Ans.* (*a*) 1.08 m; (*b*) 1.18 m

20 A mercury manometer is connected to a vessel as shown in Fig. 7.29. (*a*) What is the (gauge) pressure in the vessel? (*b*) What is the absolute pressure in the vessel, assuming atmospheric pressure is $1.01 \times 10^5$ N/m²? (*c*) If the

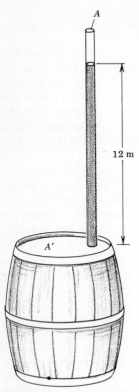

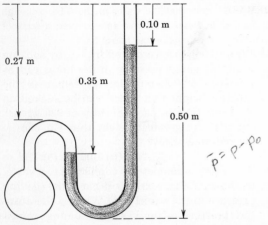

FIGURE 7.28   Problem 18.

absolute pressure in the vessel is doubled, what is the gauge pressure?

21 A mercury manometer is connected to a vessel as shown in Fig. 7.30. (*a*) If the height $d_A$ of

the left-hand column is 0.22 m, what is the height $d_B$ of the right-hand column when the gauge pressure inside the vessel is $0.16 \times 10^5$ N/m²? (*b*) What are the heights $d_A$ and $d_B$ when the gauge pressure is $0.32 \times 10^5$ N/m²?

*Ans.* (*a*) 0.34 m; (*b*) 0.16 and 0.40 m

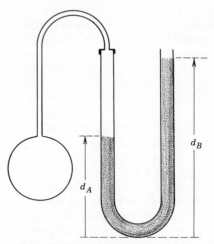

FIGURE 7.30   Problem 21.

22 A cylinder of cross-sectional area $A = 4 \times 10^{-4}$ m² is connected by a tube to one side of a mercury manometer (Fig. 7.31). What is the difference in the heights of the two columns when a 3-kg mass is placed on the piston of the cylinder?

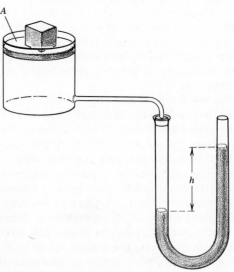

FIGURE 7.31   Problem 22.

FIGURE 7.29   Problem 20.

$\bar{p} = p - p_0$

23 How tall would a barometer filled with glycerin have to be?      *Ans.* 8.18 m

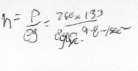

24 A hemispherical object is immersed in a fluid (Fig. 7.32). Prove that the total force **F** on the curved portion of the hemisphere, i.e., the vector sum of the forces acting on every point of this surface, has the magnitude $F = \pi r^2 p$, where $r$ is the radius of the sphere and $p$ is the pressure in the fluid. (*Hint:* First find the total force on the flat surface and then use Newton's first law. Neglect variations of pressure with depth.)

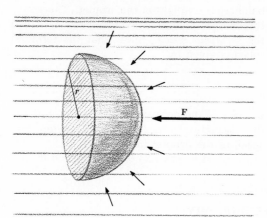

FIGURE 7.32    Problem 24.

25 In 1654 Otto von Guericke gave a demonstration in Magdeburg of the effect of air pressure. He used an air pump he had invented to evacuate the air from between two brass hemispheres. Teams of eight horses, pulling on each hemisphere, were then unable to pull them apart. If the radius of each hemisphere was 0.3 m and the pressure between them was 0.1 atm, what force would each team of horses have to exert to pull the hemispheres apart? (Use the result of Prob. 24.)

         *Ans.* $2.57 \times 10^4$ N

26 What fraction of an iceberg is below the surface of the water?

27 A "bubble" of warm air (30°C), formed near the ground, rises into the cold air (10°C) above the ground. (*a*) If the volume of the bubble is 8 m³, what is the total force on it? (*b*) What is the upward acceleration of the bubble, neglecting air resistance?

         *Ans.* (*a*) 7.06 N; (*b*) 0.76 m/s²

28 What is the upward acceleration of a block of wood released from the bottom of a lake?

29 A 2-kg block of aluminum is suspended in

FIGURE 7.33    Problems 29 and 30.

water by a cord attached to a scale (Fig. 7.33). What does the scale read?    *Ans.* 12.3 N

30 When a weight $W$ is suspended in water by a cord attached to a scale (Fig. 7.33), the scale reads $W'$. Show that the density $\rho$ of the suspended object is

$$\rho = \frac{W}{W - W'} \rho_w$$

where $\rho_w$ is the density of water.

31 The speed $v_m$ of blood through the center of a capillary is 0.066 cm/s. The length $L$ of the capillary is 0.1 cm, and its radius $r$ is $2 \times 10^{-4}$ cm. (*a*) What is the fluid flow $Q$ through the capillary? (*b*) Estimate the total number of capillaries in the body from the fact that the fluid flow through the aorta is 83 cm³/s.

     *Ans.* (*a*) $4.14 \times 10^{-9}$ cm³/s; (*b*) $2 \times 10^{10}$

32 (*a*) Calculate the resistance (to blood) of the capillary described in Prob. 31. (*b*) Calculate the resistance when the radius of the capillary dilates to $2.5 \times 10^{-4}$ cm.

33 (a) What is the resistance to water of a glass capillary of length 20 cm and radius 0.06 cm? (b) What is the fluid flow through the capillary when the pressure difference between its ends is 15 cm $H_2O$? (c) What pressure difference gives a fluid flow of 0.5 cm³/s?
*Ans.* (a) $3.93 \times 10^9$ N-s/m⁵; (b) 0.374 cm³/s; (c) 20 cm $H_2O$

34 (a) What is the resistance to water of a hypodermic needle of length 8 cm and internal radius 0.04 cm? (b) The needle is connected to a syringe with a plunger of area 3.5 cm². What force must be applied to the plunger to get water to flow from the syringe into a vein with a flow rate of $Q = 2$ cm³/s? Assume the pressure in the vein is 9 mm Hg.

35 It is found experimentally that the flow of a fluid of density $\rho$ and viscosity $\nu$ through a pipe of radius $r$ is laminar as long as the *Reynolds number*

$$\mathrm{Re} = \frac{\overline{v} r \rho}{\nu}$$

is less than 1000. Here $\overline{v}$ is the average speed of the fluid in the pipe. From the data given in Sec. 7.5, calculate the Reynolds number for the flow of blood (a) through the aorta and (b) through a typical capillary.
*Ans.* (a) 780; (b) $1.73 \times 10^{-4}$

36 (a) Show that the Reynolds number defined in Prob. 35 can also be written

$$\mathrm{Re} = \frac{Q\rho}{\pi r \nu}$$

(b) Show that the flow of water through a capillary is laminar as long as $Q$ (in cubic centimeters per second) is less than the diameter of the capillary (in millimeters).

37 A fluid exerts a viscous force $F_v$ on any object moving through it. For a small sphere of radius $r$ moving slowly with speed $v$, the force is given by Stokes' law

$$F_v = 6\pi \nu r v$$

(a) What is the viscous force on a water droplet of radius $r = 0.02$ cm moving through air with the speed $v = 2$ m/s? (b) A falling droplet increases its speed until the viscous force balances the weight of the droplet. Thereafter the droplet falls at constant speed $v_t$, called the *terminal speed*. Show that the terminal speed is given by

$$v_t = \frac{2\rho r^2 g}{9\nu}$$

where $\rho$ is the density of the droplet and $\nu$ is the viscosity of air. (c) What is the terminal speed of the droplet in part (a)?
*Ans.* (a) $1.38 \times 10^{-7}$ N; (c) 4.76 m/s

**BIBLIOGRAPHY**

ALEXANDER, R. M.: "Functional Design in Fishes," Hutchinson University Library, London, 1967. The role of the swim bladder in the buoyancy of fish is discussed in chap. 3.

———: "Animal Mechanics," University of Washington Press, Seattle, 1968. The principles of mechanics are applied to a variety of biological problems. The application of fluid properties to animals is given in chaps. 5 and 6.

GUYTON, ARTHUR C.: "Textbook of Medical Physiology," W. B. Saunders Company, Philadelphia, 1971. The physics of blood flow is described in chaps. 18 and 19.

# gases CHAPTER 8

A given quantity of gas does not have a definite volume or density but expands to fill any closed container in which it is placed. There is, however, a definite relation between the volume, pressure, and temperature of a gas, so that a given quantity of gas occupies a definite volume at any given pressure and temperature. Although this relationship is very complex for a dense, or *real*, gas, it is very simple for a dilute, or *ideal*, gas. The properties of an ideal gas are of practical importance because the gases involved in respiration are approximately ideal. These properties are also of theoretical interest because they can be completely understood in terms of the motions of the molecules of the gas. The study of real gases is necessary in order to understand the transformation between the gas and liquid states.

## 8.1 PARTICLE DENSITY

The mass of an individual atom or molecule is usually given in *atomic mass units* (u). This unit is defined in terms of a normal carbon atom, $^{12}C$, which is assigned the mass of 12.00000 u. Since a hydrogen atom, H, has about one-twelfth the mass of a carbon atom, its mass is 1.0 u. The mass of a hydrogen molecule, $H_2$, is 2.0 u, because each molecule is composed of two hydrogen atoms bonded together. Similarly, the mass of an oxygen atom, O, is 16.0 u, whereas the mass of an oxygen molecule, $O_2$, is 32.0 u.

**REMARK** Carbon is composed of two different types of atoms that have the same chemical properties but different masses. These carbon *isotopes* are denoted $^{12}C$ and $^{13}C$, respectively, the superscript being the integer nearest to the mass of the atom (Sec. 20.1). Table 20.1 gives the exact masses of the isotopes of carbon and other elements, whereas the table of the elements inside the back cover gives the average mass of the atoms that compose each element. For example, 99 percent of the atoms in carbon are $^{12}C$ and 1 percent are $^{13}C$; therefore the average mass of carbon is

$$m_C = \frac{99 \times m_{^{12}C} + 1 \times m_{^{13}C}}{100}$$

$$= \frac{99 \times 12 \text{ u} + 1 \times 13 \text{ u}}{100}$$

$$= 12.01 \text{ u}$$

In this chapter we use the average masses given in the table of the elements unless we specify a particular isotope by using a superscript.

The absolute mass of a $^{12}$C atom is found experimentally to be

$$m_{^{12}C} = 19.92637 \times 10^{-27} \text{ kg}$$

This determines the conversion between the atomic mass unit and the kilogram because, by definition,

$$m_{^{12}C} = 12.00000 \text{ u}$$

Therefore the conversion is

$$1 \text{ u} = \frac{19.92637 \times 10^{-27} \text{ kg}}{12}$$

$$= 1.660531 \times 10^{-27} \text{ kg} = 1.660531 \times 10^{-24} \text{ g}$$

For many purposes it is convenient to measure the quantity of a substance in *moles*, a unit specifying the number of molecules in a substance, rather than in grams or kilograms, which are units of mass. This is particularly true in chemistry because the amount of one substance that reacts with a given amount of another substance depends on the number of molecules in each substance, not on their masses. It is also true in the study of gases, because the properties of a gas depend on the number of molecules in the gas, not on its mass.

The *mole* (abbreviated mol) is a standard number of molecules, equal to the number $N$ of atoms in 12 g of carbon. The mass of $N$ carbon atoms is clearly $Nm_C$, so

$$Nm_C = 12 \text{ g}$$

or

$$N = \frac{12 \text{ g}}{m_C} = \frac{12 \text{ g}}{19.9 \times 10^{-24} \text{ g}}$$

$$= 6.02 \times 10^{23}$$

This is *Avogadro's number*. A mole of any substance is a quantity of matter that contains $N$ molecules of that substance. If the mass of an individual molecule is written

$$m = M \text{ u} = (1.66 \times 10^{-24} M) \text{ g}$$

then the mass $\mathfrak{M}$ of 1 mol of these molecules is

$$\mathfrak{M} = Nm$$
$$= (6.02 \times 10^{23})(1.66 \times 10^{-24})(M) \text{ g}$$
$$= M \text{ g}$$

That is, the mass $M$ g of a mole of a substance in grams is equal to the mass $M$ u of an individual molecule in atomic mass units.

Since $\mathfrak{M}$ is the mass (in grams) of 1 mol of a substance, $n$ mol of the substance has the mass $n\mathfrak{M}$ (in grams). If the substance occupies a volume $V$ (in cubic centimeters), its density $\rho$ (in grams per cubic centimeter) is

$$\rho = \frac{n\mathfrak{M}}{V} \qquad\qquad 8.1$$

The particle density† $\eta$ is defined as the number of molecules per unit

---

†$\eta$ is the Greek letter eta.

volume. Since $n$ mol contains $nN$ molecules, the particle density of $n$ mol of a substance in a volume $V$ is

$$\eta = \frac{nN}{V} \qquad\qquad 8.2$$

The volume $V$ must be in cubic centimeters in Eq. 8.1, but it may be in any units in Eq. 8.2.

As an example, suppose a 1500-cm³ jar contains 8 g of oxygen. What is the particle density of the oxygen in this jar? Since the mass of 1 mol of oxygen ($M = 32$) is $\mathfrak{M} = 32$ g, there is $\frac{1}{4}$ mol of oxygen in the jar, or $\frac{1}{4}N$ molecules. Therefore the particle density is

$$\eta = \frac{\frac{1}{4}N}{1500 \text{ cm}^3}$$

$$= \frac{0.25 \times 6 \times 10^{23} \text{ molecules}}{1500 \text{ cm}^3}$$

$$= 10^{20} \text{ molecules/cm}^3$$

Because a gas expands to fill the container in which it is held, its density can be varied by varying the size of the container. When the particle density is very small (less than $10^{20}$ molecules/cm³), the molecules of the gas are so far apart that they rarely collide with each other. The properties of such a dilute gas do not depend on the type of molecules of which it is composed, since its molecules seldom interact. Thus all gases tend to have the same behavior when $\eta$ is small. This behavior, which characterizes the *ideal gas*, is described in the next three sections.

## 8.2 TEMPERATURE

The concept of temperature is rather subtle. Although everyone has a rough idea that temperature is a measure of the relative hotness or coldness of a substance, this does not take one very far without an understanding of the physical meaning of hot and cold. It turns out that it is best to begin with an operational definition of temperature, just as was done with length and time in Chap. 1. Then the fundamental physical meaning of temperature will emerge as we develop the properties of the ideal gas.

From the operational point of view, temperature is what is measured with a thermometer. A normal laboratory thermometer consists of a sealed glass tube with a bulb at one end (Fig. 8.1). The bulb and part of the tube are filled with a liquid (usually mercury), and the rest of the tube is vacuum. The temperature of a water and ice mixture is defined to be 0°C on the Celsius† scale. Therefore, to calibrate the thermometer, it is first put into an ice-water mixture, and a line is etched on the outside of the tube opposite the position of the top of the mercury column. The thermometer is then placed

† The name Celsius has officially been adopted to replace the name centigrade in international use.

FIGURE 8.1
A laboratory thermometer.

in boiling water, the temperature of which (at atmospheric pressure) is defined to be 100°C. When heated, mercury expands more than glass, and so the mercury rises in the glass tube. A second line is etched on the tube opposite the position of the top of the mercury column after it stops rising. Intermediate temperature marks are made on the thermometer by dividing the distance between the 0°C mark and the 100°C mark into 100 equal intervals. The temperature of a substance can then be measured with this thermometer by placing its bulb in contact with the substance and noting the position of the top of the mercury column. If it is at the twentieth division between the 0°C line and the 100°C line, its temperature is 20°C. Temperatures above 100°C and below 0°C are measured by extending the scale above and below the 100 and 0°C marks on the tube.

Unfortunately, the temperature measured this way depends to some extent on the liquid used in the thermometer. For instance, a mercury thermometer and an alcohol thermometer, both calibrated in the same way, give slightly different measurements for the temperature of a substance because these liquids expand differently with temperature. One could *define* temperature by specifying a particular substance to be used in the standard thermometer, but such a definition lacks fundamental physical significance. To get a more meaningful definition of temperature it is desirable to have a thermometer that does not depend on the particular properties of any one substance. Since all gases behave the same at low density, a dilute-gas thermometer is used for the standard definition of temperature.

In a standard gas thermometer, a dilute gas is contained in a glass bulb that is connected through a narrow tube to a mercury-filled manometer (Fig. 8.2). The left-hand side of the manometer is sealed at zero pressure, so that the difference in the heights of the two mercury columns is equal to the absolute pressure of the gas. The two sides of the manometer are connected by a flexible rubber tube to allow the left-hand side to be raised or lowered relative to the right-hand side. During the operation of this instrument, the left-hand side is moved up and down until the mercury in the right-hand side just reaches the top of the manometer tube. In this way the volume occupied by the gas is kept constant.

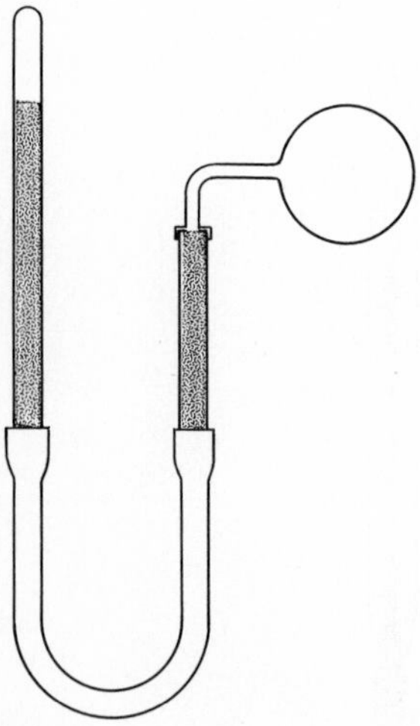

This instrument is calibrated by first measuring the absolute pressure $p_i$ in the bulb when the bulb is in a water and ice mixture (ice temperature). When the bulb is then placed in boiling water (steam temperature), the pressure of the gas increases, so that some of the mercury is pushed out of the right-hand tube. By raising the left-hand tube, the mercury level in the right-hand tube is brought back to the top of the tube. After this adjustment is made, the pressure $p_s$ of the gas at the steam temperature is measured. The values of $p_i$ and $p_s$ depend, of course, on the amount of gas in the bulb. However, if the particle density $\eta$ is sufficiently small, the ratio $p_s/p_i$ depends neither on the magnitude of $\eta$ nor on the chemical composition of the gas. It is a universal constant, independent of the properties of any one substance. Careful measurements of this ratio using very dilute gases have determined it to be

$$\frac{p_s}{p_i} = 1.3661$$

FIGURE 8.2
Standard gas thermometer. A glass bulb containing a dilute gas is connected to a mercury-filled manometer.

The only thing special about the temperatures of ice and steam is that they are easily reproduced in the laboratory, so that they are convenient calibration points. However, the ratio $p/p_i$, where $p$ is the pressure in the bulb at any other temperature, is also independent of the nature of the gas. Thus it is natural to define the fundamental temperature scale in terms of this ratio. That is, the fundamental or *absolute temperature* $T$ is defined by

$$T = a\frac{p}{p_i} \qquad\qquad 8.3$$

where $p$ is the pressure in the bulb at this temperature and $a$ is an arbitrary constant.

This constant is chosen to make the difference between the steam temperature $T_s$ and the ice temperature $T_i$ exactly 100 temperature units, or kelvins (K). By definition, these temperatures are given by Eq. 8.3 with $p$ replaced by $p_s$ and $p_i$, respectively. That is,

$$T_s = a\frac{p_s}{p_i} = 1.3661a$$

$$T_i = a\frac{p_i}{p_i} = a$$

and so their difference is

$$T_s - T_i = 1.3661a - a = 0.3661a$$

Equating this to 100 K we get

$$T_s - T_i = 100 \text{ K} = 0.3661a$$

or

$$a = \frac{100 \text{ K}}{0.3661} = 273.15 \text{ K} \qquad\qquad 8.4$$

Thus the absolute temperature $T$ of the bulb when its pressure is $p$ is

$$T = 273.15 \text{ K} \times \frac{p}{p_i}$$

Temperature is a new fundamental dimension, and its unit of measure is the kelvin. The different temperature scales in common use are distinguished by putting a letter after the unit. A temperature measured on the absolute, or *Kelvin*, scale is labeled K. For instance, the ice temperature is 273.15 K.

Since the gas pressure in the bulb cannot be less than zero, temperatures below 0 K are not defined. In fact, temperatures below absolute zero do not exist, so that the zero point on the absolute scale corresponds to the coldest temperature possible.

A temperature $t$ on the Celsius scale is related to the absolute temperature $T$ by

$$t = T - 273.15 \qquad\qquad 8.5$$

On this scale, the ice temperature is 0°C the steam temperature is 100°C, and absolute zero is $-273.15$°C. Note that the superscript ° for degree is retained for Celsius temperatures but is omitted for Kelvin temperatures.

Since the gas thermometer is not convenient for everyday laboratory use, the mercury thermometer is commonly used. It should be thought of as an instrument that approximately measures the temperature as defined by the gas thermometer. Over the range of temperatures usually encountered in the laboratory, the discrepancy between the mercury and the gas thermometers is not significant.

### 8.3  THE IDEAL GAS

**Ideal-gas law**

An *ideal gas* is a gas in which the molecules are so far apart that they rarely collide with one another. Since this is the case for any real gas at low particle density, all gases are ideal at low density. This fact was used in the last section to define the absolute temperature $T$ in terms of the pressure $p$ of an ideal gas at constant volume. The result is Eq. 8.3, which can also be written

$$p = \frac{p_i}{a} T \qquad\qquad 8.6$$

This says that at constant volume the pressure of an ideal gas is proportional to the absolute temperature.

The dependence of $p$ on the volume of the gas can be derived from Eq. 8.6 on the assumption that $p_i$, the pressure of the gas at the ice temperature, is proportional to the particle density,

$$p_i = b\eta \qquad\qquad 8.7$$

where $b$ is another constant independent of the chemical composition of the gas. This assumption is reasonable because, as will be shown in the next section, the pressure of the gas is caused by the gas molecules colliding with the wall of the container. When the particle density is doubled, there are twice as many molecules in the container. For a dilute gas, this means that twice as many molecules collide with the wall, so that the pressure is doubled.

REMARK  Equation 8.7 is not true if $\eta$ is very large. In a dense gas, so many molecules strike the wall at one time that there is little space available for additional molecules. Although the pressure increases when the particle density doubles, the pressure does not double because some of the additional molecules collide with other molecules near the wall, instead of with the wall itself.

The particle density is related to the number of moles $n$ and the volume $V$ of the gas through Eq. 8.3, so that Eq. 8.7 can be written

$$p_i = \frac{bnN}{V}$$

The dependence of $p$ on the volume is thus obtained by substituting this last expression into Eq. 8.6:

$$p = n\frac{bN}{a}\frac{T}{V} \qquad\qquad 8.8$$

The quantity $bN/a$ is the *gas constant* $R$. Its dimension is [energy/kelvin], and its magnitude is

$$R = \frac{bN}{a} = 8.314 \text{ J/K}$$

In terms of $R$, Eq. 8.8 becomes

$$p = \frac{nRT}{V}$$

or
$$pV = nRT \qquad \qquad 8.9$$

Equation 8.9 is the ideal-gas law. It gives the relation between the pressure $p$, volume $V$, and temperature $T$ of $n$ mol of an ideal gas. Care and thought must be exercised in using this equation because the quantities that appear in it must be of the proper form and in the correct units. The pressure is the absolute pressure in newtons per square meter, the volume is in cubic meters, and the temperature is the absolute temperature in kelvins.

As an example, we calculate the mass density $\rho$ of oxygen at 0°C and atmospheric pressure; this is called *standard temperature and pressure* (STP). To do this, Eq. 8.9 is used to determine the number of moles $n$ of an ideal gas in unit volume $V = 1$ m³ at a pressure of $p = 1$ atm $= 1.01 \times 10^5$ N/m² and a temperature of $T = 273$ K. Equation 8.9 first is solved for $n$, and then these values are substituted into it. The result is

$$n = \frac{pV}{RT} = \frac{(1.01 \times 10^5 \text{ N/m}^2)(1 \text{ m}^3)}{(8.31 \text{ J/K})(273 \text{ K})}$$

$$= 44.5$$

Since the mass of a mole of oxygen, $O_2$, is 32 g, the mass of 44.5 mol of oxygen is $44.5 \times 32$ g $= 1425$ g $= 1.425$ kg. Therefore the STP density of oxygen is 1.43 kg/m³, which agrees with the value given in Table 7.2.

As another example, suppose an automobile tire is filled with air at a temperature of 20°C to a (gauge) pressure of 24 lb/in.². After driving 100 mi on a highway, the temperature of the tire is 50°C. What is the tire pressure at this temperature?

The best way to proceed is to first rewrite Eq. 8.9 so that the quantities that change are on one side of the equation and the quantities that remain constant are on the other side. In this case $p$ and $T$ change, but the volume $V$ and, of course, $n$ and $R$ do not. Therefore Eq. 8.9 is rewritten in the form

$$\frac{p}{T} = \frac{nR}{V} \qquad \qquad 8.10$$

Next, the given quantities must be put into the proper forms. The Celsius temperatures must be expressed as absolute temperatures, and the gauge pressures must be expressed as absolute pressures. From Eq. 8.5 we have that the initial temperature is $T_I = 273 + 20 = 293$ K, and the final temperature is $T_F = 273 + 50 = 323$ K. Assuming that atmospheric pressure is 14.7 lb/in.², the initial absolute pressure is $p_I = 14.7 + 24.0$ lb/in.² $= 38.7$ lb/in.².

At the initial temperature, Eq. 8.10 is

$$\frac{p_I}{T_I} = \frac{nR}{V}$$

and at the final temperature it is

$$\frac{p_F}{T_F} = \frac{nR}{V}$$

so we have

$$\frac{p_I}{T_I} = \frac{p_F}{T_F}$$

Finally, we solve this for $p_F$ to get

$$p_F = \frac{T_F}{T_I} p_I$$

$$= \frac{323 \text{ K}}{293 \text{ K}} \times 38.7 \text{ lb/in.}^2$$

$$= 42.6 \text{ lb/in.}^2$$

The gauge pressure in the tire at the higher temperature is $42.6 - 14.7 \text{ lb/in.}^2 = 27.9 \text{ lb/in.}^2$, which is $3.9 \text{ lb/in.}^2$ higher than the initial gauge pressure.

**REMARK** In this last example the gas constant $R$ is not needed because only the change in pressure is calculated. By using proportions to find $p_F$, the pressure did not have to be converted into mks units. In the first example, on the other hand, $R$ is needed to find the number of moles in a gas under specified conditions. Since the value of $R$ is given in mks units, all other quantities had to be in mks units also. It is often more convenient to measure pressure in atmospheres, millimeters of mercury, or pounds per square inch and to measure volume in liters, cubic centimeters, or cubic feet. Instead of converting these units into mks units, the value of $R$ appropriate to these units is used. Table 8.1 gives the value of $R$ to be used in Eq. 8.9 when $p$ and $V$ are measured in these other units.

### Dalton's law of partial pressure

A mole of dry air consists of $6.02 \times 10^{23}$ molecules, of which 78.1 percent are nitrogen molecules, 20.9 percent are oxygen, 0.9 percent are argon, and 0.03 percent are carbon dioxide. Since, as we have calculated, there are 44.5 mol of an ideal gas in 1 $m^3$ at STP, there are $44.5 \times 0.209 = 9.31$ mol of oxygen in 1 $m^3$ of air at STP. The oxygen molecules exert a partial pressure $p_{O_2}$ equal to the pressure they would exert if the other gas molecules were not present. From Eq. 8.9 this pressure is found to be

TABLE 8.1 **Values of the gas constant $R$ appropriate for various units of pressure and volume**
The temperature is always in kelvins.

| Volume | Pressure | $R$ |
|---|---|---|
| $m^3$ | $N/m^2$ | 8.314 J/K |
| l | atm | $8.205 \times 10^{-2}$ atm-1/K |
| l | mm Hg | 62.4 mm Hg-1/K |
| $cm^3$ | atm | 82.05 atm-$cm^3$/K |
| $cm^3$ | mm Hg | $6.24 \times 10^4$ mm Hg-$cm^3$/K |
| $ft^3$ | atm | $2.90 \times 10^{-3}$ atm-ft/K |
| $ft^3$ | $lb/in.^2$ | $4.26 \times 10^{-2}$ lb-$ft^3$/in.$^2$-K |

$$p_{O_2} = \frac{nRT}{V} = \frac{(9.31 \times 8.31 \text{ J/K})(273 \text{ K})}{1 \text{ m}^3}$$
$$= 0.211 \times 10^5 \text{ N/m}^2$$
$$= 0.209 \text{ atm}$$

which is 20.9 percent of the total pressure of 1 atm exerted by the air. This is an example of *Dalton's law of partial pressure*, which says that in a gas mixture each component gas exerts a partial pressure proportional to its molecular concentration. The total pressure of the mixture is equal, therefore, to the sum of the partial pressures of all the component gases.

Dalton's law is important because the physiological effects of each component of air depend on the partial pressure of the component in the lungs, not on the total pressure. For example, the amount of nitrogen that dissolves in the blood and body tissues is proportional to the partial pressure of nitrogen in the lungs. The partial pressure of nitrogen becomes dangerously large for an underwater diver at great depths.

### Underwater respiration

In normal breathing, the diaphragm is moved down, increasing the volume of the chest cavity and, in accordance with Eq. 8.9, lowering the pressure inside the lungs. During quiet inhalation this pressure falls about 3 mm Hg below atmospheric pressure, which pulls air into the lungs against the resistance to fluid flow of the air passageways (Sec. 7.5). During exhalation, the diaphragm is lifted, decreasing the volume of the chest cavity and raising the pressure inside the lung about 3 mm Hg above atmospheric pressure. This forces the air out of the lungs. Normally, then, the pressure of the air in the lungs is within a few mm Hg of atmospheric pressure, so that the partial pressures of oxygen and nitrogen in the lungs are 0.209 and 0.781 atm, respectively.†

An underwater diver can breathe normally at great depths if air is delivered to him at a pressure equal to the pressure $p_A$ of the surrounding water. Modern *self-contained underwater breathing apparatus* (scuba) is designed to supply the diver with air at the prevailing pressure on demand, i.e., as the diver breathes. The apparatus consists of a tank of air at high pressure (initially over 170 atm) connected by a flexible tube to a mouthpiece. A regulator (Fig. 8.3) between the tank and the mouthpiece allows air to enter the diver's mouth only when the pressure in chamber $A$ falls below $p_A$.

Since chamber $B$ is open to the water, it is always at the prevailing water pressure $p_A$. This chamber is separated from chamber $A$ by a flexible watertight diaphragm $D$. When the pressures in chambers $A$ and $B$ are equal, the diaphragm is positioned so that the spring and lever arrangement exerts sufficient pressure on the valve $V$ to prevent air from flowing from the high-pressure tank.

The air in the diver's lungs is normally at the prevailing pressure $p_A$, but when he inhales, he decreases the pressure in his lungs below

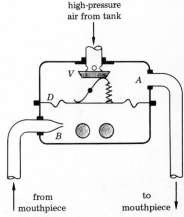

high-pressure air from tank

from mouthpiece

to mouthpiece

FIGURE 8.3
A regulator for delivering air on demand from a high-pressure tank to a scuba diver. Chamber $A$ is sealed from chamber $B$ by the flexible diaphragm $D$. The spring and lever arrangement is connected to the diaphragm in such a way that it permits air to flow through valve $V$ only when the pressure in $A$ is less than the pressure in $B$. Chamber $B$ is open to the water, so its pressure changes as the diver changes the depth at which he is swimming.

---

† The partial pressures of oxygen and nitrogen in the lungs are actually somewhat less than this because of the water vapor in the lungs (Sec. 8.5).

$p_A$. This lowers the pressure in chamber $A$, so that the diaphragm moves up, and the lever mechanism moves down, permitting air to flow from the tank. When the diver exhales, he increases the pressure in his lungs above $p_A$. This increases the pressure in chamber $A$ above $p_A$, so that the diaphragm moves down and the lever mechanism moves up, shutting off the flow of air from the tank. At the same time, another valve (not shown) routes the exhaled air into chamber $B$ and out the holes. Since the pressure in chamber $B$ is always that of the surrounding water, the regulator automatically adjusts the pressure of the air it delivers to equal that of the surrounding water.

The pressure on a diver increases about 1 atm for each 10 m of depth (Sec. 7.3). At 60 m, for instance, a diver must breathe air at a pressure of 7 atm. The partial pressure of nitrogen at this depth is $p_{N_2} = 0.78 \times 7$ atm $= 5.5$ atm. A nitrogen pressure this large causes the amount of nitrogen dissolved in the blood and body tissues to reach a toxic level. *Nitrogen narcosis* develops, the symptoms of which include light-headedness and impairment of judgment, similar to alcohol intoxication. For this reason nitrogen narcosis is frequently called *rapture of the deep*.

For man to live and work safely at great depths under the ocean, he must breathe special gas mixtures. Experiments in which men lived for weeks at depths of 200 m used a mixture of 97 percent helium and 3 percent oxygen. The helium was used instead of nitrogen to avoid nitrogen narcosis. The low oxygen concentration was necessary to avoid oxygen poisoning, which occurs at high oxygen pressure. Even with only a 3 percent oxygen concentration, the partial pressure of the oxygen is above normal levels, because the pressure at 200 m is 21 atm. Thus the partial pressure of oxygen is $p_{O_2} = 0.03 \times 21$ atm $= 0.63$ atm, which is still three times the normal oxygen pressure. At a partial pressure above 2 atm, oxygen becomes toxic enough to produce convulsions and coma.

## 8.4 KINETIC THEORY OF IDEAL GASES

### Pressure, temperature, and kinetic energy

The pressure a gas exerts on the walls of its container is the result of the collisions of the molecules of the gas with the walls. It is possible to calculate this pressure in terms of the average kinetic energy of the molecules in an ideal gas. The exact calculation is complicated, but with a few simplifying assumptions a crude approximation can be reached that still contains all the essential points. This calculation also gives a fundamental relation between temperature and the kinetic energy of the molecules of an ideal gas.

Consider a molecule of mass $m$ moving with speed $v$ directly toward a wall (Fig. 8.4). The molecule is unaffected by the wall until it gets close enough to interact with it. For simplicity we assume that the wall exerts no force on the molecule until the molecule reaches point $A$ and that inside this point the wall exerts a constant force $\mathbf{F}$ directed away from the wall. This force decelerates the molecule, bringing it to rest at a point $B$ a distance $d$ from point $A$. The

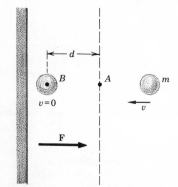

**FIGURE 8.4**
A molecule of mass $m$ moving with speed $v$ toward the wall of a container. The wall exerts a constant force $\mathbf{F}$ on the molecule once the molecule passes point $A$.

molecule is then accelerated back toward $A$ by $\mathbf{F}$, arriving at $A$ with the same speed $v$ with which it approached the wall. The result is that the molecule bounces off the wall without a net loss of kinetic energy.

While the molecule travels from $A$ to $B$, its kinetic energy does change. At $A$ its kinetic energy is $K_A = \frac{1}{2}mv^2$, while at $B$ it is $K_B = 0$. According to the work-energy theorem (Sec. 5.2), this change in kinetic energy is equal to the work $W$ done by $\mathbf{F}$. Since $\mathbf{F}$ is directed opposite to the motion of the molecule during this part of the collision, $W$ is equal to $-Fd$, and so the work-energy theorem gives the relation

$$W = -Fd = K_B - K_A = 0 - \tfrac{1}{2}mv^2$$

or

$$F = \frac{\tfrac{1}{2}mv^2}{d}$$

This force also acts on the molecule to accelerate it back from $B$ to $A$. Thus as long as the molecule is between $A$ and $B$, it is acted upon by $\mathbf{F}$.

At any one time there are many molecules close enough to the wall to experience the force $\mathbf{F}$. For instance, on a given area $A$ of the wall, all molecules in the region of volume $V = Ad$ (Fig. 8.5) experience this force. The total number of molecules in this volume is $\eta V = \eta Ad$, where $\eta$ is the particle density. Since the wall exerts the force $\mathbf{F}$ on each of these molecules, the magnitude of the total force $\mathbf{F}_{\text{total}}$ exerted by the wall on all the molecules in the volume is

$$\begin{aligned} F_{\text{total}} &= F\eta Ad \\ &= \frac{\tfrac{1}{2}mv^2}{d}\eta Ad \\ &= \tfrac{1}{2}mv^2\eta A \end{aligned}$$

which is independent of the distance $d$. By Newton's third law (Property 3, Sec. 2.1), these molecules exert a reaction force $\mathbf{R}_{\text{total}}$ on the wall that is equal in magnitude to $\mathbf{F}_{\text{total}}$ but opposite in direction. Therefore the pressure these molecules exert on the wall is

$$\begin{aligned} p = \frac{R_{\text{total}}}{A} &= \frac{F_{\text{total}}}{A} = \eta\tfrac{1}{2}mv^2 \\ &= \eta\bar{K} \end{aligned} \qquad 8.11$$

where $\bar{K}$ is the kinetic energy of a gas molecule. Actually the molecules in a gas have a wide distribution of kinetic energies, so that $\bar{K}$ must be understood as being the average kinetic energy of a molecule in the gas.

This calculation has neglected the fact that a molecule can approach the wall from any angle, not just perpendicularly. When this is properly taken into account, it is found that the pressure is related to $\bar{K}$ by

$$p = \tfrac{2}{3}\eta\bar{K} \qquad 8.12$$

Thus the exact result differs from Eq. 8.11 only by a numerical factor. Our crude calculation correctly gives the dependence of the pressure on $\eta$ and $\bar{K}$.

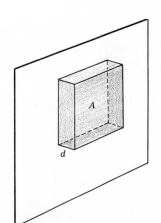

FIGURE 8.5
A region of area $A$ and depth $d$ on the wall of a container.

The pressure is also related to the temperature by Eq. 8.9:

$$p = \frac{nRT}{V}$$

By equating these two expressions for $p$ we obtain a relation between the temperature and the average kinetic energy of a gas molecule,

$$\frac{nRT}{V} = \tfrac{2}{3}\eta\bar{K}$$

or

$$\bar{K} = \frac{3}{2}\frac{nRT}{\eta V}$$

Then, using Eq. 8.2 to write $\eta V = nN$, we get

$$\bar{K} = \frac{3}{2}\frac{nRT}{nN} = \frac{3}{2}\frac{RT}{N}$$

$$= \tfrac{3}{2}kT \qquad\qquad 8.13$$

where $k$ is the *Boltzmann constant* defined by

$$k = \frac{R}{N}$$

Since the constant $R$ is the gas constant for 1 mol of gas and $N$ is the number of molecules in a mole, the Boltzmann constant $k$ is the gas constant per molecule. Its magnitude is

$$k = \frac{8.314 \text{ J/K}}{6.02 \times 10^{23}} = 1.38 \times 10^{-23} \text{ J/K}$$

The Boltzmann constant relates the absolute temperature $T$ to the average kinetic energy of the molecules of an ideal gas. In fact, temperature is just a measure of this kinetic energy. This is the fundamental physical meaning of temperature. This connection between temperature and energy is also of practical importance because, as we shall see, it explains such things as why certain gases are not found in the earth's atmosphere and why the rate of a chemical reaction depends on temperature.

## Distribution of kinetic energy

The speed of a molecule in a gas changes as it collides with other molecules, so that at any instant all the molecules in a gas have different speeds. Therefore some molecules must have a kinetic energy less than the average energy $\bar{K}$ while others have a kinetic energy greater than $\bar{K}$. The actual number of molecules with different energies can be calculated for an ideal gas, and the result is in agreement with experimental measurements of molecular energies. The calculation is beyond the scope of this book, but Table 8.2 gives the result and shows the fraction $f$ of the molecules with a kinetic energy greater than $y\bar{K}$ for various values of $y$. For example, the table gives $f = 0.112$ for $y = 2$, which means that at any instant 11.2 percent of the molecules have a kinetic energy greater than twice the average kinetic energy. This result does not depend on the temperature of the gas, though of course the magnitude of $\bar{K}$ does.

**TABLE 8.2 Fraction $f$ of the molecules of an ideal gas with a kinetic energy greater than $y\bar{K}$**

$\bar{K}$ is the average kinetic energy of the molecules in the gas.

| $y$ | $f$ |
|-----|-----|
| 0.5 | 0.685 |
| 1 | 0.392 |
| 2 | 0.112 |
| 3 | $0.29 \times 10^{-1}$ |
| 4 | $0.76 \times 10^{-2}$ |
| 5 | $0.18 \times 10^{-2}$ |
| 10 | $0.14 \times 10^{-5}$ |
| 15 | $0.94 \times 10^{-9}$ |
| 20 | $0.59 \times 10^{-12}$ |
| 30 | $0.24 \times 10^{-18}$ |
| 40 | $0.77 \times 10^{-25}$ |
| 50 | $0.31 \times 10^{-31}$ |
| 100 | $0.11 \times 10^{-63}$ |

At the temperature of 300 K the average kinetic energy of the molecules in a gas is

$$\bar{K}_{300} = \tfrac{3}{2}kT$$
$$= (1.5 \times 1.38 \times 10^{-23} \text{ J/K})(300 \text{ K})$$
$$= 6.21 \times 10^{-21} \text{ J}$$

The fraction of molecules in this gas with energy greater than $E = 31.05 \times 10^{-21}$ J is found by calculating the factor

$$y = \frac{E}{\bar{K}_{300}} = \frac{31.05 \times 10^{-21} \text{ J}}{6.21 \times 10^{-21} \text{ J}} = 5$$

and then using Table 8.2 to get $f_{300} = 0.18 \times 10^{-2}$. If the temperature of the gas is raised 25 percent to 375 K, the average kinetic energy is also increased 25 percent to $\bar{K}_{375} = 7.76 \times 10^{-21}$ J. At this temperature $y$ equals 4, so that the fraction $f_{375}$ of the molecules with energies greater than $E$ is $0.76 \times 10^{-2}$. That is, for every 10,000 molecules in the gas, 18 have energy greater than $E$ at 300 K and 76 have energy greater than $E$ at 375 K. Thus there are

$$\frac{76}{18} = \frac{f_{375}}{f_{300}} = 4.2$$

times as many molecules with energy greater than $E$ at 375 K as there are at 300 K. In this case a 25 percent temperature increase produces a 420 percent increase in the number of molecules with energy greater than $E$.

The energy distribution in a gas determines the rate of a chemical reaction because two molecules can react chemically only if one collides with the other with enough kinetic energy to overcome the repulsive force between their outer electrons. If one molecule's energy is less than a specific value $E_a$, called the *activation energy* of the reaction, it will bounce off the other molecule without reacting with it. The value of $E_a$ is characteristic of each chemical reaction.

The rate of a chemical reaction obviously depends on the fraction of molecules with energy greater than $E_a$, and this fraction, as we have just seen for a gas, is very sensitive to temperature. The behavior of the molecules of a substance dissolved in a liquid is similar to the behavior of the molecules of a gas, so that the energy distribution of these molecules is also given by Table 8.2. Hence the rates of reaction both in gases and in liquid solutions increase rapidly with temperature.

For example, if the activation energy of a certain reaction is $E_a = 31.05 \times 10^{-21}$ J, then, from the last example, it is seen that there are 4.2 times as many molecules with an energy greater than $E_a$ at 375 K as there are at 300 K. Hence the reaction will go 4.2 times as fast at 375 K as at 300 K.

The sensitivity of biological functions to temperature is a direct consequence of the temperature dependence of the rates of chemical reactions. The activity of any organism is a reflection of the rates of the thousands of different reactions that take place in its body. For example, the growth rate of bacteria and other simple organisms in culture increases by a factor of 2 or 3 for each 10°C increase between 10 and 30°C. That is, the growth rate at 30°C may be as much as nine times the growth rate at 10°C. This is, of course, why refrigeration is used to retard food spoilage.

Insect activity is also very temperature-dependent. Most insects become immobile at temperatures below 10°C (50°F), though some nocturnal moths remain active below 0°C. Warm-blooded animals are able to function over a much wider range of air temperatures because they are able to maintain a constant internal body temperature. However, should the body temperature rise, as with a fever, all the body's chemical processes are speeded up. At a body temperature of 110°F, which is 6°C above normal, the rate of metabolism is double the normal rate in man.

### Composition of the atmosphere

Although the energy distribution of the molecules in a gas does not depend on the mass of the molecules, the distribution of molecular speed does. Since the average value of $K = \frac{1}{2}mv^2$ is the same for all gases at the same temperature, the average molecular speed of molecules with small mass is greater than the average molecular speed of molecules with large mass. This has a number of consequences. For instance, light molecules are more reactive than heavy molecules because of their greater speed. The composition of the earth's atmosphere is also related to the dependence of molecular speeds on mass.

The earth's atmosphere is believed to have been formed billions of years ago out of gases that escaped from the earth's molten crust. These gases formed a primitive atmosphere with a temperature of 1000 K or more. The earth could retain only those molecules which had speeds less than the escape speed from the earth. In Sec. 5.4 this speed was found to be $v_e = 11.2 \times 10^3$ m/s. A molecule with this speed (or greater) could, if it were traveling away from the earth, escape into outer space. Of course, it might collide with another gas molecule before it escaped, but if a large number of molecules had such a speed, some would escape sooner or later.

It is easy to estimate the fraction of oxygen molecules that could have escaped from the primitive atmosphere. The mass $m_{O_2}$ of an oxygen molecule is 32 u, or

$$m_{O_2} = (32 \text{ u})(1.66 \times 10^{-27} \text{ kg/u})$$
$$= 53.1 \times 10^{-27} \text{ kg}$$

so the kinetic energy of an oxygen molecule moving with the escape speed is

$$K_e = \frac{1}{2}m_{O_2}v_e^2$$
$$= (0.5 \times 53.1 \times 10^{-27} \text{ kg})(11.2 \times 10^3 \text{ m/s})^2$$
$$= 3.33 \times 10^{-18} \text{ J}$$

The average kinetic energy of the molecules in the primitive atmosphere (assuming a temperature of 1000 K) is

$$\bar{K} = \frac{3}{2}kT$$
$$= (1.5 \times 1.38 \times 10^{-23} \text{ J/K})(1000 \text{ K})$$
$$= 2.07 \times 10^{-20} \text{ J}$$

and so the ratio $K_e/\bar{K}$ is

$$y = \frac{K_e}{\bar{K}} = \frac{3.33 \times 10^{-18} \text{ J}}{2.07 \times 10^{-20} \text{ J}} = 160$$

Table 8.2 does not give $f$ for so large a value of $y$, but it shows that $f$ is much smaller than $10^{-63}$, so that not a single oxygen molecule would have had the escape speed.†

The situation is different for the lighter gases such as hydrogen and helium. These have the same average kinetic energy as oxygen, but because of their smaller mass they have a smaller kinetic energy when moving with the escape speed. For instance, since the mass of hydrogen is one-sixteenth the mass of oxygen, the kinetic energy for escape of hydrogen is only one-sixteenth that of oxygen, so

$$K_e = \tfrac{1}{16}(3.33 \times 10^{-18} \text{ J}) = 2.07 \times 10^{-19} \text{ J}$$

This is only a factor

$$y = \frac{K_e}{\bar{K}} = \frac{2.07 \times 10^{-19} \text{ J}}{2.07 \times 10^{-20} \text{ J}} = 10$$

times $\bar{K}$. Table 8.2 shows that a fraction $f = 0.14 \times 10^{-5}$ of the hydrogen molecules have sufficient speed to escape.

As these high-energy molecules escape, the average kinetic energy of the remaining hydrogen molecules is lowered. This means the temperature of the hydrogen component of the atmosphere is lowered slightly below that of the other components. From collisions with these other molecules, the temperature of the remaining hydrogen is raised back to 1000 K, and in the process the proper number of high-energy hydrogen molecules is restored. As long as the temperature of the atmosphere remains at 1000 K, the fraction $0.14 \times 10^{-5}$ of the hydrogen molecules will have sufficient energy to escape. Thus, although this fraction is small, it remains constant regardless of the number of molecules that do escape. There is a small but constant leakage of hydrogen molecules into outer space, so that over the many millions of years the atmosphere was forming, all the hydrogen eventually escaped.

## 8.5  REAL GASES

The pressure and volume of a real gas are related by the ideal-gas law, Eq. 8.9, only when the particle density is small. As the pressure of a gas is increased, the decrease in volume produces an increase in density. At high pressure, therefore, the pressure-volume relation of a real gas can be expected to deviate from that of an ideal gas. For most gases these deviations become appreciable at particle densities above $10^{21}$ particles/cm$^3$, or about 1 mol/l. That is, at a pressure at which 1 mol of gas occupies a volume of 1 l or so, the ideal-gas law does not apply.

† We know there are far fewer than $10^{63}$ molecules in the atmosphere because it is shown in Appendix I that there are only $10^{57}$ atoms in the sun. There are obviously fewer molecules in the atmosphere than atoms in the sun. In fact there are about $10^{44}$ molecules in the atmosphere.

## $pV$ **diagrams**

In Fig. 8.6 the pressure $p$ is plotted against the volume $V$ occupied by 1 mol of gaseous $H_2O$. Each curve gives the variation of pressure with volume at a fixed temperature, so these curves are called *isotherms* (meaning "equal temperature"). The light curves are the $pV$ relations for an ideal gas, whereas the dark curves are the actual $pV$ relations of $H_2O$. As expected, the real curve deviates appreciably from the ideal curve for values of $V$ less than 1 or 2 l.

The isotherms in Fig. 8.6 are typical of any real gas. Above a certain critical temperature, the real isotherms are smooth curves which lie somewhat below the corresponding ideal isotherm. This is because in a real gas some of the molecules are prevented from striking the walls of the container by the presence of other molecules near the wall, so that the pressure of the real gas is less than the pressure of an ideal gas at the same temperature and volume. Below the critical temperature, the isotherms are no longer simple smooth curves but, like the 350°C isotherm in Fig. 8.6, consist of a horizontal section joined abruptly to a smoothly varying section.

A horizontal section in an isotherm indicates that the volume occupied by the gas can be decreased without an increase in pressure. This is possible because as the gas is compressed, some of it condenses into the liquid state. However, a gas can be liquefied by compression only if its temperature is below the critical temperature of the gas.

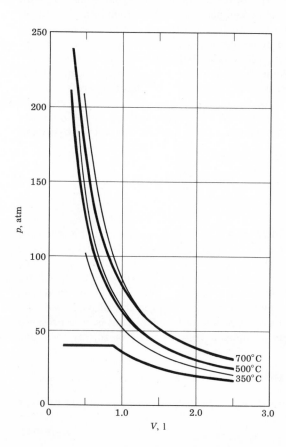

**FIGURE 8.6**
Plot of the pressure $p$ against the volume $V$ occupied by 1 mol of gaseous $H_2O$. The dark lines are the real isotherms at the temperatures indicated. The light curves are the isotherms of an ideal gas at the same temperatures.

Table 8.3 gives the critical temperature of some common substances. Note that two of the isotherms in Fig. 8.6 are at temperatures above 374.1°C, the critical temperature of $H_2O$, and one is at a temperature just below the critical temperature.

The isotherm of a substance at its critical temperature, the *critical isotherm*, is horizontal only at a single point, the *critical point*. The isotherms of $H_2O$ in the vicinity of its critical point are shown in Fig. 8.7. The scale in this figure is enlarged over that in Fig. 8.6 in order to display the necessary detail. An isotherm just above the critical isotherm has a kink in it near the critical point (point $C$ in Fig. 8.7), but the isotherm is never horizontal. An isotherm below the critical isotherm is split into two distinct sections connected by a horizontal line. The critical isotherm, which is the transition between these two types of isotherms, is horizontal only at the critical point.

In Fig. 8.7 the ends of the horizontal lines are connected by a curve, called the *saturation curve*. This curve touches the critical isotherm at the critical point, so together these two curves divide the $pV$ diagram into four regions. In the region above the critical isotherm, the substance remains a gas for all values of the pressure; this is the

TABLE 8.3 **Critical temperatures of some common substances**

| Substance | Critical temperature, °C |
| --- | --- |
| Ammonia | 132.4 |
| Carbon dioxide | 31.0 |
| Chlorine | 144 |
| Ethanol | 126.9 |
| Helium | −267.9 |
| Iodine | 512 |
| Methane | −82.1 |
| Nitrogen | −146.9 |
| Oxygen | −118.5 |
| Propane | 96.8 |
| Sulfur | 1040 |
| Trichloromethane (chloroform) | 262 |
| Water | 374.1 |

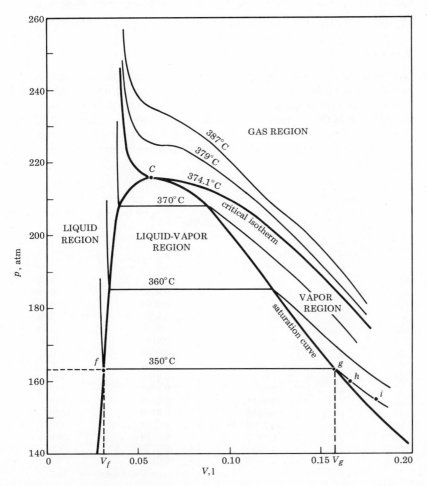

FIGURE 8.7
Isotherms of $H_2O$ in the vicinity of its critical point.

*gas region.* In the region below the critical isotherm and to the right of the saturation curve, the substance is still a gas, but it can be liquefied by compression; this is the *vapor region*. Below the critical isotherm and to the left of the saturation curve is the *liquid region*. The area under the saturation curve is the *liquid-vapor region*.

Consider point $i$ in Fig. 8.7. This represents 1 mol of gaseous $H_2O$ at a temperature of 350°C and a pressure of 155 atm. If this gas is compressed while its temperature is held constant, its pressure will increase. For example, decreasing the volume from 0.180 to 0.165 l increases the pressure from 155 to 160 atm. The substance is now represented by point $h$. As the substance is further compressed, its corresponding point in the $pV$ diagram moves along the isotherm. When the substance reaches point $g$, it begins to condense into the liquid state, so the point moves along the horizontal line $fg$. At the pressure $p_v = 163$ atm of this line, the substance can occupy any volume between $V_f$ and $V_g$. The point $g$ corresponds to the substance's being all in the gaseous state, whereas the point $f$ corresponds to the substance's being all in the liquid state. Between these two points part of the substance is liquid and part is gas. When all the substance is in the liquid state, a further decrease in its volume requires an enormous increase in pressure; hence the isotherm rises steeply to the left of $f$.

### Vapor pressure

The pressure $p_v$ of the horizontal section of an isotherm is called the *vapor pressure*. It is the pressure at which the vapor and liquid can coexist at the temperature of the isotherm. Table 8.4 gives the vapor pressure of $H_2O$ for various temperatures between 0 and 374.1°C. (The vapor pressure obviously does not exist above the critical temperature.) Also listed in Table 8.4 are the volumes $V_f$ and $V_g$ occupied

TABLE 8.4 **Vapor pressure of water at temperatures from 0 to 374.1°C**
The table also lists the volumes $V_f$ and $V_g$ occupied by 1 mol of $H_2O$ in the liquid and gaseous state, respectively.

| Temperature, °C | Pressure, atm | $V_f$, l | $V_g$, l |
|---|---|---|---|
| 374.1 | 218.3 | 0.057 | 0.057 |
| 300 | 84.8 | 0.0252 | 0.389 |
| 250 | 39.2 | 0.0224 | 0.897 |
| 200 | 15.3 | 0.0207 | 2.28 |
| 150 | 4.69 | 0.0196 | 7.04 |
| 100 | 1.00 | 0.0187 | 30.0 |
| 80 | 0.466 | 0.0185 | 61.1 |
| 60 | 0.196 | 0.0183 | 138 |
| 40 | 0.0728 | 0.0180 | 350 |
| 37 | 0.0621 | 0.0.80 | 411 |
| 30 | 0.0418 | 0.0180 | 590 |
| 20 | 0.0230 | 0.0179 | 1004 |
| 15 | 0.0168 | 0.0179 | 1400 |
| 10 | 0.0121 | 0.0179 | 1910 |
| 5 | 0.00856 | 0.0179 | 2640 |
| 0 | 0.00626 | 0.0179 | 3700 |

by 1 mol of $H_2O$ at the pressure $p_v$ when it is completely in the liquid and gaseous state, respectively.

Table 8.4 shows that 1 mol of water (liquid $H_2O$) at 20°C has a volume of 0.0179 l. If it is placed inside an evacuated 2-l container, some of the liquid will evaporate. That is, some of the $H_2O$ molecules leave the liquid, forming a gas in the empty space above the liquid. As more molecules evaporate, the pressure of this vapor increases. At the same time, some of the molecules in the vapor collide with the surface of the liquid, reentering the liquid state. The number of molecules which recondense increases as the pressure of the vapor increases, until eventually a point of *dynamic equilibrium* is reached, where the rate at which molecules evaporate from the liquid equals the rate at which molecules condense back into the liquid. The pressure of the vapor at this equilibrium point is the vapor pressure given in Table 8.4, which at 20°C is 0.0230 atm. When the pressure reaches this value, the number of molecules in the liquid and gaseous states remains constant. That is, there is no net evaporation or condensation, so the system is in equilibrium. However, there is still a constant flow of molecules from the liquid to the gas and an equal flow of molecules from the gas to the liquid, which is why the equilibrium is termed *dynamic*. The system is represented by a point in the liquid-vapor region of the $H_2O$ $pV$ diagram.

At equilibrium, the vapor occupies a volume of 2 l minus the volume occupied by the liquid, or

$$V = 2\ l - 0.018\ l = 1.982\ l$$

From Table 8.4 we see that 1 mol of gaseous $H_2O$ occupies 1004 l at this temperature and pressure, so that only

$$\frac{1.982\ l}{1004\ l} = 0.00197\ mol$$

of $H_2O$ is in the gaseous state. The fraction 0.99803 mol is still in the liquid state. If the volume of the container is increased while keeping the temperature constant at 20°C, a larger fraction of the $H_2O$ will go into the gaseous state in order to maintain the pressure of $H_2O$ at 0.0230 atm. When the volume of the container reaches 1004 l, all the $H_2O$ is in the gaseous state. A further increase in the volume beyond this produces a decrease in pressure because the number of molecules in the gaseous state cannot be increased. The system is now represented by a point in the vapor region of the $pV$ diagram. The vapor pressure is the maximum pressure the vapor can have at a given temperature.

Air contains gaseous $H_2O$ in concentrations that vary from 0 to 6 percent, depending on the weather conditions. This gas exerts a partial pressure in proportion to its concentration.

For example, in air that has an $H_2O$ content of 0.85 percent, the partial pressure of the $H_2O$ is

$$p_{H_2O} = 0.0085 p_0$$

where $p_0$ is the total atmospheric pressure. With $p_0 = 1$ atm, the partial pressure of $H_2O$ is just 0.0085 atm.

The maximum partial pressure of $H_2O$ possible at a particular air temperature is the vapor pressure of $H_2O$ at that temperature. For instance, at a temperature of 20°C the partial pressure of $H_2O$ can be at most 0.023 atm (Table 8.4). The actual $H_2O$ content of air is often given as the percentage of the maximum $H_2O$ content possible at the existing air temperature. This percentage is called the *relative humidity*.

For example, if the $H_2O$ in air at 20°C has a partial pressure of 0.0085 atm, the relative humidity is

$$\frac{p_{H_2O}}{p_v} \times 100 = \frac{0.0085 \text{ atm}}{0.0230 \text{ atm}} \times 100 = 37\%$$

The same air at 10°C has a relative humidity of

$$\frac{0.0085 \text{ atm}}{0.0121 \text{ atm}} \times 100 = 70\%$$

The temperature at which the air would have a relative humidity of 100 percent is called the *dew point*. From Table 8.4 the dew point of air with $p_{H_2O} = 0.0085$ atm is found to be about 5°C. Moisture begins to condense out of this air when its temperature falls to 5°C. This often happens at night, the moisture appearing as dew on the ground. Moisture will also condense on any surface that is 5°C or less, such as the outside of a glass containing an iced drink.

The dew point is easily measured by wrapping a piece of wet cloth around the bulb of a thermometer and then waving the thermometer rapidly in the air to increase the rate of evaporation of the water from the cloth. During evaporation, only the most energetic molecules leave the liquid, so that the temperature of the remaining liquid is lowered (Sec. 9.1). The temperature thus falls until it reaches the value at which the vapor pressure equals the partial pressure of $H_2O$ in the air. At this temperature no further evaporation takes place because water condenses on the cloth as fast as it evaporates. The thermometer, which measures the temperature of the cloth, thus registers the dew point.

For example, suppose the thermometer reads 10°C. This means that $p_{H_2O}$ is 0.0121 atm, the vapor pressure of $H_2O$ at 10°C. If the air temperature is 20°C, the relative humidity is

$$\frac{0.0121 \text{ atm}}{0.0230 \text{ atm}} \times 100 = 52.6\%$$

The *boiling point* of a liquid is the temperature at which the vapor pressure equals atmospheric pressure. At this temperature bubbles of vapor are formed in the interior of the liquid. Vapor bubbles cannot be formed at a lower temperature because the pressure inside them, which is the vapor pressure, is less than atmospheric pressure, so that the bubbles collapse. From Table 8.4 we see that the vapor pressure of $H_2O$ equals 1.0 atm at 100°C. This, of course, is not a coincidence, since 100°C was first defined as the boiling point of water. However, this is the boiling point only at a pressure of 1 atm. At elevations above sea level the air pressure is less than 1.0 atm, so that water boils at a lower temperature. In Mexico City, for instance, where the air pressure is only 0.85 atm, water boils at 95°C. On the other hand, in a pressure cooker the water boils at a temperature above 100°C because the pressure inside the cooker is greater than 1.0 atm.

*1* (*a*) What is the molecular mass of $H_2O$? (*b*) How many $H_2O$ molecules are there in 1 g of water? *Ans.* (*a*) 18 u; (*b*) $3.34 \times 10^{22}$

*2* What is the particle density of nitrogen ($M = 28$) at STP (0°C and 1 atm)? Use Table 7.2.

*3* (*a*) On the average, what is the volume occupied by a single water molecule (see Prob. 1)? (*b*) What is the average distance between molecules of $H_2O$?
 *Ans*? (*a*) $30 \times 10^{-24}$ cm³; (*b*) $3.11 \times 10^{-8}$ cm

*4* (*a*) What is the average volume occupied by a nitrogen molecule at STP (see Prob. 2)? (*b*) What is the average distance between nitrogen molecules under these conditions?

**REMARK** Problems 3 and 4 show that the molecules of a gas at STP are 10 times farther apart than the molecules in a liquid.

*5* Air in a sealed bulb has a pressure of 1 atm at 20°C. (*a*) What is the pressure in the bulb at 150°C? (*b*) At what temperature is the pressure in the bulb 2 atm? (*c*) At what temperature is the pressure in the bulb 0.5 atm?
 *Ans.* (*a*) 1.44 atm; (*b*) 313°C; (*c*) −126.5°C

*6* On the Fahrenheit temperature scale, the temperature of the ice point is 32°F, and the temperature of the steam point is 212°F. (*a*) How many Fahrenheit degrees equal one Celsius degree? (*b*) Show that the conversion from a temperature $t_F$ on the Fahrenheit scale to a temperature $t_C$ on the Celsius scale is given by

$$t_C = \tfrac{5}{9}(t_F - 32°)$$

and the conversion from the Celsius scale to the Fahrenheit scale is given by

$$t_F = \tfrac{9}{5}t_C + 32°$$

(*c*) Plot $t_F$ against $t_C$. At what temperature does $t_F = t_C$? (*d*) What is absolute zero on the Fahrenheit scale?

*7* A gas occupies a volume of 22 l when its gauge pressure is 1.5 atm. (*a*) What is the absolute pressure of the gas? (*b*) The gas is compressed until it occupies 10 l at the same temperature. What is the new absolute pressure of the gas? (*c*) What is the new gauge pressure?
 *Ans.* (*a*) 2.5 atm; (*b*) 5.5 atm; (*c*) 4.5 atm

*8* A bubble of volume $V = 15$ cm³ is formed at the bottom of a lake, 12 m below the surface. What is the volume of the bubble when it reaches the surface of the water? (Assume the temperature does not change.)

*9* A cylinder contains 0.6 ft³ of oxygen at a temperature of 25°C and a pressure of 2200 lb/in.². (*a*) What volume does this gas occupy at 25°C and atmospheric pressure? (*b*) A man is breathing pure oxygen through a face mask at the rate of 8 l/min. How long will one cylinder of oxygen last (1 l = 0.0354 ft³)?
 *Ans.* (*a*) 89.8 ft³; (*b*) 5.28 h

*10* A cylinder of compressed air will supply a scuba diver with enough air to last for 90 min on the surface of the water. How long will the same tank last when the diver is swimming 20 m below the surface? (*Hint*: The volume of air inhaled per minute does not change with depth.)

*11* From the ideal-gas law, calculate the mass density of helium ($M = 4$) at STP. Compare with Table 7.2. *Ans.* 0.178 kg/m³

*12* From the ideal-gas law, calculate the mass density of steam at 100°C and 1 atm. Compare with Table 7.2.

*13* What is the mass of 5 l of $CO_2$ at a temperature of 35°C and a pressure of 2.5 atm? (*Hint*: Use Table 8.1.) *Ans.* 21.8 g

*14* What is the mass of the oxygen in the tank described in Prob. 9?

*15* What is the volume of air that must escape from a room when the temperature of the room is raised from 15 to 25°C? The dimensions of the room are 10 by 7 by 4 m, and the pressure remains unchanged while the room is being heated. *Ans.* 9.72 m³

*16* Calculate the volume occupied by 1 mol of an ideal gas at STP.

*17* Calculate the volume occupied by 1 g of oxygen at STP. *Ans.* 0.700 l

*18* Show that at atmospheric pressure the density of air ($M = 29$) at temperature $T$ is given by

$$\rho = \frac{353 \text{ K}}{T} \text{ kg/m}^3$$

Check this against the data in Table 7.2.

*19* Air in the lungs (alveolar air) has a different composition from atmospheric air. For instance, the partial pressure of carbon dioxide in alveolar air is 40.0 mm Hg. What is the percentage of $CO_2$ in alveolar air?

*Ans.* 5.26%

20 Oxygen constitutes only 13.6 percent of the air in the lungs (alveolar air). What is the partial pressure of $O_2$ in the lungs?

21 How many moles of oxygen are there in 1 l of air at STP? *Ans.* $9.33 \times 10^{-3}$

22 A gas composed of 3.5 g of oxygen and 1.5 g of helium occupies a volume of 3 l at a temperature of 25°C. (*a*) What are the partial pressures of oxygen and helium in this gas? (*b*) What is the density of the gas mixture?

23 What is the average kinetic energy of the molecules of a gas at 37°C?

*Ans.* $6.42 \times 10^{-21}$ J

24 What fraction of the molecules of a gas at 25°C have kinetic energies greater than the average kinetic energy of the molecules of a gas at 1217°C?

25 What fraction of the molecules of a gas have kinetic energies less than the average kinetic energy of the gas? *Ans.* 0.608

26 What is the total kinetic energy of the molecules in 1 mol of gas at 0°C?

27 (*a*) What is the average of $v^2$ for oxygen molecules at 27°C? (*b*) The square root of $\overline{v^2}$ is called the *root-mean-square* (rms) speed. Calculate the rms speed of an oxygen molecule at 27°C.

*Ans.* (*a*) $23.4 \times 10^4$ m²/s²; (*b*) 484 m/s

28 At what temperature would 11.2 percent of the oxygen molecules in the atmosphere have speeds greater than the escape speed?

29 On a winter day, outside air at 0°C and relative humidity of 30 percent is taken into a house, where it is warmed to 20°C. What is the relative humidity of the air in the house?

*Ans.* 8.16%

30 A thermometer with a piece of wet cloth wrapped around its bulb is waved in the air until the reading stops falling at 15°C. The air temperature is 30°C. (*a*) What is the dew point of the air? (*b*) What is the relative humidity of the air?

31 The relative humidity is 37.2 percent on a day when the temperature is 20°C. What is the dew point? *Ans.* 5°C

32 The dew point is 30°C on a day when the temperature is 37°C. What is the relative humidity?

33 One mole of $H_2O$ occupies a volume of 0.10 l at 300°C. (*a*) What fraction of the $H_2O$ is in the liquid state? (*b*) What is the volume occupied by the liquid $H_2O$?

*Ans.* (*a*) 79.4%; (*b*) 0.020 l

## BIBLIOGRAPHY

COUNCIL FOR NATIONAL CO-OPERATION IN AQUATICS: "The New Science of Skin and Scuba Diving," Association Press, New York, 1968. Authoritative account of the equipment and techniques of scuba diving. The physics of diving, including detailed descriptions of the mechanism of the regulator, is included.

EGAN, DONALD F.: "Fundamentals of Inhalation Therapy," The C. V. Mosby Company, St. Louis, 1969. The physics and physiology of respiration are treated in detail.

GUYTON, ARTHUR C.: "Medical Physiology," W. B. Saunders Company, Philadelphia, 1971. The physics and physiology of deep-sea diving are discussed in chap. 45. Space physiology is discussed in chap. 44.

# liquids CHAPTER 9

A liquid is a fluid that has a definite volume at any given temperature. For example, 1 mol of water at 20°C occupies 18.04 cm³ (Table 8.4), regardless of the size or shape of the container in which it is held. Although its volume can be decreased slightly by applying very high pressure, a liquid is considered to be incompressible for most purposes. Thus there is no significant change in the volume of water as its pressure is varied from 0 to 10 atm. This is in contrast to the volume of a gas, which is infinite at zero pressure and which changes by a factor of 10 as the pressure is varied from 1 to 10 atm. This chapter discusses a number of properties of liquids that are of biological importance.

## 9.1 HEAT OF VAPORIZATION

The molecules in a gas are so far apart that, to good approximation, their mutual attraction does not affect the behavior of the gas. In a liquid, on the other hand, the molecules are so close that an individual molecule is always being attracted by a number of neighboring molecules. It is this attraction that holds a liquid together and prevents it from expanding. Nevertheless, the molecules in a liquid, as in a gas, are free to move anywhere in the fluid, so that a liquid cannot maintain its shape against an external force. In other words, a liquid is not rigid.

A molecule in the interior of a liquid is completely surrounded by other molecules that exert attractive forces on it in all directions. Figure 9.1 shows some of the forces on an individual molecule at one instant of time. Since the molecules in a liquid are in constant motion, the forces on this particular molecule constantly change. However, on the average, the sum of these forces is zero, so there is no net force on a molecule in the interior of a liquid.

The situation is different for a molecule near the surface of a liquid because the molecule is only partially surrounded by other molecules. Figure 9.2 shows the forces on an individual molecule at the surface. The sum of these forces is not zero. On the contrary, there is a net force **F** on the molecule directed toward the interior of the liquid. This is the force that prevents most of the molecules at the surface from escaping from the liquid.

Some molecules do evaporate from the liquid, however, because the force **F** exists only within a short distance of the surface, and so a molecule with sufficient energy to penetrate beyond the range of **F** can escape. Suppose, for simplicity, that the magnitude of **F** is

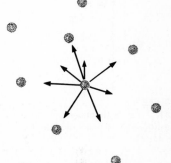

FIGURE 9.1
Forces on a molecule in the interior of a liquid. The sum of these forces is zero.

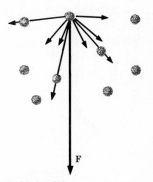

FIGURE 9.2
Forces on a molecule at the surface of a liquid. The sum of these forces is **F**.

constant in a region of width $d$ and that it is zero outside this region (Fig. 9.3). Then the force $\mathbf{F}$ does the work $-Fd$ on an escaping molecule, and so if the molecule had kinetic energy $K_A$ at point $A$ just inside the surface, its kinetic energy $K_B$ at point $B$, according to the work-energy theorem (Sec. 4.2), is given by

$$K_B - K_A = -Fd$$

Since the minimum value of $K_B$ is zero, $K_A$ must be at least equal to $Fd$. A molecule whose kinetic energy at point $A$ is less than $Fd$ will come to rest before it reaches $B$, and so it will be attracted back to the surface.

Therefore the energy $w = Fd$ is the minimum energy required to separate one molecule from a liquid. The minimum energy $H_v$ required to separate all the molecules in 1 mol of liquid is $Nw$, where $N$ is Avogadro's number. Of course, separating all the molecules in a liquid transforms the liquid into a gas. Thus $H_v$, called the *molar heat of vaporization*, is the minimum energy required to vaporize 1 mol of a substance. The values of $H_v$ for a variety of liquids are given in Table 9.1.

The energy $w$ is much greater than the average kinetic energy of a molecule in a liquid, so that only the extremely energetic molecules evaporate. As they leave the liquid, the average energy of the remaining molecules decreases. The absolute temperature $T$ of the liquid then falls because $T$ is proportional to the average molecular energy. Once the temperature of the liquid falls below the temperature of its surroundings, heat flows from the surroundings into the liquid, supplying the energy for further evaporation.

The heat supplied to an evaporating liquid is measured in a *calorimeter* (Fig. 9.4). This device consists of a small sample container immersed in a large vessel of water. The vessel is heavily insulated on the outside to prevent room heat from reaching the water. The

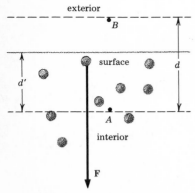

**FIGURE 9.3**
Work $-Fd$ is done by the surface of a liquid on a molecule leaving the liquid.

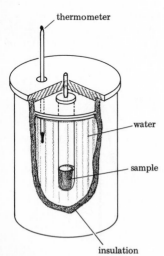

**FIGURE 9.4**
A calorimeter. The small sample container is immersed in a large vessel of water.

TABLE 9.1 **Molar heat of vaporization and surface tension of some liquids**
The liquids are listed in order of increasing molar heat of vaporization. Since the heat of vaporization and the surface tension both depend on temperature, the temperature is listed first.

| Liquid | Temperature, °C | Heat of vaporization, kcal/mol | Surface tension, N/m |
|---|---|---|---|
| Helium | −270 | 0.0275 | 0.000239 |
| Hydrogen | −255 | 0.250 | 0.00231 |
| Oxygen | −183 | 1.7 | 0.0132 |
| Chloroethane | 20 | 6.3 | 0.020† |
| Ethanol | 20 | 9.7 | 0.0227 |
| Tissue fluid | 37 | | 0.050 |
| Blood, whole | 37 | | 0.058 |
| Blood plasma | 37 | | 0.073 |
| Water | 100 | 9.70 | 0.0589 |
| | 50 | 10.24 | 0.0679 |
| | 20 | 10.55 | 0.0727 |
| | 0 | 10.75 | 0.0756 |
| Mercury | 20 | 15.1 | 0.4355 |
| Tungsten | 3410 | 180 | 2.5 |

†Estimated.

liquid to be studied is placed in the sample container, and as it evaporates, heat from the surrounding water flows into the container, lowering the temperature of the water. From a measurement of the decrease in the water temperature that results from the evaporation of a known quantity of liquid, the heat of vaporization of the liquid is determined. Of course, great care is needed to minimize heat leakage from other sources, e.g., through the top of the container.

Suppose, for example, that when 5 g of ethanol, $C_2H_5OH$, is evaporated from the container, the temperature of the surrounding water falls 0.42°C. If the mass of the water is 2.5 kg, then 2.5 kg $\times$ 0.42°C = 1.05 kcal of heat was used to evaporate the ethanol (Sec. 5.5). The heat of vaporization is thus 1.05 kcal/5 g = 0.21 kcal/g, which is the heat of vaporization per gram of substance. The molar heat of vaporization $H_v$ is found by multiplying the heat of vaporization per gram by the number of grams in a mole. A mole of ethanol contains 46 g ($M = 2 \times 12 + 6 \times 1 + 16 = 46$), so $H_v = $ 46 g/mol $\times$ 0.21 kcal/g = 9.7 kcal/mol.

The molar heat of vaporization is important because it is related to the magnitude of **F** by

$$H_v = Nw = NFd$$

Molecules differ greatly in the magnitude of the force they exert on each other and hence in the magnitude of **F**. However the range $d$ of the molecular force is approximately the same for all molecules (between $10^{-8}$ and $10^{-7}$ cm). Therefore the molar heat of vaporization is more or less proportional to the magnitude of the molecular force. Substances with very small values of $H_v$ (such as helium and oxygen) have weak molecular forces and are gases at normal temperature. They can be liquefied only at low temperature. On the other hand, substances with very large values of $H_v$ (such as tungsten) are solids at normal temperature and liquids only at high temperature. Substances with intermediate values of $H_v$ (such as ethanol and water) are liquids at normal temperature.

The heat of vaporization in kilocalories per gram is perhaps of more practical use because it gives the cooling potential of a specified mass of the liquid. It can be calculated from the molar heat of vaporization given in Table 9.1 if the molecular mass is known. For example, the heat of vaporization of water is

$$\frac{11 \text{ kcal/mol}}{18 \text{ g/mol}} = 0.61 \text{ kcal/g}$$

This means that the evaporation of 1 g of water can cool 1 kg of water by 0.61°C. Similarly, since 1 mol of chloroethane (ethyl chloride), $H_2C_5Cl$, contains 64.5 g, its heat of vaporization is

$$\frac{6.3 \text{ kcal/mol}}{64.5 \text{ g/mol}} = 0.097 \text{ kcal/g}$$

Thus the evaporation of 1 g of chloroethane can cool 1 kg of water by only 0.097°C.

Water has the largest heat of vaporization per gram of any common liquid, which makes it an ideal coolant. Many biological systems use the evaporation of water for cooling. It is well known, for instance, that birds and mammals maintain a constant body temperature by evaporating water to remove excess heat. The water is evaporated either from the surface of the skin (as in men and horses) or from the lungs (as in birds and dogs). It is less well known that the honeybee also uses evaporation to maintain a constant temperature in the brood nest, even though the temperature of an individual bee's body is not kept constant. If the temperature in a hive rises above 35.5°C, foraging bees stop bringing nectar to the hive and start bringing water. The water is deposited in cells in the brood nest, where it is fanned by other bees to increase its rate of evaporation. Even plants are kept cool by evaporation. On a hot summer day the temperature of a tree is usually several degrees cooler than the ground, because of the evaporation of water from its leaves.

The rate at which water evaporates depends on the relative humidity of the air (Sec. 8.5). If the humidity is high, molecules of $H_2O$ vapor condense on the surface of liquid $H_2O$ almost as fast as molecules of $H_2O$ evaporate from the surface. The net evaporation is then very small, and so the cooling capacity of water is small. This is why humid days are so uncomfortable. In fact, the temperature of evaporating water cannot fall below the dew point, which is the temperature at which the rate of evaporation equals the rate of condensation (Sec. 8.5).

A liquid, e.g., chloroethane, that has a small heat of vaporization is said to be *volatile*. It requires less energy for a molecule of chloroethane to evaporate than for a molecule of water; therefore at a given temperature more molecules of chloroethane are capable of evaporating. A volatile liquid thus evaporates more rapidly than water, and even though the heat removed per gram evaporated is small, the rate of heat removal is large. Chloroethane is used to anesthetize small areas of skin by freezing the skin. For instance, the evaporation of 2 g of chloroethane removes 195 cal from a small volume of tissue in a few seconds. This is sufficient to cool 5 g of tissue to 0°C. The time is important because the heat must be removed from the treated area faster than heat can flow into it from the surrounding tissue.

## 9.2 SURFACE TENSION

The surface of a liquid has special properties because of the molecular force **F** acting on it. The total surface area of a liquid, unlike its volume, can be changed by changing the shape of the liquid. But to increase the surface area, molecules from the interior of the liquid must be moved to the surface. It is seen from Fig. 9.3 that the surface does the work $-Fd'$ on a molecule passing from the interior to the surface, which means that the work $w' = Fd'$ is required to bring a molecule to the surface. Thus work has to be done to increase the surface area of a liquid.

Figure 9.5 shows a simple device for measuring this work. A thin film of liquid is suspended in a wire frame of breadth $b$ and variable length $l$. The total surface area of the film is $A = 2bl$ because the film has both a front and a back surface, each of area $bl$. Then if the length of the film is increased from $l_1$ to $l_2$ by moving the slider a distance $x = l_2 - l_1$, the total surface area of the film is increased

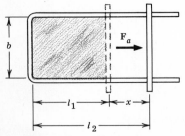

**FIGURE 9.5**
A film of liquid suspended in a wire frame.

from $A_1 = 2bl_1$ to $A_2 = 2bl_2$. The film resists having its surface area increased, so a force $\mathbf{F}_a$ must be applied to the slider to move it. The work $W = F_a x$ done in increasing the surface area is found by measuring $F_a$ and $x$.

This work is equal to the energy $nw'$ required to move $n$ additional molecules from the interior of the film to the surface. The number $n$ is proportional to the increase in surface area, $A_2 - A_1$, so we can write

$$n = a(A_2 - A_1) = a(2bl_2 - 2bl_1)$$
$$= 2ab(l_2 - l_1) = 2abx$$

where $a$ is a constant of proportionality (the number of molecules per unit area on the surface). Thus, on the one hand the work required to move $n$ molecules to the surface is

$$W = nw' = 2abxw'$$

while on the other hand the work $W = F_a x$ is done in moving the slider, so we have

$$F_a x = 2abxw'$$

We cancel the $x$'s on both sides and divide by $b$ to get

$$\frac{F_a}{b} = 2aw' \qquad\qquad 9.1$$

There are several things to note about this equation.

*1* The quantity $aw'$ is an intrinsic property of the liquid, depending on the molecular properties of the liquid, not on the particular geometry of Fig. 9.5. This quantity, called the *surface tension* $\gamma$,[†] can in principle be measured by measuring the force $\mathbf{F}_a$ required to stretch the film.

*2* The reaction to $\mathbf{F}_a$ is the force $\mathbf{R}_a = -\mathbf{F}_a$ the liquid exerts on the slider. Equation 9.1 shows that the film exerts a constant force per unit length on the slider:

$$\frac{R_a}{b} = \frac{F_a}{b} = 2\gamma \qquad\qquad 9.2$$

*3* The factor of 2 occurs in Eqs. 9.1 and 9.2 because the film has two surfaces. A liquid in a dish has only one free surface, and so the force per unit length it exerts is

$$\frac{R_a}{b} = \gamma \qquad\qquad 9.3$$

The surface tension is the force per unit length that the surface of a liquid exerts on any line in the surface. This force lies in the surface and is perpendicular to the line. Surface tension is thus analogous to pressure, which is the force per unit area that fluid exerts on any surface. Pressure exerts a force perpendicular to the surface, just as surface tension exerts a force perpendicular to the line. However, the pressure of a fluid exerts an outward force,[‡]

[†] $\gamma$ is the Greek letter gamma.
[‡] An important exception to this is discussed in Sec. 9.5.

whereas surface tension exerts an inward force. That is, pressure is directed so as to expand a volume, whereas surface tension is directed so as to shrink a surface. Surface tension tries to make the surface area of a liquid as small as possible. For instance, a freely falling drop of liquid forms as a sphere because a sphere has the smallest surface area for a given volume.

The surface tension of a liquid depends on its temperature, but otherwise it is a constant, characteristic of the liquid. In the mks system, the unit of surface tension is newtons per meter. Table 9.1 gives the surface tension of a variety of liquids at the indicated temperatures. The surface tension of water is given at several temperatures to show its temperature dependence.

**REMARK** The energy $w'$ required to move a molecule from the interior to the surface of a liquid is more or less proportional to the energy $w$ required to remove a molecule from a liquid. Therefore the surface tension $\gamma = aw'$ should increase regularly with the heat of vaporization $H_v = Nw$. Table 9.1 shows that this is the case. It is only a trend, however, and small exceptions are to be expected.

The surface tension of water is larger than that of any other common liquid (except mercury), a fact of considerable significance because of the pervasiveness of water in biological systems. The most familiar consequence of this is the ability of water to support small objects on its surface. For instance, if a razor blade is carefully placed on the surface of undisturbed water, it will not sink. The blade is supported by the surface tension, but it will sink if the surface is disturbed. Similarly, some insects are able to walk on water, and many small water animals, e.g., planaria and mosquito larvae, hang from the surface.

To understand the role surface tension plays in this phenomenon, let us first recall how the tension in a horizontal cord supports an object. Figure 9.6 shows a tightrope walker being supported by a rope with tension $T$. This means that at every point, one side of the rope pulls the other side with a force of magnitude $T$ directed parallel to the rope. Since the rope is bent where the performer is standing, the forces $\mathbf{F}_1$ and $\mathbf{F}_2$ exerted on the performer by each side of the rope are in different directions. Their horizontal components, which are in opposite directions, cancel each other, whereas their vertical components, which are in the same direction, add. Since the vertical component of each force has the magnitude $T \cos \theta$, the rope exerts an upward force of magnitude $2T \cos \theta$ on the performer.

The surface of a liquid behaves similarly. Figure 9.7 shows the small circular depression made in the surface of the water by the leg of an insect. The surface tension of the water exerts forces all around the rim of this depression. These forces, which are in the surface and perpendicular to the rim, have a magnitude of $\gamma$ per unit length. Figure 9.7 shows two of the forces on opposite sides of the rim. Since each force acts on a small section of length $s$, each has the magnitude $\gamma s$. Their horizontal components are in opposite directions, so they cancel, but their vertical components are in the same direction, so they add. Thus the magnitude of the vertical force per unit length along the rim is $\gamma \cos \theta$; therefore the total vertical force is $2\pi r\gamma \cos \theta$. This is the force that supports the weight of the leg.

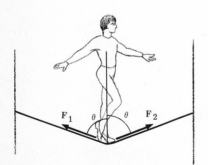

**FIGURE 9.6**
A tightrope walker standing on a tightrope.

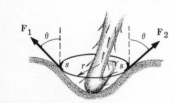

**FIGURE 9.7**
An insect's leg supported by the surface tension of water.

## Bubbles

A bubble consists of a spherical surface of liquid. The surface tension in this surface tends to shrink the bubble but is resisted by the pressure $p_i$ inside the bubble, which is greater than the pressure $p_o$ outside the bubble. This pressure difference results in an outward force on the bubble that equals the inward force of surface tension.

To be specific, consider a soap bubble of radius $r$. Imagine the bubble divided into two hemispheres joined together along a circumference of length $2\pi r$ (Fig. 9.8), and consider the forces on the upper hemisphere. Because the bubble has an inside and an outside surface, the surface tension exerts a force of $2\gamma$ per unit length on the circumference. Thus the lower hemisphere exerts a force

$$F_t = 2\gamma 2\pi r = 4\pi\gamma r$$

on the upper hemisphere, directed so as to shrink the bubble.

The pressure inside the bubble exerts an outward force of $p_i$ per unit area on the surface. It can be shown (see Prob. 24 in Chap. 7) that the total force on the upper hemisphere due to $p_i$ has the magnitude

$$F_i = Ap_i = \pi r^2 p_i$$

and is directed so as to expand the bubble. ($A = \pi r^2$ is the cross-sectional area of the circle separating the two hemispheres.) Likewise, the pressure outside the bubble exerts an inward force of $p_o$ per unit area on the surface. The force on the hemisphere due to $p_o$ has the magnitude

$$F_o = \pi r^2 p_o$$

and it is directed so as to shrink the bubble. Thus the magnitude of the net force due to pressure on the upper hemisphere is

$$F_p = F_i - F_o = (p_i - p_o)\pi r^2$$

and it tends to expand the bubble. At equilibrium, this force equals the force $F_t$ of surface tension, so

$$4\pi\gamma r = (p_i - p_o)\pi r^2$$

or
$$p_i - p_o = \frac{4\gamma}{r} \qquad\qquad 9.4$$

For a "bubble" with only one surface, such as a drop of liquid, or a gas bubble in a liquid, the corresponding formula is

$$p_i - p_o = \frac{2\gamma}{r} \qquad\qquad 9.5$$

There are two important points to note about these formulas:
1 The pressure difference is proportional to the surface tension, so that a larger pressure is required to form a bubble in a liquid with a large surface tension than in a liquid with a small surface tension.
2 The pressure difference is inversely proportional to the radius of the bubble. This means that the pressure difference is greater in a small bubble than in a large one.

A liquid boils at the temperature at which its vapor pressure equals

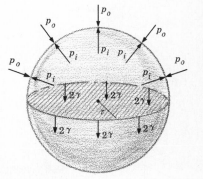

FIGURE 9.8
Forces on one hemisphere of a bubble produced by the air pressure inside and outside the bubble and by the surface tension exerted by the other hemisphere.

atmospheric pressure because at this temperature bubbles of vapor can form. Actually, the vapor pressure must be slightly greater than atmospheric pressure to overcome the tendency of the surface tension to collapse the bubble.

For example, the surface tension of water at 100°C is 0.059 N/m (Table 9.1), and so, according to Eq. 9.5, the pressure inside a vapor bubble of radius $10^{-3}$ m is

$$p_i = p_o + \frac{2\gamma}{r} = p_o + \frac{2 \times 0.059 \text{ N/m}}{10^{-3} \text{ m}}$$

$$= p_o + 0.118 \times 10^3 \text{ N/m}^2$$
$$= p_o + 0.00117 \text{ atm}$$

Larger bubbles have a still smaller inside pressure, so that the vapor pressure inside the large bubbles formed when a liquid boils vigorously is negligibly greater than atmospheric pressure. On the other hand, since the first bubbles to form in a liquid when it reaches its boiling point are very minute, the pressure inside them is very large. For instance, the pressure inside a bubble with a radius of $10^{-5}$ m is $p_o + 0.117$ atm. Normally, these first bubbles form on dust particles in the liquid, which enables them to start with a radius large enough to require only a small excess pressure. However, it is possible to heat very clean water (so-called denucleated water) above 100°C without boiling. Such water is said to be *superheated*. Its vapor pressure is large enough to form moderate sized bubbles but not large enough to start a minute bubble. Superheated water is unstable. That is, if it is agitated slightly to form a few bubbles, the entire liquid will suddenly begin to boil vigorously.

**Surfactants**

When a small quantity of detergent is dissolved in water, a layer of detergent molecules forms along the surface of the liquid. These molecules are not as strongly attracted by the water molecules in the interior as other water molecules are, and so the surface tension of the liquid is reduced. Any substance that reduces the surface tension of a liquid in this way is called a *surfactant*. A reduced surface tension is desirable for cleaning purposes because the liquid breaks more easily into tiny drops and disperses more quickly into the fibers of the material being cleaned. Also, with a surfactant, the surface of the water is attracted more strongly to the surface being cleaned than to the water itself. Consequently the water spreads out along a solid surface rather than beading up on it. For this reason a surfactant is said to make water wetter.

The body uses a surfactant to reduce the surface tension in the mucous lining of the pulmonary alveoli, the tiny sacs (about $10^{-2}$ cm in radius) at the end of the bronchial tubes in the lungs (Fig. 9.9). During normal inhalation the pressure in the alveoli is approximately 3 mm Hg below atmospheric pressure (a gauge pressure $\bar{p}_i$ of $-3$ mm Hg), which enables air to flow into them through the bronchial tubes. In these sacs, which are perfused by capillaries containing pulmonary arterial blood, the oxygen in the air is exchanged with the carbon dioxide in the blood.

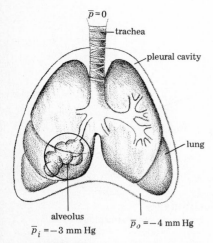

**FIGURE 9.9**
Alveoli are microscopic sacs at the termination of the bronchial tubes of the lung.

During inhalation the radius of the alveoli expands from about $0.5 \times 10^{-4}$ to $1.0 \times 10^{-4}$ m. They are lined with mucous tissue fluid, which normally has a surface tension of 0.050 N/m. With this surface tension, the pressure difference required to inflate an alveolus would be

$$p_i - p_o = \bar{p}_i - \bar{p}_o = \frac{2\gamma}{r}$$

$$= \frac{2 \times 0.050 \text{ N/m}}{0.5 \times 10^{-4} \text{ m}}$$

$$= 2 \times 10^3 \text{ N/m}^2 = 15 \text{ mm Hg}$$

This means that the gauge pressure $\bar{p}_o$ outside the alveolus would have to be 15 mm Hg less than the pressure $\bar{p}_i = -3$ mm Hg inside it. Thus $\bar{p}_o$ would be $-18$ mm Hg.

The outside pressure in this case is the pressure in the space between the lungs and the pleural cavity that holds the lungs (Fig. 9.9). The gauge pressure in this space is in fact negative. (It is this negative pressure that holds the lungs against the walls of the cavity.) However, this pressure is only about $-4$ mm Hg, so that the actual pressure difference $\bar{p}_i - \bar{p}_o$ is only about 1 mm Hg, or 15 times less than would be required to expand an alveolus with a surface tension of 0.050 N/m.

To overcome this difficulty, the walls of the alveoli secrete a surfactant that reduces the surface tension by a factor of 15. There appears to be a fixed amount of this surfactant in each alveolus, and its ability to reduce surface tension depends on its concentration. Therefore, when the alveolus is deflated, the concentration of the surfactant (per unit area) is high and the surface tension is very low, so that the alveolus is expanded without difficulty. However, as it expands, the concentration of the surfactant decreases and the surface tension increases, until a point of equilibrium is reached at maximum expansion. Upon exhaling, the increased surface tension helps to collapse the alveolus and expel the air from it.

At birth the alveoli of a baby are so completely collapsed that a pressure difference of as much as 30 mm Hg is required to expand them the first time. Thus the first breath of life requires extraordinary effort to overcome the surface tension in the alveoli.

## 9.3 CAPILLARY ACTION

When a drop of liquid is placed on a solid surface, the liquid may either spread out or bead up, depending on the relative magnitudes of the *adhesive* and *cohesive* forces. The cohesive force is the force that molecules of the liquid exert on each other; it is the force that holds the liquid together. The adhesive force is the force that the surface or any outside object exerts on the molecules of the liquid. A drop of water placed on clean glass spreads out over the glass because the attraction of the water to the glass (the adhesive force) is greater than the attraction of the water to itself (the cohesive force). At equilibrium, the surface of the water is perpendicular to the total force on it. This is illustrated in Fig. 9.10, which shows the

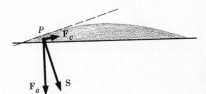

FIGURE 9.10
Adhesive and cohesive forces on a molecule at the edge of a drop of water on glass.

adhesive force $\mathbf{F}_a$ and the cohesive force $\mathbf{F}_c$ on a molecule at point $P$. The surface is perpendicular to the total force $\mathbf{S} = \mathbf{F}_c + \mathbf{F}_a$ on the molecule. Because the cohesive force of mercury is much greater than the adhesive force of glass, a drop of mercury placed on glass beads up. This is illustrated in Fig. 9.11.

In the same way, the surface of a column of liquid in a glass tube is determined by the sum $\mathbf{S}$ of the adhesive and cohesive forces. Figure 9.12$a$ shows the force $\mathbf{S}$ on a water molecule at the surface of such a column. Because the adhesive force $\mathbf{F}_a$ is greater than the cohesive force $\mathbf{F}_c$, the force $\mathbf{S}$ is directed away from the liquid. The surface of the liquid takes on a curved shape, or *meniscus*, which is concave in order to be perpendicular to $\mathbf{S}$. The angle of contact $\theta$ between the surface of the water and the glass tube is determined by the relative magnitudes of $\mathbf{F}_a$ and $\mathbf{F}_c$. Figure 9.12$b$ shows that the meniscus of mercury is convex because the cohesive force is greater than the adhesive force. The angle of contact in this case is greater than 90°.

A column of liquid can be supported in a narrow glass tube (capillary) by the adhesive and cohesive forces acting on it. This phenomenon is called *capillary action*. The capillary must have been previously immersed in the liquid so that there is a thin layer of liquid adhering to the inside wall of the tube just above the column of liquid. This layer exerts surface tension around the circumference of the liquid column. Figure 9.13 shows the forces $\mathbf{F}_1$ and $\mathbf{F}_2$ that act on sections of length $s$ on opposite sides of the capillary. Each force has the magnitude $\gamma s$ and is directed at the contact angle $\theta$ to the capillary. The horizontal components, which are in opposite directions, cancel each other, but the vertical components, which are in the same direction, add. The magnitude of the vertical force per unit length around the circumference is $\gamma \cos \theta$, so that the total vertical force is $2\pi r \gamma \cos \theta$, where $r$ is the radius of the capillary. (This analysis is similar to the analysis of the force exerted on an insect's leg in Sec. 9.2.)

At equilibrium, the upward force $2\pi r \gamma \cos \theta$ due to surface tension is equal to the downward force $mg$ due to gravity:

$$2\pi r \gamma \cos \theta = mg \qquad 9.6$$

The mass $m$ of the column is equal to $\rho V$, where $\rho$ is the density of the liquid and $V$ is the volume of the column. The volume of a column of radius $r$ and height $h$ is

$$V = \pi r^2 h$$

so

$$m = \rho V = \rho \pi r^2 h$$

Putting this expression for $m$ into Eq. 9.6, we get

$$2\pi r \gamma \cos \theta = \rho \pi r^2 h g$$

or

$$h = \frac{2\gamma \cos \theta}{\rho g r} \qquad 9.7$$

This equation gives the height $h$ of a column of liquid supported by capillary action. It is evident that $h$ is proportional to the surface

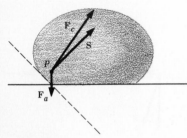

**FIGURE 9.11**
Adhesive and cohesive forces on a molecule at the edge of a drop of mercury on glass.

tension of the liquid and inversely proportional to the radius of the capillary.

**REMARK** Surface tension is caused by the cohesive forces of the liquid (Sec. 9.2). From Eq. 9.7 it might appear that adhesion plays no role in capillary action. However, this is not correct. It is the concave shape of the meniscus that causes the surface tension to exert an upward force on the column, and the shape of the meniscus depends, as we have seen, on the sum of the cohesive and adhesive forces.

If the force of adhesion is much greater than the force of cohesion, the contact angle $\theta$ is very small. In this case $\cos \theta$ is approximately 1, and Eq. 9.7 can be replaced by

$$h = \frac{2\gamma}{\rho g r} \qquad 9.8$$

This equation is valid for water in a glass capillary, where $\theta$ is 20° or less, but Eq. 9.7 must be used for water in a plastic capillary, where $\theta$ may be 45° or more.

For example, suppose one end of a glass capillary of radius $r = 0.05$ cm is dipped in water. If the capillary is dry, the water will not rise in the capillary above the surrounding water. However, if the inside of the capillary is wet, the water will rise to the height $h$ given by Eq. 9.8:

$$h = \frac{2\gamma}{\rho g r} = \frac{2 \times 0.073 \text{ N/m}}{(10^3 \text{ kg/m}^3)(9.8 \text{ m/s}^2)(5 \times 10^{-4} \text{ m})}$$
$$= 2.98 \times 10^{-2} \text{ m} = 2.98 \text{ cm}$$

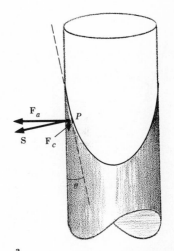

a

When the cohesive force is greater than the adhesive force, as with mercury in a glass tube, the meniscus is convex (Fig. 9.12*b*). The surface tension then exerts a downward force on the liquid so that the liquid in the tube is depressed a distance $h$ below the surface of the surrounding water (Fig. 9.14). When the angle of contact is close to 180°, $\cos \theta$ is approximately −1 and Eq. 9.7 becomes

$$h = -\frac{2\gamma}{\rho g r}$$

For example, if the glass tube in the last example is dipped in mercury, the mercury in the tube is depressed the distance

$$h = \frac{-2 \times 0.435 \text{ N/m}}{(13.6 \times 10^3 \text{ kg/m}^3)(9.8 \text{ m/s}^2)(5 \times 10^{-4} \text{ m})}$$
$$= -1.30 \text{ cm}$$

below the surface of the surrounding liquid.

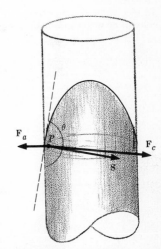

b

FIGURE 9.12
Meniscus of liquids in capillary tubes. (*a*) When the adhesive force is greater than the cohesive force, the meniscus is concave. (*b*) When the adhesive force is less than the cohesive force, the meniscus is convex.

## 9.4 OSMOSIS

When a substance such as sugar is dissolved in a liquid such as water, the molecules of the substance (the *solute*) disperse uniformly throughout the liquid (the *solvent*). The resulting mixture is called a *solution*. The molar concentration $c$ of a solution is the number of moles $n$ of solute per unit volume of solvent. For instance, suppose

2 g of sucrose, $C_{12}H_{22}O_{11}$, is dissolved in 100 cm³ of water. The molecular mass of sucrose is 342 u, so 2 g is

$$n = \tfrac{2}{342} = 5.85 \times 10^{-3} \text{ mol}$$

The concentration of the solution is

$$c = \frac{n}{V} = \frac{5.85 \times 10^{-3}}{100 \text{ cm}^3}$$

$$= 5.85 \times 10^{-5} \text{ mol/cm}^3 = 5.85 \times 10^{-2} \text{ mol/l} \qquad 9.9$$

The molecules of the solute move about randomly in the solvent, much like the molecules of a gas. Because of this motion, the solute distributes itself uniformly throughout the solvent, just as a gas fills the volume available to it. Furthermore, if the volume of the solvent is increased, the solute molecules will "expand" into the new volume, just as a gas expands to fill an increase in its volume.

This is illustrated in Fig. 9.15, which shows a sugar-water solution separated by a partition from an equal amount of pure water. The sugar molecules in the left-hand compartment are constantly colliding with the partition, so as soon as the partition is removed (Fig. 9.15b), sugar molecules begin to diffuse into the right-hand compartment. As the concentration in the right-hand compartment increases, some sugar molecules will start to diffuse back into the left-hand compartment (Fig. 9.15c). However, the number of molecules that flow from right to left will be less than the number that flow from left to right as long as the concentration on the right is less than the concentration on the left. Thus there is a net flow of molecules from left to right until the concentration is uniform throughout the solvent (Fig. 9.15d), which is the equilibrium condition. At equilibrium, as many molecules flow from right to left as flow from left to right, so there is no net flow of molecules in the solution.

Suppose now that the sugar solution is separated from pure water by a cellophane or animal membrane (Fig. 9.16). These membranes have the property of permitting water molecules to slowly diffuse through them but preventing the larger sugar molecules from doing so. Membranes permeable to some molecules and impermeable to others are called *semipermeable membranes*. The wall of a living cell is semipermeable, allowing some substances to diffuse across it into and out of the cell while preventing other substances from crossing it. The semipermeable nature of the cell wall is essential for the maintenance of the proper chemical concentrations in the cell.

The water molecules, shown as dots in Fig. 9.16, are able to diffuse back and forth across the semipermeable membrane, while the sugar molecules, shown as circles, cannot. However, the number of water molecules in the sugar solution that collide with the membrane on the left is less than the number of water molecules in the pure water that collide with the membrane on the right. This is because the sugar molecules occupy space along the membrane and thus prevent some water molecules from reaching the membrane. As a consequence, more molecules diffuse from the pure water into the sugar solution than diffuse from the solution into the pure water. Thus there is a net flow of water into the sugar solution, and so the level of liquid rises in the left-hand compartment and falls in the right-hand com-

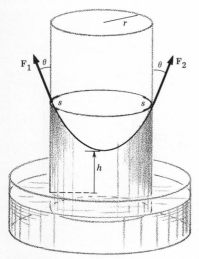

FIGURE 9.13
Surface tension supporting a column of water in a capillary.

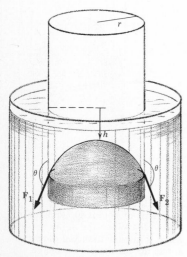

FIGURE 9.14
Surface tension depressing a column of mercury in a capillary.

partment (Fig. 9.16b). The diffusion of water through a semipermeable membrane from a region of low concentration to a region of high concentration is called *osmosis*.

**REMARK** Diffusion of water through a semipermeable membrane (osmosis) looks very different from the free diffusion of solute molecules shown in Fig. 9.15. However, both processes operate on the same general principle. In both cases the net diffusion is in the direction that tends to equalize the concentration in the two compartments. In free diffusion, sugar molecules diffuse into the pure water, increasing the sugar concentration in the right-hand compartment and decreasing the concentration in the left-hand compartment. In osmosis, sugar molecules are prevented from diffusing, so water molecules diffuse into the sugar solution, decreasing the sugar concentration in the left-hand compartment. Free diffusion quickly reaches equilibrium because the concentrations in the two compartments soon become equal. Equilibrium can also be achieved in osmosis if there is initially a small sugar concentration in the right-hand compartment (Fig. 9.17). As water flows from right to left, the solution in the left-hand compartment becomes less concentrated while the solution in the right-hand compartment becomes more concentrated. At equilibrium, when the concentrations in the two compartments are equal, osmosis ceases. However, if initially there is pure water in the right-hand compartment, the concentrations in the two compartments can never equalize, and so all the water eventually diffuses from the right-hand compartment into the left-hand one.

## Osmotic pressure

In Fig. 9.18 a piston has been fitted onto the left-hand compartment. By applying a force to this piston, the pressure $p$ of the sugar solution can be varied. It is found that as $p$ is increased, the rate of flow of water into the sugar solution is decreased. At a certain pressure $p_{os}$ the flow ceases altogether, and as the pressure is increased beyond $p_{os}$, reverse osmosis takes place: water is forced out of the sugar solution and diffuses back into the right-hand compartment. The pressure $p_{os}$ at which no osmosis occurs is called the *osmotic pressure* of the solution. It is a measure of the tendency for water to diffuse into the sugar solution.

The sugar molecules in Fig. 9.18 are trapped between the piston and the membrane, whereas the water molecules move freely through the membrane. The sugar molecules act like an enclosed gas, and the osmotic pressure can be thought of as the pressure $p_s$ of this solute "gas." If the solution is dilute, the solute pressure is given by the ideal-gas law (Eq. 8.9):

$$p_s = \frac{nRT}{V}$$

Here $V$ is the volume occupied by the solution, and $n$ is the number of moles of solute in the solution. From Eq. 9.9 this can be written

$$p_s = cRT$$

It is the pressure of the solute "gas" that resists the movement of the piston. When the pressure $p$ exerted by the piston is less than the solute pressure $p_s$, the solute "expands": water diffuses into the left-hand compartment, increasing the volume of fluid in this com-

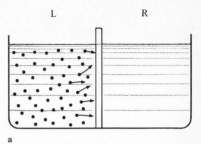

a

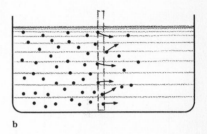

b

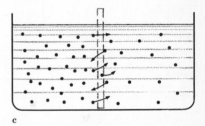

c

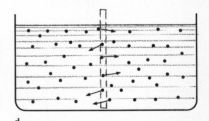

d

**FIGURE 9.15**
(a) A sugar solution separated by a partition from pure water. (b) When the partition is removed, some sugar molecules diffuse into the right-hand compartment. (c) As the concentration of sugar increases in the right-hand compartment, some sugar molecules diffuse back into the left-hand compartment. (d) At equilibrium the concentration is the same in both compartments.

partment and diluting the solution. When the pressure $p$ is greater than $p_s$, the solute is "compressed": water diffuses out of the left-hand compartment, decreasing the volume of the fluid in this compartment and concentrating the solution. When $p$ equals $p_s$, there is no flow of water in or out of the solution. Thus $p_s$ is equal to the osmotic pressure of the solution:

$$p_{os} = p_s = cRT \qquad 9.10$$

**REMARK** Often the water in the right-hand compartment is at atmospheric pressure $p_0$. To prevent osmosis in this case, the absolute pressure of the solution must be $p_0 + p_{os}$. That is, osmosis ceases when the pressure of the solution is greater than the pressure of the water by the osmotic pressure $p_{os}$.

### Osmolality

The osmotic pressure of a liquid with more than one substance dissolved in it is also given by Eq. 9.10. However, in this case, $c$ must be the total concentration of all nonpermeating particles in the solution. This concentration is called the *osmolality* of the solution. In calculating it, all nonpermeating particles contribute equally, whether they are large protein molecules or small $Na^+$ ions. In fact, a molecule, like NaCl, that dissociates into separate ions ($Na^+$ and $Cl^-$) when dissolved in water, contributes two particles to the osmolality. Osmolality is usually measured in osmoles per liter, an *osmole* being 1 mole of nonpermeating molecules and ions. The term osmole is used to emphasize that only nondiffusible particles are counted.

For example, the osmolality of the fluid inside a cell is 0.30 osmol/l. This means that in each liter of intracellular fluid there are

$$0.30N = 1.8 \times 10^{23}$$

nonpermeating molecules and ions, where $N = 6 \times 10^{23}$ is Avogadro's number. At body temperature (37°C) the osmotic pressure of this fluid is

$$p_{os} = (0.30 \text{ osmol/l})(0.082 \text{ atm-l/K})(310 \text{ K})$$
$$= 7.9 \text{ atm} = 6000 \text{ mm Hg}$$

The actual pressure inside a cell depends on the osmolality of the fluid surrounding the cell and on the rigidity of the cell wall.

When a cell with a rigid cell wall, e.g., a red blood cell, is placed in pure water, water tries to diffuse into the cell; but because the cell cannot expand, very little water enters. Instead, the internal fluid pressure increases until either it reaches 7.9 atm above the external pressure or the cell bursts. Red blood cells normally burst open when placed in pure water.

Likewise, when a cell with a nonrigid cell wall is placed in a solution with an osmolality somewhat less than 0.30 osmol/l, water diffuses into the cell; but because the cell wall is nonrigid, the internal fluid pressure does not increase. Instead, the cell expands as water diffuses into it. This decreases the concentration of the intracellular fluid, and osmosis stops when the osmolality of the intracellular fluid becomes equal to the osmolality of the extracellular fluid.

Similarly, when a cell is placed in a solution with an osmolality greater than 0.30 osmol/l, water diffuses out of the cell, which in-

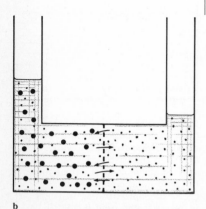

**FIGURE 9.16**
(a) A sugar solution separated by a semipermeable membrane from pure water. (b) Water diffuses into the sugar solution.

creases the osmolality of the intracellular fluid. The cell shrinks until the osmolalities of the two fluids are equal.

Thus osmosis keeps the osmolalities of the intracellular and extracellular fluids of the body equal. If the osmolality of the extracellular fluid suddenly decreases, water diffuses into the cells until equality is reached. If the osmolality of the extracellular fluid suddenly increases, water diffuses out of the cells.

An *isotonic* solution is one that has the osmolality of intracellular fluid. A cell placed in an isotonic solution will neither expand or shrink. Fluids administered intravenously to a patient are usually isotonic so that the fluid balance in the body will not be upset. For instance, an isotonic glucose, $C_6H_{12}O_6$, solution used for intravenous feeding must have an osmolality of 0.30 osmol/l. Since the molecular mass of glucose is 180 u, an isotonic solution contains 180 g $\times$ 0.30 = 54 g of glucose per liter of water.

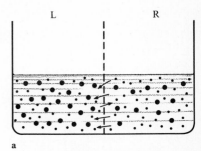

### Reverse osmosis

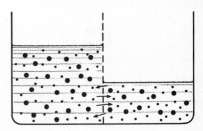

Reverse osmosis may be one practical method for desalinating seawater. The osmolality of seawater is 1.08 osmol/l, so the osmotic pressure of seawater at 20°C is

$$p_{os} = (1.08 \text{ osmol/l})(0.082 \text{ atm-1/K})(293 \text{ K})$$
$$= 25.9 \text{ atm}$$

If seawater on one side of a semipermeable membrane is subjected to a pressure greater than this, pure water will diffuse through the membrane (Fig. 9.19). At present, membranes suitable for desalination are expensive and deteriorate rapidly. When cheaper and more durable membranes are developed, reverse osmosis may be an economical process for large-scale desalination.

**FIGURE 9.17**
(*a*) A concentrated sugar solution separated by a semipermeable membrane from a dilute sugar solution. (*b*) At equilibrium the concentrations are the same in both compartments.

This process is attractive because it requires relatively little energy. If the pressure of the seawater is $p$, the force on a membrane of area $A$ is $F = pA$. Figure 9.19 shows that to purify a volume $V = Ad$ of water, the fluid must be moved the distance $d = V/A$. Thus the work done in desalinating a volume $V$ of water is

$$W = Fd = pA\frac{V}{A} = pV$$

The volume of 1 mol of water is $V = 18 \text{ cm}^3 = 18 \times 10^{-6} \text{ m}^3$. If the pressure is $p = 26 \text{ atm} = 26.3 \times 10^5 \text{ N/m}^2$, the work done is

$$W = pV = (26.3 \times 10^5 \text{ N/m}^2)(18 \times 10^{-6} \text{ m}^3)$$
$$= 47.3 \text{ J}$$

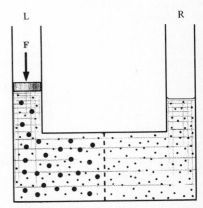

Thus it requires only 47.3 J to desalinate 1 mol of seawater by reverse osmosis. This is to be compared to the energy required to desalinate water by evaporation. From Table 9.1 it is seen that the heat of vaporization of water at 20°C is 10.55 kcal/mol = $44.2 \times 10^3$ J/mol. That is, the energy required to convert water into vapor at 20°C is nearly 1,000 times the energy required to purify water by reverse osmosis.

**FIGURE 9.18**
A piston fitted onto the left-hand compartment to adjust the pressure in this compartment.

**REMARK** Nevertheless, evaporation is a practical alternative to reverse osmosis because heat is an unavoidable by-product of electric power generation. At present this heat is dumped into the environment, creating a thermal-pollution problem. A desalination plant that uses the heat output of a power-generating plant to vaporize seawater is a practical method of solving two problems at one time.

## 9.5 NEGATIVE PRESSURE

Pressure is usually considered a positive quantity, equal to the outward force per unit area that a fluid exerts on its surroundings or to the inward force per unit area that the surroundings exert on the fluid. It is reasonable, therefore, to call the pressure in a fluid negative if the fluid pulls inward on its surroundings and the surroundings pull outward on the fluid. A fluid with negative pressure would be in a state of tension, like a stretched cord.

To understand this better, consider the fluid in the tube in Fig. 9.20 (the tube is in a vacuum to eliminate the effects of atmospheric pressure). If an inward force **F** is applied to the piston in the tube, the fluid will exert an opposite force −**F** outward on the piston. Consequently the piston will be in equilibrium. The positive pressure $p$ of the fluid is $p = F/A$, where $A$ is the area of the piston.

Suppose, instead, that an outward force **F** is applied to the piston, as shown in Fig. 9.20. Initially, the forces exerted on the piston by the fluid and by the applied force **F** are both directed outward, so that the piston accelerates to the right.

If the fluid is a gas, it will expand into the volume that becomes available as the piston moves. The gas pressure will decrease, in accordance with the ideal-gas law, but it will never fall below zero. That is, the gas always exerts a force in the direction of **F,** so that the piston is never in equilibrium. A gas cannot exert a negative pressure.

If the fluid is a liquid, the piston will normally break away from it and move freely to the right while the liquid stays behind at zero pressure. However, if it were possible to prevent the piston from breaking away, the piston would pull on the liquid, trying to increase its volume. Since a liquid has a more or less fixed volume, the liquid would resist expansion by pulling back on the piston. Equilibrium is established when the magnitude of the inward force exerted by the liquid on the piston is equal to the magnitude of the outward force **F**. In this circumstance the liquid is in a state of tension, with a negative pressure $p = -F/A$.

In practice, negative pressure is obtained using a closed-tube manometer, as shown in Fig. 9.21. The vertical tube is sealed at the top and is connected at the bottom through a reservoir to a vacuum pump and pressure gauge. Initially the reservoir is open to the atmosphere, and the liquid is held in the vertical tube by the atmospheric pressure $p_A$ at $A$. The pressure $p_B$ at the top of the liquid (point $B$) is given by Eq. 7.6:

$$p_B = p_A - \rho g h \qquad 9.11$$

where $h$ is the vertical height of $B$ above $A$ and $\rho$ is the density of the fluid.

Normally, when the pressure $p_A$ in the reservoir is decreased by

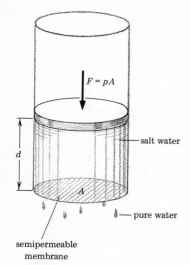

**FIGURE 9.19**
Reverse osmosis. When the pressure of the salt water exceeds its osmotic pressure, pure water diffuses through the semipermeable membrane.

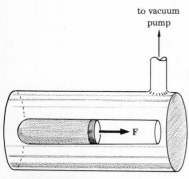

**FIGURE 9.20**
A fluid subjected to a tensive force.

means of the vacuum pump, the liquid falls out of the tube. However, if the liquid and the tube are exceptionally clean and free of impurities,† it is possible to reduce $p_A$ to zero without the liquid's falling. In this case Eq. 9.11 shows that the pressure at $B$ is negative. For example, it is possible to maintain a column of mercury 2.5 m high in this way. This corresponds to a negative pressure at $B$ of $-2500$ mm Hg, or $-3.3$ atm.

With refined techniques, pressures of nearly $-300$ atm have been obtained in water. The maximum possible negative pressure is not known, but presumably at some value the liquid will pull apart, just like a wire that breaks when its tension is very large.

**REMARK** A liquid under negative pressure is very unstable. It is like a pencil balanced on its point. At the slightest disturbance, the liquid shatters into tiny drops which fall out of the tube.

### Water transportation in trees

Evidence now definitely supports the theory that water is raised up into trees by negative pressure. Thus, while specialized techniques are required to produce negative pressure in the laboratory, it appears that nature routinely produces such pressures with ease.

Water is conducted upward in plants through the *xylem*, a system of capillaries formed from dead cells which have lost their cytoplasm. These capillaries are from 0.05 to 0.50 mm in diameter and may rise to a height of 75 m or more. How is the water raised to these incredible heights?

The method of raising water from a well is to reduce the pressure in the pipe at the top of the well, thus allowing the pressure of the atmosphere to force water up the pipe. But the maximum height water can be lifted in this way is only about 10 m, the height that a column of water can be supported by a pressure of 1 atm. This method is not sufficient to raise the water in a tall tree.

It might be thought that the water is raised by means of capillary action. However, from Eq. 9.8 the height $h$ that water rises in a capillary of radius $r = 0.02$ mm $= 2 \times 10^{-5}$ m, is found to be

$$h = \frac{2\gamma}{\rho g r} = \frac{2 \times 0.0727 \text{ N/m}}{(10^3 \text{ kg/m}^3)(9.8 \text{ m/s}^2)(2 \times 10^{-5} \text{ m})}$$

$$= 0.74 \text{ m}$$

which is far short of the needed height.

Osmosis is another mechanism to be considered. It is effective when the concentration of substances dissolved in the sap is greater than that of the groundwater. This occurs in maple trees in early spring, when the sap first starts to rise. At that time, the roots have a large concentration of sugar produced during the preceding summer. As the snow melts, the groundwater diffuses into the cells of the roots, forcing the sap up the tree. The maximum height $h$ that the sap can be lifted this way is related to the osmotic pressure by $p_{os} = \rho g h$. To get the sap up a maple tree, for instance, the osmotic pressure must be at least

† Special techniques are required to achieve the necessary purity. For instance, the liquid is boiled under vacuum to remove all the gas dissolved in it.

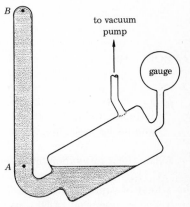

FIGURE 9.21
A closed-tube manometer connected to a vacuum pump and a pressure gauge.

$$p_{os} = \rho g h = (10^3 \text{ kg/m}^3)(9.8 \text{ m/s}^2)(30 \text{ m})$$
$$= 2.94 \times 10^5 \text{ N/m}^2 = 2.9 \text{ atm}$$

From Eq. 9.10, this requires a molar concentration of at least

$$c = \frac{p_{os}}{RT} = \frac{2.9 \text{ atm}}{0.082 \text{ atm-1/K} \times 293 \text{ K}}$$
$$= 0.12 \text{ osmol/l}$$

Since the concentration of sugar in the sap is greater than this in early spring, osmotic pressure forces the sap up the xylem in this case. In summer, however, the concentration of sugar in the sap is too low for osmotic pressure to force the sap up the tree.

Furthermore, water can rise in trees against an osmotic pressure barrier. This occurs most spectacularly in mangroves, which grow in seawater. The sap inside the xylem is essentially pure water, while the roots are in salt water. Thus an osmotic pressure of 25.9 atm (Sec. 9.4) tends to force pure water out of the trees into the seawater. In fact, however, reverse omsosis takes place, and pure water diffuses from the sea into the mangrove. This requires that the pressure in the sea be at least 25.9 atm larger than the pressure in the tree. But the absolute pressure of the seawater is 1 atm. Therefore, the absolute pressure inside the xylem of the tree must be $-24.9$ atm or less.

It is generally believed that the pressure of the water in the xylem of most trees and plants is negative; i.e., the water in the xylem is in tension. It is pulled up the plant like a rope raised from above. As water molecules evaporate from the surface of the leaves, neighboring molecules diffuse to the surface to replace them. These molecules are replaced in turn by molecules further down the xylem. The long unbroken column of water in the xylem is held together by the cohesive forces between the molecules of the water.

Direct measurements of the pressure in the xylem of tall trees [Scholander et al. (1965)] show that the pressure is most negative at the top of a tree and increases, i.e., becomes less negative, with distance below the top, in accordance with Fluid Property 3 (Sec. 7.3, Eq. 7.6). For instance, if the pressure is $-30$ atm at the top of a tree, it is $-27$ atm at a point 30 m below the top.

**PROBLEMS**

*1* The heat of vaporization of tetrachloromethane, $CCl_4$, is 0.05087 kcal/g at 20°C. What is its molar heat of vaporization?
*Ans.* 7.82 kcal/mol

*2* The heat of vaporization of diethyl ether, $C_4H_{10}O$, is 6.72 kcal/mol at 30°C. What is its heat of vaporization in kilocalories per gram?

*3* The evaporation of 50 g of trichloromethane (chloroform), $CHCl_3$, from the inner vessel of a calorimeter lowers the temperature of the 1.6 kg of water surrounding the vessel by 1.9°C. What is the heat of vaporization of trichloromethane? *Ans.* 0.0608 kcal/g

*4* By how many degrees does the evaporation of 65 g of chloroethane, $C_2H_5Cl$, from the inner vessel of a calorimeter lower the temperature of the 2.3 kg of water surrounding the vessel?

*5* The human body dissipates approximately 2500 kcal of heat per day. If all this heat were lost by evaporating water from the skin, how much water would be evaporated per day?
*Ans.* 4.27 kg

*6* How much energy is required to transform 1 l of water at 100°C into vapor (use Table 7.2)?

*7* (a) How much energy $w$ is required to remove

one molecule from water at 20°C? (*b*) If the water molecules have the same energy distribution as the molecules in an ideal gas (Table 8.2), approximately what fraction of the molecules of water at 20°C have sufficient energy to escape?      *Ans.* (*a*) $7.33 \times 10^{-20}$ J; (*b*) $10^{-7}$

8  Plot the surface tension of water against temperature. By fitting a straight line through these points, estimate the temperature at which the surface tension is zero.

**REMARK**  The surface tension of a liquid is zero at the critical temperature (Sec. 8.5), where the distinction between liquid and gas vanishes. Compare your estimate of the critical temperature from Prob. 8 with the value given in Table 8.3  A considerable discrepancy is to be expected because the surface tension does not vary linearly with temperature over such a wide range of temperature.

9  Figure 9.22 shows a common method for

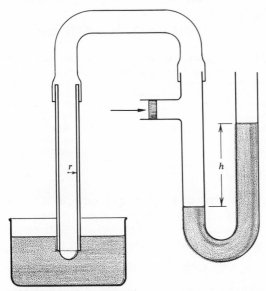

FIGURE 9.22   Problem 9.

measuring the surface tension of a liquid. A capillary of radius $r$ is dipped into the liquid, and the pressure in the capillary is increased until a bubble is formed. The maximum pressure occurs just before the bubble breaks free from the capillary. At this stage the bubble is a hemisphere, as shown in the figure. (*a*) Show that at this stage the gauge pressure $\bar{p}$ in the capillary is

$$\bar{p} = \frac{2\gamma}{r}$$

(*b*) A capillary of radius 0.2 mm is dipped into tetrachloromethane, and bubbles are formed when the height difference in the two arms of the water-filled manometer is 2.75 cm. What is the surface tension of tetrachloromethane? (*c*) What is the height difference in the arms of the manometer when the capillary forms a bubble in water at 20°C?
      *Ans.* (*b*) 0.0265 N/m; (*c*) 7.42 cm

10  Two soap bubbles of different radii $r_1$ and $r_2$ are formed on opposite ends of a glass tube closed in the middle by a stopcock (Fig. 9.23).

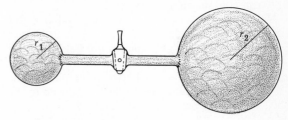

FIGURE 9.23   Problem 10.

What happens to the size of the bubbles when the stopcock is opened so that air passes freely between the bubbles?

11  (*a*) What is the (gauge) pressure inside a soap bubble of radius 2 cm made from a solution whose surface tension is 0.06 N/m? (*b*) If the surface tension of the solution is measured by the method described in Prob. 9, what is the maximum pressure in a capillary of radius 0.02 cm? [*Hint:* The bubble in (*a*) is a two-sided bubble, so Eq. 9.4 applies, whereas the bubble in (*b*) is a one-sided bubble, so Eq. 9.5 applies.]      *Ans.* (*a*) 12 N/m²; (*b*) 600 N/m²

12  What is the pressure inside an alveolus inflated to a radius of 0.08 mm if the surface tension of the fluid lining it is 0.04 N/m?

13  (*a*) The leg of an insect standing on water makes a depression (Fig. 9.7) of radius $r = 2$ mm and angle $\theta = 40°$. How much weight is this depression supporting? (*b*) What is the mass of the insect, assuming it is being supported equally on six legs?
      *Ans.* (*a*) $7.0 \times 10^{-4}$ N; (*b*) 0.43 g

14  A water spider of mass 2 g is supported on the surface of water. Assuming each leg sup-

ports one-eighth the spider's weight, what is the radius of the depression made by each leg (Fig. 9.7)? Take $\theta$ to be 45°.

15 (a) To what height $h$ will ethanol rise in a capillary of radius 0.5 mm if the contact angle is zero? (b) In an experiment with a capillary of a certain material, the alcohol is found to rise to a height of 1.09 cm. What is the contact angle between alcohol and the capillary material?          *Ans.* (a) 1.17 cm; (b) 20°

16 Trichloromethane rises to a height of 2.48 cm in a capillary of radius 0.15 mm. What is the surface tension of trichloromethane assuming the contact angle is zero?

17 Water rises 5.0 cm in a capillary. What is the radius of the capillary?          *Ans.* 0.296 mm

18 When two wet glass plates are held together and dipped into water, the water rises to a height $h$ in the space between the plates (Fig. 9.24). Show that the height $h$ is related to the distance $d$ between the plates by

$$h = \frac{2\gamma \cos \theta}{d\rho g}$$

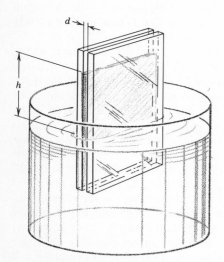

FIGURE 9.24   Problems 18 and 19.

19 If water rises 9.5 cm in the space between two glass plates dipped in water, what is the distance $d$ between the plates? Take $\theta$ to be zero (see Prob. 18).          *Ans.* 0.156 mm

20 A capillary blood vessel has a radius of $2 \times 10^{-6}$ m. How high can blood rise in such a vessel if the contact angle is zero?

21 (a) The molecular mass of sugar (sucrose) is 342.3 u. What is the osmolality of a 1% sugar solution (1 g of sugar dissolved in 100 g of water)? (b) What is the osmolality of a 1% salt (NaCl) solution, assuming that every NaCl molecule dissociates in solution into $Na^+$ and $Cl^-$ ions?

*Ans.* (a) 0.029 osmol/l; 0.34 osmol/l

22 The molecular mass of urea is 60 u. What is the concentration (in grams per liter) of a 0.20 osmol/l solution of urea in water?

23 The walls of blood capillaries are permeable to most small molecules but impermeable to proteins. Table 9.2 gives the concentrations (in grams per liter) and average molecular masses of the major protein groups found in blood plasma. Calculate the osmolality of each protein group and the total osmolality of plasma.

*Ans.* Total osmolality = 0.837 mosmol/l

TABLE 9.2 **Concentration and average molecular mass of the major protein groups in blood plasma**

| Protein group | Concentration, g/l | Molecular mass |
|---|---|---|
| Albumin | 45 | 69,000 |
| Globulin | 25 | 140,000 |
| Fibrinogen | 3 | 400,000 |

24 Show that at body temperature (37°C), the osmotic pressure (in millimeters of mercury) of a solution is related to the osmolality of the solution (in milliosmoles per liter) by

$$P_{os} \text{ (mm Hg)} = 19.3c \text{ (mosmol/l)}$$

25 What is the osmotic pressure of blood plasma caused by the proteins dissolved in it (see Probs. 23 and 24)?          *Ans.* 16.2 mm Hg

**REMARK** The normal osmotic pressure of blood plasma is 28 mm Hg. The difference between this pressure and the 16.2 mm Hg pressure due to dissolved proteins is caused by the positive ions in the plasma. The protein molecules, which are negatively charged, attract positively charged ions and prevent them from diffusing through the capillary wall. These positive ions contribute to the osmotic pressure even though the capillary wall is permeable to them. This phenomenon is known as the *Donnan equilibrium.*

26 The kidneys remove about 180 l of fluid per

day from the blood (99 percent of this fluid is returned to the blood, and the other 1 percent is secreted as urine). The removal of fluid from the plasma is reverse osmosis, because the plasma has an osmotic pressure of 28 mm Hg (see Prob. 25 and the remark following it). How much work must the kidneys do each day in filtering fluid from the blood?

## BIBLIOGRAPHY

GUYTON, ARTHUR C.: "Textbook of Medical Physiology," W. B. Saunders Company, Philadelphia, 1971. Chapter 33 contains a thorough discussion of the role of osmosis in maintaining balance between the intracellular and extracellular fluids of the body.

HAYWARD, ALAN T.: Negative Pressure in Liquids: Can It Be Harnessed to Serve Man?, *American Scientist*, **59:** 434 (1971). Semitechnical account of negative pressure and its present and future industrial applications.

PROBSTEIN, RONALD F.: Desalination, *American Scientist*, **61:**280 (1973). Survey of the various methods used for desalination and their relative economic merits. The requirements for an efficient semipermeable membrane for reverse osmosis are discussed.

SCHOLANDER, P. F., H. T. HAMMEL, E. D. BRADSTREET, and E. A. HEMMINGSEN: Sap Pressure in Vascular Plants, *Science*, **148:** 339 (1965). Report of an experiment that measured the pressure in the xylem of tall trees.

SCHOLANDER, P. F.: Tensile Water, *American Scientist*, **60:** 584 (1972). Semitechnical discussion of the relation between osmotic pressure and negative pressure.

ZIMMERMANN, MARTIN H.: How Sap Moves in Trees, *Scientific American*, **208:**132 (March 1963). A description of the upward transport of water in the xylem and the downward transport of the products of photosynthesis in the phloem.

# CHAPTER 10 solids

A solid is a rigid object that tends to maintain its shape when external forces are applied to it. Because of this rigidity, solid materials are used in building all complex structures that have a fixed shape. The engineer studies the mechanical properties of materials, such as steel and concrete, used in the structures he builds. Similarly, it is of interest to the biologist to know something about the properties of materials, such as wood and bone, that serve as the rigid components of plants and animals. This chapter discusses certain general properties of all solids and some specific properties of biological materials.

## 10.1 CRYSTALLINE SOLIDS

The molecules of a solid, like those of a liquid, are so close together that they exert strong attractive forces on each other. However, in a liquid the molecules are free to move about, whereas in a solid the molecules have fixed positions. It is the fixed positions of the molecules that give a solid its rigidity.

In a *crystalline* solid, furthermore, the molecules are arranged in a regular three-dimensional pattern, which extends for millions of molecules in all directions. Figure 10.1 shows one possible pattern, a simple cubic lattice. If identical molecules occupy every point in this lattice, the molecule at any one lattice point has the same surroundings as the molecule at any other point (except, of course, for molecules located near the surface of the lattice). Thus all the molecules in the interior of a crystalline solid are subject to identical conditions, and so they behave identically. It is the identical behavior of the molecules in a crystalline solid that gives these solids their special characteristics. Most of the chemical elements (including all metals) and most simple organic and inorganic compounds form crystalline solids.

A *crystal* is a crystalline solid in which the ordered arrangement of the molecules persists throughout and controls the external shape of the solid. The faces of a crystal are plane surfaces that meet at definite angles. These angles are characteristic of the molecular order in the solid. Figure 10.2 shows some quartz crystals, which are naturally formed crystals of silicon dioxide, $SiO_2$. Figure 10.3 is a photomicrograph of a snowflake, which is a complex crystal of $H_2O$. The regularity and symmetry of a crystal is the visible manifestation of the regularity and symmetry of the arrangement of the molecules in the solid.

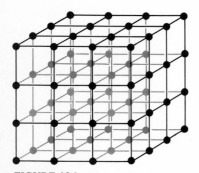

FIGURE 10.1
Molecules arranged in a simple cubic lattice.

More commonly, a solid is formed under conditions that do not favor the formation of a single crystal. Instead, as the solid forms, thousands of minute crystals grow together, forming a multi-crystalline solid. Within each microcrystal the molecules have an ordered arrangement, but the microcrystals are randomly arranged with respect to each other. Metals generally have this kind of structure. Because each microcrystal contains $10^{16}$ or more molecules, the vast majority of molecules are in the interior of a large ordered array of molecules. Only the relatively few molecules on the boundary between two microcrystals are out of order (Fig. 10.4). The term crystalline solid is used to mean either a single crystal or a multi-crystalline solid. This section discusses properties common to all crystalline solids.

**REMARK** In a noncrystalline solid the molecules are randomly arranged, so that the forces on one molecule are different from the forces on another. Glass is perhaps the most common example of a noncrystalline solid. Chemically it is predominantly $SiO_2$, the same as quartz. Its physical properties differ from those of quartz, however, because its molecules are arranged differently.

The most characteristic property of a crystalline solid is the existence of a definite temperature $T_m$, called the *melting point*. When a solid is heated, its temperature $T$ increases until it reaches $T_m$. Thereafter, further heating transforms the solid into a liquid. The molar *heat of fusion* $H_f$ is the energy required to melt one mole of a solid. At the melting point all the heat added to a solid goes into supplying the energy necessary to melt it, so that the temperature does not increase during the melting process. However, after all the solid has melted, the temperature of the liquid is increased by further heating. Table 10.1 gives the melting point and the molar heat of fusion of some common substances.

At the melting point a substance can coexist in the solid and liquid states. For example, a mixture of ice and water at 0°C is stable if no heat is added or removed. However, if heat is added, the ice will

FIGURE 10.2
Quartz crystals.

melt, and if heat is removed, the water will freeze. At temperatures above $T_m$ the substance cannot exist as a solid, and at temperatures below $T_m$ the substance cannot normally exist as a liquid.†

**REMARK**  A noncrystalline solid, such as glass, has no definite melting point. Since the molecules of the solid are randomly arranged, some molecules can break away from their positions more easily than others. As the temperature of glass is increased, the glass becomes softer and softer, until eventually it flows freely, but there is no sharp transition from the solid to the liquid state. Conversely, as molten glass cools, its viscosity increases steadily, until no perceptible flow occurs. Butter and margarine are also examples of noncrystalline solids that have no definite melting point.

The melting point of a crystalline solid depends slightly on pressure. For example, the melting point of $H_2O$, which is 0°C at a pressure of 1 atm, decreases to −1.5°C at a pressure of 200 atm. The conventional melting point $T_m$ is the melting point at a pressure of 1 atm. A plot of melting point against pressure gives the melting curve shown in Fig. 10.5. The curve is nearly vertical because $p$ has to change by many atmospheres to change the melting point by a fraction of a degree. (The melting curve in Fig. 10.5 slopes to the right because the melting point of most substances, unlike that of water, increases with increasing pressure.) A substance is a solid for values of $p$ and $T$ to the left of the melting curve and a liquid for values of $p$ and $T$ to the right of the curve. Points on the melting curve give the values of $p$ and $T$ for which the liquid and solid can coexist.

Similarly a vaporization curve is a plot of the vapor pressure of a liquid (Sec. 8.5) against temperature. As shown in Fig. 10.5, a substance is a liquid for values of $p$ and $T$ above this curve and a

† Under certain circumstances a liquid can be cooled below the melting point without solidifying. This is an unstable situation, however, and the entire liquid immediately solidifies once any part of it starts to solidify.

FIGURE 10.3
A snowflake. [*Vermont Life Magazine.*]

FIGURE 10.4
Only the molecules on the boundary between two microcrystals are out of order.

TABLE 10.1 **Melting point, heat of fusion, and boiling point of some crystalline solids in order of increasing melting point**

| Substance | Melting point, °C | Heat of fusion, kcal/mol | Boiling point, °C |
|---|---|---|---|
| Oxygen | −218.8 | 0.106 | −183 |
| Ethanol | −114.5 | 0.026 | 78.5 |
| Carbon dioxide | −56.6† | 1.90 | † |
| Mercury | −38.9 | 0.56 | 357 |
| Water | 0 | 1.436 | 100 |
| Sodium | 97.8 | 0.63 | 892 |
| Sulfur | 119 | 0.29 | 445 |
| Lead | 327 | 1.22 | 1744 |
| Aluminum | 659 | 2.55 | 2467 |
| Sodium chloride | 800 | 7.22 | 1413 |
| Copper | 1083 | 3.11 | 2595 |
| Silicon dioxide (quartz) | 1470 | 3.40 | 2230 |
| Iron | 1530 | 3.56 | 3000 |
| Tungsten | 3387 | 8.42 | 5927 |

† This is the triple-point temperature. The triple-point pressure of $CO_2$ is 5.11 atm, so that $CO_2$ does not exist as a liquid at atmospheric pressure.

gas for values of $p$ and $T$ below the curve. A point on the vaporization curve gives the pressure $p$ of the vapor that is in equilibrium with the liquid at the temperature $T$. The vaporization curve terminates at the critical temperature $T_c$, which is the highest temperature at which the substance can exist as a liquid (Sec. 8.5). The boiling point $T_b$ of a substance is the temperature at which the vapor pressure is 1 atm. Table 10.1 gives the boiling point of some common substances.

The vaporization curve intersects the melting curve at a point called the *triple point*. The temperature $T_t$ and pressure $p_t$ at this point are characteristic of the substance. For instance, at the triple point of $H_2O$ they are 0.01°C and 0.0060 atm (4.58 mm Hg). At this temperature and vapor pressure, the solid, liquid, and gaseous states of a substance all coexist. For most substances the temperatures $T_m$ and $T_t$ differ by only a few hundredths of a degree, which is inconsequential for most purposes.†

A third curve, called the *sublimation curve*, also intersects the vaporization curve and melting curve at the triple point (Fig. 10.5). This curve gives the vapor pressure of the gas that is in equilibrium with the solid at various temperatures. A substance is a solid for values of $p$ and $T$ above the curve and a gas for values of $p$ and $T$ below the curve. For example, the vapor pressure of ice at $-10°C$ is 1.95 mm Hg. On a cold, dry winter day, when the temperature is $-10°C$ (14°F) and the partial pressure of $H_2O$ in the air is less than 1.95 mm Hg, the values of $p$ and $T$ correspond to a point below

† The only important exception is carbon dioxide; see Table 10.1.

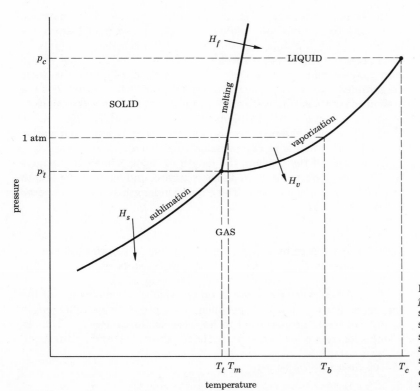

**FIGURE 10.5**
$pT$ diagram of a crystalline substance. The melting curve separates the solid and liquid states, the vaporization curve separates the liquid and gas states, and the sublimation curve separates the gas and solid states.

the sublimation curve. Under these conditions the only stable state of $H_2O$ is the gaseous state. This means that any snow or ice on the ground is not in equilibrium with the vapor but is slowly being transformed directly into vapor, a process called *sublimation*. No liquid $H_2O$ is formed because it does not exist below the triple point.

The energy required to transform a substance directly from the solid to the gaseous state is called the *heat of sublimation* $H_s$. The heat of sublimation of $H_2O$ is 12.1 kcal/mol, so that 12.1 kcal/18 = 0.672 kcal of energy must be absorbed to transform 1 g of ice into vapor. Not all the snow sublimates immediately because it takes time to absorb this energy from the sun and the surrounding air. Eventually, however, all the snow would sublimate if the temperature remained below 0°C and no fresh snow fell.

## 10.2  MECHANICAL PROPERTIES OF SOLIDS

A solid is not absolutely rigid; its size and shape are modified slightly when it is subjected to large forces. These changes are not very noticeable under ordinary circumstances, but they can be measured by instruments designed to test solid materials. The purpose of such tests is to obtain quantitative measurements of the mechanical properties of different materials.

### Young's modulus

In the simplest test, the ends of a cylindrical sample of the material are connected to movable plates. The plates are pulled apart (or pushed together), subjecting the specimen to tension (or compression). Figure 10.6*a* shows a specimen in tension, and Fig. 10.6*b* shows it in compression. In each case the forces $\mathbf{F}_1$ and $\mathbf{F}_2$ have the same magnitude because the object is in equilibrium. The tension $T$ is the magnitude of these forces (Sec. 2.2). Since compression is the opposite of tension, they can be distinguished by the sign of $T$. The tension $T$ is positive when the specimen is in tension (Fig. 10.6*a*) and negative when the specimen is in compression (Fig. 10.6*b*).

As a result of the tension, the length of the specimen is changed. Let $L_0$ be the length when the tension is zero, and let $L$ be the length when the tension is nonzero. The change in length $L - L_0$ is denoted by the symbol $\Delta L$:†

$$\Delta L = L - L_0$$

Materials are tested by measuring the change of length produced by a given tension $T$. The change of length $\Delta L$ is positive when $T$ is positive and negative when $T$ is negative (Fig. 10.6).

The change of length depends on the size and composition of the specimen. For a cylindrical specimen, $\Delta L$ is proportional to its length $L_0$ and inversely proportional to its cross-sectional area $A$. That is, a long specimen is stretched more than a short one, and a thin speci-

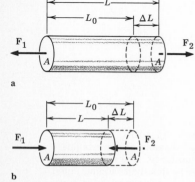

**FIGURE 10.6**
A specimen subjected to (*a*) tensile stress and (*b*) compressive stress.

† $\Delta$ is the Greek letter capital delta.

men is stretched more than a thick one. Thus the effect of size on $\Delta L$ can be written

$$\Delta L = \frac{1}{E} \frac{L_0}{A} T \qquad\qquad 10.1$$

where $E$ is a constant, called *Young's modulus*, that depends only on the composition of the specimen, not on its size. This equation also assumes that $\Delta L$ is proportional to $T$. This is true, as we shall see, provided $T$ is not too large. Table 10.2 gives the Young's modulus of various solids.

It is convenient to rewrite Eq. 10.1 in terms of the normal *strain* $\epsilon$ and the normal *stress* $\sigma$.[†] Strain is the ratio of the change of length to the original length,

$$\epsilon = \frac{\Delta L}{L_0}$$

Stress is the ratio of the tension to the cross-sectional area,

$$\sigma = \frac{T}{A}$$

In terms of these quantities, Eq. 10.1 can be written

$$\epsilon = \frac{\sigma}{E}$$

or $\qquad\qquad\qquad \sigma = E\epsilon \qquad\qquad 10.2$

The mechanical properties of solids are always described in terms of

[†] $\epsilon$ is the Greek letter epsilon, and $\sigma$ is the Greek letter sigma. Normal here means *perpendicular*, rather than *standard* or *customary*. It is used in distinction to shear, which is defined below. Used without qualification, strain and stress mean normal strain and normal stress.

TABLE 10.2 **Young's modulus, elastic limit, and strength of some common solids**
The values listed here are representative of each material. The actual values for a particular specimen can differ greatly from the listed values.

| Substance | Young's modulus, $10^9$ N/m$^2$ | Elastic limit $\sigma_e$, $10^7$ N/m$^2$ | Tensile strength, $10^7$ N/m$^2$ | Compressive strength, $10^7$ N/m$^2$ |
|---|---|---|---|---|
| Aluminum | 70 | 18 | 20 | |
| Bone:[†] | | | | |
|   Tensile | 16 | | 12 | |
|   Compressive | 9 | | | 17 |
| Brick | 20 | | | 4 |
| Copper | 120 | 20 | 40 | |
| Glass, fused quartz | 70 | | 5 | 110 |
| Granite | 50 | | | 20 |
| Iron, wrought | 190 | 17 | 33 | |
| Marble | 60 | | | 20 |
| Polystyrene | 3 | | 5 | 10 |
| Quartz | 70 | | | |
| Steel | 200 | 30 | 50 | |
| Wood | 10 | | | 10 |

[†] For more information on bone, see Table 10.5.

stress and strain, rather than tension and change of length, because the relation between stress and strain is independent of the size of the object.

**REMARK**   Strain is a dimensionless quantity, and stress has the dimension of force per unit area. Therefore, from Eq. 10.2, the dimension of Young's modulus $E$ is also force per unit area.

The stress on the specimen in Fig. 10.6$a$ is equal to the magnitude of the force $\mathbf{F}_1$, divided by the area $A$. This is similar to the definition of pressure (Sec. 7.2) except that pressure is positive when the force is directed inward and stress is positive when the force is directed outward. Thus stress is simply negative pressure: $\sigma = -p$. However, the term pressure is usually reserved for fluids, where the force per unit area is the same on all areas (Pascal's law). The term stress is reserved for solids, where the force per unit area can be different on different areas. For instance, the force per unit area in the solid in Fig. 10.6$a$ is $F_1/A$ on the ends of the cylinder but zero on the sides of the cylinder.

As an example, consider a steel wire of length $L_0 = 0.5$ m and diameter $d = 2 \times 10^{-3}$ m. How much does it stretch when a tension of 450 N is applied to it? The cross-sectional area $A$ of the wire is

$$A = \frac{\pi d^2}{4} = 3.14 \times 10^{-6} \text{ m}^2$$

so the stress is

$$\sigma = \frac{T}{A} = \frac{450 \text{ N}}{3.14 \times 10^{-6} \text{ m}^2} = 143 \times 10^6 \text{ N/m}^2$$

From Table 10.2 the Young's modulus of steel is found to be $200 \times 10^9$ N/m$^2$, so the strain is

$$\epsilon = \frac{\sigma}{E} = \frac{1.43 \times 10^8 \text{ N/m}^2}{2 \times 10^{11} \text{ N/m}^2}$$

$$= 0.715 \times 10^{-3} = \frac{\Delta L}{L_0}$$

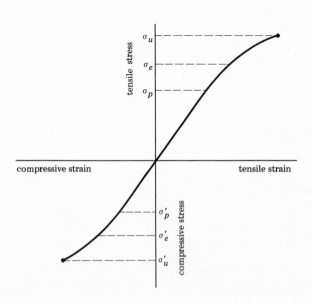

FIGURE 10.7
Typical stress-strain curve for a crystalline solid.

Therefore the change in length is

$$\Delta L = \epsilon L_0 = 0.715 \times 10^{-3} \times 0.5 \text{ m}$$
$$= 0.357 \times 10^{-3} \text{ m} = 0.357 \text{ mm}$$

The tension increases the length of the wire by 0.357 mm. When the tension is removed, the wire returns to its unstretched length.

Figure 10.7 is a plot of stress against strain for a typical crystalline solid. Equation 10.2 is valid only for the straight-line portion of this curve, where the stress is proportional to the strain. The limits $\sigma_p$ and $\sigma'_p$ of this region are different for tensile and compressive stresses. The elastic limits $\sigma_e$ and $\sigma'_e$ are the maximum tensile and compressive stresses that can be applied to the solid such that it returns to its original size when the stress is removed. If the tensile stress exceeds $\sigma_e$ or the compressive stress exceeds $\sigma'_e$, the solid is permanently deformed. This means that it will not return to its unstressed condition when the stress is removed. The solid breaks when the stress exceeds the tensile strength $\sigma_u$ or the compressive strength $\sigma'_u$. Table 10.2 gives the values of $\sigma_e$, $\sigma_u$, and $\sigma'_u$ for various solids.

If the stress on a supporting member of a mechanical structure exceeds the elastic limit, the member is permanently deformed and its physical properties are changed. In general the deformed solid is weaker than the original material. For instance, a sheet of metal can be flexed back and forth indefinitely as long as it is not flexed beyond its elastic limit. But once bent beyond this limit, the sheet is easily broken by a few more twists. Thus the members of any structure, whether animate or inanimate, are designed so that the greatest stress ever applied to them does not exceed their elastic limit.

For example, suppose a crane is required to lift weights of 20,000 lb or less. What is the minimum diameter of the steel cable that can be used? The maximum tension in the cable will be

$$T = 20,000 \text{ lb} = 8.9 \times 10^4 \text{ N}$$

From Table 10.2, the elastic limit of steel is $30 \times 10^7$ N/m². If the stress is not to exceed this, the cross-sectional area of the cable must be greater than

$$A = \frac{T}{\sigma_e} = \frac{8.9 \times 10^4 \text{ N}}{30 \times 10^7 \text{ N/m}^2} = 3.0 \times 10^{-4} \text{ m}^2$$

and the diameter must be greater than

$$d = \sqrt{\frac{4A}{\pi}} = 1.95 \times 10^{-2} \text{ m} = 1.95 \text{ cm}$$

Of course, the diameter would be made larger than this to provide a margin of safety.

The maximum elastic strain that can be produced in an object is found by substituting values of $E$ and $\sigma_e$ from Table 10.2 into Eq. 10.2.† Steel, for instance, can have a maximum strain of

$$\epsilon = \frac{\sigma_e}{E} = \frac{30 \times 10^7 \text{ N/m}^2}{2 \times 10^{11} \text{ N/m}^2}$$
$$= 1.5 \times 10^{-3}$$

† Equation 10.2 is not strictly valid for stresses as large as $\sigma_e$ (Fig. 10.7). However, the stress-strain curve does not deviate much from a straight line at $\sigma_e$, so that Eq. 10.2 is a good approximation for stresses as large as $\sigma_e$.

This means that if a steel rod is stretched by more than 0.15 percent, it will be stretched beyond its elastic limit. Except for elastomers (Sec. 10.3), most solids cannot be elastically deformed by more than a few percent.

**REMARK** A contact force (Sec. 2.2) is the force a solid exerts as a result of its deformation. Because the Young's modulus of most solids is very large, the contact force exerted by a solid can vary over a wide range without a perceptible change in the shape of the solid. Thus in many problems, such as the ones discussed in Chaps. 2 and 3, the deformation involved in producing the contact force can be neglected.

### Shear modulus

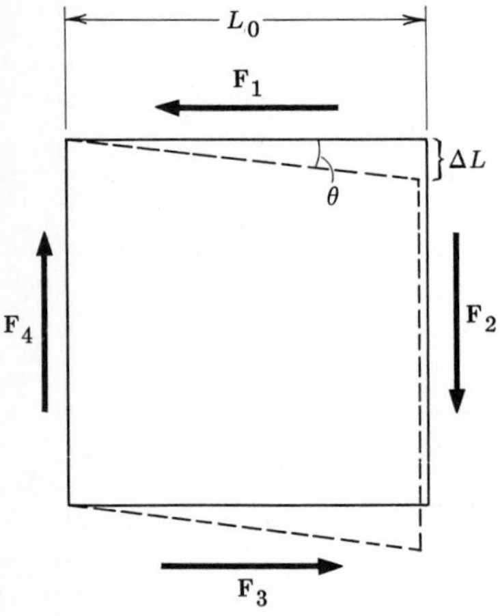

FIGURE 10.8
A specimen subjected to shear stress.

Figure 10.8 shows four forces of equal magnitude applied parallel to four faces of a cubical solid. The forces are arranged so that the total force and torque on the solid are zero. Consequently the solid is in equilibrium and remains at rest. However, it is deformed slightly into the shape shown by the dotted lines. The magnitude of the deformation is given by the *shear strain* $\gamma$, which is defined as the tangent of the angle $\theta$. From Fig. 10.8 this is seen to be related to the size of the cube by

$$\gamma = \tan \theta = \frac{\Delta L}{L_0}$$

The forces apply a *shear stress* $\tau$ to the solid which is defined as the magnitude of the parallel force on any face divided by the area of the face:

$$\tau = \frac{F_1}{A}$$

If the shear stress is not too large, it is related to the shear strain by

$$\tau = G\gamma \qquad 10.3$$

where $G$ is a constant, called the *shear modulus*, characteristic of the solid. Table 10.3 gives the values of $G$ for various solids. Generally, $G$ is between $\frac{1}{2}E$ and $\frac{1}{3}E$.

**REMARK** Recall that in a fluid at rest, the forces are always perpendicular to a surface (Sec. 7.2). This means that a fluid at rest cannot sustain a shear stress; the shear modulus of a fluid is zero. When a shear stress is applied to a fluid, the fluid flows.

TABLE 10.3 **Shear modulus of some common solids**
The values listed here are representative of each substance; actual values for a particular specimen can differ greatly from the listed values.

| Substance | Shear modulus $G$, $10^9$ N/m$^2$ |
|---|---|
| Aluminum | 25 |
| Copper | 40 |
| Glass, fused quartz | 30 |
| Iron | 50 |
| Quartz | 30 |
| Steel | 80 |
| Tungsten | 140 |
| Wood | 10 |

### Bulk modulus

The volume of a gas decreases when its pressure is increased. To a lesser extent, the volume of any substance—gas, liquid, or solid—decreases when the pressure on it increases. To subject a solid to a uniform pressure, the solid must be immersed in a fluid, as shown in Fig. 10.9. The piston applies a pressure to the fluid, which subjects the specimen to uniform pressure. (This arrangement should be contrasted to the arrangement in Fig. 10.6$b$, where the stress is applied only on the ends of the specimen.) When the pressure is increased from $p_0$ to $p_0 + \Delta p$, the volume is decreased from $V_0$ to $V_0 - \Delta V$.

The ratio of $\Delta p$ to the relative change of volume $\Delta V/V_0$ is called the *bulk modulus B*:

$$B = \frac{\Delta p}{\Delta V/V_0} \qquad\qquad 10.4$$

The reciprocal of the bulk modulus is the *compressibility K*,

$$K = \frac{1}{B} = \frac{\Delta V/V_0}{\Delta p}$$

For liquids and solids the bulk modulus is a constant, independent of pressure, provided the pressure is not exceptionally large. Therefore the change in volume is related to the change in pressure by

$$\frac{\Delta V}{V} = \frac{\Delta p}{B} = K\,\Delta p \qquad\qquad 10.5$$

Table 10.4 gives values of $B$ for various liquids and solids. Generally, the bulk modulus of a solid is between $\frac{1}{3}E$ and $E$. It is shown in Sec. 13.2 that the bulk modulus of an ideal gas is equal to the pressure of the gas.

Each modulus is a measure of a solid's resistance to a specific stress. There are other tests that measure the resistance of a solid to tearing, scratching, impact, and other types of stress. Information about these tests can be found in the books in the bibliography for this chapter.

## 10.3  NONCRYSTALLINE SOLIDS

Many important solids, such as glass, rubber, and bone, are noncrystalline. The molecules of these solids are not arranged in a regular three-dimensional array but are randomly scattered with respect to each other. In this section we discuss a few of the more important kinds of noncrystalline solids.

### Glass

Ordinary glass is a noncrystalline solid composed principally of silicon dioxide, with barium oxide, sodium monoxide, and other oxides present in varying amounts. However, the word "glass" is used technically to refer to any inorganic oxide that has the same random arrangement of molecules as ordinary glass. Although most oxides occur only as crystalline solids, a few form a glass if they are cooled very quickly while still molten. For this reason, minerals formed from molten rock that has cooled slowly in the interior of the earth are always crystalline, whereas minerals formed from molten rock that has cooled rapidly on the earth's surface may be glass. Lava from a volcano forms a glass (obsidian) if it is cooled very quickly.

The chief characteristic of glass, as mentioned in Sec. 10.1, is its lack of a definite melting point. As the temperature of glass is increased, more and more molecules break away from each other and the substance becomes more and more fluid. At sufficiently high temperature, the substance behaves like an ordinary liquid. For prac-

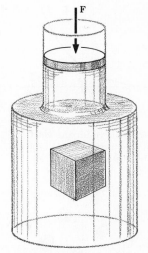

FIGURE 10.9
A specimen subjected to uniform pressure.

TABLE 10.4 **Bulk modulus of some common liquids and solids**

| Substance | Bulk modulus $B$, $10^9$ N/m² |
|---|---|
| *Liquids* | |
| Ethanol | 0.9 |
| Mercury | 25 |
| Water | 2.2 |
| *Solids* | |
| Aluminum | 70 |
| Copper | 120 |
| Glass, fused quartz | 36 |
| Granite | 47 |
| Iron | 80 |
| Marble | 70 |
| Steel | 158 |

tical purposes a glass transition temperature is often given instead of a melting point. This is the temperature at which the substance is midway between being completely rigid and being completely fluid. The transition temperature of ordinary glass is about 650°C.

### Elastomers

Elastomers are substances, like rubber, that can be elastically stretched to twice their original length or more. An elastomer is composed of long molecules that are randomly arranged and loosely connected to each other. Normally these molecules are coiled up, but when tension is applied to an elastomer, the molecules uncoil, greatly increasing the length of the substance. When the tension is removed, the molecules return to their original configuration and the substance returns to its original shape.

The elasticity of the soft connective tissue of the body is due to the presence of elastomeric fibers in the tissue. These fibers, which are composed of the protein *elastin*, are about $10^{-5}$ m in diameter. They fill the space between the cells, together with other fibers (composed of collagen) that give the tissue its strength. Figure 10.10 shows the stress-strain curve of the elastic tissue of the aorta, which is similar to the curve of a pure elastomer, like rubber. There is no straight-line portion of the curve, so that Eq. 10.2 does not hold, even for small stress. The elastic limit $\sigma_e$ is 95 percent of the tensile strength $\sigma_u$. This means that the tissue can be stretched almost to the breaking point without producing permanent deformation. (Note that a strain of 1.0 doubles the length of the tissue.)

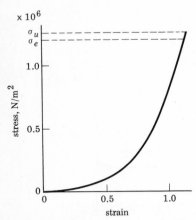

**FIGURE 10.10**
Stress-strain curve for the elastic tissue of the aorta. This is typical of the stress-strain curve of an elastomer.

### Gels

An ordinary mixture of solid and liquid consists of particles of the solid dispersed throughout the liquid. The particles will settle out of the liquid in a time that depends on their size; large particles, like sand, settle quickly, whereas small particles settle slowly. *Colloidal* particles are so small (less than 10 μm) that they remain suspended in the liquid indefinitely. A *sol* is a mixture of colloidal particles in a liquid. It is not a solution because each particle still contains billions of molecules, but it is not an ordinary mixture either because the particles are permanently suspended in the liquid. Laundry starch mixed with water is a familiar sol.

A sol is fluid because each particle is completely surrounded by liquid. Under certain circumstances, however, the particles may adhere together to form a continuous solid network throughout which the liquid is dispersed as colloidal-sized droplets. This is a *gel*, which is relatively rigid because the liquid is surrounded by solid. Figure 10.11 shows schematically how a sol, in which the liquid surrounds the solid, is transformed into a gel, in which the solid surrounds the liquid. Gelatine dessert, asphalt, and cement are examples of gels.

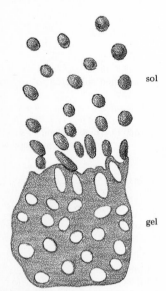

**FIGURE 10.11**
A sol, in which the liquid surrounds colloidal-sized particles, is transformed into a gel, in which the solid surrounds colloidal-sized droplets.

### Heterogeneous solids

Most solids of biological interest, such as bone and cartilage, are complex structures consisting of living cells, nerves, and blood vessels, embedded in a solid extracellular framework. Although nothing is known about the general physical properties of biological solids, empirical studies have been made to determine the specific properties of specific materials. Some of these properties are discussed in the next section.

## 10.4  BIOLOGICAL MATERIALS

Living organisms produce a variety of solid and semisolid materials, such as bone, tooth, horn, shell, nail, and cartilage. Most of these are complex heterogeneous substances. The compact part of a bone, for instance, consists of living cells embedded in a solid framework composed largely of a mixture of collagen fibers and hydroxyapatite crystals. Collagen is a protein found in all connective tissue, and hydroxyapatite is an inorganic salt consisting of calcium and phosphate, $PO_4$. The hydroxyapatite crystals, which are only about $3 \times 10^{-8}$ m long, bind the collagen fibers together.

The mechanical properties of bone and other biological materials are tested by the same methods used for engineering materials [Yamada (1970)]. The principal difficulty is obtaining fresh specimens and preserving them long enough for study. Figure 10.12 shows the stress-strain curve for compact bone. The measurement was made

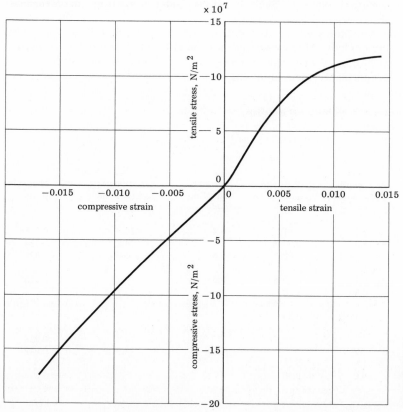

FIGURE 10.12
Stress-strain curve for compact bone. The different behavior of bone under compressive and tensile stress is typical of heterogeneous substances.

using a small sample of compact bone taken from a freshly dissected femur.

The most striking feature of this curve is the difference of the slope for tensile and compressive stress. This is characteristic of a heterogeneous solid, because the different constituents of the solid have different mechanical properties. In the case of bone, for instance, the tensile strength is due to the collagen and the compressive strength is due to the hydroxyapatite. Consequently the Young's modulus of bone and other heterogeneous substances is different for tensile and compressive stress.

Table 10.5 gives the tensile and compressive properties of some biological materials. The Young's modulus of human bone is seen to be nearly twice as large for tensile stress as for compressive stress. This means that a compressive stress produces twice as much strain as a tensile stress of equal magnitude. The mechanical properties of bones from different animals are remarkably similar, considering how dissimilar the animals are in other respects.

A comparison of Tables 10.5 and 10.2 shows that the tensile strength of bone is one-fourth that of steel and the compressive strength of bone is nearly that of granite. Considering that bone is much lighter than steel or granite, it compares very favorably with them as a structural material.

**REMARK** Bone is constructed on the same principle as reinforced concrete. Concrete alone has great compressive strength, but it lacks tensile strength. Steel rods are embedded in reinforced concrete to give it tensile as well as compressive strength. Similarly, collagen adds tensile strength to the compressive strength of hydroxyapatite.

Our knowledge of the strength of biological materials is still very incomplete. Much research is needed to understand the properties of biological materials in terms of their structure and to relate their properties to their function.

**TABLE 10.5 Tensile and compressive properties of some biological materials**
The maximum strain is the strain just before the material breaks.

| Material | Tension | | | Compression | | |
|---|---|---|---|---|---|---|
| | Young's modulus, $10^9$ N/m$^2$ | Tensile strength, $10^7$ N/m$^2$ | Maximum strain | Young's modulus, $10^9$ N/m$^2$ | Compressive strength, $10^7$ N/m$^2$ | Maximum strain |
| Bone, compact: | | | | | | |
|     Human femur | 16.0 | 12.1 | 0.014 | 9.4 | 16.7 | 0.0185 |
|     Horse femur | 23 | 11.8 | 0.0075 | 8.3 | 14.2 | 0.024 |
|     Ostrich femur | 12.6 | 7.0 | 0.0065 | 4.8 | 11.8 | 0.021 |
| Spongy: | | | | | | |
|     Human vertebra | 0.17 | 0.12 | 0.0058 | 0.088 | 0.19 | 0.025 |
| Cartilage, human ear | | 0.30 | 0.30 | | | |
| Eggshell | 0.06 | 0.12 | 0.20 | | | |
| Tooth, human: | | | | | | |
|     Crown | | | | | 14.6 | 0.023 |
|     Dentin | | | | 6.8 | 18.2 | 0.042 |
| Nail, thumb | 0.15 | 1.8 | 0.16 | | | |
| Hair | | 19.6 | 0.40 | | | |

SOURCE: Data from Yamada (1970).

1 How much energy does it take to melt 50 g of ice?      *Ans*. 3.99 kcal

2 An ice cube is placed in a glass containing 250 ml of water. If the temperature of the water was originally 18°C, what minimum mass of ice is required to lower the temperature of the water to 0°C?

3 How much energy is required to transform 25 kg of ice at 0°C into water at 20°C? (*Hint*: First calculate the energy required to melt the ice and then calculate the energy required to raise the temperature of the water.)      *Ans*. 2500 kcal

4 A 20-g ice cube is placed in a glass containing 250 ml of water. If the temperature of the water was originally 18°C, what is its temperature after the ice melts?

5 Table 10.1 gives the heat of fusion in kilocalories per mole. Calculate the heat of fusion of (*a*) water, (*b*) sodium chloride, (*c*) copper, and (*d*) tungsten in calories per gram.
*Ans*. (*a*) 79.7 cal/g; (*b*) 124 cal/g; (*c*) 49.0 cal/g; (*d*) 45.8 cal/g

6 (*a*) Show that at the triple point, the heats of sublimation, vaporization, and fusion are related by

$$H_s = H_v + H_f$$

(*b*) Calculate the heat of sublimation of water at 0°C from the data in Tables 9.1 and 10.1. Compare with the value given in Sec. 10.1.

7 A wire 13.500 m long is stretched to a length of 13.507 m. (*a*) What is the strain of the stretched wire? (*b*) If the wire is copper, what is the stress required to produce this strain? (*c*) If the cross-sectional area of the wire is $4 \times 10^{-5}$ m², what is the tension of the stretched wire?
*Ans*. (*a*) $5.18 \times 10^{-4}$; (*b*) $6.22 \times 10^7$ N/m²; (*c*) 2488 N

8 A marble column with a cross-sectional area of 25 cm² supports a weight of $7 \times 10^4$ N. (*a*) What is the stress in the column? (*b*) What is the strain in the column? (*c*) If the column is 2 m high, how much is its length changed by the weight? (*d*) What is the maximum weight the column can support?

9 An aluminum wire with a cross-sectional area of $7 \times 10^{-5}$ m² is stretched to the elastic limit. (*a*) What is the tension in the wire? (*b*) What is the strain of the wire? (*c*) What tension is required to break the wire? (*d*) What tension is required to break an aluminum wire with twice the diameter?
*Ans*. (*a*) $1.26 \times 10^4$ N; (*b*) $2.86 \times 10^{-3}$; (*c*) $1.40 \times 10^4$ N; (*d*) $5.6 \times 10^4$ N

10 The quantity $l = \sigma_u/\rho g$ is used as a measure of the tensile strength of a material relative to its density. Calculate $l$ for aluminum, bone, and steel using Tables 10.2 and 7.2.

11 Calculate the relative compressive strengths $l' = \sigma'_u/\rho g$ of (*a*) bone, (*b*) glass, (*c*) granite, and (*d*) wood.
*Ans*. (*a*) $1.1 \times 10^4$ m; (*b*) $4.3 \times 10^4$ m; (*c*) $0.76 \times 10^4$ m; (*d*) $1.5 \times 10^4$ m

12 Show that the quantity $l$ defined in Prob. 10 is equal to the maximum length of material that can hang together under its own weight. That is, show that if a length of material greater than $l$ is hung from one end, it will break apart under its own weight.

13 If the volume of a block of iron is normally 100 cm³, what is the volume when the block is subjected to a uniform pressure of $10^8$ N/m²?      *Ans*. 99.875 cm³

14 A cubic meter of seawater has a mass of 1025 kg at sea level (Table 7.2). (*a*) What is the volume occupied by 1025 kg of seawater at a depth of 10,000 m? (*b*) What is the density of seawater at a depth of 10,000 m?

15 What pressure is required to decrease the volume of glass by 1 percent?
     *Ans*. $3.56 \times 10^4$ atm

16 Tests have shown that the femoral shaft (leg bone) breaks under a compression of $5 \times 10^4$ N for men, $4 \times 10^4$ N for women, and $10 \times 10^4$ N for horses. (*a*) What is the effective cross-sectional area of the femur in men, women, and horses? (*b*) A horse weighs about six times as much as a man, yet its legs are only twice as strong. Why?

**REMARK** The center of the femur contains bone marrow, a substance without compressive strength, so that the effective cross-sectional area is the total area minus the area containing marrow. The effective area is about 0.8 times the total area.

17 Hair breaks under a tension of 1.2 N. What is the cross-sectional area of a hair?
     *Ans*. 0.006 mm²

18 Estimate the compression required to break the crown of a molar.

## BIBLIOGRAPHY

ALEXANDER, R. M.: "Animal Mechanics," University of Washington Press, Seattle, 1968. The principles of mechanics are applied to a variety of biological problems. The mechanical properties of biological materials are discussed in chaps. 3 and 4.

EVANS, FRANCIS GAYNOR: "Stress and Strain in Bones," Charles C Thomas, Publisher, Springfield, Ill., 1957. Classic monograph on the mechanical properties of bones and the relation of stress-strain phenomena to fractures and osteogenesis.

FRANKEL, VICTOR H., and ALBERT H. BURSTEIN: "Orthopaedic Biomechanics," Lea & Febiger, Philadelphia, 1970. Mechanical-engineering concepts as applied to the musculoskeletal system. Detailed description of the viscoelastic properties of biological material.

HAYDEN, H. W., WILLIAM G. MOFFATT, and JOHN WULFF: "The Structure and Properties of Materials," vol. III, "Mechanical Behavior," John Wiley & Sons, Inc., New York, 1965. Descriptions of the mechanical tests used to measure the properties of materials.

MOFFATT, WILLIAM G., GEORGE W. PEARSALL, and JOHN WULFF: "The Structure and Properties of Materials," vol. I, "Structure," John Wiley & Sons, Inc., New York, 1964. Excellent treatment of the atomic structure of crystalline and noncrystalline solids.

YAMADA, HIROSHI: "Strength of Biological Materials," ed. by F. Gaynor Evans, The Williams & Wilkins Company, Baltimore, 1970. The results of over 25 years of research by Professor Yamada and his colleagues at the Kyoto Prefectural University of Medicine. Data are given for the mechanical properties of the various bones and tissues of the body. All these data are from tests made on fresh material from Japanese cadavers. Nothing is known at present about racial differences in the mechanical properties of bone and tissue, but dietary habit is expected to have considerable effect. The data on biological materials given in this chapter are from this remarkable book.

# thermodynamics CHAPTER 11

All living organisms do work. Plants do work when raising water from their roots to their branches; animals do work when swimming, crawling, and flying. Work is also done in pumping blood through the vessels of the body (Sec. 7.5) and in pumping ions across cell walls (Sec. 17.5). All this work is obtained at the expense of the chemical energy stored in the food consumed by the organism.

Thermodynamics is the study of the relation of heat, work, and energy and, in particular, of the conversion of heat into work. The first and second laws of thermodynamics were formulated in the nineteenth century by scientists concerned with increasing the efficiency of steam engines. In spite of this, however, these laws are as fundamental as any other law of physics. They limit the efficiency of an amoeba or a whale as surely as they limit the efficiency of an automobile or a nuclear power plant.

## 11.1 THERMODYNAMIC STATES

Consider a fixed amount of gas contained in a cylinder that is fitted with a piston and a thermometer, as shown in Fig. 11.1. By moving the piston and by heating or cooling the cylinder, the pressure $p$, volume $V$, and temperature $T$ of the gas can be varied. The *thermodynamic state* of the gas is specified by giving the values of the *thermodynamic variables* $p$, $V$, and $T$. When these variables are changed, the state of the gas is changed.†

For an ideal gas these variables are related by the equation of state (Eq. 8.9)

$$pV = nRT \qquad 11.1$$

where $n$ is the number of moles of gas and $R = 8.314$ J/K is the gas constant (Table 8.1). This equation shows that if any two variables are given, the third variable is determined. This means that only two variables are required to specify the state, because the third variable can be found from Eq. 11.1. Even if the gas is not ideal, only two variables are required, because there always exists an equation of state relating these variables. Of course the equation of state of a nonideal gas is more complicated than Eq. 11.1.

If $p$ and $V$ are chosen to specify the state, the state is represented

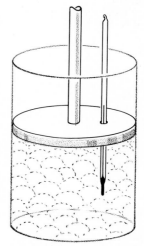

**FIGURE 11.1**
A fixed amount of gas contained in a cylinder with a piston. With this arrangement the temperature, pressure, and volume of the gas can all be varied.

† Previously (Sec. 7.1) we spoke of the three states of matter (gas, liquid, and solid). It is better to speak of the three *phases* of matter, and to reserve state for a particular set of values of the thermodynamic variables.

by a point on a graph of $p$ against $V$ (Fig. 11.2). For instance, the state $A$, with pressure $p_A$ and volume $V_A$, is represented by the point $A$ in Fig. 11.2. The temperature of this state is determined from the equation of state.

All states with the same temperature lie on a curve called the *isotherm*.† Figure 11.2 shows several isotherms for different temperatures; the temperature of a particular isotherm is greater than the temperature of all isotherms lying below it and less than the temperature of all isotherms lying above it. At high temperature the isotherms are smooth curves given by Eq. 11.1, but at lower temperature the shape of the isotherm is more complex because the gas is no longer ideal. This is discussed more fully in Sec. 8.5.

A substance such as the gas in the cylinder in Fig. 11.1 is in a definite thermodynamic state only if the temperature and pressure are the same throughout the substance. We know from Pascal's law (Fluid Property 2, Sec. 7.2) that the pressure is the same everywhere if the gas is at rest but that the pressure can differ from point to point if the gas is in motion. For instance, suppose the volume of the gas is suddenly increased by moving the piston. The gas will rush into the new volume, and for a time the system will not be in *mechanical equilibrium* because of the turbulent motion of the gas. Similarly, if the gas is heated from below, the system will not be in *thermal equilibrium* because different parts of the gas will be at different temperatures. A system that is in both mechanical and thermal equilibrium is said to be in *thermodynamic equilibrium;* only systems in thermodynamic equilibrium are in a definite thermodynamic state and can be represented on a $pV$ diagram. A thermodynamic state is thus the same as an equilibrium state.

When the state of a system is changed, as by moving the piston or heating the cylinder, the system may not be in equilibrium for a time. For example, suppose the gas is in state $A$ (Fig. 11.3), and its volume is suddenly increased from $V_A$ to $V_B$. After a time the gas settles into the equilibrium state $B$, but during the transition from $A$ to $B$ the gas is not in equilibrium, so that it is not represented by points on the diagram. This is an *irreversible* transformation, indicated by drawing a dotted line between $A$ and $B$.

On the other hand, the volume or temperature of the gas can be changed so slowly that the gas is always in equilibrium. Then, at every step of the transformation from one state to another the system is in a thermodynamic state, and the entire process can be represented by a solid line connecting the initial and final states and passing through all the intermediate states. This is called a *reversible* transformation because the system can be transformed from the final state back to the initial state through the same intermediate states. Each of the infinite number of lines that can be drawn between two states represents a different reversible transformation. A few special cases of particular importance are given special names.

In an *adiabatic* transformation no heat is allowed to enter or leave the system. This is accomplished by surrounding the cylinder with an insulating material, such as asbestos or Styrofoam (Fig. 11.4a).

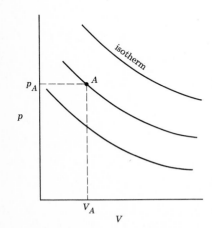

FIGURE 11.2
The state $A$ of a gas is represented by a point on a $pV$ diagram. States with the same temperature lie on isotherms.

---

† *Iso-* is a common prefix meaning equal; thus isotherm means equal temperature.

It is found that when an ideal gas is expanded adiabatically, its temperature and pressure both decrease. The system follows the solid line $AB$ in Fig. 11.3.

In an *isothermal* transformation the temperature is held constant. This is accomplished by placing the cylinder in contact with a large reservoir of water that is at the desired temperature (Fig. 11.4$b$). The cylinder has thin walls made of a heat-conducting material, such as copper, so that heat can readily flow back and forth between the reservoir and the gas. The reservoir is large enough for its own temperature to be unaffected by the amount of heat exchanged with the gas. During an isothermal expansion heat flows into the gas to keep its temperature constant. (Remember, the temperature of a gas decreases if heat is prevented from flowing into it while it expands.) The system follows the isotherm in going from the initial state $A$ to the final state $B'$ (Fig. 11.3).

In an *isochoric* transformation the volume of the system is held constant. This is accomplished by clamping the piston in a fixed position. The state of the gas is changed by heating the gas (Fig. 11.4$c$). Since the piston is held fixed, no work is done by the system during the transformation. The system follows the line $AA'$ in Fig. 11.3.

In an *isobaric* transformation the pressure of the system is held fixed. This is accomplished by applying a constant external pressure to the piston (Fig. 11.4$d$). The state of the gas is changed by heating the gas. The system follows the line $AB''$ in Fig. 11.3.

There are countless other transformations in which nothing is held constant, but it is sufficient for our purposes to consider only these special cases. There is no need to learn these transformations now, but you should review this discussion whenever a particular transformation is mentioned later in this chapter.

## 11.2  THE FIRST LAW OF THERMODYNAMICS

The first law of thermodynamics is the conservation of energy, introduced in Sec. 5.5. Briefly, it states that energy can neither be created nor destroyed but only transformed from one form to another. For the purposes of thermodynamics it is necessary to be more specific and to state the law more quantitatively.

Thermodynamics considers the relation between a *system* $\mathcal{S}$, such as the gas in the cylinder in Fig. 11.1, and the *environment* $\mathcal{E}$ surrounding it. The environment is everything outside the system that can affect the system, which in most cases includes only the immediate surroundings of the system. The system and the environment together constitute the *universe* $\mathcal{U}$.

The energy $E_\mathcal{S}$ of the system is the sum of the kinetic energies of the molecules of the system (thermal energy) and the potential energy of the atoms in the molecules (chemical energy). The energy $E_\mathcal{S}$ depends on the state of the system, changing when the state changes. For instance, in the isobaric transformation in Fig. 11.4$d$, the heat source increases the thermal energy of the system. Since

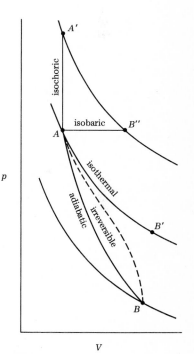

FIGURE 11.3
A gas initially in state $A$ is transformed to states $A'$, $B$, $B'$, and $B''$ by isochoric, adiabatic, isothermal, and isobaric transformations, respectively. An irreversible transformation from $A$ to $B$ is indicated by a dotted line because the system does not pass through any of the intermediate states between $A$ and $B$.

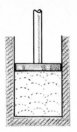

a

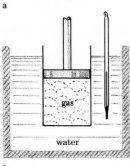

gas

water

b

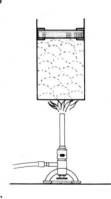

c

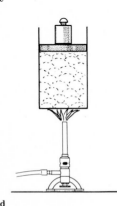

d

FIGURE 11.4
Schematic diagrams representing (*a*) an adiabatic, (*b*) an isothermal, (*c*) an isochoric, and (*d*) an isobaric transformation.

the heat source is part of the environment, the energy $E_\varepsilon$ of the environment also changes. The first law of thermodynamics says that the energy $E_u$ of the universe

$$E_u = E_s + E_\varepsilon$$

does not change. That is, if $E_s$ and $E_\varepsilon$ are the energies of the system and the environment when the system is in one state and $E'_s$ and $E'_\varepsilon$ are the energies when the system is in another state, then

$$E'_s + E'_\varepsilon = E_s + E_\varepsilon$$

or
$$(E'_s - E_s) + (E'_\varepsilon - E_\varepsilon) = 0 \qquad \text{11.2}$$

As before, delta is used as a prefix to mean "difference in" or "change of." Specifically, $\Delta E_s$ is the energy of the final state of the system minus the energy of the initial state,

$$\Delta E_s = E'_s - E_s$$

and $\Delta E_\varepsilon$ is the final energy of the environment minus the initial energy

$$\Delta E_\varepsilon = E'_\varepsilon - E_\varepsilon$$

In terms of these symbols Eq. 11.2 can be written

$$\Delta E_s + \Delta E_\varepsilon = 0$$

or
$$\Delta E_s = -\Delta E_\varepsilon \qquad \textit{first law} \qquad \text{11.3}$$

This is a convenient mathematical expression of the first law of thermodynamics. It is used to calculate the change of energy of the system when the change of energy of the environment is known, and vice versa.

## Heat and work

Energy can be transferred between the system and the environment in two essentially different ways.

*1 Heat* is energy that flows from one object to another as the result of the random motion of the molecules of the objects. The molecules of an object with a temperature $T_1$ have, on the average, greater kinetic energy than the molecules of an object with a lower temperature $T_2$. If two such objects are in contact, their molecules collide with each other where the objects touch. In each collision the more energetic molecule of the hotter object loses energy, while the less energetic molecule of the colder object gains energy. Through a succession of billions upon billions of such collisions, energy is transferred from the hotter object to the colder object.

*2 Work* is done by or on a system when there is a change of volume. For instance, the expansion of the gas in Fig. 11.5 does work in moving the piston. The gas exerts a force $F = pA$ on the piston, where $p$ is the gas pressure and $A$ is the cross-sectional area of the piston. Therefore, if the piston is moved a distance $d$, the work done is

$$W = Fd = pAd = p\,\Delta V \qquad \text{11.4}$$

where
$$\Delta V = V_{\text{final}} - V_{\text{initial}} = Ad$$

is the change of volume of the gas.† The work $W$ is produced at the expense of the energy of the system and goes to increase the energy of the environment. Thus by doing work, energy is transferred from the system to the environment. Likewise, if the gas is compressed, work is done on the system and the energy of the system is increased at the expense of the energy of the environment.

If, during a transformation, a quantity of heat $Q$ enters a system while the work $W$ is done by the system, the change of energy of the system is

$$\Delta E_s = Q - W \qquad \textit{first law} \qquad\qquad 11.5$$

This equation says that the change of energy is equal to the heat that enters the system minus the work done by the system. It is just another form of the first law. When heat leaves the system, $Q$ is negative, and when work is done on the system, $W$ is negative.

### Specific heat

When heat is added to a system, the temperature of the system increases. For a given quantity of heat, the temperature change $\Delta T$ depends on whether the pressure or the volume of the system is kept constant during the process. In an isochoric (constant-volume) transformation, the temperature change is related to the heat absorbed by

$$Q = C_v \, \Delta T \qquad \textit{isochoric} \qquad\qquad 11.6a$$

where $C_v$ is the *heat capacity* of the system at constant volume. The *specific heat* $c_v$ of a substance is its heat capacity divided by its mass:

$$c_v = \frac{C_v}{m} \qquad\qquad 11.6b$$

The specific heat is a characteristic property of a substance. It depends on temperature, but over a narrow temperature range it can be treated as a constant. Since no work is done when $\Delta V = 0$, the heat $Q$ absorbed is equal to the change of energy. Combining Eqs. 11.6a and 11.6b, we get

$$\Delta E_s = Q = mc_v \, \Delta T$$

Most transformations of biological interest occur at constant pressure rather than constant volume. In an isobaric (constant-pressure) transformation, the temperature change is related to the heat absorbed by

$$Q = mc_p \, \Delta T \qquad \textit{isobaric} \qquad\qquad 11.7$$

where $c_p$ is the specific heat at constant pressure. This is the specific heat most commonly used. Table 11.1 gives the values of $c_p$ for some common substances. Its unit is kcal/kg-°C or J/kg-°C.

From Eqs. 11.4, 11.5, and 11.7 the change of energy in an isobaric transformation is found to be

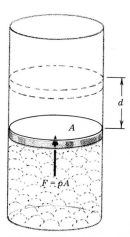

FIGURE 11.5
The expansion of a gas does work.

---

† Equation 11.4 is exact if the pressure of the gas is constant during the expansion. If the pressure is not constant, this equation is only an approximation, valid for small changes of volume.

$$\Delta E_\text{S} = Q - W$$
$$= mc_p \, \Delta T - p \, \Delta V \qquad \qquad \text{11.8}$$

**REMARK**  When a liquid or solid is heated at constant pressure, there is only a slight increase in volume, so that the term $p \, \Delta V$ is very small in Eq. 11.8. Consequently there is little difference between a constant-volume and a constant-pressure transformation for a liquid or solid, and $c_v$ is equal to $c_p$ for all practical purposes.  A gas, on the other hand, undergoes considerable expansion when heated, so that $c_v$ is very different from $c_p$.  The ratio $c_p/c_v$ of the specific heats of a gas is between 1.0 and 1.67, depending on the gas (see Table 13.2).

### Calorimetry

The heat absorbed during a transformation is measured in a *calorimeter*, a large vessel of water surrounding a smaller container (Fig. 11.6). The outside of the vessel is thermally insulated to prevent heat from entering or leaving during the transformation.  The container, on the other hand, is made of copper or some other heat-conducting material to permit heat to be readily exchanged between the water and the container.  The container encloses the system to be measured, and the water surrounding the container is the environment.

For example, suppose 1 mol of ice (18 g) is placed in the container. A thermometer in the water shows that the temperature of the water decreases as the ice melts.  Specifically, if there is 2.50 kg of water in the calorimeter, its temperature changes by $\Delta T = -0.58°C$ when

**TABLE 11.1  Specific heat at constant pressure of 1 atm of some common substances**

| Substance | Temperature, °C | $c_p$ kcal/kg-°C | J/kg-°C |
|---|---|---|---|
| *Gases* | | | |
| Air | 100 | 0.240 | 1000 |
| Carbon dioxide | 15 | 0.199 | 833 |
| Oxygen | 15 | 0.218 | 913 |
| Nitrogen | 15 | 0.248 | 1040 |
| Water (steam) | 100 | 0.482 | 2020 |
| *Liquids* | | | |
| Ethanol | 25 | 0.581 | 2430 |
| Mercury | 20 | 0.0332 | 139 |
| Water | 0 | 1.0074 | 4218.1 |
| | 14 | 1.0000 | 4186.8 |
| | 50 | 0.9985 | 4180.7 |
| | 100 | 1.0070 | 4216.0 |
| *Solids* | | | |
| Aluminum | 20 | 0.214 | 899 |
| Brass | 20 | 0.0917 | 384 |
| Copper | 20 | 0.0921 | 386 |
| Glass, crown | 20 | 0.161 | 674 |
| Flint | 20 | 0.117 | 490 |
| Granite | 20 | 0.192 | 804 |
| Human body (avg.) | 37 | 0.83 | 3500 |
| Iron | 20 | 0.115 | 481 |
| Water (ice) | 0 | 0.492 | 2060 |
| Wood | 20 | 0.42 | 1760 |

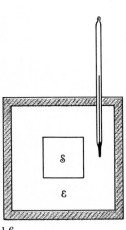

**FIGURE 11.6**
A calorimeter, consisting of a large vessel of water $\mathcal{E}$ surrounding a small container $\mathcal{S}$.

18 g of ice melts. Using Eq. 11.7, together with the specific heat of water from Table 11.1, we find the heat absorbed by the environment to be

$$Q_\mathcal{E} = mc_p \, \Delta T$$
$$= (2.50 \text{ kg})(1.0 \text{ kcal/kg-}°\text{C})(-0.58°\text{C})$$
$$= -1.45 \text{ kcal}$$

The minus sign indicates that the heat is lost by the environment and absorbed by the system (ice). That is, 1 mol of $H_2O$ absorbs 1.45 kcal of heat when it is transformed from the solid phase (ice) to the liquid phase (water). This is the heat of fusion, discussed in Sec. 10.1.

As another example, suppose 50 g of ethanol at a temperature of 30°C is placed in the container. The surrounding water initially has a temperature of 15.00°C, but as the alcohol cools, the temperature of the water increases. After a time the water and the alcohol reach the same temperature, which is found experimentally to be 15.17°C. What is the specific heat of alcohol?

The temperature change of the environment (water) is

$$\Delta T = 15.17°\text{C} - 15.00°\text{C} = 0.17°\text{C}$$

so the heat absorbed by the environment is

$$Q_\mathcal{E} = mc_p \, \Delta T$$
$$= (2.50 \text{ kg})(1.0 \text{ kcal/kg-}°\text{C})(0.17°\text{C})$$
$$= 0.425 \text{ kcal}$$

The temperature change of the system (alcohol) is

$$\Delta T = 15.17°\text{C} - 30°\text{C} = -14.83°\text{C}$$

so the heat absorbed by the system is

$$Q_\mathcal{S} = mc_p \, \Delta T$$
$$= 0.050 \text{ kg} \times c_p \times (-14.83°\text{C})$$
$$= -0.74 \text{ kg-}°\text{C} \times c_p$$

where $c_p$ is now the specific heat of the alcohol. (The minus sign indicates that the system lost heat.) Since the heat absorbed by the environment is equal to the heat lost by the system, we have

$$0.74 \text{ kg-}°\text{C} \times c_p = 0.425 \text{ kcal}$$

From this the value of $c_p$ is found to be

$$c_p = \frac{0.425 \text{ kcal}}{0.74 \text{ kg-}°\text{C}} = 0.57 \text{ kcal/kg-}°\text{C}$$

These examples show how a calorimeter is used to determine the thermodynamic properties of substances.

## 11.3  THE SECOND LAW OF THERMODYNAMICS

An object of mass $m$ released from a height $h$ spontaneously falls to the ground, where it comes to rest. In this situation the energy of the universe is the sum of the thermal energy of the object, the thermal energy of the ground, and the mechanical energy of the object. Before it is released, the object has mechanical energy just equal to its potential energy $U = mgh$, and after it comes to rest on the ground, its mechanical energy is zero. In this process, therefore, the mechanical energy of the universe decreases from $mgh$ to zero. Since the total energy of the universe does not change (the first law of thermodynamics), the thermal energy of the universe must increase by $mgh$. This increase in thermal energy manifests itself as a slight increase in the temperature of the object and the ground.

**REMARK** Such a temperature rise is normally too small to be noticed, so that the validity of the first law is not evident from everyday experience. The first law was established only after Joule's carefully executed paddle-wheel experiment (Sec. 5.5) showed that work is equivalent to heat in raising the temperature of water. This result was unexpected at the time because it conflicted with the currently held concepts of heat.

It is well known from everyday experience that an object initially at rest on the ground never spontaneously jumps into the air. On first thought such an event seems to be impossible because it violates the first law. If an object were to jump into the air, there would clearly be an increase in the mechanical energy of the universe. This would not be a violation of the first law, however, if there were a corresponding decrease in the thermal energy of the universe. The first law, therefore, does not explain why objects never spontaneously jump into the air.

The process "object spontaneously jumps into the air" is the reverse of the process "object spontaneously falls to the ground." The one process occurs with ease, whereas the reverse process never occurs at all. There are many other irreversible processes that can proceed in only one direction. For example, when a cold object and a hot object are placed in contact, heat always flows from the hot object to the cold object, and never from the cold object to the hot object. Consequently, the temperature of the hot object decreases, while the temperature of the cold object increases. If the reverse process were to occur, the cold object would get colder while the hot object got hotter. As another example, ink dropped into a glass of water diffuses until it is evenly distributed throughout the water. The reverse process, in which a uniform mixture of water and ink spontaneously separates into pure water and pure ink, never occurs.

### Kelvin and Clausius statements of the second law of thermodynamics

The second law of thermodynamics specifies those processes which cannot occur spontaneously. There are many equivalent statements of the second law, and we shall consider a number of them in this chapter. We begin with the form of the second law due to Lord Kelvin.

**Kelvin's statement of the second law of thermodynamics** *It is impossible to construct a device that can without other effect lift one object by extracting thermal energy from another.* This form of the second law is essentially the assertion that objects do not spontaneously jump into the air. But it goes beyond this by asserting that even with the help of special machinery, no matter how complex, an object cannot be lifted solely by extracting thermal energy. In other words, thermal energy cannot be completely converted into potential energy without other effect, i.e., without something else happening.

There is another form of the second law due to Clausius.

**Clausius's statement of the second law of thermodynamics** *It is impossible to construct a device that can without other effect transfer heat from a cold object to a hotter object.* This form of the second law is essentially the assertion that heat does not spontaneously flow from a cold object to a hotter object. But it goes beyond this by asserting that even with the help of special machinery, no matter how complex, heat cannot be transferred from a cold object to a hot object without other effect.

These two forms of the second law are statements of well-known facts, which makes them easy to understand. The deduction of further implications from these forms of the second law, however, requires novel and subtle reasoning. A few examples of this are given in the remainder of this section. Other forms of the second law, which are easier to apply, are given in the following sections.

The Kelvin and Clausius statements are equivalent. This means that if it were possible to construct a device that violated the Kelvin statement, it would be possible to construct a device that violated the Clausius statement, and vice versa. For example, suppose a device existed that could lift a mass solely by extracting thermal energy from object $A$, in violation of the Kelvin statement. Then, with suitable mechanical connections, this device could be made to lift the mass repeatedly and drop it onto another object $B$ (Fig. 11.7). Each time the mass was lifted, the thermal energy and temperature of object $A$ would decrease, and each time the mass fell, the thermal energy and temperature of object $B$ would increase. Consequently, thermal energy would be transferred from $A$, which would become colder, to $B$, which would become hotter, in violation of the Clausius statement. This proves that a device that violates the Kelvin statement can be used to construct a device that violates the Clausius statement. The reverse is also true, but we shall not prove it here.

### Efficiency of engines

An important application of the second law is the evaluation of the efficiency of engines. An engine is a device, such as a gasoline motor or a muscle, that converts energy into work. The engines we shall be considering all contain a substance that changes its size when it absorbs energy.† For example, the combustion of gasoline releases energy that causes the gas in the cylinder of a gasoline motor to expand, and the energy released when adenosine triphosphate (ATP) is converted into adenosine diphosphate (ADP) causes a muscle fiber to contract. These are both very complex processes which are difficult

---

† An electric motor, which converts electric energy into work, does not contain a size-changing substance. From the viewpoint of thermodynamics, electric energy, like potential energy, is equivalent to work because it can be completely converted into work without violating the second law. Therefore the production of electric energy from thermal energy in a power-generating station is equivalent to the production of work from thermal energy. This process does use engines (steam turbines) containing a size-changing substance (steam).

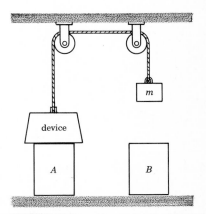

FIGURE 11.7
A device that violates the Kelvin statement by lifting a mass $m$ using thermal energy extracted from $A$.

to analyze in detail. Therefore, in order to understand the basic thermodynamic principles involved in converting energy into work, we shall analyze a much simpler system.

The simplest engine consists of an ideal gas confined to a cylinder that is fitted with a movable piston (Fig. 11.5). Initially the gas is in state $A$, with pressure $p_A$, volume $V_A$, and temperature $T_1$. To hold the piston in place, an external pressure $p_A$ must be exerted on the piston. When this external pressure is decreased slightly, the gas expands and the piston moves outward, doing work. By suitably connecting the piston to a mechanical device, the piston can be made to lift a weight as it moves. The change of potential energy of the weight is equal to the work $W = p_A \Delta V$ done by the piston.†

Suppose the gas is allowed to expand gradually while the cylinder is in thermal contact with an environment at temperature $T_1$. This means that the temperature of the gas is kept constant during the expansion by absorbing heat from the environment. During the expansion the state of the system moves along the isotherm from state $A$ to state $B$ (Fig. 11.8). From the first law (Eq. 11.4), the change of energy of the gas is

$$\Delta E_\text{S} = Q_1 - W_1$$

where $Q_1$ is the heat extracted from the environment and $W_1$ is the work done by the piston. Although this process converts thermal energy into work, it does not violate the second law (Kelvin's statement) because there has been another effect: the gas has been transformed from state $A$ to state $B$.

All practical engines operate in a cycle, the system returning repeatedly to its initial state. To transform the system from state $B$ back to state $A$, work $-W_2$ must be done to compress the gas. (The minus sign indicates that this work is done on the system.) Figure 11.8 shows a hypothetical cycle in which the gas is first expanded isothermally along path 1 and then compressed along path 2. Because the pressure is lower (on the average) along path 2 than path 1, less work is required to compress the gas than was obtained by expanding it. Thus a net amount of work

$$W = W_1 - W_2$$

is done by the engine in one complete cycle. The change of energy of the gas is zero in one cycle because the system is restored to its original state. Therefore, applying Eq. 11.5 to one cycle, we get

$$\Delta E_\text{S} = 0 = Q - W$$

or

$$Q = W$$

where $W$ is the net work done and $Q$ is the net heat absorbed.

If no heat were exchanged along path 2, the net heat $Q$ absorbed in one cycle would just be the heat $Q_1$ absorbed along path 1. Thus, without other effect, all the heat absorbed from the environment would be converted into work, in violation of the second law. Some

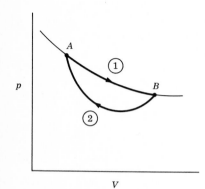

FIGURE 11.8
A cycle for converting thermal energy into work.

† This assumes that there is no friction in the system. In every real engine, some work is used to overcome friction and is unavailable to lift an external weight. However, because friction can in principle be completely eliminated, we discuss only ideal frictionless engines.

heat, therefore, must be exchanged along path 2, and in fact, the heat $-Q_2$ is ejected from the system into the environment. Only the net heat absorbed in one cycle

$$Q = Q_1 - Q_2$$

is converted into work, and this is less than the heat absorbed along path 1. It is a consequence of the second law that only a fraction of the heat absorbed along path 1 can be converted into work; the remainder is ejected along path 2 as heat. The efficiency $e$ of an engine is the ratio of the work done to the heat absorbed:

$$e = \frac{W}{Q_1} = \frac{Q_1 - Q_2}{Q_1} = 1 - \frac{Q_2}{Q_1} \qquad 11.9$$

This can be made clearer by considering the special cycle shown in Fig. 11.9. Along path 1 the system first expands isothermally from $A$ to $B$, and then it expands adiabatically from $B$ to $C$, which lowers its temperature. Work $W_1$ is done by the engine during this first half of the cycle, and heat $Q_1$ is absorbed at temperature $T_1$ during the isothermal portion of path 1. Along path 2 the system is first compressed isothermally from $C$ to $D$, and then it is compressed adiabatically from $D$ to $A$, which raises its temperature. Work $-W_2$ is done on the engine during this second half of the cycle, and heat $-Q_2$ is ejected at temperature $T_2$ during the isothermal portion of path 2. This is a *Carnot cycle*, and an engine that operates on such a cycle is a Carnot engine.

Starting with the second law, a number of remarkable things can be proved about a Carnot engine.

*1* A Carnot engine is the most efficient engine that operates between two temperatures $T_1$ and $T_2$.

*2* The efficiency of a Carnot engine is independent of the working substance. That is, the efficiency is the same whether the substance is an ideal gas, a real gas, or even a liquid.

*3* The efficiency of a Carnot engine operating between the absolute temperature $T_1$ and $T_2$ is

$$e = \frac{T_1 - T_2}{T_1} = 1 - \frac{T_2}{T_1} \qquad 11.10$$

For example, during the first half of its cycle, a steam engine uses steam, which has been heated to a temperature of 250°C in a boiler, to move a piston. During the second half of the cycle, the steam is condensed into water at a temperature of 38°C. Thus the engine operates between a high temperature $T_1 = 523$ K and a low temperature $T_2 = 311$ K. The efficiency of a Carnot engine operating between these temperatures is

$$e = 1 - \frac{311 \text{ K}}{523 \text{ K}} = 0.405$$

This means that for every 100 J of heat absorbed at $T_1$, 40.5 J is converted into work and 59.5 J is ejected as heat at $T_2$. The efficiency of an ideal steam engine is only about 0.32, because it does not operate on the Carnot cycle.

**REMARK** Equation 11.10 is valid only when $T_1$ and $T_2$ are absolute temperatures.

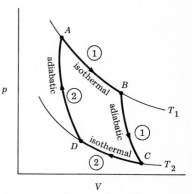

FIGURE 11.9
A Carnot cycle.

All heat engines convert only a fraction of the heat they absorb into useful work; the remainder must be dumped into the environment. Rivers and lakes are used to absorb the enormous amount of heat that accompanies the generation of electric power. Thermal pollution results when enough heat is dumped into a river to raise its temperature significantly. With our ever increasing production of electric energy, it is becoming more difficult to dispose of heat without ecological damage. The direct conversion of energy into work, as in a battery, is a more efficient method of generating electric power, but it has not yet been accomplished using inexpensive fuels.

## 11.4  STATISTICAL FORMULATION OF THE SECOND LAW

Although the second law applies to all irreversible processes, it is not easy to apply the Kelvin or Clausius statements to processes that do not involve a transfer of energy. Therefore it is useful to introduce a third statement of the second law.

**Statistical statement of the second law of thermodynamics**  *The total disorder of the universe never decreases.*  The concept of disorder will be clarified shortly, but for now it is sufficient to use the commonsense meaning of the word.

For example, when ink is dropped into water, it diffuses until it is evenly distributed throughout the water. Water and ink are in a more disordered state when mixed than when separated, because their molecules are randomly mixed together in one state and not in the other. Consequently the disorder of the universe increases during the diffusion process. The reverse process, in which a uniform mixture of water and ink spontaneously separates into pure water and pure ink, is forbidden by the second law because it would decrease the disorder of the universe.

As another example, consider again an object of mass $m$ held at a height $h$ above the ground. Its potential energy $mgh$ is distributed equally among all its molecules, since each molecule has the same potential energy $m_{mol}gh$, where $m_{mol}$ is the mass of an individual molecule. When the object is dropped, its potential energy is ultimately converted into kinetic energy of the molecules of the ground. This energy is distributed randomly among these molecules, because each molecule can have a different energy. Consequently the disorder of the universe is increased when potential energy is converted into thermal energy. The reverse process, in which thermal energy is completely converted into potential energy, is forbidden by the second law because it would decrease the disorder of the universe. This shows how the statistical statement of the second law is related to the Kelvin statement, though we shall not give a complete proof of their equivalence.

The concept of disorder can be made quantitative by considering the relation between a thermodynamic state and a molecular state of a system. A molecular state (*microstate*) is specified by giving the position and velocity of every molecule in the system, whereas a thermodynamic state (*macrostate*) is specified by giving a few average molecular properties, such as temperature and pressure. Different microstates with the same average properties correspond to the same macrostate. The number $D$ of distinct microstates corresponding to a macrostate is the *disorder* of the macrostate.

As the molecules of the system change their positions and velocities, the system changes from one microstate to another. The equilibrium macrostate of the system is the macrostate of maximum disorder, i.e., the state with the greatest number of microstates. At equilibrium the system does not change macrostates, even though its microstates are constantly changing, because almost all the microstates correspond to the equilibrium macrostate. This abstract discussion will become clearer after considering the following example.

A tray is filled with 100 plastic chips. Each chip is white on one side and black on the other. When the tray is viewed from a large distance, the individual chips cannot be seen. All that is seen is the average color of the tray, which can be any shade of gray or pure white or pure black, depending on how many chips are turned white side up and how many are turned black side up. The color of the tray is thus an average property of the system, analogous to a thermodynamic variable. A quantitative description of the color is given by the *grayness* $g$, which is defined as the fraction of chips turned black side up. The grayness varies from 0 (pure white) to 1 (pure black). The value of $g$ specifies the macrostate of the tray, just as the values of $p$ and $T$ specify the macrostate of a thermodynamic system.

A microstate of the tray is specified by giving the color of each chip. For instance, one microstate may have the first chip black side up, the second chip white side up, the third white, the fourth black, and so on. This microstate is represented by the sequence

$$bwwb \cdots$$

Another microstate may have the first chip black, the second black, the third white, the fourth white, and so on. It is represented by the sequence

$$bbww \cdots$$

Since each chip can have either one of two colors (black or white), the total number $N$ of distinct microstates is

$$N = 2 \times 2 \times 2 \times 2 \cdots = 2^{100}$$

Using logarithms,[†] we can rewrite this as a power of 10. The result is

$$N = 10^{30}$$

This shows that even a simple system with only a 100 particles in it possesses an immense number of microstates.

Of even greater importance is the number of microstates $D(g)$ corresponding to a given macrostate $g$. The all-white macrostate ($g = 0$) has only one microstate because every chip must be white side up; hence $D(0) = 1$. The $g = 0.01$ macrostate has 99 chips white

[†] From the definition of logarithm we can write

$$2 = 10^{\log 2} = 10^{0.30}$$

and so

$$2^{100} = (10^{0.30})^{100} = 10^{30}$$

side up and one chip black side up. Since the macrostate is the same regardless of which one of the 100 chips is black, there are 100 different microstates corresponding to $g = 0.01$; hence $D(0.01) = 100$. The $g = 0.02$ macrostate has 98 white chips and 2 black. There are 4950 different microstates in this case; hence $D(0.02) = 4950$. Table 11.2 gives the disorder $D(g)$ of various macrostates between 0 and 1. This shows that $D(g)$ is very different for different macrostates and that it is a maximum for the state $g = 0.50$, in which 50 chips are white and 50 black.

When the tray is shaken up and down, the chips randomly flip from one side to another. With each shake, the tray makes a random transition from one microstate to another. Although any one microstate is as likely to occur as any other, some macrostates are much more probable than others. The probability $P(g)$ of the tray's being in macrostate $g$ after any one shake is equal to the number of microstates $D(g)$ corresponding to $g$ divided by the total number $N$ of microstates of the system:

$$P(g) = \frac{D(g)}{N} \qquad 11.11$$

For example, of the $10^{30}$ microstates of the tray, only one corresponds to $g = 0$. Therefore the probability of the all-white macrostate occurring after any one shake of the tray is

$$P(g) = \frac{D(0)}{N} = \frac{1}{10^{30}} = 10^{-30}$$

This is an incredibly small probability; the tray could be shaken a billion times a second for 100 billion years without the all-white macrostate's occurring. An event with such a low probability of occurrence is, for all practical purposes, impossible. That is, we can safely say that the all-white macrostate cannot be produced by randomly shaking the tray. There is no mechanical law that prevents its occurrence, but its probability is so small that it can be considered zero.

The macrostate of maximum disorder ($g = 0.50$) occurs with the greatest probability:

$$P(0.50) = \frac{1.0 \times 10^{29}}{10^{30}} = 0.10$$

This means that on an average the tray is in the $g = 0.50$ state once in every 10 shakes of the tray. The macrostates with $g$ close to 0.50 also occur frequently. In fact, it can be shown that 98 percent of all the microstates of the tray correspond to macrostates between $g = 0.40$ and $g = 0.60$. Therefore, although the microstate changes with each shake of the tray, the macrostate is always close to 0.50.

If the number of chips in the tray is increased, the probability of the tray's being in a macrostate other than 0.50 decreases. For instance, in a tray with a million chips, over 99.99 percent of the microstates correspond to macrostates between $g = 0.499$ and $g = 0.501$. Suppose that initially these chips are all turned white side up, so that the tray is started in the $g = 0$ state. When the tray is

TABLE 11.2 **Number of microstates** $D(g)$ **corresponding to various macrostates** $g$ **of a tray containing 100 black-and-white chips**

| Macrostate $g$ | Disorder $D(g)$ |
|---|---|
| 0.00 | 1 |
| 0.02 | $4.9 \times 10^3$ |
| 0.05 | $7.5 \times 10^7$ |
| 0.10 | $1.7 \times 10^{13}$ |
| 0.20 | $5.5 \times 10^{20}$ |
| 0.40 | $1.4 \times 10^{28}$ |
| 0.50 | $1.0 \times 10^{29}$ |
| 0.60 | $1.4 \times 10^{28}$ |
| 0.80 | $5.5 \times 10^{20}$ |
| 0.90 | $1.7 \times 10^{13}$ |
| 0.95 | $7.5 \times 10^7$ |
| 0.98 | $4.9 \times 10^3$ |
| 1.00 | 1 |

shaken, it randomly changes its microstate and will, with overwhelming probability, make a transition to a microstate corresponding to a macrostate close to $g = 0.50$. Thereafter, with each shake of the tray, the microstate changes while the macrostate remains close to 0.50.

In a thermodynamic system, which consists of billions upon billions of molecules, the number of microstates corresponding to a state of greater disorder is overwhelmingly larger than the number of microstates corresponding to a macrostate of lesser disorder. Consequently the system never spontaneously changes to a state of lesser disorder. The second law of thermodynamics is fundamentally a statistical law. There is no physical law that prevents a uniform mixture of ink and water from spontaneously separating into pure ink and pure water. However, there are so many more ways to arrange ink and water molecules in the mixed system than there are in the separated system that the probability of the molecules of the mixture ever being, by pure chance, in a microstate of the separated system is essentially zero.

**REMARK**  It is important to realize that the second law says that the disorder of the *universe* never decreases. The disorder of a part of the universe can decrease if there is a corresponding increase in the disorder of another part. Such a decrease is not spontaneous, however, since it is produced by an interaction with another part of the universe. For instance, the heat $Q_1$ removed from the hot environment of a Carnot engine decreases the disorder of the hot environment. But the heat $Q_2$ added to the cold environment increases the disorder of the cold environment. The second law requires that the increase of disorder of the cold environment be at least as large as the decrease of disorder of the hot environment.

It is sometimes said that evolution violates the second law because in the course of hundreds of millions of years more and more complex forms of life have developed. The more complex forms are clearly more ordered than the lower forms. Thus evolution, the argument goes, is a process in which life progresses from a state of less order to a state of more order, in contradiction to the second law. Some people even maintain that this proves that life has unique properties that are beyond the laws of physics.

This, of course, is just ignorant nonsense. It is true that evolution has, on the whole, produced forms of increased order. Even the growth of an individual organism, from egg to maturity, is a process in which order increases. However, these are not violations of the second law because the earth is not an isolated system. The earth receives energy from the sun during the day and radiates energy into outer space at night. The energy-releasing reactions that occur on the sun increase the disorder of the sun. Thus the energy used on earth to produce ordered life from disordered matter is produced at the expense of increased disorder of the sun. The total disorder of the earth and sun decreases, in compliance with the second law.

The earth can be thought of as a giant heat engine. Some of the heat that is absorbed at high temperature from the sun during the day is released at low temperature into outer space at night. The rest is converted into the work that drives the winds, lifts water vapor into the clouds, and contracts muscles.

## 11.5   ENTROPY, FREE ENERGY, AND ENTHALPY

### Entropy

Entropy $S$ is a thermodynamic variable that measures the disorder of a thermodynamic state. This means that every state has a definite entropy and that the entropy of one state is greater than the entropy of another if the disorder of the first state is greater than the disorder of the second. But unlike disorder, the entropy of a composite system is simply the sum of the entropies of the component subsystems.† In particular, the entropy $S_\mathfrak{u}$ of the universe is the sum of the entropies of a system and its environment:

$$S_\mathfrak{u} = S_\mathfrak{s} + S_\mathcal{E} \qquad\qquad 11.12$$

Let $S_\mathfrak{s}$ and $S_\mathcal{E}$ be the entropies of the system and the environment when the system is in one state, and let $S_\mathfrak{s}'$ and $S_\mathcal{E}'$ be the entropies after the system has transformed to another state. According to the statistical form of the second law, the entropy of the universe never decreases, so that $S_\mathfrak{u}'$ must be greater than or equal to $S_\mathfrak{u}$:

$$S_\mathfrak{u}' \geq S_\mathfrak{u}$$

From Eq. 11.12 we have

$$S_\mathfrak{s}' + S_\mathcal{E}' \geq S_\mathfrak{s} + S_\mathcal{E}$$

or $\qquad\qquad (S_\mathfrak{s}' - S_\mathfrak{s}) + (S_\mathcal{E}' - S_\mathcal{E}) \geq 0$

Using $\Delta S$ to represent these entropy changes, we can write this as

$$\Delta S_\mathfrak{u} = \Delta S_\mathfrak{s} + \Delta S_\mathcal{E} \geq 0$$

or $\qquad\qquad \Delta S_\mathcal{E} \geq -\Delta S_\mathfrak{s} \qquad second\ law \qquad 11.13$

This last equation expresses the second law in a form analogous to the first law in Eq. 11.3.

It is always possible, in principle, to transform a system from one state $A$ to another state $B$ without changing the entropy of the

---

† The disorder $D$ of a system composed of two subsystems is the product $D_1 D_2$ of the disorders of the two subsystems. This is because for every microstate of subsystem 1, the composite system has $D_2$ microstates. Since subsystem 1 has $D_1$ different microstates, the composite system has a total of $D_1 D_2$ microstates.

The entropy $S$ of a state is proportional to the logarithm of its disorder. Thus the individual entropies of subsystems 1 and 2 are

$$S_1 = k \log D_1 \qquad and \qquad S_2 = k \log D_2$$

and the entropy of the composite system is

$$S = k \log D = k \log D_1 D_2$$

where $k$ is the constant of proportionality. From the rules of logarithms we have

$$\log D_1 D_2 = \log D_1 + \log D_2$$

so

$$S = k \log D_1 + k \log D_2 = S_1 + S_2$$

Thus the entropy of a composite system is equal to the sum of the entropies of its subsystems, while the disorder of the composite system is equal to the product of the disorders of its subsystems.

universe ($\Delta S_\mathrm{u} = 0$). Such a transformation is said to be reversible because the system can be transformed from $B$ back to $A$ through the same intermediate states. Normally, however, $\Delta S_\mathrm{u}$ increases during a transformation, so that the reverse process is impossible. For example, $\Delta S_\mathrm{u}$ increases when ink spontaneously diffuses in water. This is an irreversible process:† the spontaneous separation of a mixture of ink and water cannot occur.

Transformations are classified as irreversible, reversible, or impossible depending on whether $\Delta S_\mathrm{u}$ is greater than, equal to, or less than zero:

$$\Delta S_\mathrm{u} > 0 \qquad \text{irreversible}$$

$$\Delta S_\mathrm{u} = 0 \qquad \text{reversible}$$

$$\Delta S_\mathrm{u} < 0 \qquad \text{impossible}$$

Reversible transformations are idealized processes in which there is no increase in disorder. They are like frictionless processes in mechanics, which are idealized processes with no decrease in mechanical energy. Only a reversible transformation can be represented by a solid line on a $pV$ diagram.

Entropy, being a thermodynamic variable like temperature and pressure, has a definite value for every state of a system. Consequently the entropy difference $\Delta S_\mathrm{s}$ between two states of a system depends only on these states, and not on the particular process used to transform the system from one state to another. The entropy difference between states can be calculated from the following rules:

*1* In a reversible adiabatic transformation the entropy change of a system is zero:

$$\Delta S_\mathrm{s} = 0 \qquad adiabatic \qquad 11.14$$

*2* In a reversible isothermal transformation the entropy change of a system is

$$\Delta S_\mathrm{s} = \frac{Q}{T} \qquad isothermal \qquad 11.15$$

where $Q$ is the heat absorbed at the absolute temperature $T$. The unit of entropy is thus calories per kelvin or joules per kelvin.

**REMARK** In the transformation of a system from state $A$ to state $B$, the entropy change of the universe $\Delta S_\mathrm{u}$ depends on the nature of the transformation, while the entropy change of the system $\Delta S_\mathrm{s}$ depends only on the states $A$ and $B$. Thus different transformations between the same two states produce different entropy changes in the environment. The second law (Eq. 11.13) gives the relation between these entropy changes.

**Carnot cycle**

As an example, we calculate the entropy change of a system in the course of a Carnot cycle (Fig. 11.9). The system, starting in state $A$, is transformed isothermally to state $B$ at the temperature $T_1$.

---

† Ink and water can be mixed reversibly using a piston and a semipermeable membrane.

During the transformation the system absorbs the heat $Q_1$, so by rule 2 the entropy change is

$$S_B - S_A = \frac{Q_1}{T_1}$$

The system is then transformed adiabatically from state $B$ to state $C$, so by rule 1 the entropy change is zero:

$$S_C - S_B = 0$$

During the isothermal transformation from state $C$ to state $D$, the system loses the heat $-Q_2$ at the temperature $T_2$, so the entropy change is

$$S_D - S_C = \frac{-Q_2}{T_2}$$

Finally, the system is transformed adiabatically from $D$ to $A$, so there is again no entropy change:

$$S_A - S_D = 0$$

The sum of these entropy changes is

$$\Delta S_S = \frac{Q_1}{T_1} - \frac{Q_2}{T_2}$$

which is the total entropy change of the system in one cycle. But since the final state is the same as the initial state, $\Delta S_S$ is zero. Consequently we have

$$\frac{Q_1}{T_1} - \frac{Q_2}{T_2} = 0 \qquad \text{or} \qquad \frac{Q_1}{Q_2} = \frac{T_1}{T_2}$$

This equation, together with Eq. 11.9, shows that the efficiency of a Carnot engine is

$$e = 1 - \frac{Q_1}{Q_2} = 1 - \frac{T_1}{T_2}$$

This is the result given in Eq. 11.10.

**Free energy and enthalpy**

It is useful to rewrite the second law in still another form, which involves only variables of the system. To be specific, consider a chemical reaction, such as

$$2C + O_2 \longrightarrow 2CO$$

that takes place at constant temperature and pressure. For every mole of $O_2$ consumed, 2 mol of CO is produced. Therefore, in order for $T$ and $p$ to remain constant, the volume of the system must increase during the reaction, and so the system must do the work $p \, \Delta V$.

The chemical energy of carbon monoxide, CO, is less than that of carbon and oxygen separately; i.e., the change of energy $\Delta E_s$ in this reaction is negative. By the first law

$$\Delta E_s = Q - W$$

where $W$ is the total work done by the system and $Q$ is the heat absorbed by the system. In this case $Q$ is negative because the system loses heat to the environment. The environment absorbs the heat

$$Q_\varepsilon = -Q = -(\Delta E_s + W)$$

so, from Eq. 11.15, the entropy change of the environment is

$$\Delta S_\varepsilon = \frac{Q_\varepsilon}{T} = \frac{-(\Delta E_s + p\,\Delta V)}{T}$$

The second law (Eq. 11.13) can therefore be written

$$\Delta S_\varepsilon = -\frac{\Delta E_s + p\,\Delta V}{T} \geq -\Delta S_s$$

or
$$\Delta E_s + p\,\Delta V - T\,\Delta S_s \leq 0 \qquad \textit{second law} \qquad 11.16$$

This form of the second law involves only variables of the system. The Gibbs *free energy*

$$G = E + pV - TS$$

is another thermodynamic variable that depends only on the state of the system. In a chemical reaction at constant pressure and temperature, the free energy changes to

$$G' = E' + pV' - TS'$$

Therefore the change of free energy is

$$\begin{aligned}
\Delta G = G' - G \\
= (E' + pV' - TS') - (E + pV - TS) \\
= (E' - E) + p(V' - V) - T(S' - S) \\
= \Delta E + p\,\Delta V - T\,\Delta S \qquad 11.17
\end{aligned}$$

which is identical to the left-hand side of Eq. 11.16. Thus in an isothermal, isobaric reaction, the free energy of the system never increases

$$\Delta G \leq 0$$

The *enthalpy*

$$H = E + pV$$

is still another useful thermodynamic variable. The change of enthalpy in an isobaric transformation is

$$\Delta H = \Delta E + p \, \Delta V \qquad\qquad 11.18$$

From the first law, $\Delta E = Q - W$, so

$$\Delta H = Q - (W - p \, \Delta V) = Q - W_0 \qquad\qquad 11.19$$

where $W_0 = W - p \, \Delta V$ is the work, other than $p \, \Delta V$ work, done by the system. Thus when $W_0$ is zero, the change of enthalpy is equal to the heat absorbed (or lost if $\Delta H$ is negative) by the system in an isobaric transformation.

The enthalpy difference between two states, like the entropy difference, depends only on the states, and not on the particular transformation used to take the system from one state to the other. In some transformations between two states, $W_0$ is zero, and the heat absorbed is equal to $\Delta H$. In other transformations between the same states, $W_0$ is not zero, and $Q$ is not equal to $\Delta H$. The maximum heat absorbed is equal to $T \, \Delta S$, the heat absorbed in a reversible isothermic transformation (Eq. 11.15). Less heat is absorbed if the transformation is irreversible, so in general

$$Q \leq T \, \Delta S$$

This, together with Eq. 11.19, gives us the inequality

$$W_0 = Q - \Delta H \leq T \, \Delta S - \Delta H$$

But from Eqs. 11.17 and 11.18 we have

$$\Delta G = \Delta H - T \, \Delta S \qquad\qquad 11.20$$

so

$$W_0 \leq -\Delta G$$

In other words, $-\Delta G$ is the maximum work that can be obtained from a particular transformation. This could be the work generated by a battery or by the contraction of a muscle.

### Determination of $\Delta S$, $\Delta G$, and $\Delta H$

The values of $\Delta S$, $\Delta G$, and $\Delta H$ have been measured for many reactions. As an example of such a measurement, let us consider the isobaric transformation of 1 mol of $H_2O$ from the liquid to the gas phase. From calorimetry measurements it is found that at a pressure of 1 atm and a temperature of 100°C, the heat of vaporization $H_v$ is 9.7 kcal (Table 9.1). This is the heat absorbed by the system during an isobaric transformation, and so it is equal to the enthalpy difference between steam and water:

$$\Delta H = Q = H_v = 9.7 \text{ kcal}$$

Since the transformation is also isothermal, the entropy change is

$$\Delta S = \frac{Q}{T} = \frac{9.7 \text{ kcal}}{373 \text{ K}} = 0.026 \text{ kcal/K}$$

The entropy change is positive, which signifies that the molecules of $H_2O$ are more disordered in the gas phase than in the liquid phase. The free energy is found from Eq. 11.20 to be

$$\Delta G = \Delta H - T \Delta S$$
$$= 9.7 \text{ kcal} - 373 \text{ K} \times 0.026 \text{ kcal/K}$$
$$= 0$$

Thus no work, other than $p \Delta V$ work, is obtainable from this transformation.

From Table 8.4 we find that the change of volume in this transformation is

$$\Delta V = V_g - V_l = 30,000 \text{ cm}^3 - 18.7 \text{ cm}^3$$
$$= 3.0 \times 10^4 \text{ cm}^3 = 3.0 \times 10^{-2} \text{ m}^3$$

The pressure is $p = 1$ atm $= 1.0 \times 10^5$ N/m$^2$, so the work done is

$$W = p \Delta V = (1.0 \times 10^5 \text{ N/m}^2)(3.0 \times 10^{-2} \text{ m}^3)$$
$$= 3.0 \times 10^3 \text{ J} = 0.72 \text{ kcal}$$

Therefore, from Eq. 11.18, the change of energy is

$$\Delta E = \Delta H - p \Delta V = 9.7 \text{ kcal} - 0.7 \text{ kcal}$$
$$= 9.0 \text{ kcal}$$

This means that of the 9.7 kcal of heat absorbed during the transformation, 9.0 kcal goes into the thermal energy of the system and 0.7 kcal goes into the $p \Delta V$ work done by the system.

**Biochemical reactions**

The free energy is important in the study of biochemical reactions. For instance, the oxidation of glucose

$$C_6H_{12}O_6 + 6O_2 \longrightarrow 6CO_2 + 6H_2O$$

is the primary source of energy in animals. The changes of free energy and enthalpy produced by the oxidation of 1 mol of glucose are

$$\Delta H = -673 \text{ kcal} \quad \text{and} \quad \Delta G = -686 \text{ kcal}$$

The change of enthalpy $\Delta H$, also called the *heat of combustion*, is equal to the change of energy $\Delta E$ in this case because there is no change of volume in this reaction. The maximum work obtainable from this reaction is $-\Delta G = 686$ kcal, which is larger than the energy released. This is because the entropy change is positive, and so, in addition to the energy released by the reaction, energy can be absorbed from the environment.

Unfortunately, at present no practical mechanism is known for converting this $-\Delta G$ directly into work. Work can be obtained indirectly, by burning the glucose and using the heat released to run a heat engine. The work obtained in this way is $W = e(-\Delta H)$, where $e$ is the efficiency of the engine.

In animals glucose is broken down by a complex series of biochemical reactions, which include the tricarboxylic acid cycle (Krebs cycle). For every mole of glucose metabolized, 38 mol of ATP is formed from ADP, in the reaction

$$\text{ADP} + \text{phosphate} \longrightarrow \text{ATP}$$

The change of free energy of this reaction is $+8$ kcal per mole of ATP formed. Thus in the overall reaction

$$\text{Glucose} + 6O_2 + 38\text{ADP} + 38 \text{ phosphate} \longrightarrow$$
$$38\text{ATP} + 6CO_2 + 6H_2O$$

the change of free energy is

$$\Delta G = -686 \text{ kcal} + 38 \times 8 \text{ kcal} = -382 \text{ kcal}$$

This free energy is lost, in the sense that it is no longer available for work.

There is still $38 \times 8$ kcal $= 304$ kcal of free energy available from the reaction

$$\text{ATP} \longrightarrow \text{ADP} + \text{phosphate}$$

This is the principal source of work-producing free energy in animals. This reaction, for instance, produces muscular contraction, with an efficiency of about 0.50. That is, the 304 kcal available for work produces about 150 kcal of work. The rest is lost as heat. The overall efficiency of the muscle as an engine, starting with glucose as the fuel, is

$$e = \frac{W}{Q} = \frac{W}{-\Delta H} = \frac{150 \text{ kcal}}{673 \text{ kcal}} = 0.22$$

The tension in a muscle can be increased with or without an accompanying change of the length of the muscle. If the length does not change (isotonic tension), the muscle does no work, since a force is not moved through a distance. Nevertheless, energy is required to maintain isotonic tension. This is a peculiarity of muscle fibers, and it is not a consequence of the laws of thermodynamics. The energy is supplied by ATP, which is continuously converted into ADP as long as the muscle is tensed. In this case, of course, none of the free energy is converted into work.

When a tensed muscle changes its length, it does work, and so there must be a decrease of free energy. That is, the second law requires that there be some reaction, such as the conversion of ATP to ADP, that supplies free energy for muscle contraction. For example, the work done by an 80-kg man walking up a flight of stairs 5 m high is

$$W = mgh = (80 \text{ kg})(9.8 \text{ m/s}^2)(5 \text{ m})$$
$$= 3.9 \times 10^3 \text{ J} = 0.94 \text{ kcal}$$

If the efficiency of the conversion of free energy into work is 0.5, then 1.88 kcal of free energy must be released. This requires the conversion of $1.88/8.0 = 0.23$ mol of ATP into ADP. A much smaller quantity of ATP is used to maintain the isotonic tension in the leg of a man standing still.

The Gibbs free energy is of great importance in the study of bio-chemical reactions because it can be used to determine the equilibrium constant of chemical reactions and thus to study the possibility of different metabolic pathways. However, a discussion of this application is beyond the scope of this book.

## PROBLEMS

1 What is the heat capacity of a 350-g aluminum pot? *Ans.* 0.075 kcal/°C

2 How much heat is required to raise the temperature of 15 kg of granite by 5°C?

3 It takes 210 kcal to raise the temperature of 350 g of lead from 0 to 20°C. What is the specific heat of lead? *Ans.* 0.03 kcal/kg-°C

4 What is the heat capacity of a system consisting of 7.5 kg of water in a 0.75-kg aluminum bucket?

5 A 400-g aluminum teakettle contains 2 kg of water at 15°C. How much heat is required to raise the temperature of the water (and kettle) to 100°C? *Ans.* 177 kcal

6 The inner vessel of a calorimeter contains 100 g of trichloromethane at 35°C. The vessel is surrounded by 1.75 kg of water at 18°C. After a time the trichloromethane and the water reach the common temperature of 18.22°C. What is the specific heat of trichloromethane?

7 A 600-g iron bar at a temperature of 300°C is plunged into 4 kg of water at 20°C. The iron and water soon come to the same temperature. What is this temperature? (This problem requires a little algebraic manipulation.) *Ans.* 24.7°C

8 A 250-g iron ball falls 3 m onto an aluminum plate. The mass of the plate is 1.50 kg. If the initial potential energy of the ball is converted entirely into the thermal energy of the ball and plate, what is the temperature increase of the plate? Assume that the ball and plate have the same initial temperature and the same temperature increase.

9 A 20-g iron nail is being pounded by a 2.2-kg (1-lb) hammer. The speed of the hammer is 3 m/s when it strikes the nail. If half the kinetic energy of the hammer is converted into thermal energy of the nail, how many times must the nail be struck to raise its temperature 25°C? *Ans.* 49

10 The molar heat capacity is the heat capacity of 1 mol of a substance. Calculate the molar heat capacities (at constant pressure) of carbon dioxide, $CO_2$, oxygen, $O_2$, and nitrogen, $N_2$.

11 The molar heat capacity (at constant pressure) of an ideal diatomic gas can be shown to be $\frac{7}{2}R$, where $R = 8.32$ J/K is the gas constant. Check this for (*a*) oxygen and (*b*) nitrogen by converting the answers to Prob. 10 to joules per kelvin. *Ans.* (*a*) 3.51R; (*b*) 3.49R

12 The law of Dulong and Petit says that the molar heat capacities of solid elements are approximately equal. Check this for aluminum, copper, and iron.

**REMARK** This "law" was used in the nineteenth century to determine the atomic mass of an element from a measurement of its specific heat.

13 What is the atomic mass of an element whose specific heat is 0.0308 kcal/kg-K? (See Prob. 12 and the Remark following it.) *Ans.* approximately 195 u

14 One mole of oxygen is heated at a constant pressure of 1 atm from 10 to 25°C. (*a*) How much heat is absorbed by the gas? (*b*) Using the ideal-gas law, calculate the change of volume of the gas in this process. (*c*) What is

the work done by the gas during this expansion? (*d*) From the first law, calculate the change of energy of the gas in this process. (*e*) What is the heat capacity $C_v$ of 1 mol of oxygen at constant volume? (*f*) What is the difference $C_p - C_v$ between the molar heat capacities at constant pressure and constant volume?

**REMARK** Note that $C_p - C_v$ is equal to the gas constant $R$. You can show that this is a general result for any ideal gas by repeating the last problem using algebraic symbols instead of numbers.

15 The United States generates about $2 \times 10^{16}$ J of electric energy a day. This energy is equivalent to work, since it can be converted into work with 100 percent efficiency by an electric motor. (*a*) If this energy is generated by power plants with an average efficiency of 0.30, how much heat is dumped into the environment each day? (*b*) How much water is required to absorb this heat if the water temperature is not to increase more than 2°C?

> *Ans.* (*a*) $4.66 \times 10^{16}$ J; (*b*) $5.05 \times 10^{12}$ kg

16 (*a*) Calculate the efficiency of a Carnot engine that operates between the temperatures 600 and 300°C. (*b*) If the engine absorbs 100 kcal of heat at the high temperature, how much heat does it release at the low temperature?

17 Consider a tray with six chips, each of which is white on one side and black on the other. (*a*) How many microstates does this system have? (*b*) How many macrostates does the system have? (*c*) List all the distinct microstates of the system and determine the disorder *D* of each macrostate. (*d*) After shaking the tray, what is the probability that the system will be in the $g = 0.5$ macrostate?

> *Ans.* (*a*) 64; (*b*) 7; (*d*) 0.31

18 What is the probability that a tray with 100 black-and-white chips will be in the $g = 0.10$ macrostate after shaking?

19 (*a*) What is the entropy change of 1 mol of $H_2O$ when it changes from ice to water at 0°C (see Table 10.1)? (*b*) If the ice is in contact with an environment at a temperature of 10°C, what is $\Delta S_u$ when the ice melts? (*c*) If the temperature of the environment is $-10$°C,

what would $\Delta S_u$ be if the ice melted? (*d*) What must be the temperature of the environment for $\Delta S_u$ to be zero?

> *Ans.* (*a*) 5.26 cal/°C; (*b*) $+0.19$ cal/°C; (*c*) $-0.20$ cal/°C; (*d*) 0°C

20 What is the entropy change of 850 g of water that is heated from 20 to 50°C?

**REMARK** Equation 11.15 is exact only for an isothermal process. However, it can be used to good approximation whenever the temperature change $\Delta T$ is small compared to the initial temperature $T_1$ of the system. In such a case the average temperature of the system

$$T_{av} = \tfrac{1}{2}(T_1 + T_2)$$

should be used in the denominator of Eq. 11.15.

21 A 500-g block of iron at a temperature of 60°C is placed in contact with a 500-g block of iron at a temperature of 20°C. (*a*) The blocks soon come to a common temperature of 40°C. What is the entropy change in this case? (*b*) Calculate the entropy change, assuming the temperature of the hotter block increased to 80°C while the temperature of the colder block decreased to 0°C. (Since the entropy of the universe decreases in this case, this process is impossible.)

> *Ans.* (*a*) 0.023 cal/°C; (*b*) $-0.71$ cal/°C

22 What is the entropy change of the universe in the transformation described in Prob. 8? Assume the initial temperature of the block and plate is 20°C.

# BIBLIOGRAPHY

BENT, HENRY A.: "The Second Law: An Introduction to Classical and Statistical Thermodynamics," Oxford University Press, New York, 1965. An excellent introduction to thermodynamics that uses no advanced mathematics. The book contains extensive quotations from the work of the founders of thermodynamics, which provide fascinating glimpses of the history of the subject.

GUYTON, ARTHUR C.: "Textbook of Medical Physiology," W. B. Saunders Company, Philadelphia, 1971. The details of carbohydrate metabolism are given in chap. 67. Chapter 7 contains an interesting discussion of the current theories of muscle contraction.

HUXLEY, H. E.: The Mechanism of Muscular Contraction, *Scientific American,* **213:** 18 (June 1965). The author discusses his model of muscle contraction.

INGRAHAM, LLOYD L., and ARTHUR B. PARDEE: Free Energy and Entropy in Metabolism, in David M. Greenberg (ed.), "Metabolic Pathways," 3d ed., vol. 1, Academic Press, Inc., New York, 1967. This article reviews the role free energy and entropy play in the analysis of biochemical reactions. Other articles in the book discuss the details of glucose metabolism and other metabolic processes.

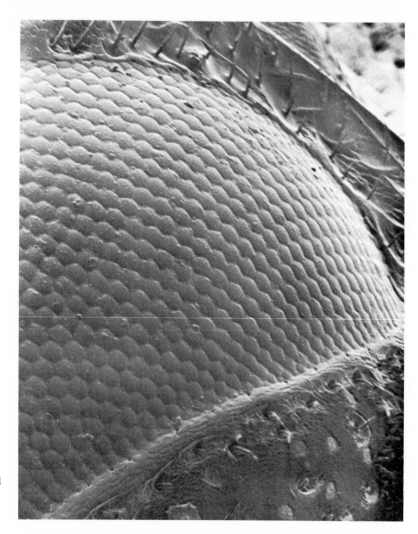

Scanning electron photomicrograph of the compound eye of a beetle. The size of the facets (ommitidia) of insect eyes has evolved to give the best vision consistent with limitations imposed by the wave nature of light (Sec. 14.2). [*Courtesy of H. F. Howden, Carleton University, Ottawa.*]

# III wave phenomena

The physical world can be considered to consist of only two basic kinds of entities: material objects and waves. Roughly speaking, a wave is pure energy, without any mass. Sound, ultrasonic waves, light, x-rays, infrared radiation, ultraviolet radiation, gamma rays, radar, and radio waves are all examples of waves, although the physical nature of some of these waves is very different. Thus sound and ultrasonic waves are both mechanical vibrations in a material medium, whereas the other waves are electromagnetic in nature. However, regardless of their fundamental nature, all waves have the same general characteristics.

Most animals obtain information about their environment by detecting some types of waves, and they communicate with each other by generating other types of waves. For instance, man detects light with his eyes, sound with his ears, and infrared radiation with his skin, and he generates sound with his speech organs. Other animals detect and generate other waves, such as ultrasonic waves and ultraviolet radiation. As we shall see, animals differ in the attributes of the waves they can detect. For instance, man detects the frequency of light as color, whereas a bee can detect both the frequency and polarization of light. Thus knowledge of the physical attributes of different types of waves is required to understand the different senses which animals have evolved to meet their special needs.

# CHAPTER 12 waves

This chapter introduces the basic properties of waves, using as an example the waves on a stretched cord. These waves are well suited for this purpose because they are easy to visualize, while at the same time they exhibit most of the general properties common to all waves.

Waves on a cord, like sound waves, consist of the mechanical vibrations of elements of a material medium. Consequently, the behavior of mechanical waves can be completely understood in terms of the laws of mechanics governing the motion of material objects.

## 12.1 WAVES ON A STRETCHED CORD

Energy can be transferred from one point to another by various means. The most obvious, perhaps, is to send an object from one point to another with speed $v$. The energy the object carries is its own kinetic energy

$$K = \tfrac{1}{2}mv^2$$

where $m$ is its mass. For instance, when gunpowder is burned in a gun, some of the energy released is transformed into the kinetic energy of a bullet. The bullet carries this energy to the target, transferring energy from the gunpowder to the target.

With the use of a material object to transfer energy there is always a corresponding transfer of mass. Is it possible to transfer energy without transferring mass? At first this does not seem likely because one tends to think of energy in terms of moving objects, and so the transfer of energy seems to involve the transfer of the objects themselves. However, it is easy to demonstrate an instance in which energy is transfered without any transfer of mass.

One end $B$ of a long cord† is attached to a fixed support, and the other end $A$ is held by a student (Fig. 12.1). When the student gives his end a sharp flip up and down, a pulse is formed that moves along the cord to the other end with constant speed (Fig. 12.2). Upon reaching $B$ the pulse hits the support and bounces back toward $A$. To demonstrate that the pulse carries energy, end $B$ is attached to a lever and weight arrangement, as shown in Fig. 12.3. Now when the pulse reaches $B$, the weight is momentarily moved upward, proving that the pulse can do work. The work it can do is equal to the work done in producing the pulse at $A$. Thus the cord allows energy to flow along it from one end to the other. In the process there is no transfer of mass, since the cord remains where it started.

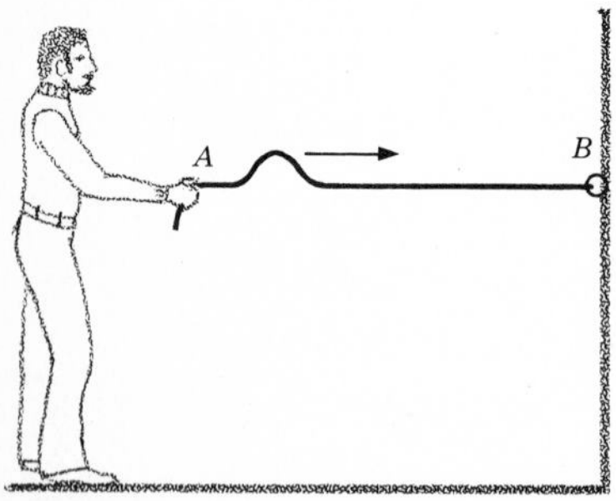

FIGURE 12.1
A pulse traveling with constant speed along a cord.

† For demonstration purposes a long spring is often used.

The disturbance that moved along the cord is called a *wave*. If the end *A* is shaken steadily, a steady wave pattern is set up on the cord (Fig. 12.4), which results in a steady oscillation of the weight. We shall often come across such wavy patterns, but the word wave is used generally for any disturbance that propagates in this manner, regardless of its shape.

To be more precise, a wave is a disturbance in a medium that propagates through the medium at a constant speed v characteristic of the medium. In the present example the cord is the medium, and the disturbance is the displacement of points of the cord from their undisturbed, or equilibrium position. It is important to realize that every medium has an equilibrium configuration in which all the points of the medium are at rest. Figure 12.5*a* shows the equilibrium configuration of a stretched cord. A ruled line is drawn through this equilibrium configuration so that each point on the cord can be labeled according to its equilibrium position. For instance, point *B* is the point that is 3.0 cm from point *A* at equilibrium.

Figure 12.5*b* shows the cord with a wave on it. Point *B* is no longer on the axis but is displaced from the axis by the distance $y_{3.0} = 0.7$ cm. The subscript 3.0 indicates that $y_{3.0}$ is the displacement of the point *B* whose equilibrium position is 3.0 cm. This displacement is the measure of how much the medium is disturbed at this point. As the wave moves to the right, the displacement of point *B* changes. In Fig. 12.5*c* the wave has moved so that now point *B* has the displacement $y_{3.0} = 0.9$ cm, and point *C* has the displacement $y_{4.0} = 0.7$ cm. Note that each point of the cord moves only up and down while the disturbance itself moves with constant speed v along the cord. This speed is not the same as the up-and-down speed of a particular point on the cord. In fact, the up-and-down speed is not constant at all, as we shall see.

To be specific, suppose the time interval between Fig. 12.5*b* and *c* is 0.2 s. In this time the point on the cord with the displacement 0.7 cm has changed from point *B* (at 3.0 cm) to point *C* (at 4.0 cm), so that the speed of the wave is

$$v = \frac{4.0 \text{ cm} - 3.0 \text{ cm}}{0.2 \text{ s}} = 5 \text{ cm/s}$$

On the other hand, during this same time the displacement at point *B* changed from 0.7 to 0.9 cm, so that the (average) up-and-down, or *transverse*, speed of this point during this time interval is only

$$v = \frac{0.2 \text{ cm}}{0.2 \text{ s}} = 1.0 \text{ cm/s}$$

The transverse speed $v$ is related to the energy carried by the wave (Sec. 12.4), whereas the wave speed v is the speed with which this energy is transferred along the cord. For a given medium the wave speed is a constant, independent of the shape of the wave, whereas the transverse speed varies from wave to wave, and even from point to point within a given wave.

Since the points of the cord move perpendicular to the direction in which the wave moves, these waves are called *transverse*. There also exist *longitudinal* waves, in which the points of the medium move back and forth in the same direction the wave moves. They can be

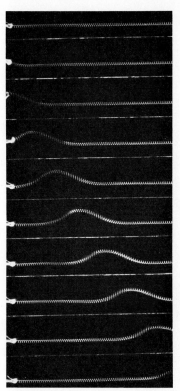

FIGURE 12.2
Photograph of the formation and motion of a pulse on a cord. [*From "PSSC Physics," 3d ed., by Education Development Center, Inc., reprinted by permission of D. C. Heath and Co.*]

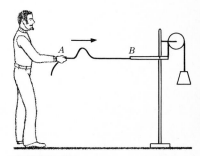

FIGURE 12.3
A lever and weight arrangement that converts the energy of a pulse into the potential energy of a weight.

demonstrated on a long spring by squeezing together a few of the coils at one end, and then releasing them. As these coils move back to their equilibrium position, neighboring coils are compressed, which in turn return to their equilibrium positions, compressing coils still further along the spring. The result is a longitudinal compression wave that moves along the spring with constant speed.

Figure 12.6a shows a spring in its equilibrium position, and Fig. 12.6b shows the displacement $y_{2.0}$ of the point whose equilibrium position is 2.0 cm from the end of the spring. Even though this point moves parallel to the motion of the longitudinal wave, there is still no net flow of mass along the spring because the point only moves back and forth about its equilibrium position.

Both transverse and longitudinal waves can propagate through solid media. For example, after an earthquake (or a moonquake) both transverse and longitudinal *seismic* waves travel outward from the quake site through the earth's (or moon's) crust. These two types of waves have different speeds which depend on the properties of the rocks through which they travel. Measurements of the speed of seismic waves provide valuable information about the nature of the rocks inside the earth and moon.

As a consequence of Fluid Property 1 (Sec. 7.2), only longitudinal waves can propagate through fluid media. This is discussed more fully in Sec. 13.1, where it is shown that sound is a longitudinal compression wave in air. Waves which are disturbances in material media, whether solid or fluid, are called *mechanical* waves, to distinguish them from electromagnetic waves, such as light.

## 12.2 THE MATHEMATICAL THEORY OF WAVES

All the properties of mechanical waves can be mathematically derived from the principles of mechanics (especially Newton's second law),

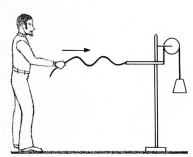

FIGURE 12.4
A train of pulses traveling on a cord.

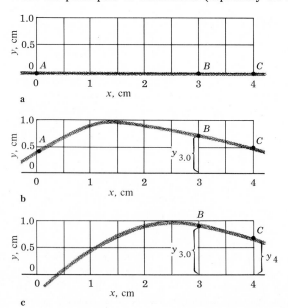

FIGURE 12.5
Portion of a stretched cord. (*a*) Equilibrium position of the cord. (*b*) Displacement of the cord produced by a wave at one instant. (*c*) Displacement of the cord 0.2 s later.

so that no new physical principles are required to understand these properties. In fact, the ability of the principles of mechanics to explain all aspects of such complex phenomena is strong evidence for the universal truth of these principles. Although the details of the mathematical theory of waves are beyond the scope of this book, some discussion of the results of this theory is useful for understanding the mechanics of waves.

The mathematical theory considers only small displacements in the medium. A bit of matter that is displaced slightly from its equilibrium position is pulled back to equilibrium by forces exerted on it by neighboring bits of matter. At the same time, by Newton's third law, the displaced matter exerts forces on its neighbors, so that as the displaced matter is pulled back to equilibrium, it pulls its neighbors away from equilibrium. These neighboring bits of matter are in turn acted upon by their neighbors, and so the process repeats, propagating the original disturbance through the medium.

**REMARK** The displacements must be so small that the medium is not permanently deformed. For example, a piece of metal that is flexed slightly will spring back to its original shape. However, if it is flexed beyond a certain amount, it will be permanently deformed. Mechanical waves are limited to displacements that do not exceed the material's ability to spring back to its original shape.

Figure 12.7 shows the successive positions of a pulse traveling to the right along a long cord. A small region $R$ of the cord is marked so that its motion can be studied as the pulse passes it.

In Fig. 12.7$a$ the pulse is just reaching $R$. Since $R$ has not yet started to move, its vertical speed is zero. At the same time, the cord exerts the forces $\mathbf{F}_1$ and $\mathbf{F}_2$ on either end of $R$. The magnitudes of $\mathbf{F}_1$ and $\mathbf{F}_2$ are both $T$ (the tension in the cord), but because the region $R$ is curved, the sum $\mathbf{S} = \mathbf{F}_1 + \mathbf{F}_2$ of these forces is not zero. Figure 12.8$a$ shows that since the angle $\theta$ is small, $\mathbf{S}$ is directed upward, perpendicular to the cord. Therefore, by Newton's second law, the region $R$ has an upward acceleration $a = S/m$, where $m$ is the mass of the cord in $R$. It is this acceleration that causes $R$ to start to move upward as the pulse moves along the cord.

A short time later $R$ has acquired a considerable upward speed and is in the position shown in Fig. 12.7$b$. But now the total force $\mathbf{S}$ on $R$ is directed downward, as shown in Fig. 12.8$b$. This means that $R$ will continue to move upward but with decreasing speed.

When $R$ reaches its maximum displacement (Fig. 12.7$c$), it again has zero speed. In this position the tension in the cord exerts a downward force $\mathbf{S}$ on $R$ (Fig. 12.8$c$), and so $R$ starts to move downward with increasing speed. Figure 12.7$d$ shows $R$ in its position of greatest downward speed, while Fig. 12.8$d$ shows that the total force on it is now upward. Thus $R$'s downward speed now decreases until $R$ returns to its equilibrium position (Fig. 12.7$e$), where it has zero speed and zero acceleration (Fig. 12.8$e$).

This discussion shows that while the pulse as a whole moves to the right with constant speed v, each point on the cord moves with continually changing speed in the vertical (transverse) direction. Figure 12.9 shows the speed of different points of the pulse. The arrows point in the direction of motion, and their lengths are proportional to the speed at that point. The points in the leading half of the pulse

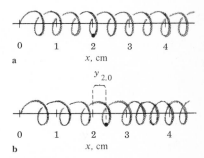

FIGURE 12.6
Portion of a spring. ($a$) Equilibrium configuration of the spring. ($b$) Longitudinal displacements of points of the spring.

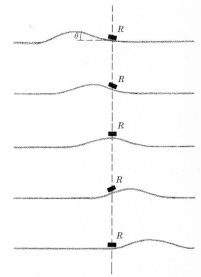

FIGURE 12.7
Successive positions of a pulse traveling to the right along a cord. A small portion $R$ of the cord is marked so that its motion can be studied.

(the right-hand side) are moving upward, while points in the trailing half of the pulse (the left-hand side) are moving downward.

This qualitative discussion indicates how the displacement itself generates the forces that keep it moving. The mathematical theory shows in general that a wave of any shape will propagate in a similar fashion. In particular it can be shown that the speed v of a wave on a cord of mass $M$ and length $L$ is

$$v = \sqrt{\frac{T}{M/L}}$$

where $T$ is the tension. The quantity $M/L$ is just the mass per unit length, or *linear density*, of the cord. Thus, if we let

$$\mu = \frac{M}{L} \qquad\qquad 12.1$$

the speed of the wave is

$$v = \sqrt{\frac{T}{\mu}} \qquad\qquad 12.2$$

**REMARK** It is not possible to give an elementary derivation of this result. However, in Appendix V the speed of a pulse is estimated by considering the forces acting on it. This calculation is worth studying in order to appreciate how the principles of mechanics are applied to a mechanical wave.

According to Eq. 12.2, the speed of a wave on a stretched cord increases with the square root of the tension and decreases with the square root of the mass per unit length. For example, suppose the tension in a guitar string is $T = 30$ N and its mass per unit length is $\mu = 0.015$ kg/m. Then the speed of transverse waves on this string is

$$v = \sqrt{\frac{T}{\mu}} = \sqrt{\frac{30 \text{ N}}{0.015 \text{ kg/m}}}$$
$$= \sqrt{2000 \text{ N-m/kg}} = \sqrt{2000 \text{ m}^2/\text{s}^2} = 44.7 \text{ m/s}$$

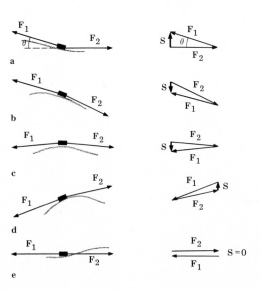

**FIGURE 12.8**
The forces $\mathbf{F}_1$ and $\mathbf{F}_2$ that act on the ends of region $R$ in Fig. 12.7 and the sum $\mathbf{S} = \mathbf{F}_1 + \mathbf{F}_2$.

If the tension were quadrupled, the speed would be doubled, and if the mass per unit length were quadrupled, the speed would be halved.

## 12.3 SUPERPOSITION

More than one wave can exist at the same point in a medium at the same time. This property of waves clearly distinguishes them from material objects. For example, the two pulses that are heading toward each other in the cord in Fig. 12.10$a$ pass through each other and continue on their way undisturbed (Fig. 12.10$b$). Material objects do not interpenetrate this way. A wave, however, is not a material object. It is a self-propagating pattern of displacements of points in a material medium.

**REMARK** A wave is a real physical entity, even though it is not a material object. It moves, carries energy, and interacts with material objects. In fact waves and material objects are the two basic physical concepts in terms of which the physical world is described.

What is the displacement of the cord while the two pulses in Fig. 12.10 are interpenetrating? The mathematical theory of waves answers this by proving the following remarkable principle.

**Superposition principle** *If at any instant two or more waves simultaneously exist at a point, the displacement of the point is the sum of the displacements the point would have with each wave separately.* In applying this principle, displacements to one side of the equilibrium position are taken to be positive, and displacements to the other side are taken to be negative.

Figure 12.11 shows how this principle applies to the pulses in Fig. 12.10 as they are interpenetrating. In each part of the figure, the gray lines represent the displacements the cord would have with each pulse separately. Pulse 1 moves steadily from right to left while pulse 2 moves steadily from left to right. The actual displacements of the cord at each instant are represented by the black lines. These displacements are found by adding the displacements produced by each pulse separately, in accordance with the superposition principle.

For instance, consider the point $A$ on the cord. In Fig. 12.11$a$ the peak of pulse 2 and the leading edge of pulse 1 are simultaneously at point $A$. Since pulse 1 alone would have displaced $A$ by 5 cm and pulse 2 alone would have displaced it by 10 cm, the actual displacement of point $A$ at this instant is 15 cm, as shown. The rest of the black curve in Fig. 12.11$a$ is calculated similarly.

In Fig. 12.11$b$ the peak of pulse 1 and the trailing side of pulse 2 are simultaneously at $A$. Again the actual displacement of $A$ is the sum of the displacements produced by each of these pulses separately. All the points on the black curves are calculated in the same way. These black curves represent the actual displacement of the cord, which can be quite complicated, even when it is just the sum of two simple pulses.

Interpenetration of a positive and a negative pulse is shown in Fig. 12.12. Figure 12.12$c$ shows that at the instant when these pulses completely overlap, the cord has zero displacement everywhere. This is because, by the superposition principle, the sum of a positive and

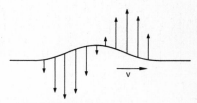

FIGURE 12.9
Speed of different points of a pulse.

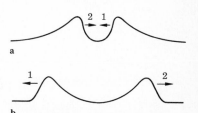

FIGURE 12.10
($a$) Two pulses heading toward each other on a cord. ($b$) The same pulses a moment later, after passing through each other.

a negative displacement of equal magnitude is zero. (Study Figs. 12.11 and 12.12 carefully until you are sure you understand how the black waves are constructed from the two gray pulses.)

It is surprising that the cord can evolve from the apparent equilibrium configuration shown in Fig. 12.12c into the two traveling pulses shown in Fig. 12.12d and 12.12e. Of course the cord is not really in equilibrium in Fig. 12.12c, because the points in the region of superposition are not at rest. Figure 12.13 shows that the points on the leading side of the positive pulse are moving away from equilibrium while the points on the trailing side are moving toward equilibrium; the points of the negative pulse behave similarly. Furthermore, the superposition principle applies to speed as well as to displacement. That is, the actual speed of each point of the cord is the sum of the speeds it would have with each pulse separately. Thus when the two pulses overlap, they combine momentarily to give a wave with zero displacement for each point but not zero speed. Each point in the wave is moving through its equilibrium position with the sum of the speeds it would have if each pulse was present alone (Fig. 12.13).

The superposition principle makes it possible to express a complex wave as the sum of several simpler waves. It is, in fact, possible to express any wave, no matter how complex, as the sum of waves of

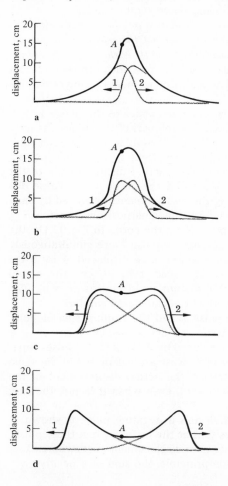

FIGURE 12.11
The superposition of the two pulses in Fig. 12.10 as they pass through each other.

an especially simple form, called *sine waves*. These are discussed in
the next section.

## 12.4 SINE WAVES

### Wavelength and frequency

A sine wave is a particular wave pattern in which positive and negative pulses alternate successively (Fig. 12.14). The distance between
neighboring positive peaks is a constant, called the *wavelength* λ† of
the wave. In fact, the displacements of the wave are the same at
any two points separated by the distance λ, such as $x$ and $y$ in Fig.
12.14. This kind of wave pattern is said to be *periodic* because the
pattern repeats itself exactly at intervals of λ.

The maximum displacement of a sine wave, called its *amplitude*
$A$, is the same for both positive and negative displacements. The
actual shape of a sine wave is related to the trigonometric sine. To
understand this consider 1 wavelength of a sine wave of amplitude
$A$, as shown in Fig. 12.15. The interval between $x = 0$ and $x = λ$ has
been divided into 360 equal divisions, corresponding to the 360° in
a circle. Therefore any point $x$ on the $x$ axis can be related to an
angle $θ$ by the proportion

$$\frac{x}{λ} = \frac{θ}{360°}$$

or

$$θ = \frac{x}{λ} 360°$$

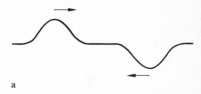

a

b

The displacement $y$ of a sine wave at point $x$ is given in terms of
the sine of $θ$ by

$$y = A \sin θ$$
$$= A \sin\left(\frac{x}{λ} 360°\right) \qquad 12.3$$

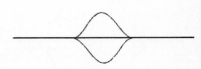

c

For example, if a sine wave has an amplitude $A = 0.5$ cm and a
wavelength $λ = 30$ cm, its displacement at $x = 6$ cm is

$$y = (0.5 \text{ cm}) \sin\left(\frac{6 \text{ cm}}{30 \text{ cm}} 360°\right)$$

$$= 0.5 \text{ cm} \times \sin 72° = 0.5 \text{ cm} \times 0.951 = 0.475 \text{ cm}$$

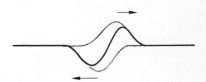

d

**REMARK** It is natural to associate the points in 1 wavelength of a sine wave
with points on a circle because a sine wave repeats after 1 wavelength just
as a circle repeats after 360°.

Figure 12.14 shows a sine wave at one particular instant of time.
If the wave is traveling to the right with speed v, the whole pattern
moves with time, as shown in Fig. 12.16. The open circles in this figure
show how a characteristic feature of the wave moves, whereas the
solid circles show how a fixed material point in the medium moves.
The peak of the wave moves to the right with speed v, whereas a
point in the medium moves down and up.

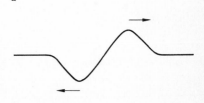

e

FIGURE 12.12
Superposition of a positive and a
negative pulse as they pass
through each other.

†λ is the Greek letter lambda.

After the wave has moved a distance λ, the pattern looks the same again. In the time $\tau = \lambda/v$ that the wave takes to move the distance λ, a fixed point in the medium returns to its initial position, completing one *cycle* of motion. The time $\tau$ is called the *period* of the wave. The relation

$$\tau = \frac{\lambda}{v}$$

or

$$v = \frac{\lambda}{\tau} \qquad \qquad 12.4$$

between the speed, wavelength, and period of a sine wave is of fundamental importance in the study of wave motion.

The period $\tau$ is also the time it takes for one whole wavelength of a wave to pass a given point in space. If $\tau = 0.25$ s, 1 wavelength passes a given point every 0.25 s, so that $1/0.25 = 1/\tau = 4$ wavelengths pass each second. The quantity

$$f = \frac{1}{\tau} \qquad \qquad 12.5$$

called the *frequency* of the wave, is the number of wavelengths that pass a point in 1 s. Its unit is $1/s = s^{-1}$. The frequency is also the number of cycles that a point in the medium executes in 1 s, so that $f$ is often quoted as so many cycles per second (cps). More recently the name *hertz* (Hz) has been adopted for this unit. Thus the frequency of a wave with the period $\tau = 0.25$ s is

$$f = \frac{1}{\tau} = 4 \text{ s}^{-1} = 4 \text{ cps} = 4 \text{ Hz}$$

An alternative form of Eq. 12.4 is obtained by combining it with Eq. 12.5 to get

$$v = \lambda f \qquad \qquad 12.6$$

Only Eqs. 12.5 and 12.6 need be remembered because Eq. 12.4 is easily derived from them.

In any particular physical situation the wave speed v is a constant. This means, for example, that the speed of a wave on a cord does not depend on the wavelength or frequency of the cord. The same is true for the speed of sound (Sec. 13.2) and the speed of light† (Sec. 14.1), though, of course, the speeds of these different kinds of waves all differ. But for a given kind of wave in a given medium, Eq. 12.6 states the relation between the wavelength and frequency of a wave.

† The speed of light in materials such as glass and water does depend slightly on its wavelength, but this can be neglected for most purposes.

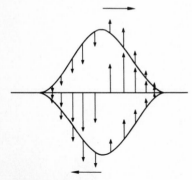

FIGURE 12.13
Superposition of the transverse speeds of the two pulses in Fig. 12.12c

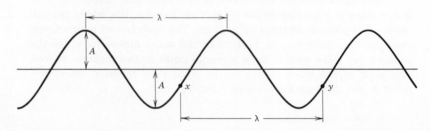

FIGURE 12.14
A sine wave.

The frequency is of importance because for sound and light it has direct psychophysiological meaning. The tone or pitch of a sound depends on the frequency of the sound wave, and the color of an object depends on the frequency of the light wave coming from it. For instance, the musical tone A is a sound wave with a frequency of about 440 Hz, and the blue color in a rainbow is a light wave with a frequency of about $7.5 \times 10^{14}$ Hz.

### Fourier's theorem

This connection between the sine wave and the psychophysiological qualities of tone and color is certainly remarkable. However, sine waves have an even more remarkable mathematical property that justifies giving them special attention.

**Fourier's theorem** *Any wave, of whatever shape, can be uniquely expressed as the superposition (sum) of sine waves of definite wavelengths and amplitudes.*

For example, according to this theorem, the rather complex wave pattern (dark curve) in Fig. 12.17 is equal to the superposition of sine waves. These sine waves, which are called the *Fourier components* of the complex pattern, are shown as gray curves in Fig. 12.17. These components are unique; i.e., only sine waves with these wavelengths and amplitudes combine to equal the given pattern. Any pattern, no matter how complex, is similarly equal to a sum of sine waves, though many sine waves may have to be superimposed to equal a very complex pattern.

The longest-wavelength Fourier component in Fig. 12.17 has a wavelength of 5 cm and an amplitude of 1.5 cm. If the wave speed is 12 m/s, the frequency of this component is

$$f = \frac{v}{\lambda} = \frac{12 \text{ m/s}}{0.05 \text{ m}} = 240 \text{ s}^{-1} = 240 \text{ Hz}$$

This component is represented on a graph of amplitude against frequency (Fig. 12.18) by a vertical line 1.5 cm high, positioned at 240 Hz on the horizontal axis. The next longest wavelength component has a frequency of

$$f = \frac{12 \text{ m/s}}{0.0167 \text{ m}} = 720 \text{ s}^{-1} = 720 \text{ Hz}$$

and an amplitude of 0.6 cm, so it is represented by a vertical line 0.6 cm high, positioned at 720 Hz on the horizontal axis. The third component is represented by a line 0.6 cm high, positioned at 1200 Hz.

A representation of the amplitudes and frequencies of the components of a particular wave, such as Fig. 12.18, is called a *spectrum*.

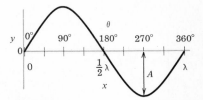

FIGURE 12.15
One wavelength of a sine wave divided into 360°.

FIGURE 12.16
A sine wave at four successive instants of time. The open circles mark a characteristic feature of the wave (a peak), whereas the solid circles mark a physical point in the medium.

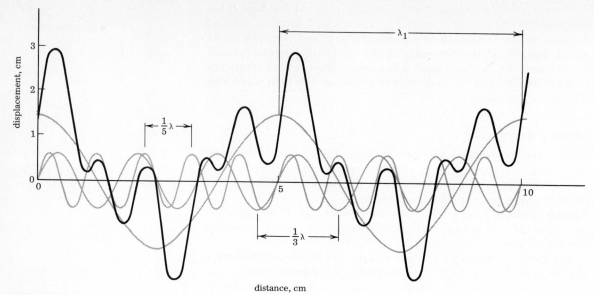

FIGURE 12.17
A complex periodic wave produced
by the superposition of three sine
waves of different amplitude and
wavelength.

It is a convenient way to specify a wave uniquely. Figure 12.19 shows
several complex waves with their corresponding spectra.

Complex wave patterns are either *periodic* or *aperiodic*. A periodic
wave, such as a single sine wave, repeats its pattern exactly at intervals of length $\lambda_1$. An aperiodic wave never repeats its pattern. The
wave in Fig. 12.19$c$ is aperiodic, and the waves in Figs. 12.17 and 12.19$a$
and $b$ are periodic. The spectrum of an aperiodic wave contains a
continuous distribution of component frequencies, whereas the spectrum of a periodic wave is discrete.

The spectrum of a periodic wave is discrete because in order for
a wave to be periodic with an interval length $\lambda_1$ the wavelengths of
its Fourier components must be members of the sequence

$$\lambda_1, \tfrac{1}{2}\lambda_1, \tfrac{1}{3}\lambda_1, \ldots, \tfrac{1}{n}\lambda_1, \ldots$$

where $n$ is an integer ($n = 1, 2, 3, \ldots$). For example, the components
of the wave in Fig. 12.17 are the first, third, and fifth members of
such a sequence. In the interval $\lambda_1$ there is exactly 1 wavelength of
the component equal to $\lambda_1$, exactly 3 wavelengths of the component
equal to $\tfrac{1}{3}\lambda_1$, and exactly 5 wavelengths of the component equal to
$\tfrac{1}{5}\lambda_1$. Thus, since all the component waves are the same at the end
of the interval as they are at the beginning, their sum repeats after
this interval. Clearly there are exactly $n$ wavelengths of the component equal to $(1/n)\lambda_1$, so that a sum of such components repeats in
the interval $\lambda_1$.

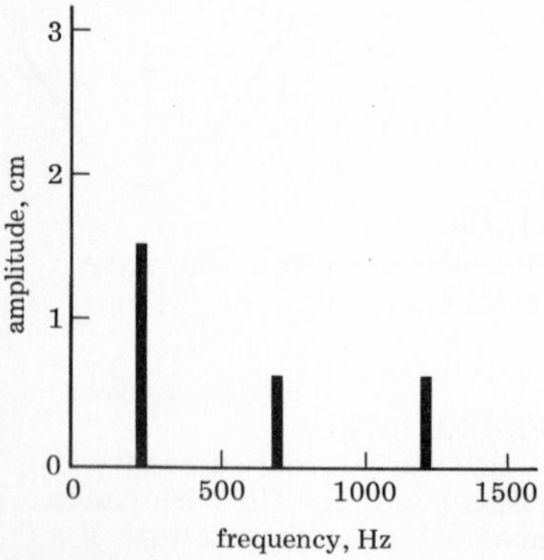

Let $f_1$ be the frequency of the component whose wavelength is
$\lambda_1$:

FIGURE 12.18
Spectrum of the complex wave in
Fig. 12.17.

$$f_1 = \frac{v}{\lambda_1}$$

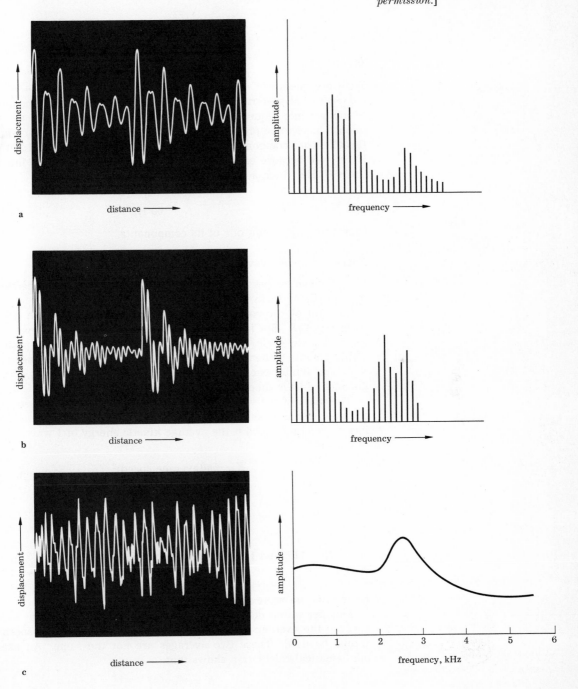

FIGURE 12.19
Spectra of some complex waves:
(a), (b) periodic waves; (c)
aperiodic. [*After P. B. Denes and
E. N. Pinson, "The Speech
Chain," Bell Telephone
Laboratories, Incorporated, used by
permission.*]

The frequencies of the components of a periodic wave then must be members of the sequence

$$f_1 = \frac{v}{\lambda_1}$$

$$f_2 = \frac{v}{\frac{1}{2}\lambda_1} = \frac{2v}{\lambda_1} = 2f_1$$

$$\cdots\cdots\cdots\cdots$$

$$f_n = \frac{v}{(1/n)\lambda_1} = \frac{nv}{\lambda_1} = nf_1$$

or                        $f_1, 2f_1, 3f_1, \ldots, nf_1, \ldots$

Such a sequence of evenly spaced frequencies is called a *harmonic sequence*. The spectrum of a periodic wave consists only of members of a harmonic sequence, although, not all the members of the sequence have to be present, not even $f_1$. For instance, a wave whose three Fourier components have the frequencies 120, 180, and 300 Hz is periodic because each of these frequencies is equal to an integer times 60 Hz. Such a wave repeats in an interval of length

$$\lambda_1 = \frac{v}{f_1} = \frac{v}{60 \text{ Hz}}$$

even though $f_1$ is not one of its components.

**Energy in a sine wave**

A sine wave carries energy as it moves, which can be understood by considering 1 wavelength of a sine wave on a cord. At any instant, each point of the cord is moving vertically with a different speed, as in Fig. 12.9. But in the time $\tau$ of one period, each point executes one complete cycle, moving from zero displacement to a positive displacement $A$, then back to zero, then to a negative displacement $-A$, and finally back to zero again. Thus a point travels a total distance of $4A$ in time $\tau$, so that its average vertical speed is $\overline{v} = 4A/\tau$.

The mass of a 1-wavelength-long piece of cord is $\mu\lambda$, where $\mu$ is the mass per unit length of the cord. Since each point on the cord has the average speed $\overline{v}$, the average kinetic energy of 1 wavelength of cord is

$$K = \tfrac{1}{2}m(\text{average speed})^2$$

$$= \tfrac{1}{2}\mu\lambda \left(\frac{4A}{\tau}\right)^2$$

$$= \frac{8\mu\lambda A^2}{\tau^2}$$

Using Eqs. 12.5 and 12.6, we can also write this as

$$K = 8\mu vfA^2$$

where $v$ is the wave speed and $f$ is the frequency.

This derivation of the kinetic energy of a sine wave is not exact because the average value of speed$^2$, rather than (average speed)$^2$, should be used. These two averages are not the same. An exact mathematical calculation shows that

$$K = \pi^2\mu vfA^2$$

That is, the factor of 8 obtained above should really be $\pi^2$. (Since $\pi^2 = 9.87$, our simple calculation is not too far from the correct result.)

Each point of the wave also has potential energy resulting from the work done in displacing it from equilibrium. It can be shown that the potential energy in 1 wavelength is equal to the kinetic energy, so that the total energy in 1 wavelength is

$$E = K + V = 2K$$
$$= 2\pi^2 \mu v f A^2$$

The number of wavelengths that pass a fixed point in 1 s is $f$. Therefore the total energy that passes this point each second is

$$P = fE = 2\pi^2 \mu v f^2 A^2$$

The symbol $P$ is used for this because it has the same dimension as power [energy/time]. The importance of this result is that it shows that a sine wave traveling in a given direction carries energy and that the rate of energy transport is proportional to the square of the amplitude of the wave.

## 12.5 STANDING WAVES

A wave pattern of particular interest is formed by the superposition of two sine waves with the same wavelength and amplitude traveling in opposite directions. At the instant ($t = 0$) when two such sine waves are in the positions shown in Fig. 12.20$a$, their superposition (dark curve) results in zero displacement of the medium everywhere. A quarter of a period later ($t = \frac{1}{4}\tau$) one sine wave has moved $\frac{1}{4}$ wavelength to the right while the other has moved $\frac{1}{4}$ wavelength to the left. This causes the two waves to coincide, as shown in Fig. 12.20$c$, so that their superposition is a sine wave of amplitude $2A$.

Figure 12.20$b$ shows that the superposition of the waves at an intermediate time ($t = \frac{1}{8}\tau$) is also a sine wave, but with an amplitude less than $2A$. Furthermore, the positions where the resulting displacement of the medium are zero are the same at both times. Figure 19.20$d$ and $e$ shows that the situation is similar at $t = \frac{1}{2}\tau$ and $t = \frac{3}{4}\tau$. Thus, as the two sine waves move through each other, their superposition remains a sine wave with *changing amplitude and fixed zeros*.

The resulting displacement of the medium at various times is shown in Fig. 12.21. This kind of wave pattern is called a *standing wave* because although the pattern changes with time, it does not move along the cord. The zeros, or *nodes*, of the pattern are fixed points that are evenly spaced at intervals of $\frac{1}{2}$ wavelength. A point of the medium located at a node has no vertical oscillation, while a point midway between two nodes (called an *antinode*) oscillates up and down between $+2A$ and $-2A$. A point between a node and an antinode has an amplitude between 0 and $2A$.

All points in a standing wave complete one oscillation in the time $\tau$ equal to the period of the sine waves of which it is composed. This is seen in Fig. 12.20, where the point $O$ in the medium is shown going through one oscillation in the time $\tau$. (The pattern at time $t = \tau$ is identical to the pattern at time $t = 0$.)

**WAVES**

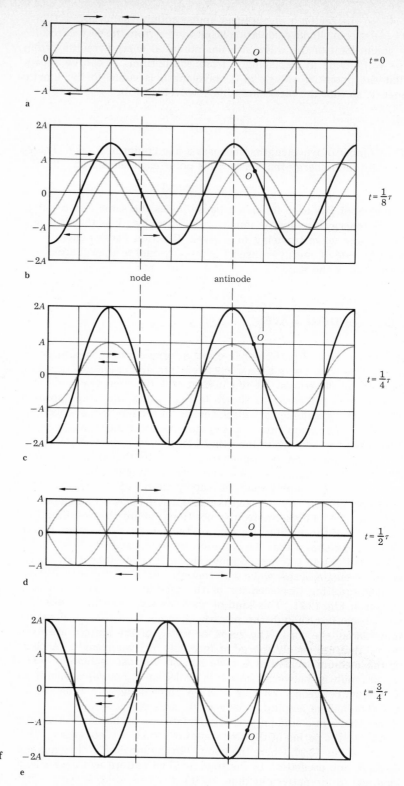

**FIGURE 12.20**
Superposition of two sine waves of
equal amplitude and wavelength
traveling in opposite directions.

Standing waves are important because they can be confined in a limited space. They are thus the natural waves to consider when studying the wave motion of a finite medium, e.g., a guitar string, which is just a cord of length $L$ fixed at both ends. A single sine wave moving in one direction cannot exist on the cord, because such a wave requires the end points to oscillate. However a standing wave can exist on the cord if the ends of the cord coincide with nodes of the wave. Figure 12.22 shows several possible standing waves that can exist on a cord of length $L$.

The condition that the end points be at nodes puts a severe restriction on the wavelengths of the possible standing waves. Since the nodes occur at intervals of $\frac{1}{2}\lambda$, and since there must be exactly a whole number of these half-wavelengths on the cord, the wavelength of a standing wave is related to the length $L$ of the cord by

$$n\tfrac{1}{2}\lambda = L$$

where $n$ is an integer. Thus only wavelengths that satisfy the condition

$$\lambda_n = \frac{2L}{n} \qquad n = 1, 2, 3, \ldots \qquad 12.7$$

can exist as standing waves on the cord. Figure 12.22 shows standing waves with $n$ equal to 1, 2, and 5.

The frequency of oscillation of the cord is related to the wavelength and the speed of the wave by Eq. 12.6. Thus a cord fixed at both ends can oscillate only with the frequencies $f_n$ given by

$$f_n = \frac{v}{\lambda_n} = \frac{nv}{2L} \qquad 12.8$$

The lowest frequency, which occurs when $n = 1$, is

$$f_1 = \frac{v}{2L}$$

Called the *fundamental*, it corresponds to $L = \frac{1}{2}\lambda$ (Fig. 12.22a). Equation 12.8 shows that all the other frequencies are integer multiples of $f_1$:

$$f_n = nf_1$$

The possible frequencies of an oscillating cord thus form a harmonic sequence.

When a guitar string is plucked, the resulting oscillation is the superposition of many different standing waves. Of course, the frequency of each standing-wave component satisfies Eq. 12.8. Usually the fundamental frequency dominates; i.e., it has an amplitude much larger than the other components. The oscillating cord causes the air to oscillate, which generates a sound wave with the same frequencies. The ear hears these waves as a musical tone of frequency $f_1$. The other frequencies (called *overtones*) contribute to the quality of the tone.

All stringed musical instruments, such as the violin, guitar, harp, and piano, produce tones in this way. The differences in the quality of the sound produced by these instruments when playing the same note, i.e., oscillating with the same fundamental frequency, is caused

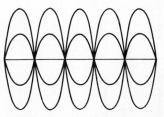
FIGURE 12.21
Displacements of a standing wave at several instants of time.

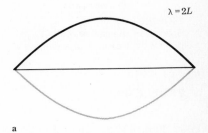
a

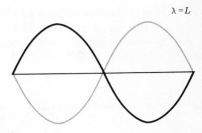
b

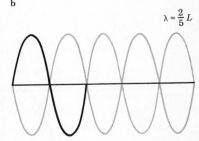
c
FIGURE 12.22
Some possible standing waves that can exist on a cord of length $L$ held fixed at both ends.

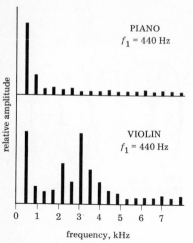

by the differences in the amplitudes of the overtones. Figure 12.23 shows the spectrum of the sound produced by a violin and a piano playing the same note ($f_1 = 440$ Hz). Each instrument produces the same sequence of harmonic frequencies but with different amplitudes.

A stringed instrument is tuned by adjusting the tension in its strings. From Eq. 12.2 this is seen to change the speed v of waves on the string, and so, from Eq. 12.8, this changes the frequencies at which the string oscillates. Once in tune, a violin or guitar is made to play different notes by pressing its strings at different points, thus changing the length $L$ that is free to oscillate. Other instruments, such as a harp and a piano, have a large number of strings of different lengths and densities, each of which is adjusted to produce a different note.

FIGURE 12.23
Spectra of the same note (A)
played by a piano and a violin.

## PROBLEMS

*1* Figure 12.24 shows a pulse on a cord at times $t = 0$ and $t = 0.2$ s. (Note that the vertical and horizontal scales are different.) (*a*) What is the speed of the pulse? (*b*) What is the vertical speed of point $A$ on the cord during this time?    *Ans.* (*a*) 1.5 m/s; (*b*) 30 cm/s

*2* Figure 12.25 shows a pulse on a cord at time $t = 0$. The pulse is moving to the right with the speed 2.5 m/s. (*a*) Draw the shape of the cord at the time $t = 0.6$ s. (*b*) What is the vertical distance that point $A$ moves between $t = 0$ and $t = 0.6$ s? (*c*) What is the (average) vertical speed of point $A$ during this time?

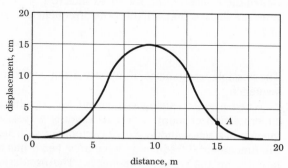

FIGURE 12.25   Problem 2.

*3* (*a*) What is the position of the peak of the pulse in Fig. 12.24 at time $t = 3.0$ s (see Prob. 1)? (*b*) At what time will the peak of the pulse arrive at the 4-m position?

*Ans.* (*a*) 6.0 m; (*b*) 1.66 s

*4* Figure 12.26 shows a complex wave moving

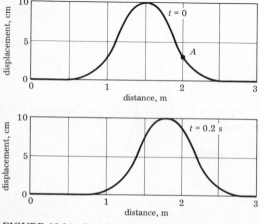

FIGURE 12.24   Problems 1 and 3.

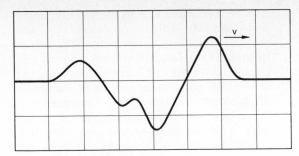
FIGURE 12.26  Problem 4.

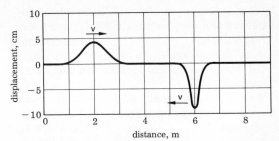
FIGURE 12.28  Problem 10.

to the right along a cord. Draw the shape of the cord an instant later, and determine which parts of the cord are moving upward and which parts are moving downward.

5 A metal guitar string has a linear density $\mu = 3.2 \times 10^{-3}$ kg/m. What is the speed of transverse waves on this string when its tension is 90 N?                   *Ans.* 168 m/s

6 The speed of transverse waves on a cord is 200 m/s. If the linear density of the cord is $7 \times 10^{-3}$ kg/m, what is the tension of the cord?

7 When the tension in a cord is 75 N, the wave speed is 140 m/s. What is the linear density of the cord?          *Ans.* $3.83 \times 10^{-3}$ kg/m

8 When the tension in a cord is 100 N, the wave speed is 120 m/s. What is the wave speed when the tension is 200 N?

9 Figure 12.27 shows two pulses on a cord at time $t = 0$. The pulses are moving toward each other, and the speed of each pulse is 40 cm/s. Sketch the shape of the cord at 0.15, 0.25, and 0.30 s.

11 Figure 12.29 shows a sine wave traveling to the right on a cord. The solid curve represents the shape of the cord at time $t = 0$, and the dotted curve represents the shape of the cord at time $t = 0.12$ s. (Note that the horizontal and vertical scales are different.) What is (*a*) the amplitude and (*b*) the wavelength of this wave? (*c*) What is the speed of the wave? What is (*d*) the frequency and (*e*) the period of the wave?
*Ans.* (*a*) 2.6 cm; (*b*) 15 m; (*c*) 12.5 m/s; (*d*) 0.833 Hz; (*e*) 1.2 s

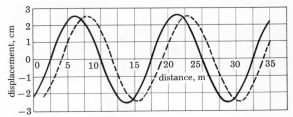
FIGURE 12.29  Problem 11.

12 Draw a sine wave with a wavelength of 10 cm and an amplitude of 6 cm. Plot at least 20 points accurately on graph paper, using the table of trigonometric functions inside the back cover.

13 The equation of a sine wave is

$$y = 3.5 \sin (60° \ x)$$

where $x$ and $y$ are both in centimeters. What is (*a*) the amplitude and (*b*) the wavelength of this wave?          *Ans.* (*a*) 3.5 cm; (*b*) 6.0 cm

14 Write the equation of a sine wave whose amplitude is 12 cm and whose wavelength is 30 cm.

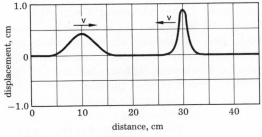
FIGURE 12.27  Problem 9.

10 Figure 12.28 shows two pulses on a cord at time $t = 0$. The pules are moving toward each other, and the speed of each pulse is 2.5 m/s. Sketch the shape of the cord at 0.6, 0.8, and 0.9 s.

15 What is the speed of a sine wave whose frequency and wavelength are 500 Hz and 0.5 m, respectively? *Ans.* 250 m/s

16 What is the wavelength of a sine wave whose speed and period are 75 m/s and 0.005 s, respectively?

17 What is the frequency of a sine wave whose speed and wavelength are 120 m/s and 30 cm, respectively? *Ans.* 400 Hz

18 The spectrum of a periodic wave contains 12, 24, 48, and 96 Hz. Which harmonics are missing?

19 A periodic wave is composed of three sine waves whose frequencies are 36, 60, and 84 Hz. If the speed of the wave is 180 m/s, what is the shortest distance in which the wave pattern repeats? *Ans.* 15 m

20 (*a*) What are the ratios of the energies (per wavelength) of the Fourier components of the wave in Fig. 12.17? (*b*) What are the ratios of the energies (per centimeter) of these components?

21 A guitar string 0.75 m long has a fundamental frequency of 440 Hz. (*a*) What is the speed of a wave on this string? (*b*) To produce other frequencies, the effective length $L$ of the string is shortened by pressing it at a point below the end of the string. What length is needed to produce a fundamental frequency of 660 Hz? *Ans.* (*a*) 660 m/s; (*b*) 0.50 m

22 Show that the total energy in a standing wave of amplitude 2A is $4\pi^2 L\mu f^2 A^2$, where $L$ is the total length of the cord. (*Hint:* The energy of a standing wave is the sum of the energies of its two sine-wave components.)

23 A guitar string 0.75 m long has a fundamental frequency of 440 Hz. The linear density of the string is $2.2 \times 10^{-3}$ kg/m. (*a*) When the string is plucked hard, the fundamental harmonic vibrates with a maximum displacement of 0.2 cm. What is the energy of the fundamental (see Prob. 22)? (*b*) What is the energy of the third harmonic if its amplitude is 0.05 cm? (*Hint:* The amplitude of a standing wave is 2A, where A is the amplitude of its sine-wave components.) *Ans.* (*a*) 0.0126 J; (*b*) 0.00709 J

24 A 20-cm cord fixed at both ends, is oscillating simultaneously with its fundamental $f_1$ and second-harmonic $2f_1$ frequencies. The amplitude of the fundamental is 5 cm, and the amplitude of the second harmonic is 3 cm. At time $t = 0$ both harmonics have maximum displacement. (*a*) Draw the displacement produced by each wave separately at $t = 0$, and then draw the shape of the cord with both waves present simultaneously. (*b*) Draw the shape of the cord at time $t = 1/2f_1$.

25 A cord of length 1.5 m is fixed at both ends. Its mass per unit length is $1.2 \times 10^{-3}$ kg/m. (*a*) If the cord has a tension of 12 N, what is the frequency of the fundamental oscillation? (*b*) What tension is required if the third harmonic $f_3$ is to have a frequency of 500 Hz? *Ans.* (*a*) 33.3 Hz; (*b*) 300 N

26 A string of length 55 cm is held fixed at both ends. When the tension is 25 N, the fundamental frequency is 40 Hz. (*a*) What is the wavelength of the fundamental? (*b*) What is the speed of a wave on this string? (*c*) What is the mass per unit length of the string?

27 The lowest note on a piano has a frequency of 27.5 Hz. The piano wire is 2.0 m long and has a tension of 300 N. What is the total mass of the wire? *Ans.* 49.6 g

## BIBLIOGRAPHY

BACKUS, JOHN: "The Acoustical Foundation of Music," W. W. Norton & Company, Inc., New York, 1969. A readable account of wave motion and its relation to music. This has always been a fascinating subject because of its unique combination of physics, psychology, and aesthetics, apparent especially in the discussion of musical scales.

JOSEPHS, JESS J.: "The Physics of Musical Sound," D. Van Nostrand Company, Inc., Princeton, N.J., 1967. Except for the integral on page 1, this is an elementary account of the physics of music. The contents are similar to those of the previous reference.

# sound CHAPTER 13

Sound is a longitudinal mechanical wave that propagates through air, water, and other material media. It is of vital importance in the life of all higher animals, which have specialized organs to produce and detect these waves. By means of sound, animals (especially man) are able to communicate with each other and to obtain information about their environment.

This chapter discusses the basic physics of longitudinal mechanical waves and the relation of their physical properties to the psycho-physiological sensations they produce in a listener. The physics of speech production, which is given in the last section, provides an opportunity to review most of the general principles of wave motion and sound treated here and in Chap. 12.

## 13.1 LONGITUDINAL MECHANICAL WAVES

Chapter 12 discussed wave motion in general, using as an example a transverse wave on a stretched cord. However, since a cord is only a one-dimensional medium, these examples must be extended so that the nature of a wave in a three-dimensional medium can be understood.

Ripples on the surface of water are a familiar example of a wave in two dimensions. Figure 13.1 shows a three-dimensional transverse wave that is moving parallel to the $z$ axis while the medium is oscillating parallel to the $x$ axis. This is called a *plane wave* because all

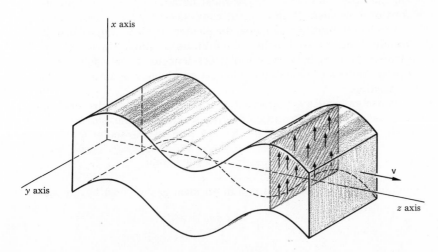

FIGURE 13.1
A sinusoidal plane wave.

points in a plane perpendicular to the $z$ axis have the same displacement at the same time. Plane waves are also characterized by having a single direction of propagation. If the displacement of a plane wave varies as the sine along its direction of propagation, the wave is a *sinusoidal plane wave*. This is the analog of a one-dimensional sine wave.

When a stone is dropped into a pond, the ripples are concentric circles that spread in all directions from the stone. The analogous wave in three dimensions is a *spherical wave* that spreads out in concentric spheres from a central point (Fig. 13.2). In a transverse spherical wave, each point on a given spherical shell oscillates parallel to the shell, whereas in a longitudinal spherical wave, the points on a shell oscillate radially in and out, about their equilibrium positions. If the displacement of a spherical wave varies as the sine along every radial line, the wave is a *sinusoidal spherical wave*.

Longitudinal mechanical waves can propagate through solid, liquid, and gaseous media, but transverse mechanical waves can propagate only through solids. This is because the points in a transverse wave oscillate parallel to a plane (Fig. 13.1), which requires that the medium exert forces parallel to the plane. A solid medium can exert such forces, but a fluid cannot (Fluid Property 1, Sec. 7.2). Consequently, only longitudinal waves can propagate in fluid media such as water and air.

We have been using point to mean a tiny bit of a medium, but a better word would be "element," because we are not using point in the mathematical meaning of the word. Here we mean a small region of the medium that is considered to move as a single unit. Clearly, the size of such a region, or *element*, must be small compared to the wavelengths of the waves that propagate through the medium because if the element were too large, different parts could have different displacements at the same time and the element would not move as a single unit. For the cord illustrated in Fig. 13.3, the region $R'$ is too big to be an element because different parts of it have appreciably different displacements. Region $R$, however, is small enough to be an element because the difference in the displacements of its parts is negligible.

On the other hand, an element must be large enough to contain billions of molecules. If an element contained only a few molecules, their random motion would cause the element itself to jiggle, so that it would not have a well-defined equilibrium position. However, the random motions of billions of molecules tend to cancel each other, and so if the element is large enough, it can have a definite equilibrium position.

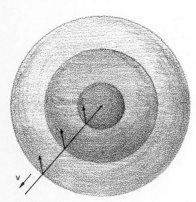

**FIGURE 13.2**
A spherical wave.

**FIGURE 13.3**
A region $R$ of a cord that is small enough to be an element and a region $R'$ that is not.

Because molecules are so small, it is possible to have elements of a size that satisfies both these conditions. For example, since 1 mol contains $6 \times 10^{23}$ molecules, an element with $10^9$ molecules is

$$n = \frac{10^9}{6 \times 10^{23}} = 1.6 \times 10^{-13} \text{ mol}$$

The volume $V$ occupied by $n$ mol of an ideal gas is given by Eq. 8.9 as

$$V = \frac{nRT}{p}$$

Thus in a gas at a pressure of 1 atm ($p = 1.01 \times 10^5$ N/m$^2$) and a temperature of 27°C ($T = 300$ K), these $10^9$ molecules occupy the volume

$$V = \frac{(1.6 \times 10^{-13})(8.3 \text{ J/K})(300 \text{ K})}{1.01 \times 10^5 \text{ N/m}^2}$$

$$= 3.9 \times 10^{-15} \text{ m}^3$$

This is the volume of a cube with sides of length $1.6 \times 10^{-5}$ m $= 1.6 \times 10^{-3}$ cm, which is much smaller than any wavelengths we shall be considering. Therefore a volume of air of this size satisfies the conditions for an element.

As a longitudinal wave propagates through air, elements of the air are displaced back and forth about their equilibrium position. Figure 13.4$a$ plots the displacements in a sine wave, and Fig. 13.4$b$ shows the directions in which elements of the air move. A positive displacement in Fig. 13.4$a$ corresponds to a longitudinal displacement of an element to the right, and a negative displacement corresponds to a displacement of an element to the left.

Notice that elements on either side of the displacement node $A$ in Fig. 13.4 are displaced toward the node, so that the density and pressure of the air at this point are above normal. Likewise the elements on either side of the displacement node $B$ are displaced away from the node, so that the density and pressure at this point are below normal. In fact, the pressure oscillates about its normal value just as the elements oscillate about their equilibrium positions. If $p_0$ is the normal pressure and $p$ is the pressure at a particular point in a wave, the pressure variation $y_p$ is defined to be

$$y_p = p - p_0$$

A longitudinal wave is as much a pattern of pressure variations as it is a pattern of displacements.

The pressure variation for the wave in Fig. 13.4$b$ is shown in Fig. 13.4$c$. It is a sine wave, like the displacement wave in Fig. 13.4$a$, except that its antinodes occur at the nodes of the displacement wave, and vice versa. The pressure wave can be written

$$y_p = A_p \sin\left(\frac{x}{\lambda} 360°\right) \qquad 13.1$$

where the pressure amplitude $A_p$ is the maximum difference between the pressure in the wave and the normal pressure. It is often useful to think of a longitudinal wave as a pressure wave rather than a displacement wave because it is the pressure variation that is usually detected.

The human ear can detect longitudinal mechanical waves with frequencies between 20 and 20,000 Hz. The speed of such waves in air is 343 m/s (Table 13.1), so the wavelength of audible sound is between

$$\lambda = \frac{343 \text{ m/s}}{20,000 \text{ Hz}} = 1.71 \times 10^{-2} \text{ m} = 1.71 \text{ cm}$$

and $$\lambda = \frac{343 \text{ m/s}}{20 \text{ Hz}} = 17.1 \text{ m} = 1710 \text{ cm}$$

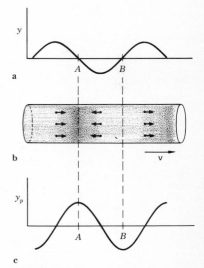

FIGURE 13.4
Displacements in a longitudinal wave. ($a$) Plot of the longitudinal displacements of the elements of a gas against their equilibrium positions. ($b$) Displacements of elements of the gas. ($c$) Plot of the pressure variation of the gas against position.

Waves in this audible range are called *sound*. Waves with frequencies above 20,000 Hz are called *ultrasonic*, and waves with frequencies below 20 Hz are called *infrasonic*. Since this classification depends on the properties of the human hearing mechanism, it has no fundamental physical significance. It is of practical significance, however, because special equipment is needed to detect ultrasonic and infrasonic waves.

## 13.2 THE SPEED OF SOUND

The propagation of longitudinal waves through material media is completely explainable in terms of the laws of mechanics. Hence the study of sound (*acoustics*) is a subspecialty of mechanics. The connection between sound and mechanics was first shown by Newton, who used the second law of motion to calculate the speed of sound. We shall give a simplified version of this derivation in order to exhibit the mechanism of sound propagation. (A similar derivation for the speed of a transverse wave on a cord is given in Appendix V.)

For simplicity we shall consider a pulse moving through a cylindrical medium (Fig. 13.5), such as the air inside a tube. The pulse consists of a localized region $R$ in which the pressure $p$ is greater than the undisturbed pressure $p_0$ of the medium. As the pulse moves along the cylinder, the elements of the air only oscillate about their equilibrium positions; they do not move with the pulse. Thus the pulse moves from region $R$ in Fig. 13.5$a$ to region $R'$ in Fig. 13.5$b$ by compressing the air in region $R_0$ rather than by moving the air in region $R$.

The time $t$ required for the pulse to move from $R$ to $R'$ is related to the wave speed $v$ and the length $L_0$ of $R_0$ by

$$t = \frac{L_0}{v}$$

In this time the gas in $R_0$ gets compressed into $R'$ by moving some of the molecules in $R_0$ to the right. The length $L$ of $R'$ is smaller than $L_0$ by the amount $L_0 - L$. Some of the molecules in $R_0$ are moved this full amount, and some are not moved at all, so that on the average the total mass of air in $R_0$ is moved the distance

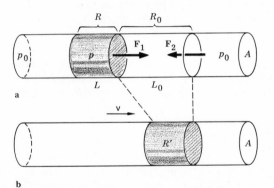

FIGURE 13.5
A sound pulse consisting of a region $R$ in which the pressure $p$ is above atmospheric pressure $p_0$. As the pulse moves to the right, the air in region $R_0$ is compressed into region $R'$.

$$d = \tfrac{1}{2}(L_0 - L)$$

From these two equations for the time and distance, the acceleration of the gas in region $R_0$ can be calculated from Eq. 4.3. The result is

$$
\begin{aligned}
a &= \frac{d}{\tfrac{1}{2}t^2} = \frac{\tfrac{1}{2}(L_0 - L)}{\tfrac{1}{2}(L_0/\mathsf{v})^2} \\
&= \frac{(L_0 - L)\mathsf{v}^2}{L_0{}^2}
\end{aligned}
\qquad\text{13.2}
$$

From Newton's second law, this acceleration is equal to the total force on $R_0$ divided by the mass $m$ of $R_0$:

$$a = \frac{F}{m} \qquad\text{13.3}$$

There are two forces on $R_0$. The pulse exerts the pressure $p$ on the left side of $R_0$, and the gas in the rest of the tube exerts the pressure $p_0$ on the right side. Thus, if the cylinder has a cross-sectional area $A$, there is a force

$$F_1 = pA$$

to the right, and a force

$$F_2 = p_0 A$$

to the left. The total force on $R_0$ is

$$F = pA - p_0 A = (p - p_0)A \qquad\text{13.4}$$

and is directed to the right. Putting Eqs. 13.4 and 13.2 into Eq. 13.3, we get

$$\frac{(L_0 - L)\mathsf{v}^2}{L_0{}^2} = \frac{(p - p_0)A}{m}$$

This can be solved for $\mathsf{v}^2$ to yield

$$
\begin{aligned}
\mathsf{v}^2 &= \frac{(p - p_0)A/m}{(L_0 - L)/L_0{}^2} \\
&= \frac{(p - p_0)L_0{}^2 A}{(L_0 - L)m}
\end{aligned}
\qquad\text{13.5}
$$

Since $L_0 A$ is the volume $V_0$ of $R_0$ and $LA$ is the volume $V$ of $R$, by multiplying the numerator and denominator of Eq. 13.5 by $A$ we get

$$\mathsf{v}^2 = \frac{(p - p_0)V_0{}^2}{(V_0 - V)m} = \frac{B}{\rho}$$

or

$$\mathsf{v} = \sqrt{\frac{B}{\rho}} \qquad\text{13.6}$$

Here $\rho = m/V_0$ is the density of the medium and

$$B = \frac{p - p_0}{(V_0 - V)/V_0}$$

is its *bulk modulus* (Sec. 10.2). The bulk modulus is the ratio of the change in pressure, $p - p_0$, to the relative change in volume, $(V_0 - V)/V_0$, produced by the pressure change. For reasonably small pressure changes, $B$ is a constant that is characteristic of the medium. The bulk modulus of some common liquids is given in Table 10.4.

**REMARK** Equation 13.6 is valid only for liquids and gases. For solids, the bulk modulus is replaced by Young's modulus $E$ (Sec. 10.2), so that

$$v = \sqrt{\frac{E}{\rho}}$$

Table 10.2 gives the Young's modulus of some common solids.

Young's modulus and the bulk modulus are measures of the stiffness, or incompressibility, of a substance. A hard substance like iron or granite requires an enormous stress to change its length a small amount, so that its Young's modulus is large. Gases, on the other hand, are easily compressed, and so their bulk moduli are small. Consequently the speed of sound in granite and other solids is larger than the speed of sound in gases. This is seen in Table 13.1, which gives the speed of sound in various substances.

Equation 13.6 is of value only in calculating the speed of sound $v$ in a fluid if its bulk modulus $B$ is known. However, it is usually easier to measure $v$ than $B$, so that in practice Eq. 13.6 is used to determine $B$ in terms of $v$, rather than vice versa.

There is one important exception: the bulk modulus of an ideal gas can be determined from the ideal-gas law, and so the speed of sound in a gas can be determined from first principles. If it is assumed that the temperature of the gas does not change while it is being compressed (an *isothermal* compression), the volumes and pressures in regions $R$ and $R_0$ of Fig. 13.5 are related by the ideal-gas law (Eq. 8.9):

$$Vp = V_0 p_0$$

TABLE 13.1 **Speed of sound in various substances**
The temperature is given for substances in which the speed varies significantly with temperature. The speed in a solid depends on the composition and structure of the solid, so only approximate values can be given.

| Substance | Temperature, °C | Speed, m/s |
|---|---|---|
| *Gases* | | |
| Carbon dioxide | 0 | 259 |
| Oxygen | 0 | 316 |
| Air | 0 | 331 |
| | 20 | 343 |
| Nitrogen | 0 | 334 |
| Helium | 0 | 965 |
| *Liquids* | | |
| Mercury | 25 | 1450 |
| Water | 25 | 1498 |
| Seawater | 25 | 1531 |
| *Solids* | | |
| Rubber | | 1800 |
| Lead | | 2100 |
| Lucite | | 2700 |
| Gold | | 3000 |
| Iron | | 5000–6000 |
| Glass | | 5000–6000 |
| Granite | | 6000 |

Subtracting the quantity $V_0 p$ from both sides of this equation gives

$$Vp - V_0 p = V_0 p_0 - V_0 p$$

or

$$p(V - V_0) = V_0(p_0 - p)$$

Dividing both sides by $V - V_0$, we get

$$p = \frac{V_0(p_0 - p)}{V - V_0} = B \qquad \text{isothermal} \qquad 13.7$$

Thus the isothermal bulk modulus of an ideal gas is equal to the pressure of the gas. Then, from Eq. 13.6, the speed of sound in an ideal gas is

$$v = \sqrt{\frac{p}{\rho}} \qquad\qquad 13.8$$

**REMARK** The pressure $p$ in Eq. 13.8 is the pressure of the gas when it is compressed. However, this differs only minutely from the undisturbed pressure (Sec. 13.3), so that the undisturbed value can be used to calculate v with sufficient accuracy.

Equation 13.8 was first derived by Newton, but it does not give the correct value for v. For example, at atmospheric pressure ($p = 1.01 \times 10^5$ N/m$^2$) and 0°C, the density of air is 1.30 kg/m (Table 7.2). With these values, the speed of sound calculated from Eq. 13.8 is

$$v = \sqrt{\frac{p}{\rho}} = \sqrt{\frac{1.01 \times 10^5 \text{ N/m}^2}{1.30 \text{ kg/m}^3}}$$
$$= \sqrt{7.77 \times 10^4 \text{ N-m}^2/\text{kg}}$$
$$= 279 \text{ m/s}$$

which is 15 percent less than the measured value of 331 m/s.

The error is in the assumption that the temperature of the gas remains constant. Actually, when a sound wave passes through a region, the gas pressure varies so rapidly that there is not enough time for heat to enter or leave the region. Under such conditions the temperature oscillates rapidly with the pressure, while no heat is exchanged. This is called an *adiabatic* process. For the adiabatic compression of an ideal gas, the bulk modulus can be shown to be

$$B = \gamma p \qquad \text{adiabatic} \qquad 13.9$$

where $\gamma$ is a constant that is characteristic of each gas. It varies from near 1.0 for very complex polyatomic gases to $5/3 = 1.67$ for monatomic gases such as helium. Table 13.2 gives the values of $\gamma$ for some common gases.

If Eq. 13.9 is used in Eq. 13.6 instead of Eq. 13.7, we get the correct equation for the speed of sound in a gas:

$$v = \sqrt{\frac{\gamma p}{\rho}} \qquad\qquad 13.10$$

For example, the $\gamma$ of air is 1.40, so the calculated speed of sound in air is

$$v = \sqrt{1.40 \times 7.77 \times 10^4 \text{ m}^2/\text{s}^2}$$
$$= 330 \text{ m/s}$$

which is in good agreement with the measured value.

TABLE 13.2 $\gamma$ **for various gases**

| Gas | $c_p/c_v = \gamma$ |
|---|---|
| Argon, Ar | 1.67 |
| Helium, He | 1.67 |
| Mercury, Hg | 1.67 |
| Oxygen, $O_2$ | 1.40 |
| Nitrogen, $N_2$ | 1.40 |
| Air | 1.40 |
| Chlorine, $Cl_2$ | 1.34 |
| Carbon dioxide, $CO_2$ | 1.29 |
| Sulfur dioxide, $SO_2$ | 1.29 |
| Ethane, $C_2H_6$ | 1.19 |
| Dimethyl ether, $C_2H_6O$ | 1.16 |

**REMARK** The quantity $\gamma$ is equal to $c_p/c_v$, the ratio of the specific heat of the gas at constant pressure to the specific heat of the gas at constant volume (Sec. 11.2). It can be shown that this quantity is $\frac{5}{3}$ for monatomic gases and $\frac{7}{5}$ for diatomic gases. Although it would take us too far afield to go into details, it is interesting to note that the speed of sound in air can be completely understood in terms of the principles of mechanics and thermodynamics.

Equation 13.10 seems to imply that the speed of sound depends on the pressure, but actually, the ratio $p/\rho$ is independent of $p$. This can be seen from the ideal-gas law,

$$pV = nRT$$

by rewriting it in the form

$$p = \frac{nRT}{V} = \frac{nNRT}{NV}$$

where $N$ is Avogadro's number. In Chap. 8 the Boltzmann constant $k$ is defined as

$$k = \frac{R}{N}$$

and the number density $\eta$ is defined as

$$\eta = \frac{nN}{V}$$

Therefore, the ideal-gas law can be written

$$p = \eta k T$$

But the mass density $\rho$ is just $\eta$ times the mass $m$ of an individual molecule:

$$\rho = \eta m$$

Therefore we have

$$p = \frac{\rho k T}{m}$$

so that Eq. 13.10 can be written

$$v = \sqrt{\frac{\gamma k T}{m}} \qquad\qquad 13.11$$

This shows that the speed depends only on the temperature of the gas and the mass of its molecules. For instance, at a given temperature, the speed of sound in helium ($m = 4$ u) is much faster than in air (average mass $= 29$ u).

## 13.3   INTENSITY

The loudness of a sound is related to the energy carried by the sound wave. Loudness is a subjective impression that a listener attributes to a particular sound, whereas the energy of the sound wave is an objective physical quantity. The relation between these quantities is studied in the branch of psychology called *psychophysics*.

## Definition of intensity

The *intensity* $I$ of a wave is the energy that crosses a unit area in unit time. It is determined experimentally by measuring the energy $E$ incident on a detector (such as a microphone) in time $t$. The intensity is then equal to this energy divided by the time and by the area $A$ of the detector:

$$I = \frac{E}{At} \qquad 13.12$$

In the mks system the unit of intensity is $J/m^2$-s, or watts per square meter ($W/m^2$).

For example, suppose that during a 5-s interval a microphone receives $1.5 \times 10^{-11}$ J of energy. If the microphone has an effective area of 3 cm$^2$ ($3 \times 10^{-4}$ m$^2$), the intensity of the sound is

$$I = \frac{E}{At} = \frac{1.5 \times 10^{-11} \text{ J}}{(3 \times 10^{-4} \text{ m}^2)(5 \text{ s})}$$
$$= 10^{-6} \text{ J/m}^2\text{-s} = 10^{-6} \text{ W/m}^2$$

This is about the intensity of the sound in a normal conversation.

It can be shown (Appendix VI) that in a sine wave the intensity is related to the pressure amplitude $A_p$ by

$$I = \frac{A_p{}^2}{2\rho v} \qquad 13.13$$

where $\rho$ is the density of the medium and $v$ is the speed of the wave in the medium. At 20°C the density of air is 1.2 kg/m$^3$, and its speed is 343 m/s. Therefore, in a wave with an intensity of $10^{-6}$ W/m$^2$, the square of the pressure amplitude is

$$A_p{}^2 = 2\rho v I$$
$$= (2 \times 1.2 \text{ kg/m}^3)(343 \text{ m/s})(10^{-6} \text{ J/m}^2\text{-s})$$
$$= 8.2 \times 10^{-4} \text{ N}^2/\text{m}^4$$

so that the pressure amplitude is

$$A_p = 2.86 \times 10^{-2} \text{ N/m}^2$$

This is the difference between the maximum pressure in the wave and the undisturbed air pressure. Since normal air pressure is about $10^5$ N/m$^2$, the pressure in the wave changes by only 0.286 parts per million. (This is why in Eq. 13.8 we can replace the disturbed pressure $p$ by the undisturbed pressure $p_0$ without affecting the value of $v$.)

## Decibel scale

Although the perceived loudness of a sound increases with its intensity, the relation between loudness and intensity is far from linear. For instance, in a lecture room the intensity of a speaker's voice may be 100 times greater at the front of the room than at the back, but a listener moving from the front to the back of the room experiences only a slight decrease in loudness.

A young adult can detect sounds with an intensity as small as $10^{-12}$ W/m$^2$ and as large as 1 W/m$^2$. By convention the intensity

$$I_0 = 10^{-12} \text{ W/m}^2$$

is taken to be the zero point on an intensity level scale called the *decibel* (dB) scale. On this scale, an increase in intensity by a factor of 10 corresponds to an increase in the intensity level $\beta$ by 10 dB. Thus, since $I_0 = 10^{-12}$ W/m² corresponds to $\beta = 0$ dB, $10^{-11}$ W/m² = $10I_0$ corresponds to $\beta = 10$ dB, $10^{-10}$ W/m² = $10^2 I_0$ corresponds to $\beta = 20$ dB, and $10^{-6}$ W/m² = $10^6 I_0$ corresponds to $\beta = 60$ dB. Mathematically, the intensity level $\beta$ (in decibels) of a sound of intensity $I$ is defined in terms of the logarithm by

$$\beta = 10 \log\frac{I}{I_0} \qquad\qquad 13.14$$

where $I_0 = 10^{-12}$ W/m².

At intensities above 1 W/m², the sensation of a wave changes from that of sound to that of pain (feeling). That is, waves with intensity greater than 1 W/m² are felt rather than heard. The intensity level of a wave at the threshold of pain is found from Eq. 13.14 to be

$$\beta = 10 \log\frac{1 \text{ W/m}^2}{10^{-12} \text{ W/m}^2} = 10 \log 10^{12}$$

$$= 120 \text{ dB}$$

The range of human hearing is thus between 0 and 120 dB, which is a factor of $10^{12}$ in the intensity and a factor of $10^6$ in the pressure amplitude. Table 13.3 gives the intensity levels $\beta$ and the intensities $I$ of some common sounds. To convert from decibels to watts per square meter you can use either this table or Eq. 13.14.

### Psychophysics of hearing

Although the ratio of the intensity of the loudest detectable sound to the softest is $10^{12}$, we are not aware of anything like a factor of $10^{12}$ in the loudness of the sounds we hear. This is because the relation between intensity, which is a physical attribute of sound, and loudness, which is a subjective attribute, involves physiological and psychological processes in the ear and brain. Psychophysical experi-

TABLE 13.3 **Sound levels and intensities of some common sounds**

| Sound level, dB | Intensity, W/m² | Sound |
|---|---|---|
| 0 | $10^{-12}$ | Threshold of hearing |
| 10 | $10^{-11}$ | Rustle of leaves |
| 20 | $10^{-10}$ | Whisper (1 m away) |
| 30 | $10^{-9}$ | Quiet home |
| 40 | $10^{-8}$ | Average home, quiet office |
| 50 | $10^{-7}$ | Average office |
| 60 | $10^{-6}$ | Normal conversation, average traffic |
| 70 | $10^{-5}$ | Noisy office |
| 80 | $10^{-4}$ | Busy traffic, inside car in traffic |
| 90 | $10^{-3}$ | Inside subway train |
| 100 | $10^{-2}$ | Machine shop |
| 120 | $10^{0}$ | Pneumatic chipper (2 m away), threshold of pain |
| 140 | $10^{2}$ | Jet airplane (30 m away) |

ments are designed to measure the relation between the physical attribute of a stimulus and the subjective attribute as perceived by an individual.

As an example, consider the following psychophysical experiment. A subject compares a standard sound, which has (say) a frequency of 1000 Hz and an intensity level of 60 dB to a test sound which has a frequency $f$. The intensity of the test sound is varied until the subject judges it to have the same loudness as the standard 1000-Hz sound. This is repeated for a range of frequencies until a curve, such as the one labeled 60 in Fig. 13.6, is obtained. Each point on the curve gives the intensity level at which a signal of the corresponding frequency is judged to have the same loudness as the standard signal (1000 Hz at 60 dB). For instance, a 100-Hz signal must have an intensity level of 72 dB to be judged as loud as a 1000-Hz signal at 60 dB. Figure 13.6 also shows curves produced when the 1000-Hz standard signal is given at different intensity levels. From these curves it is seen that hearing is most acute for sounds with frequencies between 3000 and 4000 Hz.

In another test, the intensity of a test signal of frequency $f$ is decreased until the subject no longer hears it. By repeating this for

**FIGURE 13.6**
Psychophysical relation of loudness to intensity and frequency. Each curve gives the intensities at which sounds of various frequencies have the same apparent loudness. The loudness level is arbitrarily taken equal to the intensity level of the 1000-Hz signal. [*After P. B. Denes and E. N. Pinson, "The Speech Chain." Copyright, 1963, Bell Telephone Laboratories, Incorporated, used by permission.*]

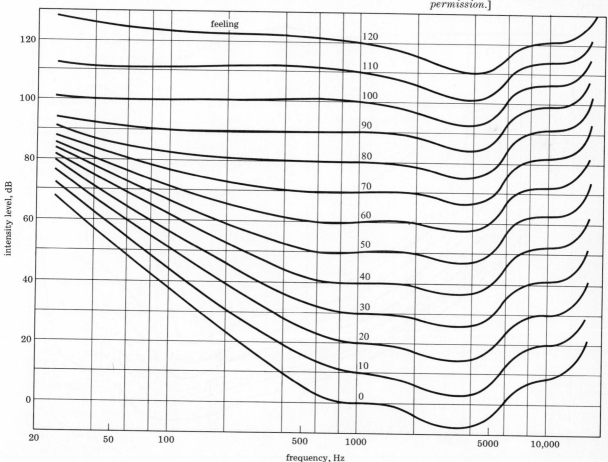

a range of frequencies, a curve of the individual's hearing acuity is obtained. Figure 13.7 summarizes the results of thousands of these tests given by the U.S. Public Health Service. Each curve is labeled by the percentage of people in the population whose hearing acuity was below the curve. Thus, 99 percent of the population could hear signals below the 99 percent curve, whereas only 10 percent could hear signals below the 10 percent curve. (The 1 percent of the population that cannot hear below the 99 percent level require hearing aids, since they cannot hear a normal conversation at 60 dB.) The 1 percent curve is often taken as the standard of unimpaired, or "normal," hearing. Only children and young adults can hear at this level, because acuity decreases with age. These curves have a dip between 3000 and 4000 Hz, which again indicates that hearing is most acute in this frequency range.

## Variation of intensity with distance

The intensity of the sound produced by a source decreases with the distance from the source. If the source is small, the sound spreads out from it in spherical waves (Fig. 13.8) in which the intensity $I_1$

**FIGURE 13.7**
Hearing-acuity profiles of the United States population. The number to the right of each curve is the percentage of the population that can hear all signals below the curve.

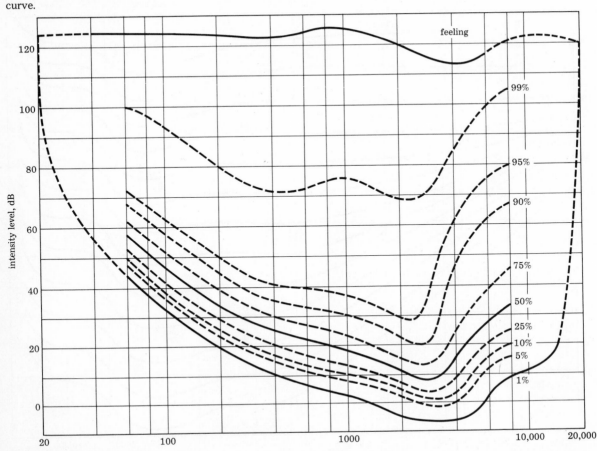

is the same at all points the same distance $d_1$ from the source. Since intensity is the energy that crosses a unit area in unit time, the energy per second, or power $P_1$, that passes through the sphere of radius $d_1$ centered at the source is the product of the intensity $I_1$ at this distance and the area $A_1 = 4\pi d_1^2$ of the sphere:

$$P_1 = A_1 I_1 = 4\pi d_1^2 I_1$$

Similarly, the power $P_2$ that passes through a sphere of radius $d_2$ is

$$P_2 = 4\pi d_2^2 I_2$$

In going from $d_1$ to $d_2$, some of the energy of the wave is absorbed by the air, but this energy is small if the distance from $d_1$ to $d_2$ is not too large. If we neglect this energy loss entirely, all the energy that passes through sphere 1 must also pass through sphere 2. Therefore the energy per second $P_1$ that passes through sphere 1 is equal to the energy per second $P_2$ that passes through sphere 2, so that from the last two equations we get

$$4\pi d_1^2 I_1 = 4\pi d_2^2 I_2$$

or
$$I_2 = \frac{d_1^2}{d_2^2} I_1 \qquad\qquad 13.15$$

**REMARK**  Equation 13.15 shows that the intensity $I_2$ decreases inversely as the square of the distance $d_2$ from the source. Thus intensity obeys an inverse-square law similar to the inverse-square laws of the gravitational (Sec. 5.4) and electric (Sec. 16.2) forces.

Equation 13.15 is used to find the intensity at a distance $d_2$ given the intensity $I_1$ at $d_1$. For example, Table 13.3 shows that the intensity of a jet airplane is $I_1 = 10^2$ W/m$^2$ at a distance $d_1 = 30$ m. The intensity $I_2$ at a distance $d_2 = 3000$ m from the airplane is

$$I_2 = \frac{d_1^2}{d_2^2} I_1 = \left(\frac{30 \text{ m}}{3000 \text{ m}}\right)^2 (10^2 \text{ W/m}^2)$$

$$= 10^{-2} \text{ W/m}^2$$

Table 13.3 shows that this corresponds to an intensity level of 100 dB. Thus, nearly 2 mi from the airplane, its noise level is equal to that of a subway train.

## 13.4  STANDING WAVES AND RESONANCE

The air inside a cavity, like a cord fixed at both ends, can vibrate only at certain discrete frequencies that are characteristic of the cavity. In Sec. 12.4 it was shown that a cord fixed at both ends vibrates only with frequencies that correspond to standing waves on the cord. Likewise, standing waves of only certain wavelengths and frequencies can exist in a cavity. The frequencies of these waves depend on the size and shape of the cavity, but they are easy to calculate only for cavities of very simple geometry.

An important special case is that of a cylindrical cavity of length $L$ open at both ends (open pipe). By blowing air across one end of the cylinder, waves are produced that travel down the cylinder. If

FIGURE 13.8
Sound from a point source spreads out radially in all directions.

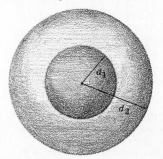

the diameter of the cylinder is small compared to the wavelength of the waves, they will be reflected back into the cylinder when they reach the other end. The superposition of waves of equal amplitude and frequency traveling in opposite directions produces a standing-wave pattern (Sec. 12.4). The pressure at each end is fixed at atmospheric pressure because the ends are open to the atmosphere. Therefore only standing waves that have pressure nodes at each end can exist in the cylinder (Fig. 13.9). The situation is identical to that of a cord fixed at both ends: only standing waves are possible that have wavelengths given by

$$L = n\frac{\lambda_n}{2} \qquad \text{or} \qquad \lambda_n = \frac{2L}{n}$$

where $n$ is an integer. The frequencies of these waves are

$$f_n = \frac{v}{\lambda_n} = \frac{nv}{2L} \qquad\qquad 13.16$$

Like the frequencies of a vibrating cord, this set includes the fundamental, $f_1 = v/2L$, and all the harmonics $f_n = nf_1$.

Organ pipes and other wind instruments, including the human voice (Sec. 13.5), produce tones by establishing standing waves in cavities. However, the characteristic frequencies are a harmonic sequence only if the cavity is a perfect cylinder that is open at both ends (open pipe). For cavities of different shape, the frequencies do not form a harmonic sequence.

For example, some organ pipes are closed at one end and open at the other (closed pipe). Such a pipe has a pressure antinode at the closed end because the air has zero displacement there. (Figure 13.4 shows that the pressure has an antinode wherever the displacement has a node.) The characteristic standing waves, then, have a node at the open end and an antinode at the closed end. Figure 13.10 shows the first three standing waves that satisfy this condition. These waves have $\frac{1}{4}, \frac{3}{4}, \frac{5}{4}, \ldots$ wavelength in the pipe, so that

$$L = m\frac{\lambda_m}{4} \qquad \text{or} \qquad \lambda_m = \frac{4L}{m}$$

where $m$ is an odd integer ($m = 1, 3, 5, 7, \ldots$). The characteristic frequencies are

$$f_m = \frac{mv}{4L} \qquad\qquad 13.17$$

The fundamental frequency is $f_1 = v/4L$, which is half the fundamental frequency of an open pipe of the same length. The overtones of the closed pipe consist only of the odd harmonics: $f_1, 3f_1, 5f_1, \ldots$.

To establish a standing wave in a cavity, it is only necessary that the air in the cavity be vibrated with a frequency near one of the characteristic frequencies of the cavity. A tuning fork is convenient for this purpose because when it is struck, it vibrates with a fixed frequency $f$ and produces sound waves of the same frequency. However, when the tuning fork is vibrating freely, only a small fraction of its mechanical energy is transformed into sound and the rest is dissipated as heat. Consequently, a freely vibrating tuning fork produces a low-intensity sound wave.

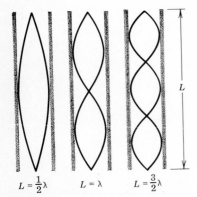

$L = \frac{1}{2}\lambda$ $\qquad L = \lambda$ $\qquad L = \frac{3}{2}\lambda$

**FIGURE 13.9**
Some standing waves in an open pipe. The curves show the pressure variation inside the tube.

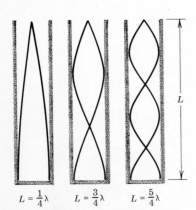

$L = \frac{1}{4}\lambda$ $\qquad L = \frac{3}{4}\lambda$ $\qquad L = \frac{5}{4}\lambda$

**FIGURE 13.10**
Some standing waves in a closed tube. The curves show the pressure variation inside the tube.

If, however, a vibrating tuning fork is placed near the open end of a cavity that has a characteristic frequency nearly equal to $f$, a large fraction of its mechanical energy is transformed into a large-amplitude standing wave in the cavity. This phenomenon, called *resonance*, is easily demonstrated by means of a cylinder connected to a reservoir of water, as shown in Fig. 13.11. A vibrating tuning fork is placed over the open end of the cylinder, while the length $L$ of the closed pipe is adjusted by raising and lowering the reservoir. When the length of the closed pipe is such that it has a characteristic frequency equal to the frequency of the tuning fork, the sound becomes noticeably louder because at resonance a larger fraction of the mechanical energy of the tuning fork is converted into sound energy.

String instruments, such as guitars and violins, have a resonating cavity in back of the strings to intensify the sound. Of course, the characteristic frequencies of the cavity cannot be equal to all the frequencies played on the strings, but fortunately resonance can occur even when the exciting frequency $f$ is not exactly equal to a characteristic frequency $f_n$. When a freely vibrating string produces a sound with an amplitude $A_p$ at frequency $f$, the cavity vibrates at the same frequency $f$ with some other amplitude $A'_p$. The ratio

$$r = \frac{A'_p}{A_p}$$

is called the *response factor*. When $r$ is greater than 1, the cavity amplifies the sound.

The response factor $r$ is much greater than 1 when $f$ is close to a characteristic frequency $f_n$ of the cavity, and $r$ is small when $f$ is far from $f_n$. This is shown by the response curves in Fig. 13.12, which are plots of $r$ against $f/f_n$ for two typical cases. These curves peak when $f/f_n = 1$, that is, when the exciting frequency $f$ equals the characteristic frequency $f_n$. By definition, the quantity $\Gamma$† is the width of a response curve at seven-tenths its peak value. That is, $r$ is within 70 percent of its peak value when $f$ is between $f_n - \frac{1}{2}\Gamma$ and $f_n + \frac{1}{2}\Gamma$. It can be shown that the peak value of $r$ is equal to $f_n/\Gamma$, so that a broad response curve (large $\Gamma$) has less amplification than a narrow response curve (small $\Gamma$).

In the design of a stringed instrument, the frequencies $f_n$ and the widths $\Gamma_n$ of the cavity must be such that all the notes will resonate. The frequencies $f_n$ depend on the size and shape of the cavity, and the widths $\Gamma_n$ depend on the material of which the cavity is made. However, only recently have the details of the relation between the design and function of musical instruments been studied in detail. By bringing modern technology to bear on the old art of instrument making, physicists hoped to improve the design of these instruments, but they found that most of the traditional instruments have evolved by trial and error into such nearly perfect forms that very little improvement is possible.

The next section discusses the most fascinating of all instruments, the human voice.

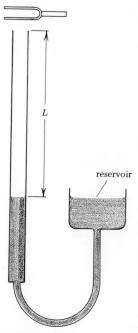

**FIGURE 13.11**
A cylinder connected to a reservoir filled with water. By varying the height of the reservoir, the length $L$ of the air column in the cylinder is varied.

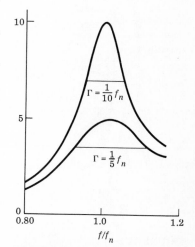

**FIGURE 13.12**
Two typical response curves.

† $\Gamma$ is capital gamma.

Through evolution, the breathing and eating organs of man have developed the additional function of producing a rich array of sounds. The ability to produce any particular sound is learned in early childhood, but the organs capable of producing sound are innate. These sounds are used to communicate symbolically with other people who share the same language. Man's ability to speak and to use symbolic language is one of the few unique characteristics that distinguish him from other animals.

The production of speech can be divided into two distinct steps: (1) the production of an audible sound and (2) the control of this sound to produce a definite *phoneme*. A phoneme is the smallest sound unit that has a functional role in a language. Each language has its own set of phonemes, and an adult speaker of one language may have difficulty learning to pronounce an unfamiliar phoneme in another language.

The alphabet is only a crude and rather inaccurate representation of the English phonemes. Since there are actually 38 phonemes in general American English, some letters must represent several phonemes while some phonemes are represented by more than one letter. For instance, the letter a represents four different vowel phonemes: the "ah" sound in f*a*ther, the "ae" sound in h*a*d, the "aw" sound in c*a*ll, and the "ei" sound in t*a*ke. On the other hand, the letters c and k sometimes represent the same phoneme, as in *c*ap and *k*ey.

All English phonemes are produced while exhaling. Air that is forced out of the lungs travels up the *trachea*, through the *larynx*, into the *pharynx*, and out the nose and mouth (Fig. 13.13). The pharynx is the part of the air passageway above the larynx and in back of the oral and nasal cavities. The larynx is a stack of cartilages between the pharynx and the trachea; it controls the flow of air through the trachea and prevents food from entering the lungs. (One cartilage in the larynx forms the Adam's apple in the front of the neck.) When a person swallows, the *epiglottis* closes the opening of the larynx to prevent food from passing into it.

The airflow is controlled by a pair of folded ligaments inside the larynx, called the *vocal cords*. During normal breathing these cords are relaxed so that air can pass freely through the larynx. A *voiced* sound† is produced when the vocal cords close off the larynx. Then, as air is exhaled, pressure builds up below the cords. When the pressure gets large enough, some air is forced through the cords, reducing the pressure in back of them. As soon as the pressure is reduced, the vocal cords close again, so that the pressure builds up once more and the sequence is repeated.

In this way a periodic series of sound pulses is produced with a frequency that depends on the tension and mass of the vocal cords. In general, the frequency is lower in men than in women because men have more massive cords. However, by changing the tension of the vocal cords, a person can vary the frequency by a factor of 3.

For a given tension, the sound produced by the vocal cords is a

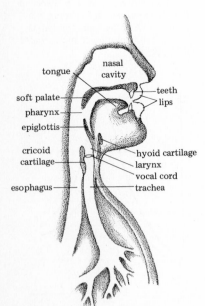

FIGURE 13.13
Anatomy of the vocal organs of man.

tongue
nasal cavity
soft palate
pharynx
epiglottis
teeth
lips
cricoid cartilage
esophagus
hyoid cartilage
larynx
vocal cord
trachea

---

† The larynx is not involved in the production of *unvoiced* phonemes, such as "f," "s," and "sh." A whisper is also unvoiced.

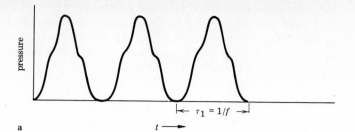

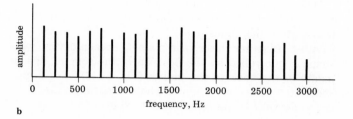

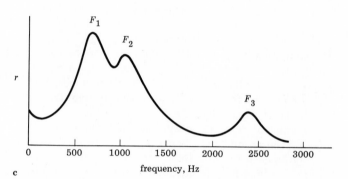

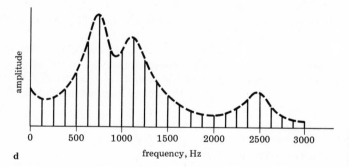

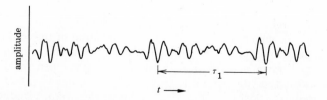

**FIGURE 13.14**
Steps in the production of a voiced phoneme. (*a*) Periodic sequence of sound pulses produced by the larynx. (*b*) Spectrum of the periodic pulses. (*c*) Response curve of the mouth and pharynx when the tongue is positioned forward and down in the mouth. (*d*) Spectrum of the sound after resonating in the oral cavity. (*e*) Final sound wave. This is the wave whose Fourier components are given in (*d*). [*After P. B. Denes and E. N. Pinson, "The Speech Chain." Copyright, 1963, Bell Telephone Laboratories, Incorporated, used by permission.*]

periodic series of positive pressure pulses, as shown in Fig. 13.14*a*. In a man the frequency $f$ of these pulses is about 125 pulses per second. Since the pattern of pulses is periodic, it can be considered to be the superposition of many sine waves (Sec. 12.3) whose frequencies form a harmonic sequence:

$$f_1 = 125 \text{ Hz}, f_2 = 2f_1, f_3 = 3f_1, \ldots$$

The spectrum of the wave contains many harmonics of nearly equal amplitude, as shown in Fig. 13.14*b*.

This is the first step in the production of speech, the production of a voiced sound. The second step, the articulation of this sound into a phoneme, takes place in the pharynx and the oral and nasal cavities. Together these form a single speech cavity, whose size and shape is controlled by the position of the tongue, lips, and the soft palate in the back of the mouth (Fig. 13.13). For a given size and shape, the cavity has its own characteristic frequencies (Sec. 13.4), called *formant* frequencies. There are usually two or three formants below 3000 Hz.

For example, to make the sound "ah," the tongue is positioned forward and down in the mouth. In this position the first three formants (in a man) are approximately 730, 1090, and 2440 Hz. As the sound wave produced by the vocal cords passes into the speech cavity, the cavity begins to resonate. Each formant has a response curve, and so although none of the frequencies in the spectrum of the sound wave is exactly equal to any formant frequency, the cavity resonates at frequencies close to the formants. Figure 13.14*c* shows the combined response curve of the three formants formed when making the sound "ah." Generally, three or four harmonics are close enough to each formant to be enhanced by resonance. This is shown in Fig. 13.14*d*, which is the superposition of Fig. 13.14*b* and *c*.

If the fundamental $f_1$ is 125 Hz, the closest harmonic to the first formant ($F_1 = 730$ Hz) in "ah" is $f_6 = 6 \times 125$ Hz $= 750$ Hz. This will have a large amplitude in the speech cavity, as shown in Fig. 13.14*d*. The neighboring harmonics, $f_5 = 625$ Hz and $f_7 = 875$ Hz, will also be enhanced but not as much as $f_6$. Consequently the spectrum of the sound is modified after entering the speech cavity. Figure 13.14*b* and *d* gives the spectra of the sound "ah" before and after entering the speech cavity, and Fig. 13.14*a* and *e* shows the corresponding wave patterns.

In summary, the first step in the production of voiced speech is the production by the vocal cords of a periodic wave with a spectrum of nearly equal amplitude harmonics. The second step is the selective enhancement of some of these harmonics by resonance in the speech cavity.

Different vowel sounds are distinguished by their formant frequencies, which are controlled by the position of the tongue and the shape of the mouth. For instance, the "ah" sound is made with the tongue down and forward in the mouth (the open front position), whereas the "oo" sound is made with the tongue up and back (the closed back position). Figure 13.15 shows the average frequencies and the relative amplitudes of the first three formants ($F_1$, $F_2$, $F_3$) of the

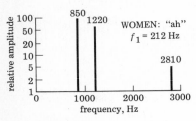

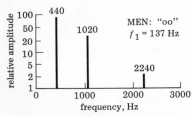

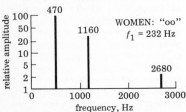

**FIGURE 13.15**
Relative amplitudes of the first three formants ($F_1$, $F_2$, and $F_3$) of the vowels "ah" and "oo" as spoken by men and women; $f_1$ is the average frequency of the fundamental in each case.

vowel sounds "ah" and "oo" as spoken by men and women. The fundamental frequency $f_1$ of the vocal cords is also given in each case.

Note that the fundamental does not differ as much between these vowels as it does between men and women, which means that the fundamental frequency plays no role in distinguishing between vowel sounds. Although the frequencies of the formants also differ between men and women, the ratio of the formant frequencies are roughly the same for a given vowel. For instance, the ratios $F_2/F_1$ and $F_3/F_1$ for "ah" are 1.49 and 3.34 for men and 1.44 and 3.31 for women. It appears that these ratios play a major role in distinguishing one vowel sound from another.

A man can speak in an unnaturally high-pitched voice (falsetto) by increasing the tension in his vocal cords. This changes the fundamental frequency of his cords but does not change the formant frequencies of his speech cavity. The falsetto voice is perfectly understandable, showing that the fundamental frequency is not used to discriminate among phonemes.

The effect of the formant frequencies on speech can be demonstrated by breathing a lungful of helium gas. The speed of sound in helium is 2.9 times the speed in air (Table 13.1). The formant frequencies of a cavity are proportional to the speed of sound in the cavity, as in Eqs. 13.16 and 13.17, so that the formant frequencies of the speech cavity are increased by a factor of 2.9 when filled with helium instead of air. A person speaking with helium in his lungs sounds like Donald Duck. Aquanauts, who breathed a gas mixture containing 97 percent helium while living 200 m under the sea, had difficulty understanding each other's speech.

## PROBLEMS

**REMARK**  Take the speed of sound in air to be 340 m/s, unless specified otherwise.

1 Bats emit ultrasonic waves with a frequency as high as $10^5$ Hz. What is the wavelength of such a wave in air?          *Ans.* 0.34 cm

2 Dolphins emit ultrasonic waves with a frequency as high as $2.5 \times 10^5$ Hz. What is the wavelength of such a wave in water?

3 (a) What is the frequency of a sound wave whose wavelength in air is 5 m? (b) What is the frequency of a wave of the same wavelength in seawater?

          *Ans.* (a) 68 Hz; (b) 306 Hz

4 What frequency must a sound wave have in seawater to have the same wavelength as a 500-Hz sound wave in air?

5 Calculate the bulk modulus of water using the data in Tables 7.2 and 13.1. Compare with the value in Table 10.4.          *Ans.* $2.24 \times 10^9$ N/m²

6 Calculate the speed of sound in carbon dioxide at 0°C and atmospheric pressure using the data in Tables 7.2 and 13.2. Compare with the value in Table 13.1.

7 Calculate the speed of sound in ethanol using the data in Tables 7.2 and 10.4.

          *Ans.* 1067 m/s

8 (a) Let $v_1$ be the speed of sound in a gas at temperature $T_1$, and let $v_2$ be the speed of sound in the same gas at temperature $T_2$. Use Eq. 13.11 to show that

$$\frac{v_2}{v_1} = \sqrt{\frac{T_2}{T_1}}$$

(b) What is the speed of sound in air at 50°C?

9 A sound wave with an intensity level of 80 dB is incident on an eardrum of area $0.60 \times 10^{-4}$ m². How much energy is absorbed by the eardrum in 3 min?          *Ans.* $1.08 \times 10^{-6}$ J

10 The sound level 25 m from a loudspeaker is 70 dB. What is the rate at which sound energy is produced by the loudspeaker?

11 (a) What is the pressure amplitude of a sound

wave with an intensity level (in air) of 120 dB? (b) What force does this exert on an eardrum of area $0.55 \times 10^{-4}$ m²?

*Ans.* (a) 28.7 N/m²; (b) 0.00158 N

12 Ultrasonic waves have many applications in medicine and technology. One of their advantages is that ultrasonic waves of large intensity can be used without danger to the ear. Consider an ultrasonic wave of intensity $I = 10^5$ W/m². (a) What is the intensity level of this wave? (b) How much energy falls on a 1-cm² surface in 1 min? (c) What is the pressure amplitude of the wave in air? (d) What is the intensity in *water* of an ultrasonic wave with the pressure amplitude found in (c)?

13 (a) What percentage of the population cannot hear a 200-Hz sound at 40 dB? (b) What is the lowest intensity level of a 1000-Hz sound that can be heard by 90 percent of the population?

*Ans.* (a) 25%; (b) 30 dB

14 What is the intensity level of a 60-Hz signal that sounds as loud as a 600-Hz signal with an intensity level of 24 dB?

15 A firework rocket explodes at an altitude of 400 m, producing an average sound intensity of $6.7 \times 10^{-2}$ W/m² for 0.2 s at a point on the ground directly below it. (a) What is the average intensity of the sound at a distance of 10 m from the rocket? (b) What is the sound level (in decibels) of the sound 10 m from the rocket? (c) What is the total sound energy radiated in the explosion?

*Ans.* (a) 107 W/m²; (b) 140 dB; (c) $2.69 \times 10^4$ J

16 The sound from an organ pipe has a sound level of 120 dB at a point 1 m from the open end of the pipe. (a) What is the sound level at a distance of 32 m from the end of the pipe? (b) What is the rate at which sound energy emerges from the open end of the pipe? (c) If the open end of the pipe has an area of 50 cm², what is the intensity of the sound inside the organ pipe?

17 At what distance from a jet airplane would the intensity level be 30 dB?

*Ans.* 9500 km

**REMARK** This result implies that a jet in London could be heard in San Francisco. Things are not quite this bad, because the sound is absorbed by the air after traveling

less than 10 km. Equation 13.15 neglects sound absorption.

18 The pressure amplitude of a sound wave is 0.04 N/m² at a distance of 12 m from its source. What is the pressure amplitude of the wave 150 m from the source?

19 (a) What are the first three characteristic frequencies of a closed pipe of length 1.5 m? (b) What are the first three characteristic frequencies if the pipe is filled with carbon dioxide at 0°C instead of air?

*Ans.* (a) 56.7, 170, and 283 Hz; (b) 43.2, 129, and 216 Hz

20 The outer ear consists of the externally visible parts of the ear and the *ear canal*, which is the passageway from the outside to the eardrum. (a) Find the fundamental frequency of this passageway if it is 2 cm long. (b) Suggest a role that this cavity might play in hearing.

21 When a tuning fork is held over the open end of the tube in Fig. 13.11, the smallest value of $L$ that gives resonance is found to be 30 cm. (a) What is the wavelength of the sound coming from the tuning fork? (b) What is the next larger value of $L$ at which resonance occurs? (c) If the frequency of the tuning fork is assumed to be 300 Hz, what value for the speed of sound in air is deduced from these data?

*Ans.* (a) 1.2 m; (b) 90 cm; (c) 360 m/s

22 When a woman pronounces the phoneme "aw," the first formant frequency of her speech cavity is 590 Hz and the fundamental frequency of her vocal cords is 216 Hz. Which harmonic is enhanced the most?

23 If the first formant frequency of a man saying "ah" is normally 730 Hz, what is it when he is breathing helium? *Ans.* 2100 Hz

## BIBLIOGRAPHY

BACKUS, JOHN: "The Acoustical Foundations of Music," W. W. Norton & Company, Inc., New York, 1969. A lucid account of the properties, production, behavior, and reproduction of musical sound. Particularly interesting is the discussion in chap. 10 of the resonance characteristics of the cavities in stringed instruments. These are compared to the formant frequencies of the human voice in chap. 11.

DENES, PETER B., and ELLIOT N. PINSON: "The Speech Chain," Bell Telephone Laboratories,

Murray Hill, N.J., 1963. A beautifully clear and concise description of the physics of speech and hearing. Recent research in artificial speech production is also discussed.

FLETCHER, HARVEY: "Speech and Hearing in Communication," D. Van Nostrand Company, Inc., New York, 1953. Classic account of the acoustic characteristics of speech. This book gives data obtained from the analysis of speech.

GRAY, GILES WILKESON, and CLAUDE MERTON WISE: "The Bases of Speech," Harper & Row, Publishers, Incorporated, New York, 1959. All aspects of speech, including the physical, physiological, phonetic, and linguistic, are treated in separate chapters. There is a detailed account of the formant theory of vowel production.

JOSEPHS, JESS J.: "The Physics of Musical Sound," D. Van Nostrand Company, Inc., Princeton, N.J., 1967. A concise account of the physics of music. Of special interest is chap. 3, which discusses the subjective nature of pitch and its relation to frequency.

OLSON, HARRY F.: "Music, Physics, and Engineering," Dover Publications, Inc., New York, 1967. A somewhat technical treatment of the physics of music. Topics include musical scales, musical instruments, properties of music, electronic reproduction, and electronic music. This book contains the spectra of many different instruments.

# CHAPTER 14 light

Light, like sound, is a wave that propagates energy without propagating mass. Light is essential for all life on earth because it is the energy of the sun, transferred to earth in the form of light, that is used by plants to synthesize carbohydrates from carbon dioxide and water (photosynthesis). Plants, in turn, are the base of the food chain for animals.

Like sound, light also supplies animals with vital information about their environment. The higher animals have evolved complex mechanisms for detecting light, though animals with divergent evolutionary histories, such as arthropods and chordates, have developed very different light-detecting apparatus. What an animal "sees" depends on the particular physical properties of light to which its eye is sensitive. Man has good color vision (frequency detection), whereas bees can detect both the color and polarization of light.

Unlike sound, however, light does not have a basis in mechanics. Since sound is the vibration of elements of the air under the action of mechanical forces (Sec. 13.2), its properties are just the consequences of the laws of mechanics applied to these elements. A light wave, on the other hand, is not the vibration of any material substance, and so its properties cannot be derived from mechanics. Light is a fundamentally different phenomenon that has no basis in mechanics. Nevertheless, a wave is a wave, and the general wave properties discussed in Chap. 12 apply as well to light as to sound. Of course, a big question is: In the absence of a material medium, what is vibrating?

## 14.1 THE NATURE OF LIGHT

### Wave nature of light

It was known to Aristotle (384–322 B.C.) that sound is caused by vibrations in the air. This knowledge was probably based on the observation that music is produced by vibrating strings. In fact, the study of the relation of musical tones to the length of the vibrating string was well developed in the ancient world. Of course, an adequate explanation of sound waves was not possible until the time of Newton.

Though nothing was known about the fundamental nature of light in Newton's time, it was natural to speculate that light was a wave, similar to sound. Christian Huygens (1629–1695), a contemporary of Newton's, developed a wave theory of light, but Newton himself favored a theory according to which light is composed of massless particles (corpuscles). Newton hypothesized that these corpuscles travel through space at constant speed and that there is a different type of corpuscle for each color. His main objection to the wave theory was that light, unlike sound, does not appear to bend around corners.

Newton's great reputation and the absence of any definitive evidence one way or another led to the general acceptance of the corpuscular theory during the eighteenth century. This did not hinder developments in optics, however, because these developments were based on empirical laws that are consistent with either a wave or corpuscular theory.

The wave nature of light was finally established by a series of experiments that demonstrated that light obeys the superposition principle (Sec. 12.3). These experiments were first performed by Thomas Young (1773-1829), the great Egyptologist, and later and more definitively by Augustin Fresnel (1788-1827). Young's experiment is discussed in the next section.

### Electromagnetic nature of light

At the time of these advances in the study of light, important discoveries were being made in the fields of electricity and magnetism (Sec. 16.1). The basic laws of electricity and magnetism, which had been discovered in the first half of the nineteenth century, were formulated into a comprehensive mathematical theory by James Clerk Maxwell (1831-1879). From this theory Maxwell deduced that there should exist electromagnetic waves consisting essentially of oscillating electric and magnetic fields that propagate through space with a definite speed. According to Maxwell's theory, this speed is given in terms of certain well-known electric constants that enter into the theory. When Maxwell calculated the speed of electromagnetic waves from these constants, he found it equal to the speed of light. This great success of Maxwell's theory established that light is a form of electromagnetic radiation.

Today physicists are familiar with electromagnetic waves with wavelengths ranging from less than $10^{-17}$ m to more than $10^4$ m. Only waves with wavelengths between $4 \times 10^{-7}$ and $7 \times 10^{-7}$ m are detected by the human eye and thus constitute visible light. Waves with longer and shorter wavelengths have special names, such as radio waves, microwaves, infrared, ultraviolet, x-rays, etc. Figure 14.1 shows the full range of electromagnetic radiation that has been studied, together with the names given to different regions. Except for the visible region, the boundaries between the regions are not sharply defined.

**REMARK**  In analogy with all other waves, electromagnetic waves were expected to be displacements in a medium. It is no ordinary medium, of course, since light and other forms of electromagnetic radiation travel to us from distant stars, through the vacuum of outer space. During the nineteenth century the medium was thought to be a special substance, called the *ether*, permeating all space. Electromagnetic waves would then be vibrations of elements of the ether about their equilibrium positions.

The ether would have to be an extraordinary substance, however, because it offers no resistance to the motion of the planets about the sun. Any ordinary material, even a very dilute gas, would apply frictional forces to the planets that would produce observable effects on their motion. A substance, like the ether, which produced no effect on ordinary matter would seem to be indistinguishable from space itself. That is, it has no physical reality other than being the medium for the propagation of electromagnetic waves. Attempts were made toward the end of the nineteenth century to detect the ether, but they failed. This led Albert Einstein (1879-1955) to abandon the idea of the ether altogether and to formulate the theory of special relativity, discussed more fully in Sec. 14.6. For the present, knowledge of "what is vibrating" is not required.

### Speed of light

In addition to the question of the nature of light, there is the question of its speed. Indeed, at one time it was not universally believed that light had a finite speed. In about the year 1600, Galileo tried and failed to measure the speed of light with a crude experiment. In 1676, Ole Römer (1644-1710) obtained the first finite value for the speed of light, based on discrepancies in the motions of the moons of Jupiter that depended on the distance between Jupiter and earth. However, his result was not generally accepted.

It was not until the middle of the nineteenth century that direct experimental measurements of the speed of light were obtained. The first measurement was made in 1849 by Armand Hippolyte Louis Fizeau (1819-1896), who used an ingenious modification of Galileo's unsuccessful measurement. Galileo had stationed two assistants, each with a lantern, some distant apart (Fig. 14.2*a*). Assistant *A* would open his lantern, and assistant *B* was instructed to open his lantern when he saw the light from *A*'s lantern. Galileo tried to measure the time between the opening of *A*'s lantern and the arrival of the light from *B*'s lantern back to *A*. Even with a distance of 5 mi between *A* and *B*, however, the time is only $5 \times 10^{-5}$ s, which is much too small for Galileo to have detected.

Fizeau replaced assistant *B* by a mirror (Fig. 14.2*b*), so that the light would be instantaneously returned to *A* without any delay caused by *B*'s reaction time. Assistant *A* was replaced by a rotating toothed wheel in front of a light source, so that as the wheel rotated, pulses of light were automatically transmitted to the mirror. The rate

**FIGURE 14.1**
Electromagnetic radiation between $10^4$ and $10^{26}$ Hz. Except for visible light, the boundaries between different types of radiation are not sharply defined.

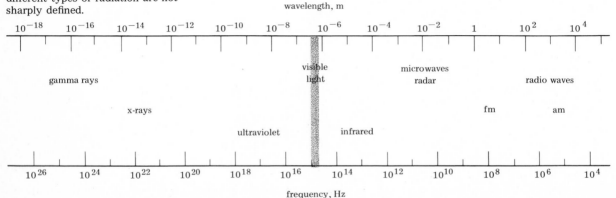

of rotation of the wheel was increased until an observer in back of the wheel did *not* see the reflected light. This happens when the time required for the light to travel from the wheel to the mirror and back again is equal to the time required for the wheel to move from a space to the next tooth. From knowledge of the distance between the wheel and the mirror and the rate of rotation of the wheel Fizeau calculated the speed of light.

Since Fizeau's first measurement, the speed of light has been measured repeatedly with increasing accuracy. Table 1.1 gives the results of some of these measurements. For our purposes the speed of light (in vacuum) is taken to be

$$c = 3.00 \times 10^8 \text{ m/s}$$

The symbol $c$ is always used for the speed of light in vacuum.

The speed of light in transparent substances, e.g., water and glass, as found from measurements is always less than the speed of light in vacuum. (The measurement of Fizeau and other early investigators was in air, of course. The speed of light in air is negligibly different from its speed in vacuum.)

For instance, the speed of light in water is

$$c_w = 2.25 \times 10^8 \text{ m/s}$$

The ratio of the speed of light $c$ in vacuum to the speed of light $c'$ in a particular substance is called the *index of refraction* $n$ of the substance:

$$n = \frac{c}{c'} \qquad\qquad 14.1$$

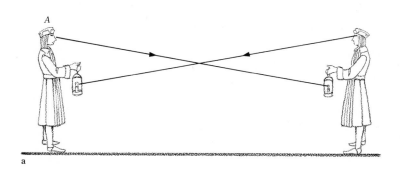

a

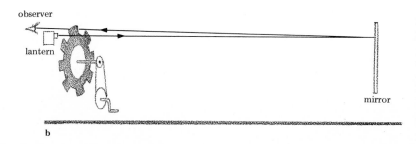

observer

lantern

mirror

b

FIGURE 14.2
Measurements of the speed of light. (*a*) Galileo tried unsuccessfully to measure the time interval between assistant $A$'s opening his lantern and $A$'s first seeing $B$ open his. (*b*) Fizeau replaced assistant $A$ by a toothed wheel rotating in front of a lantern and assistant $B$ by a mirror that instantaneously reflects light pulses back toward the wheel.

Thus, the index of refraction of water is

$$n = \frac{3.00 \times 10^8 \text{ m/s}}{2.25 \times 10^8 \text{ m/s}} = 1.33$$

Since $c'$ is always less than $c$, the index of refraction of every substance is greater than 1. For a particular substance, the value of $n$ depends somewhat on the wavelength of light. Table 14.1 gives the index of refraction of some common substances at the wavelength of yellow light, $\lambda = 5.89 \times 10^{-7}$ m $= 589$ nm.†

## 14.2 INTERFERENCE AND DIFFRACTION

### Interference

The principle of superposition (Sec. 12.3) says that when two or more waves coexist at the same point, the displacement of the medium is the sum of the displacements that each wave would produce separately. This has the surprising consequence that two waves which separately would produce equal and opposite displacements at a point superimpose to produce zero displacement at that point. This has already been seen in the case of a standing wave (Sec. 12.5), where two sine waves with the same wavelength and amplitude but traveling in opposite directions always cancel each other at certain points called nodes.

To demonstrate the wave nature of light, Thomas Young devised an experiment in which two light beams are caused to cancel each other at some points. His experiment can best be understood by first considering a series of small water waves (ripples) moving toward a screen that has two small holes in it (Fig. 14.3a). The crests of the waves are shown as solid lines, and the troughs, or minima, are shown as dashed lines halfway between the crests. Each line is called a *wavefront* because all the points on a given line have the same displacement. The wavelength $\lambda$ is the distance between the solid lines.

When the wave hits the screen, a disturbance is produced at each hole, so that each hole acts as a point source of waves. That is, semicircular waves emanate from each hole. Wherever the crest of a wave from one hole intersects the trough of a wave from the other hole, a region of *destructive interference* is formed, in which the elements of the medium do not oscillate, i.e., in which their amplitude is zero. Figure 14.3a shows that these regions are nearly straight lines that radiate from the screen. Between these lines, a region of *constructive interference* is formed, in which the crests (and troughs) of the two waves coincide, so that the elements of the medium oscillate with large amplitudes. Thus the interference of the waves from the two holes produces alternating regions of zero and large-amplitude oscillation. This interference pattern is identical to that pro-

TABLE 14.1 **Index of refraction of some common substances**
These values are for yellow light.

| Substance | Index of refraction |
|---|---|
| Acetone at 20°C | 1.3584 |
| Air at STP | 1.0002926 |
| Canada balsam | 1.530 |
| Diamond | 2.4168 |
| Ethanol at 20°C | 1.36008 |
| Glass: | |
|     Fused quartz | 1.458 |
|     Heavy flint | 1.650 |
|     Zinc crown | 1.517 |
| Quartz, crystalline | 1.553 |
| Sodium chloride | 1.544 |
| Water at 20°C | 1.33335 |

† 1 nm = 1 nanometer = $10^{-9}$ m. In the older literature a nanometer is called a millimicron (m$\mu$), but this term is no longer acceptable.

duced by circular waves emanating from two points, as shown in Fig. 14.3b.

Young demonstrated the wave nature of light by producing a similar interference pattern with light. Today this demonstration is easily performed with a laser. A screen with a pair of narrow slits in it is placed in front of a laser beam, and the light emerging from the slits is allowed to fall on a viewing screen (Fig. 14.4). The pattern on the screen consists of alternating bright and dark bands, called *fringes*. The bright central fringe corresponds to point $A$ in Fig. 14.3. This fringe is equally distant from the two slits, so that a crest from one slit always coincides with a crest from the other, resulting in constructive interference. On either side of the central fringe there are dark fringes, where the lights from the two slits cancel each other. Points in these regions correspond to points $B$ and $B'$ in Fig. 14.3. They are exactly $\frac{1}{2}$ wavelength farther from one slit than from the other, so that a crest from one slit always falls on a trough from the other, resulting in destructive interference.

Away from the central fringe, on either side of the dark fringe, are two more bright fringes. These points, which correspond to points $C$ and $C'$ in Fig. 14.3, are 1 wavelength farther from one slit than from the other, so that again a crest from one slit always coincides with a crest from the other slit, resulting in constructive interference. In general, a dark fringe occurs at a point that is a distance $(n + \frac{1}{2})\lambda$ farther from one slit than the other, where $n$ is an integer and $\lambda$ is the wavelength of the light. Similarly, a bright fringe occurs at a point that is a distance $n\lambda$ farther from one slit than the other.

Figure 14.5 shows the distances from two slits to a point $P$ in a bright fringe. The distance $d$ between the slits is usually very small compared to the distance $D$ from the slits to the screen, so that the lines $SP$ and $S'P$ are almost parallel and make approximately the same angle $\theta$ with the perpendicular to the screens. The line $ST$ is the perpendicular from slit $S$ to the line $S'P$. The distance from $T$ to $P$ is approximately equal to the distance from $S$ to $P$, so that the distance $l$ between $S'$ and $T$ is the difference between the distances from $S$ and $S'$ to $P$. From the right triangle $STS'$ we get the relation

$$l = d \sin \theta$$

The condition for a bright fringe is $l = n\lambda$. Therefore, the bright fringes occur at angles given by

$$\sin \theta = \frac{l}{d} = \frac{n\lambda}{d}$$

14.2

where $n$ is an integer.

For $n = 0$, Eq. 14.2 gives $\theta = 0°$, which is the central fringe. The first bright fringe on either side of the central fringe corresponds to $n = 1$, the second bright fringe corresponds to $n = 2$, and so on. The distance $x_n$ between the central fringe and the $n$th fringe is approximately

$$x_n = D \tan \theta$$

Because $\theta$ is very small, $\tan \theta$ is approximately equal to $\sin \theta$ (see

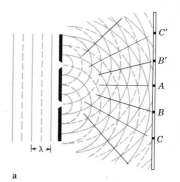

a

b

FIGURE 14.3
(a) Interference of circular waves emanating from two holes. The radial lines indicate regions of destructive interference. (b) The interference pattern of circular waves of water emanating from two points. [*From "PSSC Physics," 3d ed., copyright © 1971, by Education Development Center, Inc., reprinted by permission of D. C. Heath and Co.*]

the table of trigonometric functions). Then, using Eq. 14.2 for sin $\theta$, we get

$$x_n = D \sin \theta = \frac{Dn\lambda}{d}$$

or

$$\lambda = \frac{x_n d}{nD} \qquad\qquad 14.3$$

For example, suppose that with a slit separation of $d = 0.02$ cm a neon-helium laser forms a pattern of fringes on a screen a distance $D = 3.0$ m from the slits. In this pattern the first bright fringe ($n = 1$) is 0.95 cm from the central fringe. The wavelength of the light is then found from Eq. 14.3 to be

$$\lambda = \frac{x_n d}{nD} = \frac{0.95 \text{ cm} \times 0.02 \text{ cm}}{1 \times 3.0 \times 10^2 \text{ cm}} = 6.3 \times 10^{-5} \text{ cm}$$

The double-slit experiment provides convincing proof of the wave nature of light because the pattern of bright and dark fringes it produces can result only from superposition, which is a unique wave property. The experiment also provides a method for measuring the wavelength of a light, because the positions $x_n$ of the bright fringes depend on the wavelength. Light containing more than one wavelength is broken into its component wavelengths after passing through a double slit, because the fringes for the different wavelengths have different positions on the screen. The frequency composition, or spectrum, of light is measured this way. (In practice a grating that consists of thousands of slits is used to increase the intensity of the bright fringes.) *Spectroscopy*, the study of the spectra of light emitted by different atoms, is the basis of our present knowledge of atomic structure (Sec. 19.2).

**REMARK** Double-slit interference is very easy to demonstrate with a laser because a laser produces a single long continuous wavetrain, like the water wave in Fig. 14.3. Such light is said to be *coherent*, whereas normal light from the sun or a light bulb is *incoherent*. Incoherent light consists of billions of independent short wavetrains that have no relation to each other. To get an interference pattern with ordinary incoherent light, instead of coherent laser light, each individual wavetrain must be made to interfere with itself. This is accomplished by using a very small source of light, so that each wavetrain illuminates both slits at the same time. Because the light source must be small,

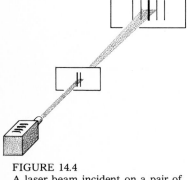

**FIGURE 14.4**
A laser beam incident on a pair of slits produces an interference pattern on a screen.

**FIGURE 14.5**
Distances from two slits $S$ and $S'$ to a point $P$ on a screen. Destructive interference occurs when the difference $l$ between these distances is $(n + \frac{1}{2})\lambda$.

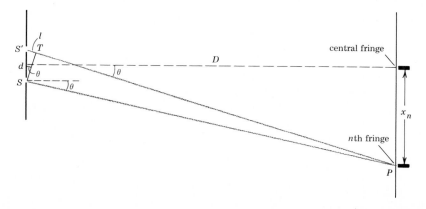

the intensity of the light will be small and the interference pattern very faint. An experiment with incoherent light must be performed in a dark room and requires considerable skill and patience.

If a large incoherent light source were used in an interference experiment, a different wave would pass through each slit at a given time. These two waves interfere, of course, but the exact position of the interference pattern on the viewing screen depends on the exact relation between the displacements of the two waves at the slits. If the crests of both waves arrive at the slits at the same time, the pattern will be in one position, whereas if the crest of one wave arrives at the same time as the trough of the other wave, the pattern will be in a different position. There will thus be millions of overlapping interference patterns on the screen, which obliterate each other, so that no interference pattern is observed. In part this explains why the wave nature of light is not ordinarily apparent.

## Diffraction

In Fig. 14.3 it is assumed that semicircular waves emanate from the narrow slits when the incident waves reach them. This in itself is an interesting wave phenomenon. Certainly, if the slits are very wide, we would not expect the incident waves to be modified very much as they pass through. However, the wavefronts of a plane wave are always bent slightly when passing through a slit, as shown in Fig. 14.6. Consequently, the wave no longer travels in one direction but instead has an angular divergence $\theta$. This phenomenon is called *diffraction*. It can be shown that $\theta$ is related to the wavelength $\lambda$ and the slit width $d$ by

$$\sin \theta = \frac{\lambda}{d} \qquad 14.4$$

If $d$ is very large compared to $\lambda$, the angle $\theta$ is very small and the incident wave is little affected by the slit. On the other hand, if $d$ is comparable to $\lambda$, the angle $\theta$ is large, and the wave radiates in all directions from the slit. This latter condition must hold for the slits in a double-slit experiment for the light from the slits to overlap on the screen.

The wavelength of light is about $5 \times 10^{-5}$ cm, which is small compared to normal-sized objects. Therefore light is not noticeably diffracted by ordinary objects but appears to travel in straight lines without bending. In contrast, the wavelengths of sound are comparable to normal-sized objects, so that sound is highly diffracted. For example, the wavelengths of the human voice are large compared to the size of the mouth, and so sound is diffracted in all directions as it leaves the mouth. If this were not the case, you would have to be directly in front of a speaker in order to hear him. Ultrasonic waves, which have much shorter wavelengths than sound, do travel in straight lines without much diffraction, just like light.

Objects smaller than a wavelength of light cannot be seen with an ordinary light microscope because the light reflected from the object is diffracted in all directions. Conversely, objects can be detected with sound if the wavelength of the sound is smaller than the object. Bats emit ultrasonic waves with wavelengths as small as 0.3 cm, which can reflect off objects as small as a moth without

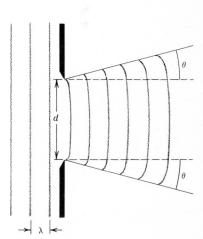

FIGURE 14.6
Diffraction of a plane wave passing through a slit. When the slit is much wider than the wavelength of the wave, the diffraction angle $\theta$ is small.

appreciable diffraction. The bat flies through dark caves and catches moths at night by detecting the echos of his ultrasonic signals.

**Insect vision**

Diffraction seems to have played a fundamental role in the evolution of the insect eye. The compound eye of insects consists of transparent fibers, called *ommatidia*, which are clustered together in a hemispherical arrangement (Fig. 14.7). Each ommatidium can receive only incident light making an angle less than $\phi$ with its central axis (Fig. 14.8). All incident light that lies inside this angle enters the ommatidium, passes along the fiber, and stimulates the nerves at its base. Light from different objects that enters the same ommatidium cannot be differentiated. That is, for an insect to see two objects, light from these objects must enter two different ommatidia. This requires that the two objects have an angular separation of at least $\phi$. Thus the angle $\phi$ of the ommatidia determines the discrimination of the insect eye.

To achieve maximum discrimination, the angle $\phi$ should be as small as possible. From Fig. 14.8, it is seen that $\phi$ can be decreased by either increasing the length $L$ or decreasing the width $d$ of the ommatidia. But if $L$ is increased, the eye will take up more space in the insect's head. Therefore it seems reasonable that the smallest possible $d$ should have resulted from evolutionary adaptations.

The ommatidia of the bee have a width $d = 2 \times 10^{-3}$ cm. Light that enters such a small opening is diffracted through an angle $\theta$ given by Eq. 14.4. Taking $\lambda = 5 \times 10^{-5}$ cm, we get

$$\sin \theta = \frac{5 \times 10^{-5} \text{ cm}}{2 \times 10^{-3} \text{ cm}} = 2.5 \times 10^{-2}$$

From the small-angle approximation given in the table of trigonometric functions inside the back cover, the diffraction angle $\theta$ is found to be

$$\theta = \frac{\sin \theta}{0.0174} = 1.4°$$

This means that light making an angle of less than 1.4° with the central axis of an ommatidium will be diffracted into the ommatidium regardless of the angle $\phi$. Thus, because of diffraction, there is no advantage to making $\phi$ less than $\theta$, and in fact measurements have found that the acceptance angle $\phi$ is between 1 and 2° in the bee. The width $d$ of the bee's ommatidia is of optimal size, since if it were smaller, the diffraction angle $\theta$ would be larger, and if it were larger, the acceptance angle $\phi$ would be larger. The insect eye has achieved, through evolution, the maximum discrimination consistent with its size.

a

b

**FIGURE 14.7**
Ommatidia in the compound eye of an insect. (*a*) Cutaway drawing showing that each ommatidium points in a different direction. (*b*) Scanning electron photomicrograph of the eye of a fly. [*Courtesy of H. F. Howden, Carleton University, Ottawa, Canada.*]

## 14.3 REFLECTION AND REFRACTION

In a homogeneous medium, a plane wave travels in a straight line with a speed $v_1$ that is characteristic of the medium. When the wave arrives at the boundary separating one medium from another, part of the wave is *reflected* from the boundary back into the first medium and part of the wave proceeds into the second medium. The part that enters the second medium is called the *refracted* wave (Fig. 14.9). The intensity of the reflected wave is $rI_0$, where $I_0$ is the intensity of the incident wave and $r$ is the fraction of the wave reflected. The intensity of the refracted wave then is $I_0 - rI_0 = (1 - r)I_0$. The value of $r$ varies between 0 and 1, depending on the circumstances. It may be exactly 1.00, in which case there is no refracted wave and the incident wave is totally reflected. However, $r$ is never exactly zero: there is always some reflection from the boundary separating two media.

Figure 14.9$a$ shows both the wavefronts and the direction of propagation of the incident, reflected, and refracted waves. Light propagates in the direction of the lines, called *rays*, that are perpendicular to the wavefronts. Physically a ray is just a narrow beam of light, as seen in Fig. 14.9$b$. This figure shows an incident beam of light reflected and refracted at the boundary between air and glass. Of particular interest are the angles that the incident, reflected, and refracted rays make with a boundary. These angles are customarily measured with respect to a line (the *normal*) drawn perpendicular to the boundary. The angles of incidence $\theta_1$, of reflection $\theta_1'$, and of refraction $\theta_2$ are shown in Fig. 14.9$a$. Note that these are also the angles between the respective wavefronts and the boundary itself.

## Laws of reflection and refraction

The relations of $\theta_1'$ and $\theta_2$ to the angle of incidence $\theta_1$ can be deduced by considering the behavior of the wavefronts. Figure 14.10 shows in detail a wavefront of the incident and reflected waves. When the lower edge $A'$ of the incident wavefront $AA'$ is just coming into contact with the boundary surface, the upper edge $A$ still has a distance $d$ to go to reach point $B$ on the surface. The edge $A$ will arrive at the surface a time $t = d/v_1$ later, where $v_1$ is the speed of the wave in the first medium. At this later time, the wavefront will have been reflected into the position $BB'$, the edge $B'$ having been reflected from $A'$. The line $A'B'$ is in the direction of propagation of the reflected light, so it is perpendicular to the wavefront $BB'$. Since both the incident and reflected waves travel in the same medium, they both have the speed $v_1$. Therefore the distance $d'$ between $A'$ and $B'$ is

$$d' = v_1 t = v_1 \frac{d}{v_1} = d$$

Thus the right triangles $A'AB$ and $BB'A'$ are congruent, so that the angles $\alpha$ and $\beta$ are equal. But $\alpha$ is equal to $\theta_1$ because the sides of angle $\alpha$ are perpendicular to the corresponding sides of angle $\theta_1$ (Theorem 3, Appendix II). Likewise, $\beta$ is equal to $\theta_1'$. Therefore we have

$$\alpha = \theta_1 \qquad \beta = \theta_2'$$

and

$$\alpha = \beta$$

from which we get

$$\theta_1 = \theta_1' \qquad \text{law of reflection} \qquad 14.5$$

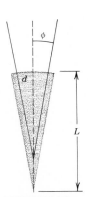

FIGURE 14.8
An ommatidium of length $L$ and width $d$. Only incident light that makes an angle less than $\phi$ with the central axis can penetrate to the base of the ommatidium.

MEDIUM 1

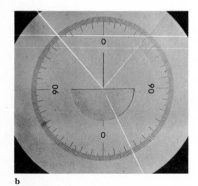

a

FIGURE 14.9
Reflection and refraction of light
at the boundary between two
media. (a) The angles of incidence
$\theta_1$, reflection $\theta_1'$, and refraction $\theta_2$
are measured with respect to the
normal. (b) Photograph of a light
ray reflected and refracted at the
boundary of air and glass. The
refracted ray is not refracted again
as it passes back into air because
its angle of incidence on the
curved glass-air boundary is zero.
[From "PSSC Physics," 3d ed.,
copyright © 1971, by Education
Development Center, Inc.,
reprinted by permission of D. C.
Heath and Co.]

The angle of incidence is equal to the angle of reflection because the two waves travel with the same speed in the same medium.

**REMARK** Only for a very smooth surface, such as a mirror, is the angle of incidence the same at all points on the surface. For a smooth surface, all the incident light coming from one direction is reflected in the same direction. This is called *specular* or *regular* reflection. Most surfaces, such as the paper of this book, are not smooth. Under high magnification, the surface is seen to consist of millions of microsurfaces, each oriented in a different direction (Fig. 14.11), so that the angle of incidence of the light illuminating the paper at any point depends on the orientation of the microsurface at that point. Consequently, light that is incident from a single direction is reflected in different directions from different points on the surface. This is called *diffuse* reflection. It is not a contradiction of the law of reflection because the law is still valid for each microsurface.

Figure 14.12 shows in detail a wavefront of the incident and refracted waves. Again, the lower edge $A'$ of the incident wavefront $AA'$ is just coming into contact with the surface, while the upper edge $A$ still has a distance $d$ to go to reach point $C$ on the surface. The edge $A$ will arrive at the surface a time $t = d/v_1$ later. At this later time, the wavefront will have been refracted into the position $CC'$, the edge $C'$ having been refracted from $A'$. The line $A'C'$ is in the direction of propagation of the refracted light, so it is perpendicular to the wavefront $CC'$. Since the refracted wave travels in the second medium, where its speed is $v_2$, the distance $d'$ between $A'$ and $C'$ is

$$d' = v_2 t = v_2 \frac{d}{v_1} = \frac{v_2}{v_1} d \qquad 14.6$$

Again we see from Theorem 3 of Appendix II that $\alpha$ is equal to $\theta_1$ and $\beta$ is equal to $\theta_2$, so that from the right triangles $A'AC$ and $CC'A'$ we have

$$\sin \theta_1 = \sin \alpha = \frac{d}{A'C}$$

$$\sin \theta_2 = \sin \beta = \frac{d'}{A'C}$$

or

$$\frac{\sin \theta_1}{\sin \theta_2} = \frac{d}{d'}$$

But according to Eq. 14.6,

$$\frac{d}{d'} = \frac{v_1}{v_2}$$

so

$$\frac{\sin \theta_1}{\sin \theta_2} = \frac{v_1}{v_2}$$

This is one form of the law of refraction.

It is convenient to rewrite this in terms of the index of refraction of the two media. From Eq. 14.1 we have

$$n_1 = \frac{c}{v_1} \qquad \text{and} \qquad n_2 = \frac{c}{v_2}$$

so the law of refraction can be written

$$\frac{\sin \theta_1}{\sin \theta_2} = \frac{n_2}{n_1}$$

or $\qquad n_1 \sin \theta_1 = n_2 \sin \theta_2 \qquad$ law of refraction $\qquad$ 14.7

This is often referred to as *Snell's law*.

**REMARKS**  (1) In 1621 Willebrord Snell van Royen (1591–1626) discovered that the ratio $(\sin \theta_1)/(\sin \theta_2)$ is a constant, independent of $\theta_1$, for light refracted between two given media.  The index of refraction of most materials is determined directly from this relation, without any assumption about the relation of $n$ to the speed of light.  It was not until the time of Newton and Huygens that $n$ was related to v.  In fact, both the corpuscular and the wave theory predict the relation in Eq. 14.7.  However, in the corpuscular theory v is related to $n$ by v $= nc$, whereas in the wave theory the relation is v $= c/n$.  For example, from Eq. 14.7 it is found that glass has an index of refraction of 1.5, assuming the index of refraction of air to be 1.0.  Therefore, light is expected to travel faster in glass than in air if it is corpuscular and to travel slower if it is a wave.  Measurements of the speed of light in different media made by Foucault (1819–1868) in 1862 showed that light travels more slowly in glass than in air, which further demonstrated the wave nature of light.  (2) Two transparent media are considered to be optically the same if they have the same index of refraction, even though they may be chemically quite dissimilar.  Light will not be reflected from the boundary between two such media, and by Eq. 14.7 the angle of incidence will equal the angle of refraction.  This causes difficulty in the microscopic examination of biological structures, such as cells, that have the same index of refraction as water.  Without special stains, a cell may be nearly invisible in water, because light passes through it without deviation or reflection.  Similarly, canada balsam is used to cement cover slips to glass microscope slides because its index of refraction is nearly the same as that of glass (Table 14.1).  This prevents distortion that might be caused by the light passing from glass into a medium with a different index of refraction.

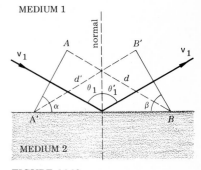

FIGURE 14.10
A wavefront reflected from a surface.

### Examples involving Snell's law

As an example of Snell's law, we consider a beam of light passing through a sheet of glass with parallel sides (Fig. 14.13).  The light travels from air ($n_1$) into glass ($n_2$) and back into air again.  Let $\theta_1$ be the angle at which the light is incident on the glass.  Then, according to Eq. 14.7, the angle of refraction $\theta_2$ of the light in the glass is given by

$$\sin \theta_2 = \frac{n_1}{n_2} \sin \theta_1$$

Since the sides of the sheet of glass are parallel, $\theta_2$ is equal to the angle $\phi_2$ at which the light is incident on the second surface.  As the light passes back into air, it is again refracted.  This angle of refraction $\phi_1$ is given by

$$\sin \phi_1 = \frac{n_2}{n_1} \sin \phi_2 = \frac{n_2}{n_1} \sin \theta_2$$

$$= \frac{n_2}{n_1} \frac{n_1}{n_2} \sin \theta_1 = \sin \theta_1$$

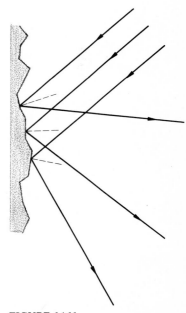

FIGURE 14.11
Enlarged view of light reflected from the microsurfaces of a rough surface.

MEDIUM 1

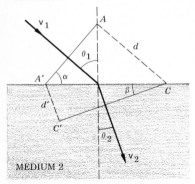

MEDIUM 2

**FIGURE 14.12**
A wavefront refracted at the
boundary between two media.

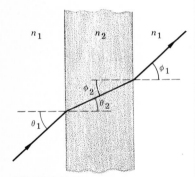

**FIGURE 14.13**
A light ray passing through a
sheet of glass with parallel sides.

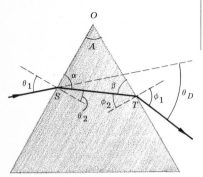

**FIGURE 14.14**
A light ray passing through two
sides of a prism.

Therefore, $\phi_1 = \theta_1$, which means that the light emerges from the glass parallel to the incident light ray. That is, a sheet of glass with parallel sides produces no net deflection of a ray of light.

The situation is different if the sides of the glass are not parallel. Figure 14.14 shows a ray of light passing through two sides of a prism at an angle $A$ with each other. In this case, the ray emerges from the prism at an angle $\theta_D$ to its incident direction.

For example, consider a beam of light incident at an angle $\theta_1 = 40°$ on a prism with an *apex angle* $A$ of 60°. With $n_2 = 1.5$ for glass, the angle of refraction $\theta_2$ is

$$\sin \theta_2 = \frac{n_1}{n_2} \sin \theta_1 = \frac{1}{1.5} \sin 40°$$

$$= 0.429$$

or $\qquad\qquad \theta_2 = 25.4°$

The angle of incidence $\phi_2$ of this ray on the other side of the prism is found by simple geometry. The sum of the interior angles of the triangle $OST$ is 180°:

$$A + \alpha + \beta = 180°$$

But $\alpha = 90° - \theta_2$, and $\beta = 90° - \phi_2$, so

$$A + (90° - \theta_2) + (90° - \phi_2) = 180°$$

or $\qquad\qquad \phi_2 = A - \theta_2$

Thus, in this example, $\phi_2 = 60° - 25.4° = 34.6°$. The angle of refraction $\phi_1$ with which the ray emerges back into air is found by using Eq. 14.7 once more:

$$\sin \phi_1 = \frac{n_2}{n_1} \sin \phi_2 = 1.5 \sin 34.6°$$

$$= 0.853$$

or $\qquad\qquad \phi_1 = 58.6°$

The *angle of deviation* $\theta_D$ can be shown by simple geometry to be

$$\theta_D = (\theta_1 - \theta_2) + (\phi_1 - \phi_2)$$
$$= \theta_1 + \phi_1 - A$$

which in this example is $\theta_D = 40° + 58.6° - 60° = 38.6°$.

The angle of deviation clearly depends on the apex angle $A$, the angle of incidence $\theta_1$, and the index of refraction $n_2$ of the glass. Since the index of refraction of glass is slightly different for different wavelengths of light, the angle of deviation depends also on wavelength. White light is a mixture of light of different wavelengths, so that when a beam of white light is passed through a prism, the different wavelengths of which it is composed are deviated through different angles. Consequently the light is broken into its component wavelengths. The eye senses these different wavelengths as different colors, and so the light emerges with its spectrum of colors. A rainbow is produced by sunlight refracting through drops of water in the sky.

## Total internal reflection

A ray of light is refracted toward the normal when it goes into a medium of higher index of refraction, and it is refracted away from the normal when it goes into a medium of lower index of refraction. This latter case is especially interesting. Suppose, for example, that a ray of light in glass ($n_1 = 1.5$) is incident on the glass-air boundary at an angle $\theta_1 = 60°$. From Snell's law the angle of refraction $\theta_2$ is given by

$$\sin \theta_2 = \frac{n_1}{n_2} \sin \theta_1 = 1.5 \sin 60° = 1.3$$

But there is no angle with a sine of 1.3, and in fact no light emerges into the air. All the light incident on the glass-air boundary is reflected back into the glass. This phenomenon, called *total internal reflection*, occurs whenever the angle of incidence $\theta_1$ produces a value of $\sin \theta_2$ that is greater than 1. Thus the condition for total internal reflection is

$$\frac{n_1}{n_2} \sin \theta_1 > 1$$

which can be satisfied only if $n_1 > n_2$. The smallest, or *critical*, angle for internal reflection occurs when

$$\sin \theta_1 = \frac{n_2}{n_1} \qquad\qquad 14.8$$

For instance, the sine of the critical angle at the glass-air boundary is

$$\sin \theta_1 = \frac{1}{1.5} = 0.667$$

so that the critical angle is

$$\theta_1 = 42°$$

The index of refraction of a substance can be determined most easily by measuring the critical angle of the substance and using Eq. 14.8 to calculate $n_1$. Optical devices for measuring the index of refraction, called *refractometers*, are used to analyze the concentration of solutions. For instance, a mixture of water and alcohol has an index of refraction between the values for pure water and pure alcohol. By accurate measurement of the index of refraction of a water-alcohol mixture, the percentage of alcohol in the mixture can be determined. Simple optical instruments, using the principle of internal reflection, are now used in garages to test the antifreeze mixture in a car's radiator.

Total internal reflection finds important application in the field of fiber optics. An *optical fiber* is a long narrow filament of glass or transparent plastic. Because it is narrow, light entering the fiber at one end strikes the inside wall of the fiber at an angle greater than the critical angle (Fig. 14.15*a*). Since the light is totally reflected inside the fiber, it travels along the fiber, making hundreds of internal reflections without escaping. Even when the fiber is bent into a complicated shape, the light will still pass from one end to the other without loss. Light will be refracted out of the fiber only when it is bent so sharply that the light strikes an internal surface at an angle less than the critical angle.

Bundles of hundreds of thousands of extremely narrow optical fibers (generally about $2 \times 10^{-3}$ cm in diameter) are used to examine otherwise inaccessible objects. Each individual fiber transmits light from a small region of the object, and the thousands of fibers together form an image of the entire object. The fibers are thus similar to the ommatidia in the compound eye of an insect.

Fiber optics is finding increasing application in medicine. For example, the inside of a patient's stomach is examined by inserting a fiber bundle into it. Light is sent down the fibers on the outside of the bundle to illuminate the stomach wall, and the reflected light passes back through the fibers inside the bundle (Fig. 14.15$b$). The presence of lesions inside the stomach can thus be determined, and they can be examined without surgery [Persyko (1969)].

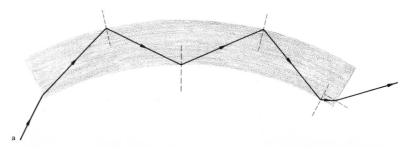

a

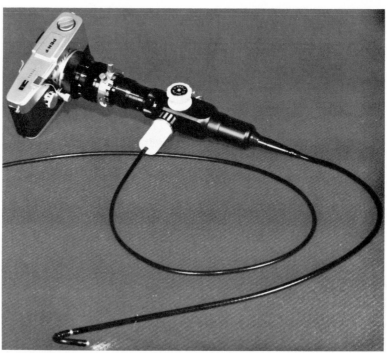

**FIGURE 14.15**
($a$) An optical fiber. Light that enters one end of the fiber is internally reflected until it reaches the other end. ($b$) A fiber gastroscope attached to a camera. The fiber bundle on the right is inserted into the stomach, which is illuminated by light that passes from the bundle on the left, through the outer fibers of the inserted bundle. Light reflected from the stomach wall then passes back through the inner fibers of the bundle and forms an image on the film in the camera. [*American Optical Corporation, Fiber Optics Division.*]

b

Fiber optics is also used to measure the blood oxygen concentration inside the heart. A fiber bundle only a few millimeters wide is passed through an artery into the heart. Light sent down one set of fibers in the bundle is scattered back from the blood and returns through another set of fibers. No image is formed, but the oxygen concentration is determined by a spectroscopic analysis of the returned light [Gamble (1965)].

## 14.4  COLOR

Light of a single wavelength is perceived by an observer with normal color vision to have one of the *spectral* colors. These colors range from bluish purple for light with a wavelength of 420 nm, through green for light with a wavelength of 520 nm, to red for light with a wavelength of 700 nm. However, most light we perceive, such as the light reflected from a blue book, consists of more than one wavelength. Spectral analysis of such a light would show that it contains light of all wavelengths, though there would probably be a greater intensity of light from the shorter-wavelength (blue) end of the spectrum than from the longer (red) end.

Color is a psychological attribute of light, and the relation of these attributes to the physical attributes of light lies in the domain of psychophysics. Unlike the other senses, however, color vision is reducible to exact mathematical laws, which were formulated in 1853 by Hermann Grassmann (1809–1877). These laws are the basis of all industrial color processes.

The relation between the color and the wavelength of light seems to be analogous to the relation between the pitch and the wavelength (or frequency) of sound. There are very important differences, however, that are due to the way these waves are detected. The ear, for instance, makes a spectral analysis of an incident sound wave. The wave causes vibration in the eardrum, which transmits these vibrations by a mechanical linkage composed of three tiny bones (the *auditory ossicles*) to a structure in the inner ear called the *cochlea* (Fig. 14.16). This spiral-shaped organ is filled with a viscous fluid (*perilymph*) and has a partition running through it. Figure 14.17

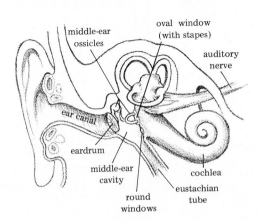

**FIGURE 14.16**
Anatomy of the inner and outer ear. [*After P. B. Denes and E. N. Pinson, "The Speech Chain." Copyright 1963, Bell Telephone Laboratories, Incorporated; used by permission.*]

shows a sagittal section of the uncoiled cochlea. Vibrations transferred to the perilymph through the *oval window* cause the partition to oscillate. Because the partition is thinnest near the entrance to the cochlea (*basal* end) and thickest at the other (*apical*) end, low-frequency waves cause maximum displacement of the partition near the apical end and high-frequency waves cause maximum displacement near the basal end. Inside the partition sensory organs convert displacements of the partition into nerve impulses, so that waves of different frequency stimulate different sensory organs. The result is that the ear effectively analyzes a complex sound into its frequency components.

The eye has much less ability to discriminate between light of different frequency compositions than the ear has to discriminate between sounds of different frequency compositions. Light incident on the eye is focused onto the retina, which is an area covered with two kinds of light-sensitive cells, *rods* and *cones*. The rods are very sensitive to light intensity, but they do not respond differently to different wavelengths. The cones, on the other hand, contain pigments that preferentially absorb light of different wavelengths. There are only three different kinds of pigments, and each cone cell contains one of them. Consequently the cones can make only a crude three-part decomposition of the light.

Grassmann's first law of color states that the normal human eye perceives only three attributes of light, often referred to as *brightness*, *saturation*, and *hue*. From this psychological fact, physiologists inferred that there must be three, and only three, kinds of pigments in the cones. From color-matching tests with subjects who have different forms of color blindness, it has even been possible to infer the color of these pigments. But only in the last two decades have these pigments been isolated and the inferences verified.

The three attributes of color can best be understood by a simple demonstration. A projector fitted with a red filter projects a red spot on a screen. *Hue* is the attribute of the spot that distinguishes it from a blue or green spot. The *brightness* of the spot depends on its intensity, which can be changed by changing the intensity of the projection bulb. Brightness is thus analogous to loudness: it is the psychological sensation that corresponds to the physical attribute of intensity. If a second projector without a filter projects a white spot over the red one, the result is a pink spot. The pink spot differs from the red one in *saturation*. Saturation describes the strength or purity of the hue.

**FIGURE 14.17**
Sagittal view of uncoiled cochlea.
[*After P. B. Denes and E. N. Pinson, "The Speech Chain." Copyright 1963, Bell Telephone Laboratories, Incorporated; used by permission.*]

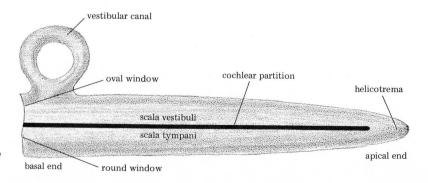

If the intensity of the white light is very small compared to the intensity of the red light, the spot has high saturation. As the relative amount of white light increases, the saturation of the spot decreases. The reason colored photographs often seem more vivid than real life is that their colors are more saturated than natural colors. The photograph reproduces the hues correctly but raises the saturation.

The word color refers collectively to hue, brightness, and saturation. That is, two images have the same color if and only if they have the same hue, brightness, and saturation. This is a psychological definition of color and makes no reference to the physical attributes of a light.

The physical attributes of a light are conveniently displayed by plotting the intensity of each of the wavelengths in it. Figure 14.18 gives an example of such a plot, called the *spectrum of the light*. In this figure, each dot represents the intensity of the light in a 10-nm-wavelength interval. (The dots are connected by a solid line for clarity.) Since each wavelength in the spectrum is a different physical attribute of the light, there is much more physical variability in a light than is detected by the eye. That is, lights with different spectra can be perceived by the eye as having the same color.

Because color has only three attributes, it is possible to match a sample color by mixing, in appropriate amounts, three standard, or *primary*, colors. The laws governing this are conveniently studied using four projectors, each one with a control for varying the intensity of its bulb. One projector is used to project the sample color $S$, and the other three are fitted with filters so they project the primary colors. The three primary colors are superimposed on a spot next to a spot of the sample color (Fig. 14.19), and the intensities of the primaries are varied in order to match the color of the primary mixture to the color of the sample. If a match is obtained, the sample color is uniquely specified by the intensities $X$, $Y$, $Z$ of the primary colors required to match it. These intensities, called the *tristimulus values*, depend, of course, on the three colors used as primaries.

The *chromaticity* of a sample color is specified by its *chromaticity coordinates* $(x, y, z)$, which are defined in terms of its tristimulus values by

$$x = \frac{X}{X + Y + Z} \qquad y = \frac{Y}{X + Y + Z} \qquad z = \frac{Z}{X + Y + Z}$$

These coordinates are the fractional amounts of each primary re-

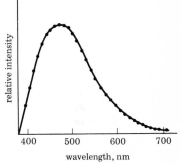

FIGURE 14.18
Spectrum of a light. Each point represents the intensity of the light in a 10-nm-wavelength interval.

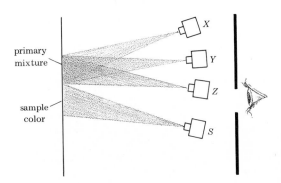

FIGURE 14.19
Color-matching experiment. The sample color is matched by varying the intensities $X$, $Y$, $Z$ of three primary colors projected onto a single spot.

quired to match the sample. Two colors with the tristimulus values $X, Y, Z$ and $2X, 2Y, 2Z$ have the same chromaticity. Since these colors differ mainly in brightness, chromaticity corresponds mainly to hue and saturation.

The sum $x + y + z$ of the chromaticity coordinates of a color always equals 1, so that only two of these coordinates are needed to specify the chromaticity of a color. Consequently, the chromaticity of every color that can be matched by the three primaries can be represented as a point in a plot of $y$ against $x$.

For example, suppose the primaries are chosen to be the spectral colors red (700 nm), green (520 nm), and blue (420 nm). Then $x$, $y$, and $z$ are the fractions of red, green, and blue required to match the sample. The chromaticity of every color that can be matched by a mixture of these colors is a point in the chromaticity diagram shown in Fig. 14.20. The primaries themselves are the corners of the triangle. For instance, green ($G$) has the coordinates ($x = 0$, $y = 1$), and so the value of $z$ at this point is zero, because $x + y + z = 1$. The coordinates of blue ($B$) are ($x = 0$, $y = 0$), because the value of $z$ at this point is 1.

Grassmann's second law of color states that any color that can be obtained by mixing two specific colors lies on the straight line that connects these colors in a chromaticity diagram. For instance, all colors that are obtained by mixing the colors represented by the points $P$ and $Q$ in Fig. 14.20 lie on the line $PQ$. Furthermore, the ratio $d_P/d_Q$ of the distances of a color $H$ on the line $PQ$ from $P$ and $Q$, is equal to the ratio $f_Q/f_P$ of the fractions of $Q$ and $P$ in the mixture. That is, if $H$ is obtained by a mixture of $\frac{1}{4}P$ and $\frac{3}{4}Q$, the point representing $H$ will be located one-fourth of the distance from $Q$ to $P$. If the color $H$ is mixed with a third color $T$, the resulting color $J$ will lie on the line $HT$. Thus, all colors represented by points inside the triangle $QPT$ can be obtained by a mixture of the colors $Q$, $P$, and $T$, just as all the colors in the chromaticity diagram itself can be obtained by a mixture of the primaries $R$, $B$, and $G$.

As shown in Fig. 14.20, the color $H$ can also be obtained by mixing the colors $P'$ and $Q'$. This $H$, when mixed with $T$, yields the same colors as were obtained by mixing $T$ with the $H$ made from $P$ and $Q$. This fact is not self-evident, because the light from the two $H$'s may have different spectral compositions. However, Grassmann's third law says that the same colors have the same effect in mixtures, even though their spectral compositions are different. Therefore color can be treated as a mathematical entity, because it obeys the axioms of addition, e.g., equal colors added to equal colors yields equal colors.

Only the colors that can be obtained from a mixture of the primaries are represented on the chromaticity diagram. Because no set of three "real" primaries will reproduce all colors, it is not possible to represent all colors this way. However, there is a sense in which any color can be matched. When the sample color in Fig. 14.19 cannot be matched by any values of the intensities of the primaries, one of the primaries is mixed with the sample, and the intensities of the three primaries are again varied in an effort to match the two colors. That is, if no values of $X$, $Y$, and $Z$ match the sample $S$, one tries to match (say) $X + Y$ and $S + Z$. This is done by simply projecting the $Z$ primary light onto the sample color. If a match is obtained, the

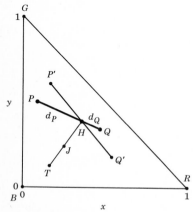

**FIGURE 14.20**
Chromaticity diagram based on three real primaries, red (700 nm), green (520 nm), and blue (420 nm).

tristimulus values are designated $X$, $Y$, $-Z$. In some cases two primaries may have to be added to $S$, so that, for instance, $X$ is matched to $S + Y + Z$. In this case the tristimulus values are $X$, $-Y$, $-Z$. The minus sign on a tristimulus value thus indicates that the corresponding primary was added to the sample rather than to the other primaries. With this extension of the notion of color matching it is found that every color can be matched by a suitable combination of three other colors. Furthermore, any three colors can be used as primaries, provided only that they do not lie on the same straight line in a chromaticity diagram, i.e., provided that none of them can be matched by a mixture of the other two.

The introduction of negative mixing enables any color to be represented by a set of tristimulus values. However, negative values prevent displaying chromaticity in a simple two-dimensional diagram. In 1931 the International Commission on Illumination (ICI) established a system that enables every color to be specified with positive chromaticity coordinates. Known as the ICI system,† it assigns positive chromaticity coordinates to the spectral colors. Since no set of real primaries can match all the spectral colors with positive coordinates, the ICI coordinates are sometimes said to be based on imaginary primaries. It is simpler, perhaps, just to accept these coordinates as an arbitrary representation of the spectral colors, sanctioned by international agreement.

Every color is represented by a point on a plot of the $x$ and $y$ ICI coordinates, as shown in Fig. 14.21 and in the colored diagram inside the front cover. The spectral colors lie on the curve, and all other colors lie inside the closed region bounded by this curve and the straight line connecting its extremities. This follows from Grassmann's laws and the fact that every color is a mixture of spectral colors, so that all such mixtures lie in the closed region. For example, all colors obtained by a mixture of the 420- and 520-nm spectral colors lie along the straight line $BG$ that connects the two colors. All colors that can be obtained by a mixture of the 420-, 520-, and 700-nm spectral colors lie inside the triangle $BGR$. These are the same colors represented by the chromaticity diagram in Fig. 14.20.

**REMARK** In Fig. 14.20 the points with the coordinates ($x = 0$, $y = 0$), ($x = 0$, $y = 1$), and ($x = 1$, $y = 0$) correspond to the primaries on which the diagram is based. In Fig. 14.21 these same points lie outside the region of real color, which is why they are said to represent imaginary primaries.

The point $C$ in Fig. 14.21 is a standard white, resembling daylight on an overcast day. The region around it includes all other whites, including those of incandescent and fluorescent lamps. Two colors are *complementary* if they can be mixed together to give white. This means they must lie on opposite sides of the neutral region. For example, $C$ can be obtained by a mixture of the 470- and 573-nm colors. All the colors that are obtained by mixing white with a spectral color lie on the line connecting $C$ to the spectral color. The spectral color has maximum saturation, and the saturation decreases along the line toward $C$. That is, saturation increases as one moves radially outward from $C$ toward the periphery of the diagram, whereas

---

†In England and the continent it is known as the CIE system, after the French name Commission Internationale d'Éclairage.

**FIGURE 14.21**
ICI chromaticity diagram. All
colors are represented by points
inside the region bounded by the
curve and the straight line that
connects its extremities. The
colors represented by the diagram
in Fig. 14.20 lie inside the triangle
*BGR* in this diagram. Note that
the points (0,0), (0,1), and (1,0),
which represent the primaries on
which the diagram is based, lie
outside the region of real color.
[*After D. B. Judd and G.
Wyszecki, "Color in Business,
Science and Industry," 2d ed.
Copyright 1963, John Wiley & Sons,
Inc., used by permission.*]

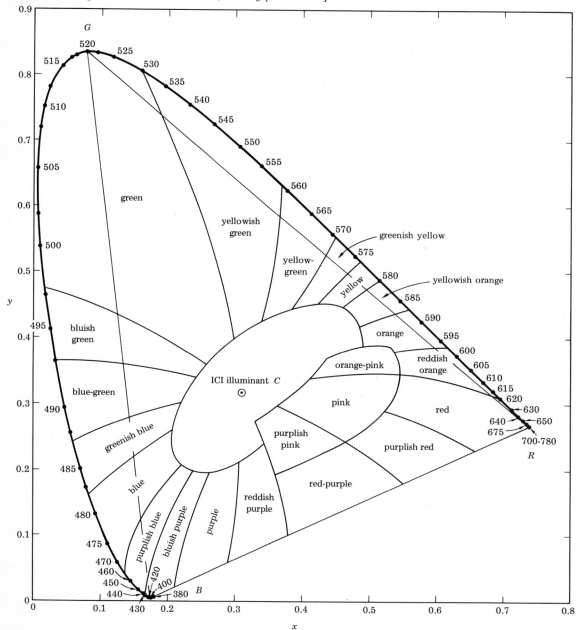

hue varies as one moves around the periphery. Since the purples are the only colors that cannot be obtained by mixing $C$ with a spectral color, they are sometimes referred to as nonspectral colors.

All color-reproduction systems, such as color television, color photography, and color printing, use three primaries. The primary colors used in a particular system can be represented by three points in an ICI diagram, and the gamut of colors the system yields lies inside the triangle that connects these points. It is evident from the ICI diagram that no set of three primaries will reproduce all colors.

The ICI diagram is also useful for describing the different types of color blindness. *Dichromatic* color blindness results from the absence of one of the three pigments in the cones. There are thus three different forms of dichromatism, called *protanopia*, *deuteranopia*, and *tritanopia*. In deuteranopia there is no distinction between green and red. This is shown in Fig. 14.22*b*, where all the colors on the same straight line are indistinguishable by a deuteranope. Similarly, a protanope cannot distinguish between colors on the same lines in Fig. 14.22*a*, and a tritanope cannot distinguish between colors on the same lines in Fig. 14.22*c*. These lines are obtained from color-mixing experiments in which a color-blind subject judges the match. The colors of the pigments in the eye were deduced from these data before they were measured directly.

## 14.5   POLARIZATION

Sound is a longitudinal mechanical wave (Sec. 13.1); i.e., in a sound wave, the elements of the medium vibrate parallel to the direction of propagation of the wave. There are also transverse waves (Sec. 12.1), such as those of a vibrating cord, in which the elements of the medium vibrate perpendicular to the direction of propagation of the wave. The question naturally arises: Is light a longitudinal or a transverse wave?

It may at first seem impossible to answer this question without probing more deeply into the nature of an electromagnetic wave. Surprisingly, however, there are simple experiments that indicate that light is a transverse wave. The easiest experiment uses two identical Polaroid sheets,† such as are used in polarizing sunglasses. Each sheet is fairly transparent, and when one sheet is placed over the other, the two together are still transparent. However, as one sheet is slowly rotated with respect to the other (Fig. 14.23), the region of overlap darkens until it is completely dark when the one sheet has been rotated 90° with respect to the other; i.e., the two sheets together are now opaque. As the one sheet is rotated further, the region of overlap starts to lighten, and when the sheet has been rotated through another 90°, the region of overlap is again transparent.

Centuries before the invention of Polaroid sheets, a similar phenomenon was observed using certain naturally occurring crystals, such as calcite. In fact, Newton considered this phenomenon to be evidence against the wave theory of light because everyone then

---

† Polaroid sheets are the patented invention of Edwin H. Land (1909–      ).

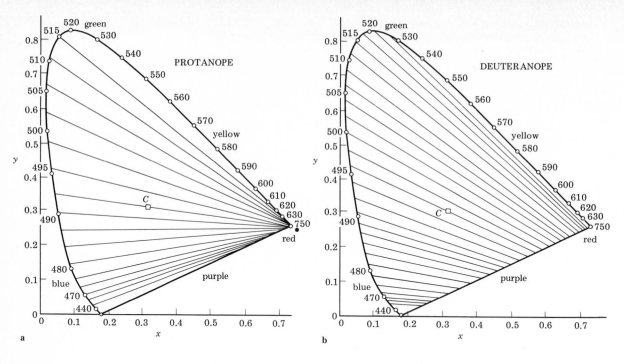

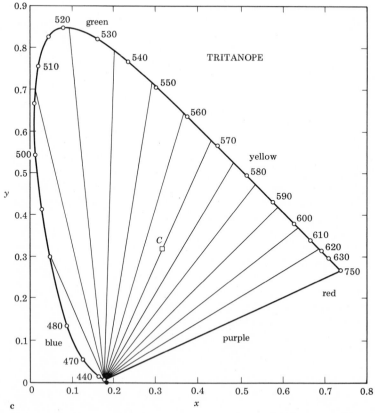

FIGURE 14.22
The three types of dichromatic color blindness. In each diagram, colors on the same straight line are indistinguishable by a person with the specified dichromatism. [*After D. B. Judd and G. Wyszecki, "Color in Business, Science and Industry," 2d ed. Copyright 1963, John Wiley & Sons, Inc., used by permission.*]

assumed that if light was a wave, it was a longitudinal wave, like sound. However, no one could explain how the intensity of a longitudinal wave could possibly be affected by rotating something around an axis parallel to the direction of motion of the wave. In 1817 Young first suggested the natural explanation of the phenomenon—that light is a transverse wave.

A longitudinal wave can vibrate in only one direction, whereas a transverse wave can vibrate in any direction that lies in a plane perpendicular to the direction of motion. In a beam of *polarized* light, all the wavetrains vibrate in the same transverse direction, so that the beam can be represented by a single amplitude **A,** as illustrated in Fig. 14.24. The amplitude of polarized light is treated as a vector quantity because it is characterized by both a magnitude and a direction.

There exist materials, such as Polaroid sheets, that transmit polarized light according to the angle between the amplitude **A** of the light and the *axis* of the sheet. This axis is a fixed direction in the sheet determined by the arrangement of the molecules in the sheet. In Fig. 12.24 a sheet of such material (called the *analyzer*) is placed in front of a beam of polarized light. When **A** is parallel to the axis of the sheet, all the light is transmitted, and when **A** is perpendicular to the axis, none of the light is transmitted.

When the amplitude **A** of polarized light makes an angle $\theta$ with the axis of an analyzer, part of the light is transmitted and part is absorbed. The amount transmitted is found by calculating the component of **A** parallel to the axis. From Fig. 14.25 it is seen that the magnitude of the parallel component is $A \cos \theta$. This component is transmitted, while the perpendicular component is absorbed. As with sound, the intensity of light is proportional to the square of the amplitude, so that the intensity of the light incident on the analyzer is proportional to $A^2$ and the intensity of the light transmitted is proportional to $(A \cos \theta)^2 = A^2 \cos^2 \theta$. Therefore, the fraction of the light transmitted is $\cos^2 \theta$. For example, when $\theta = 0°$, all the light is transmitted ($\cos^2 0° = 1^2 = 1$); when $\theta = 45°$, half the light is transmitted ($\cos^2 45° = 0.707^2 = 0.5$); when $\theta = 90°$, none of the light is transmitted ($\cos^2 90° = 0^2 = 0$). Note that the transmitted light is polarized in the direction of the axis of the analyzer and not in the direction of polarization of the incident light. The analyzer thus changes the direction of polarization as well as the intensity of the transmitted light.

Ordinary (unpolarized) light consists of billions of wavetrains, each vibrating in a different direction. This is illustrated in Fig. 14.26, where the amplitudes of several wavetrains are shown. When unpolarized light passes through a Polaroid sheet (called the *polarizer*), only the component of amplitudes parallel to the axis of the sheet are transmitted. The result is a beam of polarized light whose amplitude **A** is parallel to the axis of the polarizer. If the intensity of the incident unpolarized light is $I_0$, the resulting polarized light has the intensity $I = \frac{1}{2}I_0$, since on average only half the wavetrains of the incident light are transmitted.

For example, suppose the axis of an analyzer is oriented at 35° to the axis of a polarizer, as shown in Fig. 14.26. What is the intensity $I'$ of the light transmitted through the analyzer? Let **A** be the ampli-

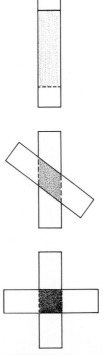

FIGURE 14.23
One Polaroid sheet rotated with respect to another.

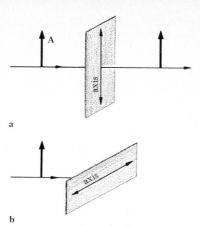

a

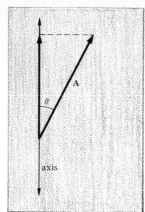

b

FIGURE 14.24
Polarized light of amplitude **A**
incident on a Polaroid sheet. (*a*) **A**
is parallel to the axis of the sheet,
so the light is transmitted. (*b*) **A** is
perpendicular to the axis of the
sheet, so the light is not
transmitted.

FIGURE 14.25
Amplitude **A** of a beam of
polarized light at an angle $\theta$ to the
axis of a Polaroid sheet.

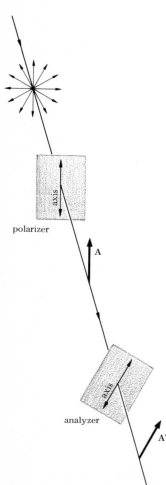

FIGURE 14.26
A beam of unpolarized light
incident on a polarizer and an
analyzer. The axis of the analyzer
is oriented at 35° to the axis of the
polarizer.

tude of the polarized light incident on the analyzer. Since **A** makes
an angle of 35° to the axis of the analyzer, the magnitude of the
amplitude **A′** of the transmitted light is

$$A' = A \cos 35° = 0.574A$$

and the intensity $I'$ of the transmitted light is

$$I' = A'^2 = (0.574A)^2 = 0.341A^2 = 0.341I$$

where $I = A^2$ is the intensity of the polarized light incident on the
analyzer. But $I = \frac{1}{2}I_0$, where $I_0$ is the intensity of the unpolarized
light incident on the polarizer, so that

$$I' = 0.1705I_0$$

The final light is 0.1705 times the intensity of the incident unpolarized
light, and it is polarized in the direction of the axis of the analyzer.

*Partially polarized* light is a mixture of polarized and unpolarized
light that can vary from 0 to 100 percent polarization. The wavetrains
in partially polarized light vibrate in all directions, but the amplitudes
of these vibrations are greatest in one direction, as illustrated in Fig.
14.27. Unpolarized light becomes partially polarized when it is reflected from a nonmetallic surface, the direction of polarization being
parallel to the reflecting surface (Fig. 14.27).

The lenses of Polaroid sunglasses consist of Polaroid sheets oriented with their axes vertical. This orientation is perpendicular to
the direction of polarization of light reflected from a horizontal surface. Consequently, these glasses eliminate much of the spectrally
reflected light that contributes to glare.

Blue skylight consists of sunlight that is reflected (or, more accurately, scattered) by the molecules in the upper atmosphere. As

a result, skylight is partially polarized, the magnitude and direction of the polarization at any point in the sky depending on the position of the point relative to the sun. Figure 14.28 is a map of the polarization of skylight when the sun is in the position indicated. The dotted lines separate regions with different percentages of polarization, and the arrows show the direction of polarization at various points. The percentage of polarization is zero near the sun and reaches a maximum value of 70 percent at angles 90° from the sun.

Through a series of brilliant experiments conducted over many years, Karl von Frisch (1886–    ) and his collaborators have demonstrated that bees navigate by reference to the sun. This is most extraordinary, because the sun continuously moves across the sky during the day. For example, suppose that in the morning a bee, in going from her hive to a particularly good source of nectar, finds her way by flying at an angle of 30° west of the sun. Then in the afternoon she must fly at (say) 40° east of the sun to reach the same source. Experiments have shown that bees do compensate for the motion of the sun, even when they are kept in a dark room between their morning and afternoon flights. Thus bees not only measure angles with respect to the sun but also have an internal clock with which they can compensate for the sun's motion.

Frisch also discovered that bees can navigate even when the sun is covered by a cloud, provided some blue sky is visible. He deter-

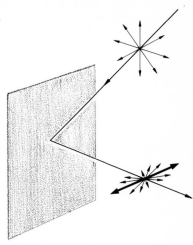

FIGURE 14.27
Unpolarized light reflected from a surface. The reflected light is polarized in a direction parallel to the surface.

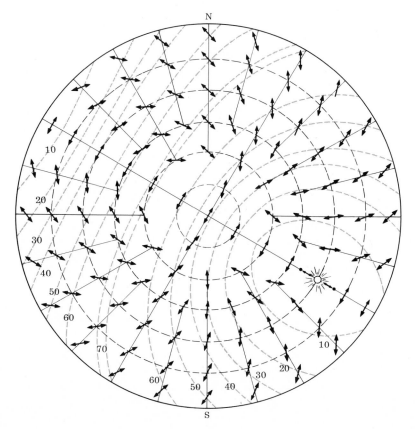

FIGURE 14.28
Polarization of the sky when the sun is in the position shown. The dotted lines separate regions with different percentages of polarization. [*After K. von Frisch, "The Dance Language and Orientation of Bees." Copyright 1967, The Belknap Press of Harvard University Press, used by permission.*]

mined that bees detect the polarization of the sky and from it infer the position of the sun. Incredible as this seems, it has been demonstrated by the most convincing experiments. For example, when the apparent polarization of the sky is changed by placing a Polaroid sheet over a hive, the bees orient themselves relative to the polarization of sky as seen through the polarizer.

To understand the mechanism for this requires a further look at an insect's eye. Figure 14.29 shows schematic drawings of the ommatidium of an insect's eye in sagittal and cross section. The body of the ommatidium consists of eight sensory cells, arranged symmetrically around the central axis. Electrophysiological experiments have shown that each cell responds to a different direction of polarization, as indicated in Fig. 14.29$b$. To see what a bee sees, Frisch constructed an octagonal analyzer consisting of eight triangles of Polaroid sheet fitted together as shown in Fig. 14.30. Figure 14.31 shows how different regions of the sky appear when viewed through this analyzer. These regions of sky, which look the same to the human eye, give very different patterns when viewed through the star-shaped analyzer because the polarization of skylight is different in different regions (Fig. 14.28).

In one of Frisch's experiments, bees were conditioned to fly to a feeding station located due west of the hive. The hive was arranged so that a bee could see only a small portion of the sky west of the hive. At the time of the experiment, this portion of the sky gave the pattern shown in Fig. 14.32$a$ when viewed through the analyzer. Bees give a little orientation dance before leaving the hive, from which an observer can infer their intended direction of flight. Initially the conditioned bees properly oriented themselves due west, but when a Polaroid sheet was placed over the hive to change the polarization of the sky as seen from the hive, the bees oriented themselves 35° south of west. With a Polaroid sheet in front of the analyzer, the sky to the west gave the pattern shown in Fig. 14.32$b$, which Frisch found was the same as the pattern given by the sky 34° north of west when viewed without the Polaroid sheet in front of the analyzer. This indicates that bees looking west through the Polaroid sheet interpreted that direction to be 34° north of west, because it gave the pattern of polarization appropriate for that direction. Since the feeding station was west, the bees reoriented themselves 35° south of what they took to be 34° north of west. The difference of 1° is within the experimental uncertainties of both man and bee.

Thus, although man has known about polarization for only a few hundred years, bees and other insects have been utilizing it for a hundred million years.

## 14.6 THE THEORY OF RELATIVITY

**REMARK** Section 4.1 should be reread before reading this section.

In the nineteenth century light waves were thought to be displacements in an extraordinary medium called the *ether*. On the one hand, the ether had to be rigid, because light is a transverse wave and transverse mechanical waves cannot propagate through a fluid me-

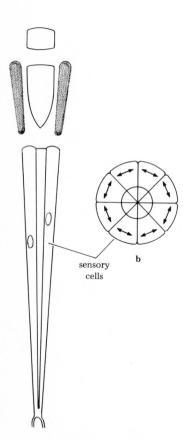

**a**

**b**

sensory
cells

FIGURE 14.29
Ommatidium of an insect's eye: (*a*) sagittal section; (*b*) cross section. Each of the eight light sensitive cells detects light of different polarization, as indicated.

dium. On the other hand, the ether had to be ethereal, i.e., very tenuous, because it offered no resistance to the motion of the planets and other objects in outer space. These contradictory properties, which follow with logical necessity from the known properties of light, were bothersome, but they did not dispel belief in the ether. The ether, after all, was not supposed to be ordinary matter, and so it could have extraordinary properties.

Since the ether permeated all space, the frame of reference of the ether would be a unique inertial frame. An observer anywhere in the universe should be able to measure his speed relative to this frame, and an observer at rest with respect to the ether could justifiably say he was at absolute rest. Of course, since the laws of mechanics are the same in all inertial frames, an observer cannot measure his speed relative to the ether using the laws of mechanics. But to nineteenth-century physicists this was a peculiarity of mechanics that did not apply to other laws of physics.

In particular, the speed of light was expected to depend on the frame in which it is measured. With respect to the ether, light would propagate with a certain speed $c$. Relative to an observer moving with respect to the ether, light would propagate with a different speed. As an analogy, consider a siren emitting a sound that travels with a speed of 343 m/s through the air. A physicist on a railroad flatcar that is moving 10 m/s toward the siren would measure the speed of this sound to be 343 m/s + 10 m/s = 353 m/s. Likewise, if the flatcar is moving away from the siren at 10 m/s, he would find the speed to be 343 m/s − 10 m/s = 333 m/s. The speed of anything is relative to the frame of reference in which it is observed.

The earth travels with a speed of about $3 \times 10^4$ m/s as it circles about the sun, so that it must be moving relative to the ether, at least some of the time. For even if the earth should, by chance, be at rest relative to the ether at one point in its orbit, 6 months later it is halfway around the sun, traveling in the opposite direction. Thus the earth cannot be at rest (with respect to the ether) at all times of the year.

In 1887 Albert A. Michelson (1852-1931) and E. W. Morley (1838-1923) performed an experiment to detect the motion of the earth relative to the ether. Assuming that the earth is moving with a speed $v$ relative to the ether, the speed of light, as observed on earth, should be $c - v$ for light moving in the direction of the earth's motion and $c + v$ for light moving in the opposite direction. For experimental reasons, Michelson and Morley compared the speeds of two light beams traveling at right angles to each other. By similar reasoning, the speeds of these light beams should be different from each other. The experiment used an interference method that was sensitive enough to detect a difference in speed of $3 \times 10^3$ m/s, much less than the expected difference. But no difference was detected. The speed of light, unlike the speed of anything else, is independent of the motion of the observer.

The Michelson-Morley experiment completely destroyed the idea of the ether. At the same time, it raised a far greater problem. To understand this problem, let us consider two inertial frames in relative motion. Let one frame be the ground, and let the other be a train moving at a constant speed $V$ relative to the ground (Fig. 14.33).

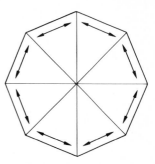

FIGURE 14.30
Frisch's star-shaped analyzer, consisting of eight triangularly shaped Polaroid sheets arranged into an octagon.

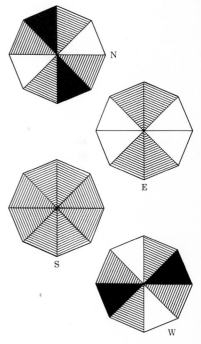

FIGURE 14.31
Different regions of the sky as viewed through Frisch's analyzer. [*After K. von Frisch, "The Dance Language and Orientation of Bees." Copyright 1967, The Belknap Press of Harvard University Press, used by permission.*]

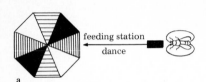

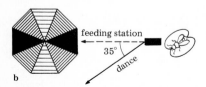

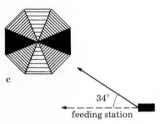

FIGURE 14.32
Frisch's experiment to test the ability of bees to navigate by the polarization of the sky. Bees conditioned to fly to the west were placed in a hive in which only a portion of the sky in the west was visible. (*a*) Western sky as seen through Frisch's analyzer. When the hive was uncovered, conditioned bees oriented to the west. (*b*) Western sky as seen through the analyzer and a Polaroid sheet placed over hive. When the hive was covered with the Polaroid sheet, bees oriented 35° south of west. (*c*) Sky 34° north of west looks the same through the analyzer as the western sky looks through the analyzer and the Polaroid sheet together. As a consequence the bees looking west through the Polaroid sheet behaved as though that direction were 34° north of west. [*After K. von Frisch, "The Dance Language and Orientation of Bees." Copyright 1967, The Belknap Press of Harvard University Press, used by permission.*]

At time $t = 0$ Smith, at one end of the railroad car, just passes Jones, who is on the ground. At this instant, Smith throws a ball toward the opposite end of the car. It hits the wall there at time $t = T$, and both Smith and Jones note the position of the ball, each in his own frame, at the instant $T$. Smith then calculates the speed of the ball (relative to the railroad car). It traveled a distance $L$ in a time $T$, so its speed $v$ relative to the train is

$$v = \frac{L}{T}$$

Jones finds it traveled the distance $D$, so its speed $v'$ relative to the ground is

$$v' = \frac{D}{T}$$

But from Fig. 14.33 it is evident that

$$D = VT + L \qquad\qquad 14.9$$

so

$$v' = \frac{VT + L}{T} = V + \frac{L}{T} = V + v$$

This formula for the addition of speeds corresponds to everyday experience. However, if the ball is replaced by a pulse of light, both Smith and Jones will observe it to travel with the same speed $c$. This is the problem raised by the Michelson-Morley experiment.

Einstein solved this problem in 1905. Starting with the assumption (demonstrated by the Michelson-Morley experiment) that the speed of light is the same in all inertial frames, he proceeded to analyze the consequences of this assumption. The result, which is known as the theory of (special) relativity, is one of the deepest and most beautiful set of ideas ever conceived by man. At the heart of Einstein's analysis is the operational principle that length and time are only what can be measured with meterstick and clock. In Sec. 1.2 the operations needed to measure length and time in one frame were discussed. Einstein went beyond this and asked, in essence, how a stationary observer, such as Jones, can measure the length of a moving object, such as the railroad car. This is important because the only length that is meaningful to Jones is the length he can measure. Because of the finite speed of light, there are unexpected difficulties with this measurement.

It is clear that Jones has to mark two points in his frame that correspond to the two ends of the car. Of course, these points must be marked simultaneously or else the car will have moved between the time the first and second marks are made. Let us make matters easy for Jones and imagine that lightning strikes the two ends of the car simultaneously, leaving marks $A$ and $B$ on both the car and the tracks. All Jones has to do is measure the distance between the marks on the tracks, in accordance with the method of Sec. 1.2, and he has the length of the car. This is correct. Surprisingly, however, this measurement will be less than the length $L$ that Smith measures. Length, like speed, is not an absolute quantity, but depends on the frame of reference in which it is measured.

To see this, let us consider the situation from Smith's point of view. By assumption, the two lightning bolts struck the railroad car simultaneously, as seen by Jones. Suppose Smith was seated in the middle of the car at that time. If these flashes occurred simultaneously for him, he would see them at the same time. But in fact, since he is moving toward $B$, the flash at $B$ reaches him before the flash at $A$. To Smith the flashes did not occur at the same time. To him, the mark at $B$ was made before the mark at $A$. Since the train moved toward $B$ during the time between the two flashes, the distance between the marks on the ground is less than the length $L$ of the car as measured by Smith.

Einstein's analysis shows that length and time intervals are relative to the frame of reference in which they are observed. This contradicts age-old notions about the absolute nature of space and time. These notions seem natural to us because the speed of light is so large compared to everyday speeds that relativistic effects are negligible. But if the speed of light were comparable to the speed of sound, relativistic effects would be an everyday experience and people would take it for granted that length and time for a moving observer are different from the length and time of a stationary observer.

With this new insight, Einstein showed that the analysis given above for the addition of speeds is wrong. The distance $D$ in Eq. 14.9

FIGURE 14.33
Two frames of reference: the ground and a railroad car moving at constant speed $V$ relative to the ground. At time $t = 0$ Smith, who is at one end of the railroad car, passes Jones, who is on the ground. At the same time, Smith throws a ball toward the other end of the car. The ball hits the other end of the car at time $t = T$.

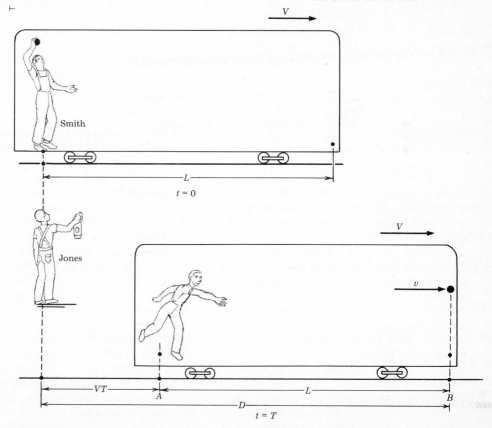

is not the distance Jones would measure, and $T$ is not the time he would observe the ball to hit the wall. Analyzing the situation in detail, Einstein found that the correct formula for the addition of speeds is

$$v' = \frac{V + v}{1 + vV/c^2}$$

Because $c$ is so large, the term $vV/c^2$ is negligibly small for all every-day speeds, including the speed of a rocketship. However, if $v$ is equal to $c$, this equation gives

$$v' = \frac{V + c}{1 + cV/c^2} = \frac{V + c}{1 + V/c} = c$$

Light travels with the same speed $c$ relative to both frames.

In summary, the principle of relativity says that the laws of physics are the same in all inertial frames. There is no ether or preferred inertial frame. Because the speed of light is the same in all frames, certain notions about space and time must be modified. In particular, the law of the addition of speeds is changed. The new law does not disagree with the old one for speeds that are small compared to the speed of light, and it has been verified with great accuracy for atomic particles traveling with nearly the speed of light.

**REMARK**   The word relativity has caused much misunderstanding in people's minds. The theory of relativity does not say that "all things are relative." Rather, it tells us which physical quantities are relative, i.e., depend on the frame of reference in which they are observed, and which quantities are abso-lute, i.e., do not depend on a frame of reference. Its unexpected result is that length and time intervals are relative quantities. However, the speed of light, rest mass, charge, and many other quantities are absolute.

**PROBLEMS**

*1* (*a*) Calculate the wavelength of the radio waves broadcast by an FM station operating at 100 MHz. (*b*) What is the wavelength of the waves broadcast by an AM station oper-ating at 100 kHz?

*Ans.* (*a*) 3.00 m; (*b*) 3.00 km

*2* What is the frequency of an electromagnetic wave that has the same wavelength as an ultrasonic wave of frequency $10^5$ Hz?

*3* (*a*) What is the speed of light in a diamond (see Table 14.1)? (*b*) The speed of light in topaz is $1.85 \times 10^8$ m/s. What is the index of refraction of topaz?

*Ans.* (*a*) $1.24 \times 10^8$ m/s; (*b*) 1.62

*4* (*a*) What is the frequency of green light with a wavelength of 525 nm? (*b*) When this light enters glass ($n = 1.50$), its frequency does not change but its speed and wavelength do. (*c*) What is the wavelength of this light in glass?

*5* What is the wavelength in water of a red light whose wavelength in air is 650 nm (see Prob. 4)?

*Ans.* 488 nm

*6* Use a compass to make an accurate drawing similar to Fig. 14.3. Place the slits 2 cm apart and let the wavelength of the incident wave be 1 cm. Find the positions of the nodes and antinodes on a screen 12 cm from the slits. Measure the distance between two antinodes and compare with Eq. 14.3. Explain any dis-crepancy.

*7* Light incident on a pair of slits produces an interference pattern on a screen 2.5 m from the slits. If the slit separation is 0.015 cm and the distance between the bright fringes in the pattern is 0.76 cm, what is the wavelength of the light?

*Ans.* $4.56 \times 10^{-5}$ cm

*8* Light from a neon-helium laser (630 nm) is incident on a pair of slits. In the interference pattern on a screen 1.5 m from the slits, the bright fringes are separated by 1.35 cm. What is the slit separation?

9 Light of wavelength 589 nm incident on a pair of slits produces an interference pattern in which the separation between the bright fringes is 0.53 cm. A second light produces an interference pattern with a separation of 0.64 cm between fringes. What is the wavelength of this second light? *Ans.* 711 nm

10 Light enters the eye through the *pupil*, a transparent aperture about 7 mm in diameter. What is the diffraction angle $\theta$ that results when a parallel beam of yellow light (589 nm) passes through the pupil?

**REMARK** Because of diffraction, two rays that make an angle with each other of less than $\theta$ will form overlapping images on the retina and cannot be distinguished. Diffraction thus limits the discrimination of the human eye as well as that of the insect eye.

11 Two mirrors are arranged at right angles to each other, as shown in Fig. 14.34. A ray of light incident on the horizontal mirror at an angle $\theta = 72°$ is reflected toward the vertical mirror. (*a*) What is its angle of incidence on the vertical mirror? (*b*) Show that after being reflected from the vertical mirror the ray emerges parallel to the incoming ray.

*Ans.* (*a*) 18°

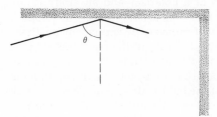

FIGURE 14.34   Problem 11.

**REMARK** During the first moon landing, an arrangement of mirrors like this, called a *corner mirror*, was left on the moon's surface. A laser beam aimed at this mirror through a telescope on earth is reflected back into the telescope. By timing the interval between the transmission of a light pulse and its return, the distance to the moon is determined with great accuracy.

12 Two mirrors, each 1.6 m long, are facing each other, as shown in Fig. 14.35. The distance between the mirrors is 20 cm. A light ray is incident on one end of one of the mirrors at

FIGURE 14.35   Problem 12.

an angle of incidence of 30°. How many times is the ray reflected before it reaches the other end? (*Hint:* Trace the ray through a scale drawing of the mirrors.)

13 A beam of light in air is incident on water at an angle of 30°. What is the angle of the beam inside the water? *Ans.* 22°

14 A ray of light is incident at an angle of 53° on a block of clear plastic. The angle of refraction of the ray is 34°. What is the index of refraction of the plastic?

15 At what angle must a ray of light be incident on ethanol to be refracted into the liquid at 25°? *Ans.* 35.1°

16 Plot the angle of refraction against the angle of incidence of light passing from air into glass ($n = 1.50$).

17 What is the critical angle for total internal reflection between water and air?

*Ans.* 48.5°

18 The laws of reflection and refraction are the same for sound as for light. The index of refraction of a medium (for sound) is defined as the ratio of the speed of sound in air (343 m/s) to the speed of sound in the medium. (*a*) What is the index of refraction (for sound) of water (use Table 13.1)? (*b*) What is the critical angle for total reflection of sound from water?

**REMARK** This means that sound must be incident on water at an angle of less than 13.2° in order for any of it to be refracted into the water. This is one reason why it is so quiet under water.

19 Figure 14.36 shows a ray of light incident at an angle $\theta_1$ on one end of an optical fiber. Its angle of refraction is $\theta_2$, and it strikes the side of the fiber at an angle $\phi_2$. If the index of refraction of the fiber is 1.30, what is the largest angle of incidence $\theta_1$ that a ray can have and still be totally reflected from the side of the fiber? *Ans.* 55.7°

20 The optical fiber in Fig. 14.36 is 2 m long and has a diameter of $2 \times 10^{-3}$ cm. If a ray of light is incident on one end of the fiber at an angle $\theta_1 = 40°$, how many reflections does it make before emerging from the other end? (The index of refraction of the fiber is 1.30.)

21 In practice, optical fibers have a coating of glass ($n_3 = 1.512$) to protect the optical sur-

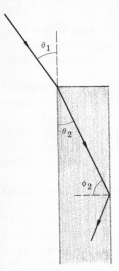

FIGURE 14.36    Problems 19 to 22.

face of the fiber. If the fiber itself has an index of refraction $n_2 = 1.700$, what is the critical angle for total reflection of a ray inside the fiber?                    *Ans.* 62.7°

22 Repeat Probs. 19 and 20 for the coated fiber described in Prob. 21.

23 A ray of light is incident at 30° on a prism with apex angle 55° and index of refraction 1.50. Calculate the angles the ray makes with the sides of the prism. With a protractor, draw the prism and trace the ray through it. What is the angle of deviation?          *Ans.* 35.5°

24 Figure 14.37 shows two plates of glass ($n_1 = 1.50$) separated by a liquid film ($n_2$). Show that if the liquid is water ($n_2 = 1.33$), a ray of light incident on the upper glass-liquid surface at

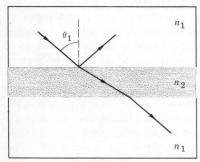

FIGURE 14.37    Problem 24.

an angle $\theta_1 = 64°$ will be totally reflected but if the liquid is alcohol ($n_2 = 1.36$), some of the light will be refracted through to the lower glass plate.

25 (*a*) What is the color $P$ with ICI chromaticity coordinates ($x = 0.400$, $y = 0.200$)? (*b*) What are the coordinates of the color which when mixed in equal proportions with $P$ gives standard white ($C$)?
      *Ans.* (*a*) red-purple; (*b*) $x = 0.22$, $y = 0.43$

26 What proportions of the spectral colors with wavelengths 480, 510, and 610 nm are required to produce standard white ($C$)?

27 What proportions of the spectral colors with wavelengths 485, 520, and 600 nm are required to produce the green color whose chromaticity coordinates are (0.25, 0.40)?
                    *Ans.* 0.45, 0.23, and 0.32

28 (*a*) What is the wavelength of the spectral color that is the complement of 485 nm? (*b*) Is there a spectral color that is the complement of 520 nm?

29 The amplitude of a beam of polarized light makes an angle of 65° with the axis of a Polaroid sheet. What fraction of the beam is transmitted through the sheet?      *Ans.* 0.179

30 The axes of a polarizer and an analyzer are oriented at 30° to each other. (*a*) If unpolarized light of intensity $I_0$ is incident on them, what is the intensity of the transmitted light? (*b*) Polarized light of intensity $I_0$ is incident on this polarizer-analyzer system. If the amplitude of the light makes an angle of 30° with the axis of the polarizer, what is the intensity of the transmitted light?

31 The axes of a polarizer and an analyzer are oriented at right angles to each other. A third Polaroid sheet is placed between them with its axis at 45° to the axes of the polarizer and analyzer. (*a*) If unpolarized light of intensity $I_0$ is incident on this system, what is the intensity of the transmitted light? (*b*) What is the intensity of the transmitted light when the middle Polaroid sheet is removed?
                    *Ans.* (*a*) $0.125I_0$; (*b*) zero

32 Polarized light of intensity $I_0$ is incident on a pair of Polaroid sheets. Let $\theta_1$ and $\theta_2$ be the angles between the amplitude of the light and the axes of the first and second sheets, respectively. Show that the intensity of the transmitted light is

$$I = I_0 \cos^2 \theta_1 \cos^2 (\theta_1 - \theta_2)$$

33 A futuristic spaceship, traveling at half the speed of light (relative to the sun), launches a projectile with half the speed of light (relative to the spaceship) in the direction of motion of the spaceship. What is the speed of the projectile relative to the sun?  *Ans.* 0.8*c*

## BIBLIOGRAPHY

EINSTEIN, ALBERT: "Relativity," Crown Publishers, New York, 1961. A simple and elegant explanation of relativity that uses no mathematics.

FEYNMAN, RICHARD P., ROBERT B. LEIGHTON, and MATTHEW SANDS: "Lectures on Physics," Addison-Wesley Publishing Company, Inc., Reading, Mass., 1963. The mechanism of color vision is discussed in chap. 35, and chap. 36 has a beautiful analysis of the optimal size of an ommatidium in an insect's eye.

FRISCH, KARL VON: "The Dance Language and Orientation of Bees," The Belknap Press of Harvard University Press, Cambridge, Mass., 1967. Comprehensive review of over 50 years' experimenting with bees. The evidence for the bee's ability to navigate by means of the polarization of skylight is well documented.

GAMBLE, WALTER J., PAUL G. HUGENHOLTZ, R. GRIER MONROE, MICHAEL POLANYI, and ALEXANDER S. NADAS: The Use of Fiberoptics in Clinical Cardiac Catherization, *Circulation*, **31**: 328 (1965). Technique for measuring blood oxygen saturation in the heart by use of fiber optics.

GAMOW, GEORGE: "Mr. Tomkins in Wonderland: Stories of *c*, *G* and *h*," Cambridge University Press, Cambridge, 1953. A famous fairy tale of relativity. Mr. Tomkins enters a world in which the speed of light is only 100 mi/h, and relativistic phenomena are commonplace occurrences.

JUDD, DEANE B., and GUNTER WYSZECKI: "Color in Business, Science and Industry," 2d ed., John Wiley & Sons, Inc., New York, 1963. Comprehensive account of all aspects of color, including measurement, specification, reproduction, and vision.

PERSYKO, LUCIEN, JOHN E. SWIFT, BEKAL VENKATACHALAM, HAROLD SEREBRO, and IVAN T. BECK: Diagnosis of Pathological Changes in the Stomach by Gastroscopy, *Canadian Medical Association Journal*, **100**: 1067 (1969). Fiberscope used to position an intragastric camera.

SHURCLIFF, WILLIAM A., and STANLEY S. BALLARD: "Polarized Light," D. Van Nostrand Company, Inc., Princeton, N.J., 1964. Elementary account of the many ways in which polarized light is produced and utilized. There are a few technical chapters on the mathematical description of polarized light.

WALD, GEORGE: Molecular Basis of Visual Excitation, *Science*, **162**: 230 (1968). Lecture delivered in Stockholm by Professor Wald when he received the 1967 Nobel prize in physiology. He describes his research on the three pigments in the cones.

# CHAPTER 15 optics

Optics is the application of lenses, mirrors, and prisms to instruments that control and manipulate light. It involves the design and manufacture of both the lenses themselves and the instruments in which they are used. This chapter does not consider how lenses are made but assumes that lenses with certain characteristics are available and discusses the principles of various optical instruments in terms of these characteristics. Since many optical instruments in common use, such as the microscope, the telescope, the camera, the projector, and the eye, are just different combinations of lenses, they can be understood in terms of the function of a single lens and the general rules for determining the effects of combinations of lenses.

FIGURE 15.1
Front and side view of a positive (convex) lens.

FIGURE 15.2
Side view of a negative (concave) lens.

## 15.1 LENSES

A simple lens is a thin circular piece of transparent material, usually glass, the thickness of which varies from its center to its edge. Figure 15.1 shows a front and a side view of a *convex* (positive) lens, which is thicker at its center than at its edge. A *concave* (negative) lens, shown in Fig. 15.2, is thinner at its center than at its edge.

A lens is similar to a prism in that a ray of light is deviated as it passes through it. It differs from a prism, however, in that the angle of deviation of the ray depends on where the ray enters the lens. Figure 15.3 shows several parallel rays incident on a convex lens. The ray $aa'$, which passes through the center of the lens, is undeviated because the front and back surfaces of the lens are parallel to each other at the points where the ray enters and leaves. The ray $bb'$ is deviated as though it passed through a prism with an apex angle $B$, where $B$ is the angle that the front and back surfaces make with each other at the points where the ray $bb'$ enters and leaves. Similarly, the ray $cc'$ is deviated as though it passed through a prism with an apex angle $C$. Because of the curvature of the lens, the angle of deviation increases with the distance from the center of the lens.

A convex, or *positive*, lens is shaped so that all parallel rays incident on it are deviated through the same point, as shown in Fig. 15.4. The line through the center $C$ of the lens, perpendicular to the plane of the lens, is called the *optic axis*. All incident rays that are parallel to the optic axis are deviated so that they pass through the point $F'$ on the optic axis. This is the *focal point* of the lens, and its distance from $C$ is the *focal length f* of the lens. The focal length is the primary characteristic of a lens.

For example, the rays of light coming from the sun are nearly parallel because the sun is so far from the earth. Therefore, all the sun's rays incident on a lens are brought to a focus at the focal point. If a piece of paper is placed one focal length from the lens, the concentrated rays at the focal point will be sufficient to burn the paper.

The deviation of a ray is the same regardless of the side of the lens on which it is incident. Therefore, parallel rays incident from the right on the lens in Fig. 15.4 intersect the optic axis at the point $F$, which is the same distance $f$ to the left of $C$ as $F'$ is to the right. Consequently a lens has two focal points, one on each side, which are the same distance $f$ from the center of the lens.

A given ray follows the same path through an optical system regardless of its direction of motion through the system. This principle of *reversibility* is important for the analysis of optical systems. For example, suppose a source of light is placed at the focal point $F'$ of the lens in Fig. 15.4. Since rays diverge in all directions from this source, some of them will pass through the lens to the left. By the principle of reversibility, these rays will emerge parallel to the optic axis, following the paths shown in Fig. 15.4 but in the reverse direction. Although we shall continue to put arrowheads on rays for clarity, it should be remembered that the light can travel in either direction along a ray.

As an immediate application of reversibility, we shall consider what happens to a beam of parallel light that is incident on a lens at an angle $\theta$ to the optic axis (Fig. 15.5). Of all the incident rays, one ray, called the *chief ray*, passes undeviated through the center of the lens. Another ray $AO$ passes through the front focal point $F$

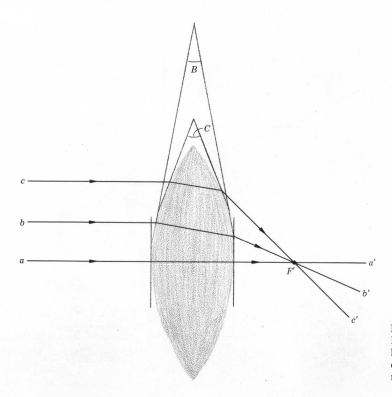

**FIGURE 15.3**
Parallel rays incident on a positive lens. The farther a ray is from the center of the lens, the more the ray is bent.

of the lens. We know that if a ray $BO$ parallel to the optic axis were incident on the lens from the right, it would follow the path $OA$ through $F$. Therefore, by reversibility, the ray $AO$ follows the path $OB$ to the right and intersects the chief ray at $I$. If the angle $\theta$ is small, all other incident rays parallel to $AO$ also intersect at $I$.

The triangles $FCO$ and $IOC$ in Fig. 15.5 are both right triangles, with a common side $OC$ and with the same angle $\theta$ at $F$ and $I$. Therefore these triangles are congruent, and so the distance $OI$ is equal to the focal length $CF$ of the lens. This means that $I$ and $F'$ lie in a plane, called the *focal plane*, that is perpendicular to the optic axis at $F'$. All parallel rays are brought to focus in this plane at a point whose position depends on the angle the rays make with the optic axis.

## 15.2 REAL IMAGES AND ONE-LENS INSTRUMENTS

A positive (convex) lens produces a *real image* of a distant object in the focal plane; i.e., if a screen is placed in the focal plane, an image of the object will be projected on the screen. This is because all the light that arrives at the lens from a single point of the object is brought to a focus at a single point in the focal plane.

This is shown in detail in Fig. 15.6. Under normal illumination each point of an object diffusely reflects light in all directions. However, if the object is very far from a lens, all the rays that reach the lens from a single point of the object are nearly parallel to each other

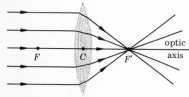

FIGURE 15.4
Rays parallel to the optic axis of a positive lens are brought to a focus at the focal point $F'$.

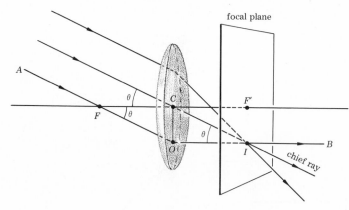

FIGURE 15.5
Parallel rays incident on a lens at an angle $\theta$ to the optic axis. The rays are brought to a focus at a point $I$ in the focal plane.

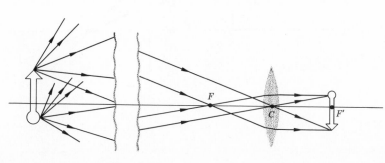

FIGURE 15.6
The real image of a distant object produced by a positive lens. The rays that reach the lens from a single point of a distant object are nearly parallel and so are focused at a single point.

and so are brought to a focus at a single point in the focal plane. Since rays from different points of the object make different angles with the optic axis, they are focused at different points in the focal plane. Thus, if a screen is placed in the focal plane, a real inverted image of the object is projected onto it. The image is called *real* to distinguish it from a *virtual* image, which is discussed in Sec. 15.3.

A lens can form a real image of an object even if the object is not distant from the lens. In fact, if the object is at any distance $s$ from the lens (called the *object distance*), there is a distance $s'$ (called the *image distance*) at which a real image will be formed, provided only that $s$ is greater than $f$. The image distance $s'$ is found by tracing a pair of rays from a point of the object through the lens. For example, consider point $B$ of the object in Fig. 15.7. Since light rays are reflected in all directions from this point, one of these rays goes toward the lens, parallel to the optic axis. Consequently, it is deviated by the lens so that it passes through the focal point $F'$. Another ray from $B$ goes through the center $C$ of the lens and is undeviated. The intersection of these two rays determines the point $B'$ where the image of $B$ is located. The image of any other point of the object is found similarly. Thus by this simple graphical procedure, called *ray tracing*, the size $h'$ and position $s'$ of the image can be determined.

Ray tracing is useful for understanding the function of a lens system, but it is often not accurate enough for calculating the image distance. It is therefore necessary to obtain an exact formula relating $s$, $s'$, and $f$. This formula is derived from Fig. 15.7 with the help of a little geometry.

*1* Since the right triangles $CAB$ and $CA'B'$ are similar, we have

$$\frac{h'}{h} = \frac{s'}{s} \qquad\qquad 15.1$$

*2* Since the right triangles $F'CP$ and $F'A'B'$ are similar, we have

$$\frac{h'}{h} = \frac{\overline{F'A'}}{\overline{CF'}}$$

But the distance $\overline{CF'}$ is the focal length $f$, and the distance $\overline{F'A'}$ is $s' - f$, so the last equation is

$$\frac{h'}{h} = \frac{s' - f}{f}$$

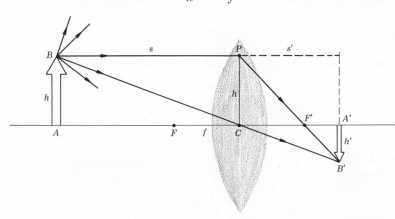

FIGURE 15.7
The real image of an object produced by a positive lens. $s$ is the object distance, and $s'$ is the image distance.

*3* Using Eq. 15.1, we get

$$\frac{s'}{s} = \frac{s' - f}{f}$$

which, with a little algebraic manipulation, can be rewritten

$$\frac{1}{s} + \frac{1}{s'} = \frac{1}{f} \qquad\qquad 15.2$$

This formula gives the exact relation between object distance $s$, image distance $s'$, and focal length $f$.

When the object distance $s$ is very large, the term $1/s$ is nearly zero and Eq. 15.2 reduces to

$$\frac{1}{s'} = \frac{1}{f}$$

or

$$s' = f$$

This is our original observation: a real image of a distant object is formed at a distance $f$ from a lens. Equation 15.2 shows that as the object is moved toward the lens, its image is formed farther from the lens.

For example, the image of an object that is 100 cm in front of a lens of focal length 10 cm is formed at a distance $s'$, given by

$$\frac{1}{100 \text{ cm}} + \frac{1}{s'} = \frac{1}{10 \text{ cm}}$$

Solving this for $s'$ we get

$$\frac{1}{s'} = \frac{1}{10 \text{ cm}} - \frac{1}{100 \text{ cm}} = \tfrac{9}{100} \text{ cm}^{-1}$$

or

$$s' = \tfrac{100}{9} \text{ cm} = 11.1 \text{ cm}$$

In this case, the image is formed 1.1 cm farther from the lens than it is when the object is very far away. The image distance increases still more as the object distance decreases. For instance, when $s = 20$ cm, the image distance is $s' = 20$ cm, and when $s = 11.1$ cm, the image distance is $s' = 100$ cm. You should verify these results using Eq. 15.2.

The relation between object and image distance can be graphically displayed by introducing the *reduced* distances $\bar{s}$ and $\bar{s}'$, which are the object and image distances divided by the focal length:

$$\bar{s} = \frac{s}{f} \qquad \text{and} \qquad \bar{s}' = \frac{s'}{f}$$

In terms of $\bar{s}$ and $\bar{s}'$, the object and image distances are

$$s = \bar{s}f \qquad \text{and} \qquad s' = \bar{s}'f$$

Substitution of these expressions into Eq. 15.2 gives

$$\frac{1}{\bar{s}f} + \frac{1}{\bar{s}'f} = \frac{1}{f}$$

The focal length $f$ is common to all the terms, so it can be canceled. The result is

$$\frac{1}{\bar{s}} + \frac{1}{\bar{s}'} = 1 \qquad\qquad 15.3$$

which is the relation between the reduced object and image distances. This relation is interesting because it does not depend on the focal length of the lens. It is a universal relation between $\bar{s}$ and $\bar{s}'$ that is valid for all positive lenses. For example, when $\bar{s} = 3$, Eq. 15.3 gives

$$\frac{1}{\bar{s}'} = 1 - \frac{1}{\bar{s}} = 1 - \tfrac{1}{3} = \tfrac{2}{3}$$

or $\qquad\qquad \bar{s}' = 1.5$

This means that for any lens, if the object is located 3 focal lengths in front of the lens, the image is formed 1.5 focal lengths in back of the lens. In particular, if $f = 10$ cm, then $s = 30$ cm and $s' = 15$ cm.

Figure 15.8 is a graph of $\bar{s}'$ versus $\bar{s}$ obtained by plotting, for each value of $\bar{s}$, the value of $\bar{s}'$ given by Eq. 15.3. Thus, for instance, at $\bar{s} = 3$, the curve is at $\bar{s}' = 1.5$.

Figure 15.8 demonstrates a number of important characteristics of a positive lens.

1 The curve is symmetric about the straight line drawn from the origin through the point $\bar{s} = 2$, $\bar{s}' = 2$. This is a consequence of the principle of reversibility, which says that if an object a distance $s$ from a lens is focused at a distance $s'$, then an object a distance $s'$ from the lens will be focused at a distance $s$.

2 The closer the object is to the lens, the farther away the image.

3 If $\bar{s} > 2$, then $\bar{s}' < 2$, and if $\bar{s} < 2$, then $\bar{s}' > 2$. This means that $s$ and $s'$ cannot simultaneously be less than $2f$, nor can they simultaneously be greater than $2f$.

4 It is not possible to form a real image if the object distance is less than $f$, and, by reversibility, no real image is ever formed at a distance less than $f$.

We can now discuss a number of one-lens optical instruments. These instruments differ from each other in the relative values of $\bar{s}$ and $\bar{s}'$ at which they operate, as indicated in Fig. 15.8.

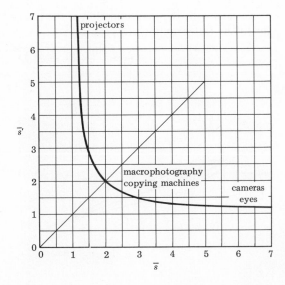

FIGURE 15.8
Graph of the reduced image distance $\bar{s}'$ against the reduced image distance $\bar{s}$. Various optical instruments operate at different values of $\bar{s}'$ and $\bar{s}$.

**The projector**

In a projector (Fig. 15.9), the object, such as a transparent slide, is located close to the lens, and the image is formed a large distance from the lens. The projector thus operates in the region of small $\bar{s}$ and large $\bar{s}'$ in Fig. 15.8.

Suppose a slide projector with a lens of focal length $f = 15$ cm projects an image on a screen a distance $s' = 4$ m from the lens. Where should the slide be positioned? With $f = 15$ cm and $s' = 400$ cm, Eq. 15.2 gives

$$\frac{1}{s} = \frac{1}{f} - \frac{1}{s'} = \frac{1}{15 \text{ cm}} - \frac{1}{400 \text{ cm}} = \frac{77}{1200} \text{ cm}^{-1}$$

so

$$s = \tfrac{1200}{77} \text{ cm} = 15.6 \text{ cm}$$

The slide is placed just a fraction of a centimeter in front of the focal point of the lens. When a projector is being focused, the position of the lens is adjusted until it is the proper distance $s$ from the slide. The projector can form an image at distances from 244 cm on up by changing the distance between the lens and the slide from 16 to 15 cm.

The *magnification* $m$ of an image is the ratio of the image size $h'$ to the object size $h$:

$$m = \frac{h'}{h}$$

According to Eq. 15.1, this is equal to

$$m = \frac{s'}{s} \qquad\qquad 15.4$$

In the case of a projector, $s$ is approximately equal to the focal length $f$ of the lens, so that the magnification is approximately

$$m = \frac{s'}{f} \qquad \text{projector} \qquad 15.5$$

For example, the magnification of the image formed by the projector in the last example is

$$m = \frac{400 \text{ cm}}{15 \text{ cm}} = 26.7$$

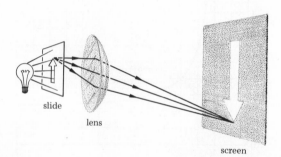

FIGURE 15.9
A slide projector. The lens forms an image of the slide on a distant screen.

slide

lens

screen

Thus, if the object is a slide of width $h = 35$ mm, the width $h'$ of the image on the screen is

$$h' = mh = 26.7 \times 3.5 \text{ cm} = 93.3 \text{ cm}$$

The size of the image can be increased either by increasing the distance between the projector and the screen or by using a lens with a shorter focal length.

### The camera

Eyes and cameras operate in the region of large $\bar{s}$ and small $\bar{s}'$ in Fig. 15.8. All the rays that reach the eye from a single point of a distant object are focused by the lens of the eye onto a single point on the retina (Fig. 15.10$a$), so that a real inverted image of the object is formed on the retina. The retina is covered with light-sensitive cells that send signals to the brain in response to the light incident on them. Since it is these signals that the brain interprets as a picture, a person "sees" only what is imaged on his retina.

Similarly, the lens of a camera focuses an image of a distant object onto light-sensitive film mounted in back of it (Fig. 15.10$b$). The focal length of a standard camera lens is about 50 mm, so the film must be at least 50 mm in back of the lens. The lens of an inexpensive fixed-focus camera is permanently mounted at the distance $s' = f$ from the film. Only distant objects are focused on the film of such a camera. For example, an object 4 m from the lens is brought to focus at a distance $s' = 50.6$ mm in back of the lens, or 0.6 mm in back of the film. The image on the film will be in fair focus in this case, but the image becomes very blurry if the object distance is much less than 4 m.

The lens of a more expensive camera is movable, so that its distance from the film can be adjusted to give a sharply focused image on the film. For instance, most 35-mm cameras† can focus on objects as close as 0.8 m. With a 50-mm lens, this requires an image distance $s'$ given by

$$\frac{1}{80 \text{ cm}} + \frac{1}{s'} = \frac{1}{5.0 \text{ cm}}$$

or

$$s' = 5.33 \text{ cm}$$

Thus, to be able to focus a camera on objects at any distance from 0.8 m on up, the distance of the lens from the film need vary only between 5.33 and 5.00 cm.

Since the image distance $s'$ in a camera is always approximately $f$, the magnification (Eq. 15.4) is approximately

$$m = \frac{f}{s} \qquad \text{camera} \qquad \qquad 15.6$$

and is always much less than 1.

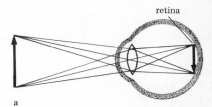

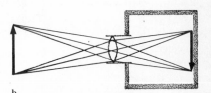

**FIGURE 15.10**
An eye and a camera. ($a$) The lens of the eye forms an image of a distant object on the retina. ($b$) The lens of a camera forms an image of a distant object on the film.

---

† The 35 mm refers to the width of the film used and not to the focal length of the lens. However, the choice of focal length depends on the film size, because a camera with a large film requires a longer-focal-length lens than a camera with smaller film if both cameras are to photograph the same scene.

For instance, the magnification of an object 20 m from a 50-mm lens is

$$m = \frac{5 \text{ cm}}{2000 \text{ cm}} = \frac{1}{400} = 0.0025$$

If the object is a 6-ft man, the height $h'$ of his image on the film is

$$h' = hm = 6 \text{ ft} \times 0.0025$$
$$= 0.015 \text{ ft} = 0.46 \text{ cm}$$

The image size can be increased either by moving the camera closer to the object or by using a lens with a longer focal length.

A telephoto lens is a camera lens with a focal length greater than 50 mm. It is invaluable in sports and news photography because it allows the photographer to get a closeup picture of distant events. It is also of great value in wildlife photography, allowing detailed close-ups to be taken of very shy animals. For instance, the magnification obtained with a 300-mm telephoto lens is 300 mm/50 mm = 6 times the magnification obtained with a 50 mm lens. This means that the image size of an animal 30 m away taken with a 300-mm lens is the same as the image size taken with a 50-mm lens 5 m away.

**REMARK** A telephoto-lens attachment is conspicuous by its great length, because the lens must be mounted in a tube at least 1 focal length long. For instance, a 300-mm lens must be mounted 300 mm in front of the film, because the image distance of a distant object is 300 mm. Some newer telephoto lenses are mounted in relatively short tubes in which the light travels the proper distance from the lens to the film by reflecting back and forth from mirrors in the tube.

A wide-angle camera lens has a focal length less than 50 mm, and so its magnification is less than that of a normal lens. Its advantage is that for a given film size it offers a wider *field of view*. The field of view is measured by the angle $\theta$ subtended by the scene that is imaged on the film. Figure 15.11 shows the chief rays that reach the extreme edges of the film for lenses with different focal lengths. Clearly, the shorter-focal-length lens gives a wider field of view.

### Macrophotography

Some optical devices operate in the middle region of Fig. 15.8, where $\bar{s}$ and $\bar{s}'$ are both approximately 2. In a copying machine, for instance, the object and image are both 2 focal lengths from the lens. This gives a magnification of 1, so that the image is the same size as the object. Similarly, in macrophotography the camera is brought close enough to a small object, such as an insect, to get a life-size image on the film. Of course, to focus on an insect that is 10 cm in front of a 50-mm lens, the lens must be located 10 cm in front of the film. Since the lens of a camera normally cannot be moved out this far, a camera must be modified to take macrophotographs. One way to do this is to mount the lens on an extension tube to place it the necessary distance from the film.

### *F* number

In addition to focal length, lenses are characterized by their diameter, or *aperture*. In cameras, the aperture is given in terms of the *F*

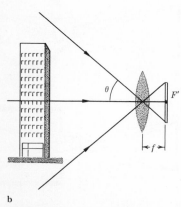

a

b

**FIGURE 15.11**
The fields of view on film of a given size given by lenses of different focal length. The long-focal-length lens (*a*) has a smaller field of view than the short-focal-length lens (*b*).

number, which is the ratio of the focal length $f$ to the diameter $d$ of the lens:

$$F = \frac{f}{d}$$

For example, the diameter of a 50-mm lens with an $F$ number of 2.8 is

$$d = \frac{f}{F} = \frac{50 \text{ mm}}{2.8} = 17.8 \text{ mm}$$

The effective aperture of a lens can be adjusted by masking off part of it with an adjustable diaphragm (Fig. 15.12). The mask does not cut off any of the image formed by the lens because the chief rays are unaffected by the mask, but the diaphragm decreases the brightness of the image. This is shown in Fig. 15.12. Rays from each point of the object pass through all points of the lens. Masking part of the lens prevents some of these rays from reaching the image point, but all image points receive some rays.

The effective $F$ number of a camera increases as the diameter of the diaphragm is decreased. For instance, with an $F$ number of 16, the diaphragm of a 50-mm lens has a diameter of only

$$d = \frac{50 \text{ mm}}{16} = 3.1 \text{ mm}$$

The smaller the $F$ number, the larger the aperture and the brighter the image. The usual values of the $F$ numbers on a camera are 2.8, 4, 5.6, 8, 11, 16, and 22. This peculiar sequence is explained by noting that the amount of light entering the lens is proportional to the area of the aperture and hence proportional to $d^2$. The squares of the $F$ numbers are roughly in the ratios $1:2:4:8: \cdots$, so that a change of the aperture by one $F$ number changes the brightness of the image by a factor of 2.

## 15.3  VIRTUAL IMAGES

A positive lens does not form a real image of an object when the object distance $s$ is less than the focal length $f$. This is seen in Fig. 15.13, which traces the chief and the parallel ray from point $B$ of an object that is located less than 1 focal length from the lens. Like all other rays, these rays are diverging as they radiate from $B$, but they are still diverging even after they pass through the lens, so that no real image is formed.

However, if the diverging rays that emerge from the lens are extended backward, they intersect at point $B'$ on the object side of the lens. This means that the rays that emerge from the lens appear to be coming from a point $B'$ located a distance $s'$ in front of the lens. Similarly, after passing through the lens, rays from other points of the object appear to be coming from corresponding points on the line $A'B'$. That is, if one looks at the light coming from the lens, the image formed on the retina of one's eyes is the same as the image formed by a real object of height $h'$ located between $A'$ and $B'$. In

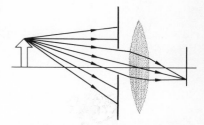

**FIGURE 15.12**
An aperture in front of a lens. The aperture prevents some light from reaching each point of the image, but it does not prevent all the light from reaching any point of the image.

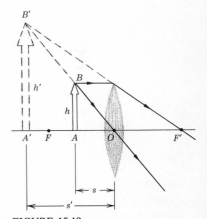

**FIGURE 15.13**
The virtual image formed by an object inside the front focal point of a positive lens. The rays from a single point $B$ of the object do not converge after passing through the lens, so no real image is formed. However, when the rays are projected backward, they appear to come from a single point $B'$ in the virtual image.

this sense, the rays that emerge from the lens are said to form a *virtual image* of the object on the object side of the lens. It is not a real image because there is no light at the position of the virtual image and no image would be formed on a screen placed there.

The virtual image in Fig. 15.13 is larger than the object itself and it is not inverted, so the eye sees an erect, magnified image of the object. The magnification $m = h'/h$ is still equal to $s'/s$, as given by Eq. 15.4. (This is seen by noting that the right triangles $OA'B'$ and $OAB$ in Fig. 15.13 are similar.) However, the relation between $s$, $s'$, and $f$ is no longer given by Eq. 15.2. When the image is virtual, the correct relation is

$$\frac{1}{s} - \frac{1}{s'} = \frac{1}{f} \qquad \text{virtual image} \qquad\qquad 15.7$$

This can be derived from Fig. 15.13 in the same way that Eq. 15.2 was derived from Fig. 15.7.

**REMARK**  Equation 15.2 can be used for both real and virtual images if the image distance $s'$ is taken to be negative for virtual images. However, it may be simpler to treat all distances as positive quantities and use different formulas for real and virtual images.

### The magnifying glass

A magnifying glass is just a simple convex lens used to examine small objects. The object is positioned inside the focal point of the lens, as in Fig. 15.13, so that an erect, magnified virtual image is formed. The magnification $m = s'/s$ depends on the distance $s$ between the lens and the object. For calculations it is usually assumed that $s$ is adjusted so that the virtual image is formed 25 cm in front of the lens, because this is the comfortable reading distance for a person with normal vision. Using $s' = 25$ cm in Eq. 15.7, we get

$$\frac{1}{s} = \frac{1}{25 \text{ cm}} + \frac{1}{f} = \frac{f + 25 \text{ cm}}{f \times 25 \text{ cm}} \qquad\qquad 15.8$$

and so

$$m = \frac{s'}{s} = \frac{25 \text{ cm}}{s} = \frac{f + 25 \text{ cm}}{f} = 1 + \frac{25 \text{ cm}}{f} \qquad\qquad 15.9$$

This gives the magnifying power of a magnifying glass in terms of the focal length of the lens. In Eq. 15.9, the focal length $f$ must be in centimeters, of course.

For example, the magnifying power of a lens of focal length 5 cm is

$$m = 1 + \frac{25 \text{ cm}}{5 \text{ cm}} = 6$$

According to Eq. 15.8, the distance $s$ from the lens to the object is given by

$$\frac{1}{s} = \frac{5 \text{ cm} + 25 \text{ cm}}{5 \text{ cm} \times 25 \text{ cm}} = 0.24 \text{ cm}^{-1}$$

or

$$s = \frac{1}{0.24 \text{ cm}^{-1}} = 4.16 \text{ cm}$$

This means that the object is held 0.84 cm inside the front focal point of the lens.

**REMARK** To achieve the theoretical magnification given in Eq. 15.9, the lens must be held directly in front of the eye. This means that with a 5-cm focal length lens, the object is held only 4.16 cm from the eye. Without a lens, an object held 4.16 cm from the eye appears 25 cm/4.16 cm = 6 times larger than when held 25 cm from the eye. Of course, without a lens, the eye cannot focus on an object only 4.16 cm away. Thus, in a sense, the magnifying glass does not magnify at all but simply enables the eye to focus on an object held very close to it.

The magnifying power of a simple lens increases with decreasing focal length. With a focal length of 1 cm, for instance, a lens has a magnifying power of 26. Since the object distance in this case is less than 1 cm, such a lens is difficult to use. For this reason, a lens with a magnifying power of more than 15 is seldom used. However, before the development of the microscope (Sec. 15.4), magnifying lenses of great power were used. Anton van Leeuwenhoek (1632-1723) discovered bacteria and protozoa using simple lenses with magnifying powers of as much as 200. This requires lenses with focal lengths as small as

$$f = \frac{25 \text{ cm}}{199} = 0.125 \text{ cm}$$

In addition to having a very short object distance, lenses of such short focal length must be very small, and they require much skill and patience to use.

### Reading glasses

The retina of the eye is a fixed distance in back of the eye's lens, and objects at different distances from the eye are focused on the retina by changing the focal length of the lens. This process is called *accommodation*. The normal adult eye can accommodate to objects that are more than 25 cm from the eye, whereas children can focus on objects as little as 10 cm from their eyes. Most people over forty no longer focus on objects held at the normal reading distance (25 cm) and require corrective lenses. The purpose of these lenses is to produce a virtual image of an object at a distance $s'$ at which the eye can accommodate.

For example, suppose that the closest distance at which a person can focus an object (the *near point*) is 100 cm. The focal length of the corrective lens that produces a virtual image at 100 cm of an object held 25 cm from the lens is found from Eq. 15.7 to be

$$\frac{1}{f} = \frac{1}{25 \text{ cm}} - \frac{1}{100 \text{ cm}} = \frac{3}{100 \text{ cm}}$$

or $\qquad f = 33.3 \text{ cm}$ $\qquad\qquad$ 15.21

With eyeglasses of this focal length, the person can read a book held at the normal reading distance because after passing through the glasses, the rays from a book held 25 cm from the eyes have the same divergence as rays coming from a book held 100 cm from the eyes.

In optometry a lens is specified by $1/f$, rather than by $f$ itself. The reciprocal of $f$ is called the *power* of the lens. If $f$ is in meters, the unit of $1/f$ is m$^{-1}$, which is termed a *diopter*. Thus, the power of the 33.3-cm focal length lens in the previous example is

$$\frac{1}{f} = \frac{1}{0.333\ \text{m}} = 3\ \text{m}^{-1} = 3\ \text{diopters}$$

### Negative lenses

A positive (convex) lens always decreases the divergence of the rays that pass through it. If an object is more than 1 focal length from a positive lens, the divergence of the rays reaching the lens is small enough for the lens to convert them into converging rays, forming a real image. A negative (concave) lens always increases the divergence of the rays that pass through it. Figure 15.14 shows that parallel rays incident on a negative lens diverge after passing through the lens. When extended backward, these diverging rays intersect at a point $F$ on the optic axis, which is the focal point of the negative lens. The distance from $F$ to the center of the lens is the focal length and is taken to be a negative quantity.

A negative lens always forms a virtual image of an object any distances $s$ from the lens (Fig. 15.15). The image distance $s'$ is given by Eq. 15.7 if a negative focal length is used. Negative lenses are used to correct myopia (nearsightedness), which is a visual defect in which the eye cannot accommodate to objects farther than some distance $d$ (the *far point*). A negative lens forms a virtual image in its focal plane of very distant objects. Thus, by wearing eyeglasses with negative lenses of focal length equal to $d$, a virtual image of a distant object is formed at a distance to which the eye can accommodate.

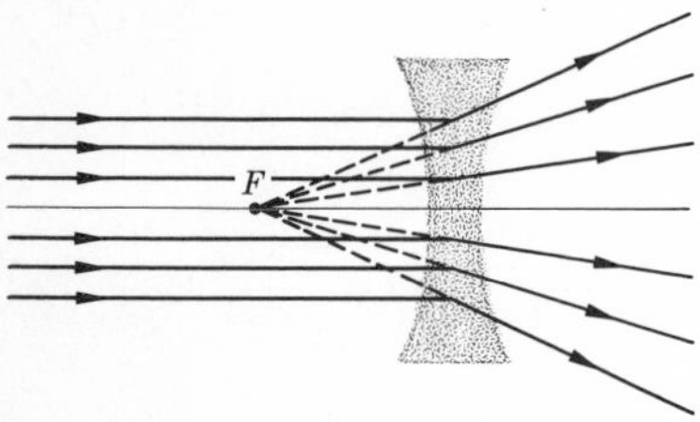

For example, if the far point is 250 cm, the lens is required to form an image at $s' = 250$ cm when $s$ is very large. Since $1/s$ is essentially zero in this case, the focal length given by Eq. 15.7 is

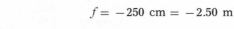

$$\frac{1}{f} = \frac{1}{s} - \frac{1}{250\ \text{cm}} = -\frac{1}{250\ \text{cm}}$$

or
$$f = -250\ \text{cm} = -2.50\ \text{m}$$

The power of the lens is

$$\frac{1}{f} = \frac{1}{-2.5\ \text{m}} = -0.4\ \text{m}^{-1} = -0.4\ \text{diopter}$$

The minus sign indicates that it is a negative (concave) lens.

## 15.4  TWO-LENS INSTRUMENTS

### The microscope

The microscope overcomes the limitation of the simple magnifying lens by using a combination of two lenses. One lens, called the *objective*, is at the lower end of the microscope tube, just above the object on the slide (Fig. 15.16). The objective forms a real image of this

**FIGURE 15.14**
Rays parallel to the optic axis diverge after passing through a negative lens. The diverging rays appear to come from the front focal point of the lens.

**FIGURE 15.15**
The virtual image of an object produced by a negative lens. After passing through the lens, all the rays from a single point of the object appear to come from a single point of the image.

object. The other lens, called the eyepiece or *ocular*, is at the upper end of the microscope tube. It acts as a magnifying lens, forming a magnified virtual image of the image formed by the objective. When the microscope is in focus, the virtual image is located 25 cm below the ocular, so that it can be focused by a normal eye looking through the ocular.

Figure 15.17 traces some rays through such a system of lenses. Notice that the object is placed just outside the front focal point $F_1$ of the objective in order to produce a real image. On the other hand, this real image is placed inside the front focal point $F_2$ of the ocular in order to produce the desired virtual image.

For example, suppose the focal lengths of the objective and ocular are $f_1 = 0.5$ cm and $f_2 = 3.0$ cm, respectively. From Eq. 15.9, the magnifying power of the ocular is

$$m_2 = 1 + \frac{25 \text{ cm}}{3.0 \text{ cm}} = 9.3$$

The ocular produces a virtual image, at a distance $s_2' = 25$ cm, of the real image that is located a distance $s_2$ from the ocular. This distance $s_2$ is found from Eq. 15.7:

$$\frac{1}{s_2} = \frac{1}{s_2'} + \frac{1}{f_2} = \frac{1}{25 \text{ cm}} + \frac{1}{3.0 \text{ cm}} = 0.373 \text{ cm}^{-1}$$

or
$$s_2 = \frac{1}{0.373 \text{ cm}^{-1}} = 2.68 \text{ cm}$$

That is, when the microscope is in focus, the real image produced by the objective must be 2.68 cm below the ocular.

Suppose, furthermore, that the distance $d$ between the ocular and

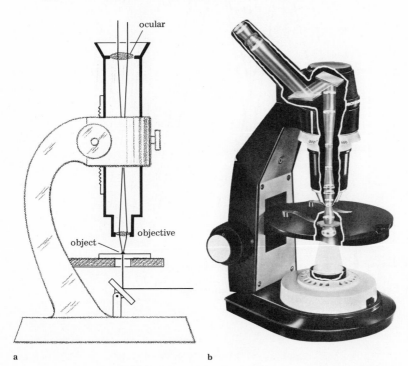

FIGURE 15.16
The microscope. (*a*) The essential features of a microscope are two lenses, the objective and the ocular, arranged so that the objective forms a real image of the object just inside the front focal point of the ocular. (*b*) In a modern microscope, both the objective and the ocular are themselves composed of several lenses. These compound-lens systems have better optical characteristics than a single lens, but they serve the same function. The microscope shown here has additional lenses for zooming and mirrors to bring out the light at a convenient angle. [*Bausch & Lomb.*]

a                                    b

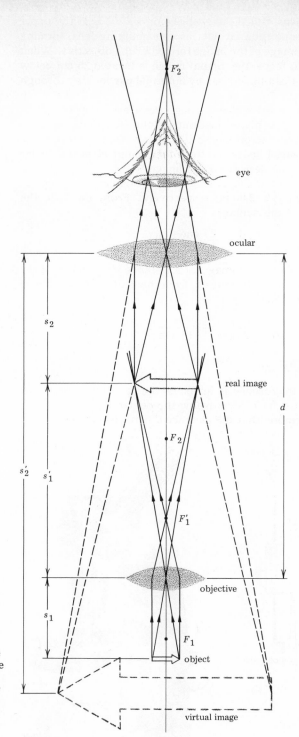

**FIGURE 15.17**
The optics of the microscope. The
objective forms a real image of the
object just inside the front focal
point $F_2$ of the ocular. The ocular
forms a virtual image of the real
image, which is located 25 cm
below the ocular.

the objective is 18 cm. Then the image distance $s_1'$ of the real image from the objective is

$$s_1' = d - s_2 = 18.00 \text{ cm} - 2.68 \text{ cm} = 15.32 \text{ cm}$$

The object distance $s_1$ can now be obtained from Eq. 15.2:

$$\frac{1}{s_1} = \frac{1}{f_1} - \frac{1}{s_1'} = \frac{1}{0.5 \text{ cm}} - \frac{1}{15.32 \text{ cm}} = 1.935 \text{ cm}^{-1}$$

or $\qquad s_1 = \dfrac{1}{1.935 \text{ cm}^{-1}} = 0.517 \text{ cm}$

The object distance $s_1$ is just a little larger than $f_1$, so that the magnification $m_1 = s_1'/s_1$ of the real image is approximately

$$m_1 = \frac{s_1'}{f_1} = \frac{15.34 \text{ cm}}{0.5 \text{ cm}} = 30.7$$

The real image is larger than the object by a factor $m_1$, and the virtual image is larger than the real image by a factor $m_2$. Therefore, the virtual image is $M$ times larger than the object, where

$$M = m_1 m_2 = \frac{s_1'}{f_1}\left(1 + \frac{25 \text{ cm}}{f_2}\right) \qquad\qquad 15.10$$

This is the magnifying power of a microscope. In the present example we have

$$M = m_1 m_2 = 30.7 \times 9.3 = 285$$

**REMARK** To give the same magnification, a single magnifying lens would have to have a focal length of

$$f = \frac{25 \text{ cm}}{284} = 0.088 \text{ cm}$$

and the object would have to be held less than 1 mm from the lens.

### Compound lenses

Two lenses placed close to each other act as a single lens. Most quality optical instruments use several lenses together this way. In a microscope, for instance, the ocular and objective are themselves composed of two or more lenses (Fig. 15.16$b$) in order to eliminate the distortions (aberrations) produced by a single lens.

Figure 15.18 shows parallel rays incident on two lenses of focal

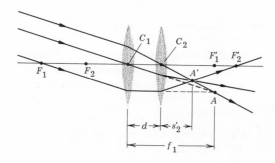

FIGURE 15.18
Parallel rays incident on two lenses. The position $A'$ of the image formed by these two lenses together can be found from the fact that the first lens alone would have formed an image at $A$.

lengths $f_1$ and $f_2$, which are separated by a distance $d$ that is less than $f_1$. Thus, the rays reach lens 2 before they are focused in the focal plane of lens 1. In the absence of lens 2, the rays passing through the center $C_1$ and the front focal point $F_1$ of lens 1, would be brought to focus at point $A$ in the focal plane of lens 1. The effect of lens 2 is to bend the rays further, so that they come to focus at a point $A'$, closer to lens 1.

To locate $A'$, first note that the ray that is parallel to the optic axis between lenses 1 and 2 will pass through the focal point $F'_2$ of lens 2. Second, in the absence of lens 2 there is a ray that goes through point $C_2$ to $A$. But this ray is not affected by the presence of lens 2 because it passes through its center. Consequently, the intersection of these two rays at $A'$ determines the position of the focal plane of this two-lens system.

This graphical construction shows that the focal length $f$ of the two lenses together is less than that of either lens alone. Equation 15.2 can be used to find $f$ by noting that the image that would have been formed by lens 1 alone acts as the object for lens 2. Since this image is a distance $s_2 = f_1 - d$ in back of lens 2, it must be treated as a negative distance in Eq. 15.2. With this understanding, the distance $s'_2$ from lens 2 at which the image is formed is given by

$$-\frac{1}{f_1 - d} + \frac{1}{s'_2} = \frac{1}{f_2}$$

or

$$\frac{1}{s'_2} = \frac{1}{f_2} + \frac{1}{f_1 - d}$$

If the distance between the two lenses is very small ($d$ approximately zero), this becomes

$$\frac{1}{f} = \frac{1}{f_1} + \frac{1}{f_2} \qquad\qquad 15.11$$

where we have set $s'_2$ equal to $f$ because the rays incident on the system are parallel.

Equation 15.11 is used by an optometrist when fitting eyeglasses. He adjusts the strength of a lens combination by adding or subtracting lenses in the system.

For instance, a 0.50-m lens in combination with a 0.75-m lens has the focal length

$$\frac{1}{f} = \frac{1}{0.50 \text{ m}} + \frac{1}{0.75 \text{ m}} = 2.0 \text{ m}^{-1} + 1.33 \text{ m}^{-1}$$

$$= 3.33 \text{ m}^{-1}$$

or $\qquad\qquad f = 0.30 \text{ m}$

The optometrist works with the power $1/f$ of a lens, rather than with the focal length $f$, because the power of a lens combination is given by simply adding the power of the individual lenses. In this example, the powers of the two lenses are 2.0 and 1.33 diopters, and their combination has a power of 3.33 diopters.

**The telescope**

A telescope is a two-lens instrument designed to magnify distant objects. The two lenses, with focal lengths $f_1$ and $f_2$, are separated by a distance $d = f_1 + f_2$, so that the back focal plane of lens 1 and the front focal plane of lens 2 coincide (Fig. 15.19). The rays from a point on a distant object are parallel when they reach the front (objective) lens and so are brought to focus at a point in its focal plane. Since this is also the focal plane of the second (ocular) lens, these rays are again parallel after emerging from lens 2. This arrangement is said to be *afocal* because the parallel rays incident on the system, emerge as parallel rays.

Figure 15.19 shows that rays entering the system from above the optic axis emerge from below the axis. Therefore, an observer looking at the parallel rays that emerge from the ocular will see an inverted image of a distant object. If $f_2 < f_1$, the angle $\alpha'$ at which these rays emerge is greater than the angle $\alpha$ at which they were incident. This has the effect of widening (magnifying) the image seen through the instrument.

The ratio $\alpha'/\alpha$ of these angles is approximately equal to the ratio $(\tan \alpha')/(\tan \alpha)$ of their tangents. But from Fig. 15.19 we have

$$\tan \alpha = \frac{\overline{DC}}{f_1}$$

$$\tan \alpha' = \frac{\overline{D'C'}}{f_2} = \frac{\overline{DC}}{f_2}$$

Consequently, the angular magnification $a$ of the telescope is

$$a = \frac{\tan \alpha'}{\tan \alpha} = \frac{\overline{DC}/f_2}{\overline{DC}/f_1} = \frac{f_1}{f_2}$$

Large magnification requires a very long focal length objective, which is why telescopes are so long.

**REMARK** It is believed that the Dutch spectaclemaker Hans Lippershey was the first to combine two lenses in the microscope and telescope configurations between 1590 and 1608. However, with the quality of lenses then available, the early microscope was not superior to the magnifying lens. The microscope did not replace the magnifying lens for another century, when it became possible to produce high-quality lenses. (All the great discoveries of Leeuwenhoek were made using a single lens of great power.) The history of the telescope is more dramatic.

Galileo happened to be in Venice about May 1609, where he heard rumors of a perspective instrument. The day after he got back to Padua Galileo made his first telescope and soon learned how to build superior instruments, with magnifications of up to 30. Galileo was the first man to look at the stars with a telescope, and the world was never the same. He saw the mountains on the moon, the spots on the sun, and the moons around Jupiter. These observa-

**FIGURE 15.19**
The optics of the telescope. The objective forms a real image of a distant object at the front focal point of the ocular. Since the rays that emerge from the ocular are parallel, they appear to come from a distant virtual image. However, since the angle $\alpha'$ is greater than $\alpha$, the distance of the image is less than the distance of the object.

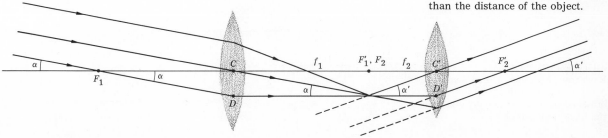

tions contradicted Aristotle's old idea that the heavenly bodies were composed of incorruptible matter, different in substance from the matter on earth. As a consequence, the authority of Aristotle was shaken, and men began to open their minds to new ideas. (Some men, that is. Others hung stubbornly to the old ideas, refusing even to look through Galileo's telescope.)

A thousand years from now man may have forgotten who first stepped on the moon, but as long as there are men who think about the nature of the world, the story of Galileo and his telescope will be remembered.

## 15.5 THICK LENSES

The optical theory discussed so far is valid only for a *thin* lens, i.e., a lens with a thickness $t$ much less than its focal length $f$. A lens for which this is not true is called a *thick* lens. The theory of thick lenses is a generalization of thin-lens theory and therefore more complicated, but since the eye consists of thick lenses, some knowledge of thick-lens theory is required to understand the optics of human vision. Furthermore, it can be shown that any multiple-lens system, such as a microscope, can be considered to be a single thick lens for the purpose of analysis. Thus, the theory of thick lenses has a wide range of application.

The basic difference between a thin lens and a thick lens is shown in Fig. 15.20. With a thin lens, rays are considered to be refracted at the single plane $H$ perpendicular to the optic axis and passing through the center $C$ of the lens. With a thick lens, rays are considered to be refracted by two planes, $H$ and $H'$, perpendicular to the optic axis and passing through the *principal points* $P$ and $P'$ on the optic axis. These planes are called the *principal planes* of the lens. Figure 15.20*b* shows that a ray that intersects the front principal plane $H$ at a point $A$ is considered to be refracted at the point $A'$ on the back plane $H'$. The point $A'$ is found by drawing a line parallel to the optic axis from $A$ to the plane $H'$.

Principal points $P$ and $P'$ exist for every lens and for every combination of lenses. They can be located either by measurement or by calculation. Once the positions of the principal points and focal points of a lens or lens system are known, the image formed by an object can be found by the usual method of ray tracing. This is shown in Fig. 15.21. A ray emerging from the object parallel to the optic axis is refracted at the back principal plane $H'$ so that it passes through the back focal point $F'$. A ray emerging from the object and passing

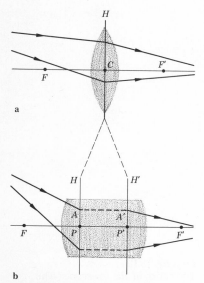

**FIGURE 15.20**
Comparison of a thin lens and a thick lens. The two principal points $P$ and $P'$ of a thick lens replace the center $C$ of a thin lens in ray diagrams.

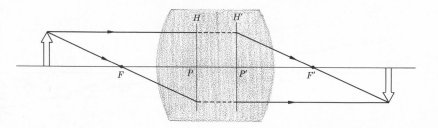

**FIGURE 15.21**
Rays traced through a thick lens. A ray incident on the front principal plane is refracted from the back principal plane.

through the front focal point $F$ emerges from the back plane traveling parallel to the optic axis. The image is determined by the intersection of these two rays. Note that a ray is refracted at a point on the back principal plane that is at the same distance from the optic axis as the point where the ray intersects the front principal plane.

The optical properties of a lens depends on the medium in which it operates. If the media in front and in back of the lens are the same, the system has the following special properties.

**Property 1** *The distance $f'$ of the rear focal point $F'$ from the rear principal plane $H'$ is equal to the distance $f$ of the front focal point $F$ from the front principal plane.*

**Property 2** *A ray that passes through the front principal point $P$ at an angle $\alpha$ to the optic axis emerges from the back principal point $P'$ at the same angle $\alpha$ to the optic axis.* (Such a ray is called a *chief ray*, in analogy to a ray that passes through the center $C$ of a thin lens.)

**Property 3** *If the object distance $s$ is measured from the front principal plane and the image distance $s'$ is measured from the rear principal plane, these distances are related by Eq. 15.2, and the magnification is given by Eq. 15.4.*

These properties are illustrated in Fig. 15.22, which shows a lens system with a focal length $f = 5$ cm and with a separation of 3 cm between the principal planes. The image of an object located a distance $s = 8$ cm in front of the front principal plane is found by tracing the rays 1 and 2, as in Fig. 15.21. The chief ray is then drawn from the object to point $P$, and from point $P'$ to the image. The angles $\alpha$ and $\alpha'$ these two lines make with the optic axis are seen to be equal, in accordance with Property 2. The distance $s'$ of the image from the rear principal plane is found from the diagram to be 13.33 cm. This value can also be obtained using Eq. 15.2. From Eq. 15.4 the magnification is $m = s'/s = 1.66$, which agrees with the ratio $h'/h$ found from the ray diagram.

The human eye, shown in Fig. 15.23, consists of two thick lenses, the cornea and the crystalline lens, embedded in a fluid (vitreous humor) with an index of refraction of 1.34. Figure 15.24 shows the location of the principal planes of the optical system of the relaxed eye. Since the media in front and in back of this lens system are different, the three special properties listed above do not apply. In particular, the front focal length $f$ is not equal to the rear focal length $f'$. Figure 15.24 shows that the front focal length is 1.70 cm and the

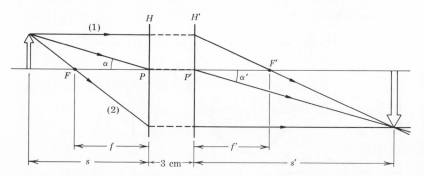

**FIGURE 15.22**
Scale diagram illustrating the properties of a thick lens. (1) The front and back focal points $F$ and $F'$ are equidistant from the front and back principal planes, respectively. (2) A ray that intersects the front principal point $P$ at an angle $\alpha$ appears to emerge from the back principal point $P'$ at the same angle. (3) Equations 15.2 and 15.4 hold for a thick lens if $s$ and $s'$ are measured from the front and back principal points, respectively.

rear focal length is 2.27 cm. The rear focal point $F'$ is located on the retina because the relaxed eye focuses distant objects on the retina.

Since the principal planes of the eye are only 0.03 cm apart, for practical purposes they can be treated as a single plane. These planes appear to coincide on the scale diagram shown in Fig. 15.25, which shows that the image of an object can be found using the normal ray-tracing method. The only differences from the method used for a thin lens are that the focal points $F$ and $F'$ are not equidistant from the principal planes and that a ray does not pass through the principal point $P$ undeviated.

There does exist a chief, or undeviated, ray for the eye, as shown in Fig. 15.25. This ray intersects the optic axis at the point $N$, called the *nodal point*, that is 0.60 cm back of the principal plane (Fig. 15.25). The nodal point, like the focal points and principal points, is a fixed point of the optical system, independent of the positions of the object and image. That is, any ray that is directed toward the nodal point emerges from the optical system undeviated.

**REMARK** A camera is adjusted to focus objects at different distances by changing the distance $s'$ between the lens and the film. In the eye, the distance between the retina and the lens is fixed. To focus on objects at different distances, the focal length of the crystalline lens is changed. This is accomplished by muscles attached to the lens that control its shape. When the eye is accommodated for nearby objects, the focal points, principal points, and nodal point of the eye's optical system are different from what they are for the relaxed eye.

Because Property 3 does not apply to the eye, the magnification of the image is not given by Eq. 15.4. The image size is determined instead from the position of the nodal point. From Fig. 15.25 it is seen that the triangles $NAB$ and $NA'B'$ are similar, so the ratio of image size to object size is

$$\frac{h'}{h} = \frac{\overline{A'N}}{\overline{AN}} = \frac{s' - 0.6 \text{ cm}}{s + 0.6 \text{ cm}}$$

**FIGURE 15.23**
Anatomy of the human eye.

**FIGURE 15.24**
The focal points and principal points of the relaxed human eye. (This diagram is not to scale.) The back focal point $F'$ is not the same as the front focal point $F$ because the medium (aqueous humor) on one side of the lens system is different from the medium (air) on the other.

**FIGURE 15.25**
Scale diagram of the focal points, principal points, and nodal points of the relaxed human eye. To the accuracy of this drawing, the two principal points coincide, and the two nodal points coincide.

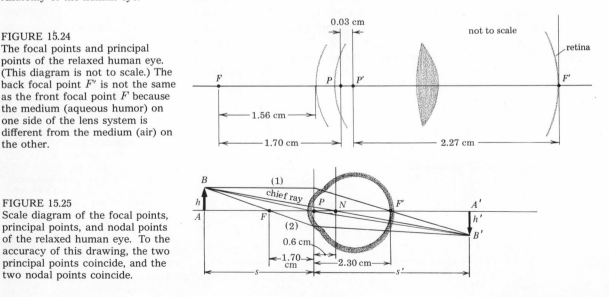

The relaxed eye is adjusted to focus distant objects on the retina, which is a distance $s' = 2.30$ cm from the principal plane. Thus the magnification is given by

$$m = \frac{h'}{h} = \frac{1.70\ \text{cm}}{s} \qquad 15.12$$

where the term 0.6 cm in the denominator has been dropped because it is very small compared to $s$.

For example, consider a man who is watching a girl 1.6 m tall who is 5 m away. What is the size of her image on the man's retina? From Eq. 15.12 we have

$$m = \frac{1.70\ \text{cm}}{500\ \text{cm}} = 0.0034$$

so

$$h' = mh = 0.0034 \times 1.6\ \text{m} = 0.00544\ \text{m} = 5.44\ \text{mm}$$

**REMARK** In the most general optical system there are two nodal points, just as there are two principal points. A ray directed toward the front nodal point $N$ emerges undeviated from the rear nodal point $N'$ (Fig. 15.26). The eye has two nodal points, but since they are only 0.03 cm apart, they are treated as a single point. The focal points, the principal points, and the nodal points are the six *cardinal* points of a general lens system. When the media in front and in back of the lens system are the same, the nodal points coincide with the principal points. In a thin lens, the principal points coincide with the center of the lens.

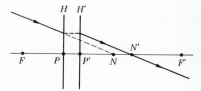

FIGURE 15.26
The cardinal points of a general lens system. The optics of any lens or combination of lenses can be described by six cardinal points: two focal points, two principal points, and two nodal points.

## PROBLEMS

1 Consider a positive lens of focal length 12 cm. Make a table of the values of the image distance $s'$ for the following values of the object distance $s$: 15, 18, 21, 24, 30, 36, 48, and 60 cm. Convert these values of $s$ and $s'$ to reduced distances $\bar{s}$ and $\bar{s}'$, and plot them on the curve in Fig. 15.8.

2 An object is 21 cm in front of a lens of focal length 14 cm. (a) Find the image distance $s'$ from the lens formula, from the curve in Fig. 15.8, and by graphical construction. (b) What is the magnification of the image?

3 A slide projector with a lens of focal length 10 cm projects an image on a screen that is 2.5 m from the lens. (a) What is the distance $s$ between the slide and the lens? (b) What is the magnification of the image? (c) What is the width of the image of a slide 35 mm wide?     *Ans.* (a) 10.4 cm; (b) 25; (c) 87.5 cm

4 A classroom slide projector is 12 m from a screen that is 1.5 m wide. What focal length lens is required for the image of a 35-mm slide to just fill the screen?

5 The distance between the lens of a projector and the slide can be varied from 20 to 30 cm. If the focal length of the lens is 21 cm, what is the smallest distance (between the lens and the screen) at which an image can be focused?     *Ans.* 70 cm

6 A slide projector with a lens of focal length 20 cm projects the image of a 35-mm slide onto a screen that is 0.8 m wide. How far should the screen be from the lens for the image to just fill the screen?

7 A camera with a lens of focal length 50 mm takes a picture of a child 1.2 m tall standing 3.0 m away. (a) What must be the distance $s'$ between the film and the lens to get a properly focused picture? (b) What is the magnification of the image? (c) What is the height of the image of the child on the film?     *Ans.* (a) 5.08 cm; (b) 0.0167; (c) 20.0 mm

8 A camera with a 50-mm lens takes a picture of a tree 25 m tall. How far must the camera be from the tree for its image on the film to be 25 mm high?

9 A camera with a telephoto lens of focal length 450 mm takes a picture of an object 60 m

away. How far from the object would a camera with a 50-mm lens have to be to get an image of the same size?        *Ans.* 6.67 m

10 A naturalist wishes to photograph a rhinoceros from a distance of 75 m. The beast is 4.0 m long, and its image is to be 1.2 cm long on the film. (*a*) What focal length lens should be used? (*b*) What would be the size of the image if a normal 50-mm lens were used?

11 A camera with a wide-angle lens of focal length 35 mm takes a picture of an object 12 m away. How far from the object would a camera with a 50-mm lens have to be to get an image of the same size?        *Ans.* 17.1 m

12 A photographer wishes to photograph a building that is 15 m tall from a distance of 20 m. What focal length lens should be used in order for the image to be 20 mm high?

13 A lens of focal length 40 mm is used to take extreme close-ups (macrophotography). (*a*) If the lens can be at most 5.20 cm from the film, what is the closest distance (from the lens) at which an object can be focused? (*b*) What is the magnification in this case? (*c*) If the lens cannot be less than 5.0 cm from the film, what is the farthest distance (from the lens) at which an object can be focused?
        *Ans.* (*a*) 17.3 cm; (*b*) 0.30; (*c*) 20.0 cm

14 A lens of focal length 30 mm is used in macrophotography to take a picture. The image on the film is to be twice the object size ($m = 2$). Find the object and image distances.

15 A camera is fitted with a 125-mm telephoto lens. The distance of this lens from the film can be varied from 125 to 130 mm. What is the closest distance that an object can be from this camera and still be focused on the film?
        *Ans.* 3.25 m

16 A camera with a 50-mm lens can focus on an object at any distance greater than 1.2 m. Find the minimum and maximum distances of the lens from the film.

17 An expensive camera lens of focal length 50 mm has an $F$ number of 1.7. What is the diameter of this lens?        *Ans.* 29.4 mm

18 The lens of the relaxed eye has an effective focal length of 17 mm. The diameter of the pupillary aperture varies from 1.5 to 8 mm, depending on the intensity of the light entering the eye. What are the corresponding $F$ numbers of the eye?

**REMARK** The $F$ number is a measure of the light-gathering capacity of a lens; the smaller the $F$ number, the greater the light-gathering capacity. The human eye has about the same light-gathering capacity as a medium-priced camera lens.

19 An object is 4 cm in front of a lens of focal length 6 cm. (*a*) Locate the position of the virtual image. (*b*) What is the magnification of the image?
        *Ans.* (*a*) 12 cm in front of lens; (*b*) 3

20 Use Eq. 15.7 to plot the reduced virtual image $\bar{s}'$ against the reduced object distance $\bar{s}$ for values of $\bar{s}$ less than 1.

21 (*a*) What is the focal length of a magnifying glass with a magnifying power of 10? (*b*) How far must the object be from this lens to obtain this magnification?
        *Ans.* (*a*) 2.78 cm; (*b*) 2.50 cm

22 What is the magnifying power of a lens with a power of 15 diopters?

23 (*a*) What is the focal length of the reading glasses required by a person whose near point is 150 cm? (*b*) What is the power (in diopters) of this lens?
        *Ans.* (*a*) 30 cm; (*b*) 3.33 diopters

24 The power of a farsighted person's reading glasses is 2.5 diopters. At what distance must he hold a book in order to read it without glasses?

25 Find the virtual image of an object 6 cm in front of a lens of focal length $-3$ cm (*a*) by tracing rays and then (*b*) by using Eq. 15.7. (*c*) What is the magnification of the image?
        *Ans.* (*a*) 2 cm; (*b*) 2 cm; (*c*) 0.33

26 Plot a curve of $\bar{s}'$ against $\bar{s}$ for a negative lens. Compare with Fig. 15.8.

27 A nearsighted person wears eyeglasses with negative lenses of focal length $-2$ m. (*a*) When he looks at an object 6 m away, where is the virtual image produced by the glasses? (*b*) Locate the image graphically.
        *Ans.* (*a*) 1.5 m

28 At age forty a man requires eyeglasses with lenses of 2 diopters power in order to read a book at 25 cm. At forty-five he finds that while wearing these glasses he must hold a book 40 cm from his eyes. What power lenses does he require at forty-five to read a book at 25 cm?

29 A microscope has an objective of focal length 0.3 cm and an ocular of focal length 2.0 cm.

(*a*) Where must the image formed by the objective be for the ocular to produce a virtual image 25 cm in front of the ocular? (*b*) If the lenses are 20 cm apart, what is the distance of the objective from the object on the slide? (*c*) What is the total magnification of the microscope? (*d*) What distance would the object have to be from a single lens that gave the same magnification?

*Ans.* (*a*) 1.85 cm in front of the ocular; (*b*) 0.305 cm; (*c*) 804; (*d*) 0.031 cm

30 A dissecting microscope is designed to have a large distance between the object and the objective. Suppose the focal length of the objective of a dissecting microscope is 5.0 cm, the focal length of the ocular is 4.0 cm, and the distance between these lenses is 17.0 cm. (*a*) What is the distance between the object and the objective? (*b*) What is the total magnification? Note that Eq. 15.10 cannot be used in this case.

31 An object is placed 12 cm in front of a lens of focal length 5 cm. Another lens of focal length 4 cm is placed 2 cm in back of the first lens. Find the image produced by this two-lens system by tracing rays. (*Hint:* First find the image produced by the front lens alone, and use it to find the image formed by the second lens.)

*Ans.* 2.5 cm is back of the 4-cm lens

32 A lens of focal length 500 mm is mounted in front of a 50-mm camera lens. (*a*) What is the focal length of the combination, assuming zero distance between the lenses? (*b*) If the lens-to-film distance can be varied from 5.00 to 5.22 cm, what are the closest and farthest object distances at which the camera can focus with this attachment?

33 A 3- and a 30-cm focal length lens are used to build a small telescope. (*a*) Which lens should be the objective? (*b*) What is the magnifying power of the telescope? (*c*) How far apart are the two lenses in the telescope?

*Ans.* (*b*) 10; (*c*) 33 cm

34 A terrestrial telescope consists of an objective of focal length $f_1$ and an ocular of focal length $-f_2$ (negative lens) separated by a distance $d = f_1 - f_2$. Trace parallel rays through such a lens system when $f_1 = 10$ cm and $f_2 = -2$ cm. Show that the magnification is $f_1/f_2$ and that the virtual image is erect.

**REMARK** This lens system is used in binoculars and spyglasses because it gives an erect image and requires a shorter distance between the objective and the ocular than the astronomical telescope does.

35 A thick lens has a focal length of 15 cm, and its principal planes are 3 cm apart. By graphical construction, locate the image of an object that is 25 cm in front of the front principal plane. Check your result with Eq. 15.2.

*Ans.* 37.5 in back of the back principal plane

36 The distances shown in Fig. 15.25 are for the relaxed eye, i.e., an eye accommodated for distant vision. By graphical construction, locate the image of an object that is 10 cm in front of the relaxed eye.

37 What is the size of the image on the retina of the print in a book held 25 cm in front of the relaxed eye if the print is 2 mm high?

*Ans.* 0.136 mm

38 Show that the relation between the object and image distance for a thick-lens system, like the eye, in which the front and back focal lengths are different, is given by Eq. 15.3, with $\bar{s} = s/f$ and $\bar{s}' = s'/f'$. Here $f$ and $f'$ are the front and back focal lengths, respectively, and $s$ and $s'$ are the object and image distances measured from the front and back principal planes, respectively.

39 Find the position of the image of an object that is 25 cm in front of the relaxed eye shown in Fig. 15.24 (see Prob. 38).

*Ans.* 0.166 cm in back of the retina

## BIBLIOGRAPHY

BUTTERFIELD, HERBERT: "The Origins of Modern Science, 1300–1800," rev. ed., Free Press, New York, 1965. The significance of Galileo's telescopic discoveries in the historic conflict between the copernican and ptolemaic world systems is told in chap. 4.

GUYTON, ARTHUR C.: "Textbook of Medical Physiology," 4th ed., W. B. Saunders Company, Philadelphia, 1971. Chapter 52 contains a concise discussion of the optics of human vision.

OGLE, KENNETH NEIL: "Optics: An Introduction for Ophthalmologists," Charles C Thomas, Springfield, Ill., 1968. Complete treatment of the optics of thin and thick lenses. The optics of the human eye is described in detail.

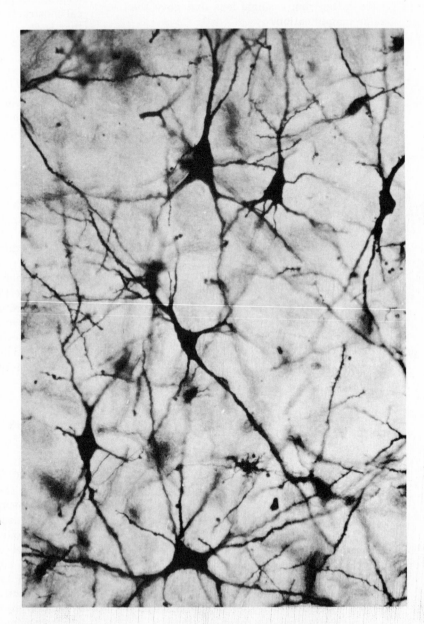

Photomicrograph of generalized
neurons of the reticular formation
of the brain stem. Signals are
transmitted between neurons by
electric impulses traveling along
their processes (Sec. 17.5).
[*Courtesy of E. Ramon-Moliner,
Université de Sherbrooke,
Sherbrooke, Quebec.*]

# IV electricity and magnetism

It has been known since ancient times that certain substances, notably amber and glass, after being rubbed with a material such as silk or fur, acquire the capacity to attract small objects such as bits of paper and cork. The substance is said to be electrified, after the Greek word for amber (*elektron*). It is now known that this phenomenon is just one manifestation of *electricity*, a fundamental force of nature. Similarly, the ability of certain iron ores, such as lodestone (*magnes*), to attract small bits of iron, has been known for thousands of years. This phenomenon is a manifestation of *magnetism*, another fundamental force of nature.

Although simple electrical and magnetic phenomena have been known for thousands of years, most of the basic quantitative laws of electricity and magnetism were discovered between 1784, when Charles Coulomb investigated the forces between charged objects, and 1831, when Michael Faraday discovered magnetic induction. The practical utilization of electricity, upon which our modern civilization is based, and the conceptual realization that light is a form of electromagnetic radiation both developed from the discoveries made during this 50-year period.

# CHAPTER 16 electricity

Electricity is a fundamental force of nature, analogous to gravity. But whereas the gravitational force between two objects depends on their mass, the electric force between two objects depends on their *charge*. Charge is a basic property of two of the elementary particles (electrons and protons) that compose all ordinary matter. In fact, it is the electric force between the protons and electrons in an atom that holds the atom together.

The practical utilization of electricity is made possible by our ability to generate and control a flow of charged particles. In this chapter we discuss the principles of electricity needed to understand devices, such as x-ray and cathode-ray tubes, that utilize the flow of high-speed electrons inside an evacuated vessel. In Chaps. 17 and 18 we discuss devices, such as heaters and motors, that utilize the flow of electrons in a wire.

## 16.1 THE FUNDAMENTAL FORCES

All the forces that have been discussed in this book, such as friction, buoyancy, and surface tension, are the observable effects of forces between the atoms of the objects involved. These observable forces are not considered to be fundamental because, in principle, they can be analyzed in terms of atomic forces.† Even the forces between atoms are not fundamental because they can be analyzed further in terms of the forces between the particles that compose the atoms.

There are three kinds of particles inside an atom: *protons*, *neutrons*, and *electrons*. The protons and neutrons are tightly packed together to form the dense central core, or *nucleus*, of the atom. The nucleus contains over 99.95 percent of the atom's mass, but it occupies only a minute fraction of the atom's volume. The electrons, which swarm in orbits about the nucleus, contain the remaining 0.05 percent of the mass and occupy most of the volume. Figure 16.1 shows a schematic diagram of a carbon atom.‡ The details of the structure of atoms and nuclei are given in Part V.

Because protons, neutrons, and electrons are not themselves composed of still smaller particles, they are said to be *elementary* particles,

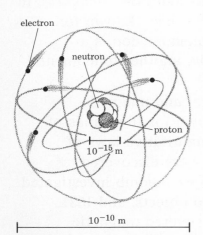

electron

neutron

proton

$10^{-15}$ m

$10^{-10}$ m

**FIGURE 16.1**
Schematic diagram of a carbon atom. Six electrons orbit about a nucleus composed of six protons and six neutrons.

† A crude analysis of surface tension in terms of atomic forces is given in Sec. 9.2.
‡ If the atom were drawn to scale, with the nucleus the size shown in Fig. 16.1, the orbits of the outer electrons would be the size of a football stadium.

and the forces between them are the fundamental forces of nature. All ordinary matter is composed of these elementary particles, and so all forces can be ultimately understood in terms of the fundamental forces between them. At present, four distinct fundamental forces are known:

*1* Gravitational force (gravity)
*2* Electromagnetic force (electricity and magnetism)
*3* Nuclear force
*4* Weak force

In a sense, the gravitational force is the weakest fundamental force because the magnitude of the gravitational force between two elementary particles is far smaller than the magnitude of any other fundamental force. The force of gravity between elementary particles is so weak, in fact, that it has no measurable effect on the behavior of these particles inside an atom. Only an object of astronomical size has sufficient mass to exert a significant gravitational force on an atom. Since this force is attractive, such an object attracts the atoms in the space around it. The accretion of these atoms increases the mass of the object, so that the gravitational force it can exert also increases. As a consequence, the object attracts still more atoms, which increases its mass even more. This is how immense aggregates of matter, such as stars and galaxies of stars, are formed from the dust and atoms scattered throughout space. Gravity is by far the most important force controlling the structure of astronomical bodies.

The electromagnetic force is the principal force determining the structure of atoms. The electrons are maintained in orbit about the nucleus by electrical attraction, just as the planets are maintained in orbit about the sun by gravitational attraction. The electrons also exert electric forces on each other, and the force between two nearby atoms is just the electric force between their electrons. All the forces discussed so far in this book, with the exception of gravity, are the result of this electric force. Thus all the forces of everyday life, with the exception of gravity, are electromagnetic in origin.

The protons and neutrons in the nucleus are held together by the nuclear force. This force is very strong when the particles are close together, but it decreases rapidly with distance. For instance, the nuclear force between two neutrons is essentially zero when they are more than $10^{-14}$ m apart. This means that the nuclear force does not extend beyond the outer electrons of the atom, which are $10^{-10}$ m from the nucleus. Thus, although the nuclear force is essential for holding the nucleus together, it plays no direct role in the interaction of atoms with each other.

The weak force is also a short-ranged force confined entirely to the nucleus. It is responsible for some forms of radioactivity, which is the spontaneous transformation of the nucleus of one kind of atom into the nucleus of another (Sec. 20.2).

The electromagnetic force is thus the principal force governing the physics and chemistry of ordinary matter. It is also of great practical importance, since our entire industrial civilization is based on electric-power generation and its utilization in motors, lighting, and heating.

## 16.2 COULOMB'S LAW

Even though the electric and gravitational forces are fundamentally distinct from each other, they have a number of similar properties. In order to understand electricity, therefore, it is helpful to briefly review some of the properties of gravity discussed in Sec. 5.4.

Gravity is an attractive force that exists between two objects with masses $m_1$ and $m_2$. This means that the force $\mathbf{F}_1$ exerted by $m_2$ on $m_1$ is directed toward $m_2$ (Fig. 16.2). By Newton's third law (Property 3, Sec. 2.1), the reaction to $\mathbf{F}_1$ is the force $\mathbf{F}_2$ exerted by $m_1$ on $m_2$. Force $\mathbf{F}_2$ has the same magnitude as $\mathbf{F}_1$ but the opposite direction, so it is directed toward $m_1$, as shown in Fig. 16.2.

Gravity is a force that acts at a distance; i.e., two objects exert their mutual attraction on each other across empty space, without any mechanical connection. The magnitude $F_g$ of the gravitational forces ($\mathbf{F}_1$, $\mathbf{F}_2$) on two objects with masses $m_1$ and $m_2$ separated by a distance $r$ is given by Newton's law of gravity,

$$F_g = G\frac{m_1 m_2}{r^2}$$

where $G$ is the universal gravitational constant ($G = 6.67 \times 10^{-11}$ N-m$^2$/kg$^2$).

It is customary to distinguish between attractive and repulsive forces by the sign of their magnitudes. An attractive force is indicated by a negative magnitude, and a repulsive force is indicated by a positive magnitude. According to this convention, the force of gravity must be written

$$F_g = -G\frac{m_1 m_2}{r^2} \qquad\qquad 16.1$$

to indicate that the force is attractive.

Electricity is a force acting at a distance between two objects with *charges* $q_1$ and $q_2$. Charge, like mass, is a basic attribute of matter. The dimension of charge is taken to be fundamental, along with mass, length, time, and degree (of temperature). The dimension of any physical quantity can be expressed in terms of these five dimensions. The unit of charge is the *coulomb* (C).

The electric force between two objects with charges $q_1$ and $q_2$ separated by a distance $r$ is given by *Coulomb's law*,

$$F_e = +K\frac{q_1 q_2}{r^2} \qquad\qquad 16.2$$

where $K$ is the universal electric constant ($K = 9.0 \times 10^9$ N-m$^2$/C$^2$). Notice the similarity between Eqs. 16.1 and 16.2. The electric force depends on the product of the charges of the two objects, just as the gravitational force depends on the product of their masses. Also, the electric and gravitational forces are both inversely proportional to the square of the distance between the objects.

The fundamental difference between gravity and electricity is that whereas gravity is always attractive, the electric force can be either attractive or repulsive because there are two kinds of charge, positive and negative. Objects with the same kind of charge repel each other,

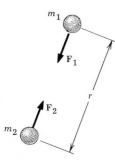

FIGURE 16.2
Gravitational attraction of two masses.

and objects with opposite charge attract each other. This is the significance of the plus sign in Eq. 16.2. When $q_1$ and $q_2$ have the same sign, either both positive or both negative, the product $q_1q_2$ is positive, and so $F_e$ is positive, indicating a repulsive force (Fig. 16.3). On the other hand, if $q_1$ and $q_2$ have opposite signs, the product $q_1q_2$ is negative, and so $F_e$ is negative, indicating an attractive force (Fig. 16.4).

The elementary particles are the ultimate carriers of charge. A proton has a positive charge $e = 1.602 \times 10^{-19}$ C, and an electron has a negative charge $-e$ of exactly the same magnitude. (A neutron has zero charge.) The charge of an object is the sum of the charges of all its protons and electrons. Thus, an object that has an equal number of protons and electrons has zero charge. Such an object is said to be uncharged or *neutral*.

A neutral object acquires charge by gaining or losing electrons. For instance, when a glass rod is rubbed with silk, electrons are transferred from the glass to the silk. If the glass loses $N$ electrons, it will have $N$ more protons than electrons, so its total charge will be $Ne$. Similarly, the silk will have $N$ more electrons than protons, so its total charge will be $-Ne$. The total charge of the rod and the silk together is

$$Ne + (-Ne) = 0$$

the same as it was before they were rubbed together.

This is an example of the law of *conservation of charge*, which states that in any physical process, the total charge is not changed. This law is obviously true for a process that involves merely the transfer of electrons from one object to another. However, the law is much more general than this and is true even for processes, such as the decay of a nucleus, in which protons and electrons are created and destroyed (Sec. 20.2).

It is interesting to compare the gravitational and electric forces between two elementary particles. Table 16.1 gives the mass and charge of the electron, proton, and neutron. Using this table, we find from Eq. 16.1 that the gravitational force between a proton and an electron separated by a distance $r$ is

$$F_g = -G\frac{m_p m_e}{r^2}$$

$$= -\frac{(6.67 \times 10^{-11} \text{ N-m}^2/\text{kg}^2)(1.67 \times 10^{-27} \text{ kg})(9.11 \times 10^{-31} \text{ kg})}{r^2}$$

$$= -\frac{1.01 \times 10^{-67} \text{ N-m}^2}{r^2}$$

FIGURE 16.3
Electrical repulsion of two positive charges.

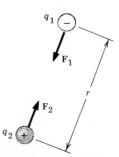

FIGURE 16.4
Electrical attraction of opposite charges.

TABLE 16.1 **Mass and charge of the elementary particles**

| Particle | Mass, kg | Charge, C |
|---|---|---|
| Proton | $1.673 \times 10^{-27}$ | $+1.602 \times 10^{-19}$ |
| Neutron | $1.675 \times 10^{-27}$ | 0 |
| Electron | $9.110 \times 10^{-31}$ | $-1.602 \times 10^{-19}$ |

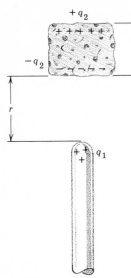

**FIGURE 16.5**
Induction of charge on a piece of cork. The cork is attracted to the inducing charge $q_1$ because the attraction between $q_1$ and the negative induced charge is greater than the repulsion between $q_1$ and the positive induced charge.

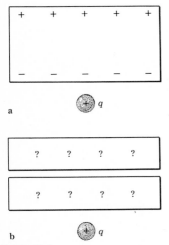

**FIGURE 16.6**
(*a*) Induction of charge on an object by an external charge $q$. (*b*) Object cut to try to isolate the induced charge.

From Eq. 16.2, the electric force between an electron and a proton is found to be

$$F_e = +K\frac{q_p q_e}{r^2}$$

$$= +\frac{(9.0 \times 10^9 \text{ N-m}^2/\text{C}^2)(1.60 \times 10^{-19} \text{ C})(-1.60 \times 10^{-19} \text{ C})}{r^2}$$

$$= -\frac{2.30 \times 10^{-28} \text{ N-m}^2}{r^2}$$

The ratio of these two forces is

$$\frac{F_e}{F_g} = \frac{2.30 \times 10^{-28}}{1.01 \times 10^{-67}} = 2.28 \times 10^{39}$$

Since the electric force is immensely larger than the gravitational force, the force of gravity can be completely neglected when calculating the force between elementary particles. However, the force between large aggregates of matter, such as astronomical bodies, is dominated by gravity because these objects tend to be electrically neutral.

### Induction

A glass rod which has been positively charged by rubbing with silk will attract a small piece of cork even though the cork is uncharged. The cork is composed of electrons and protons of opposite charge, which are normally distributed uniformly, so that the net charge is zero everywhere in the cork. When a charged object, such as the glass rod, is brought near the cork, the positive charge on the rod attracts the electrons in the cork and repels the protons, causing these particles to shift their positions slightly. The result is that negative charge accumulates on the side of the cork near the rod and positive charge accumulates on the other side. This process is called *induction*. According to the law of conservation of charge, the total charge of the cork must remain zero, since no charge has been added or removed from it. Thus there is as much negative charge induced on one side of the cork as there is positive charge induced on the other.

Suppose the glass rod has a charge $q_1$ and is a distance $r$ from the front side of the cork (Fig. 16.5). If the charge induced on the front side is $-q_2$, the rod attracts it with the force

$$F_e = +K\frac{q_1(-q_2)}{r^2} = -K\frac{q_1 q_2}{r^2}$$

At the same time, the rod repels the charge $+q_2$ induced on the back side. This charge is a distance $d$ farther from the rod, where $d$ is the thickness of the piece of cork. Therefore, the repulsive force on the positive induced charge is

$$F'_e = K\frac{q_1 q_2}{(r+d)^2}$$

which is less than $F_e$. The magnitude $F$ of the total force on the cork is the sum of $F_e$ and $F'_e$, or

$$F = F_e + F'_e = -K\frac{q_1 q_2}{r^2} + K\frac{q_1 q_2}{(r+d)^2}$$

$$= -Kq_1 q_2 \left[\frac{1}{r^2} - \frac{1}{(r+d)^2}\right]$$

$$= -Kq_1 q_2 \frac{2rd + d^2}{r^2(r+d)^2}$$

This shows that the glass rod exerts an attractive (negative) force on the cork.

**REMARK** If the rod were negatively charged, positive charge would be induced on the side of the cork near the rod and negative charge would be induced on the other side, so the cork would still be attracted by the rod. Thus any charged object, whether positively or negatively charged, exerts an attractive force on an uncharged object.

### Insulators and conductors

Substances differ in the relative freedom with which charge can move through them. A substance in which charge can move very freely is called a *conductor*, and a substance in which charge can move only slightly is called an *insulator* (or *dielectric*). Metals are all good conductors; glass, rubber, and cork are examples of good insulators.

The difference between insulators and conductors is illustrated by the following experiment. A charge $q$ is brought near an uncharged object, inducing charge on it (Fig. 16.6a). The object is then cut in half, as shown in Fig. 16.6b, and the charge $q$ is removed. Is each half of the object left with the charge that was induced on it? The answer is yes if the object is a conductor and no if the object is an insulator. To understand this, we must compare the atomic structures of insulators and conductors.

In an insulator, every electron is bound to an atom and is not free to move away. The atoms of an insulator are represented in Fig. 16.7a as neutral units composed of positive and negative charges superimposed on each other. A positive charge $q$ brought near an insulator attracts the negatively charged electrons in the atoms and repels the positively charged nuclei. This causes the negative and positive charge in each atom to separate slightly. The atom as a whole remains neutral, but one end becomes positively charged and the other becomes negatively charged. The negative end of each atom faces the inducing charge $q$, as shown in Fig. 16.7b. As a result, an excess of negative charge forms on the surface of the insulator facing $q$, and an excess of positive charge forms on the opposite surface. Of course, these are not free charges; they are merely the charged ends of neutral atoms. Figure 16.7c shows that if the insulator is cut in half, induced charge appears on the cut surfaces but the total charge of each half remains zero.

In a metal conductor, at least one electron is separated from each atom and is free to move anywhere in the conductor. Atoms from which electrons are missing are called *ions*. In a metal they are positively charged and remain in fixed positions. A metal thus consists of positively charged ions in fixed positions and negatively charged electrons free to move around. Normally the ions and electrons are uniformly distributed, so that the net charge is zero everywhere in

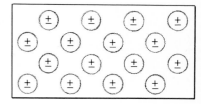

a

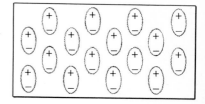

b  q

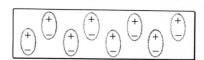

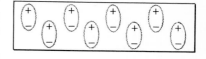

c  q

FIGURE 16.7
(*a*) An insulator. Each atom is a neutral object composed of positive and negative charges. (*b*) Induction of charge on an insulator. The external charge $q$ causes the positive and negative charge in each atom to separate slightly. (*c*) The charges induced on an insulator are not isolated when the insulator is cut in half.

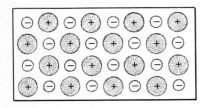

a

b

 $q$

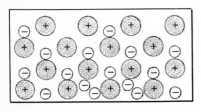

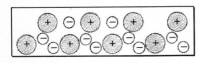

c     $q$

FIGURE 16.8
(*a*) A conductor. The positive charges are the ions, which have fixed positions, and the negative charges are the electrons, which are free to move around. (*b*) Induction of charge on an insulator. The external charge $q$ causes the free electrons to move toward one side of the insulator. (*c*) When the insulator is cut in half, an excess of electrons is trapped on one half and a deficit of electrons is trapped on the other half.

the metal (Fig. 16.8*a*). However, a charged object brought near the conductor changes this distribution. A positive charge $q$ attracts the electrons and repels the ions. Since the electrons are free to move, they accumulate on the side near $q$, which thus becomes negatively charged (Fig. 16.8*b*). The opposite side, being deficient in electrons, is positively charged. If the conductor is cut in half, an excess of electrons is trapped on the half facing $q$, leaving the other half with a shortage of electrons (Fig. 16.8*c*). Therefore each half is left charged, even after the inducing charge is removed.

To discharge these conductors it is necessary only to connect a conducting wire from one to the other. The excess electrons on the negatively charged conductor, attracted to the positively charged conductor, flow freely through the wire, until each conductor is again neutral.

### Induction machines

Figure 16.9 shows one of the many machines that use induction to produce a continuous separation of charge. Such machines are primarily used for physics demonstrations today, but in the nineteenth century they were used in electrical research. In the machine shown, metal rods are attached radially to a wheel, which is rotated by a crank in the direction indicated. The plate $P$ initially has a small positive charge on it, so that charges are induced on the rod in position 1. When this rod is rotated into position 2, it makes momentary contact with the wire $W$, which simultaneously is in contact with the rod in position 6. Negative charge flows from the rod in position 6, so that when contact is broken, this rod, now in position 7, is left positively charged. Likewise, the rod in position 3 is left negatively charged. These rods deposit their charge to the two conducting spheres by means of contacts at $A$ and $B$.

Some of the positive charge deposited at $A$ accumulates on plate $P$. As the charge on $P$ increases, the amount of charge induced on the rod in position 2 increases. This, in turn, increases the charge delivered back to $P$. Thus this machine uses positive feedback (Sec. 6.5) to rapidly build a large charge on the two conducting spheres. When this charge is sufficiently large, the air between the spheres suddenly becomes conducting and the electrons are able to flow from the negative to the positive sphere. This is seen as a spark jumping between the spheres.

### 16.3 THE ELECTRIC FIELD

Figure 16.10 shows an arbitrary configuration of charges ($q_1, q_2, \ldots, q_n$). The force $\mathbf{F}$ that this set of charges exerts on some other positive charge $q$ at point $P$ is the vector sum of the forces $\mathbf{F}_1, \mathbf{F}_2, \ldots, \mathbf{F}_n$ that each charge in the set exerts individually on $q$. This force depends jointly on the *test* charge $q$ and on the *source* charges $q_1, q_2, \ldots, q_n$ in the original configuration. It is useful to express $\mathbf{F}$ as the product of a factor that depends only on the test charge and a factor that depends only on the source charges. This is easily done because **F** is proportional to $q$, and so the ratio

$$\mathbf{E} = \frac{\mathbf{F}}{q} \qquad\qquad 16.3$$

is independent of $q$. The vector **E** is the *electric field* at point $P$ produced by the source charges. It is equal to the force that these charges would exert on a positive 1-C charge at $P$. The unit of **E** is newtons per coulomb (N/C). In terms of **E**, the force that the source charges exert on an arbitrary charge $q$ at point $P$ can be written

$$\mathbf{F} = q\mathbf{E} \qquad\qquad 16.4$$

Equation 16.4 expresses the vector **F** as the product of a number $q$ that depends only on the test charge and a vector **E** that depends only on the source charges. The product of a positive number $q$ and a vector **E** is a vector **F** that has the direction of **E** and the magnitude $qE$. If $q$ is negative, the direction of **F** is opposite the direction of **E**.

The electric field is an important concept because it allows us to think about the force that a configuration of charges would exert on a charge at a point even if no charge is actually there. Since the force that would be exerted on a test charge depends on the position of the point, the electric field varies from point to point. The effect of a charge configuration can be represented by drawing the electric field at various points. For example, Fig. 16.11 shows the electric field at various points produced by a positive point charge. The arrows representing the field are all directed away from the charge because the direction of the field is the direction of the force that would be exerted on a positive test charge. The lengths of the arrows decrease with distance from the charge because the force on a test charge decreases with distance in accordance with Coulomb's law (Eq. 16.2).

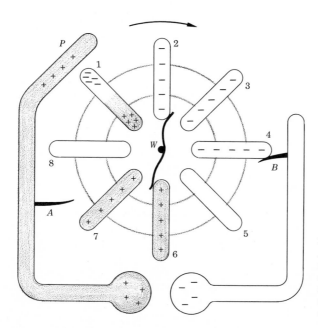

FIGURE 16.9
An induction machine. The rotation of the wheel in a clockwise direction causes positive charge to accumulate on the left and negative charge to accumulate on the right.

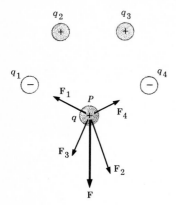

**FIGURE 16.10**
The force on a test charge $q$ due
to four source charges, $q_1$, $q_2$, $q_3$, $q_4$.

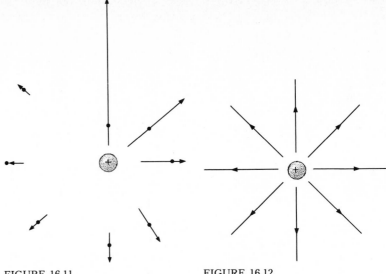

**FIGURE 16.11**
Electric field at various points
produced by a positive point
charge.

**FIGURE 16.12**
The lines of force of a positive
point charge.

Figure 16.12 shows another representation of the electric field of a positive point charge. The lines, called *lines of force*, are drawn parallel to the field at every point, giving a good picture of the direction of the electric field everywhere. Comparing Figs. 16.11 and 16.12, we see that the lines of force are closer together where $E$ is large and farther apart where $E$ is small. Thus, the spacing of the lines of force displays the relative magnitude of the field.

Figure 16.13 shows the lines of force for a negative point charge. They are identical to the lines of force of a positive point charge except that they point toward the charge. This is because a positive test charge is attracted toward the negative charge. Figure 16.14 shows the lines of force produced by two equal and opposite charges separated by a distance $d$. The lines of force all start on the positive charge and end on the negative charge. The electric field at any point $P$ is tangent to the line of force at $P$ and is equal to the vector sum of the electric fields due to each charge separately.

The charge configuration in Fig. 16.14 is called a *dipole*. Even though the total charge of the configuration is zero, it will exert an electric force on another charge. This is just the situation already encountered in discussing the force between a charged object and an uncharged piece of cork (Sec. 16.2). The charged object induced positive and negative charge on opposite sides of the cork. These induced charges form a dipole that exerts a force on the charged object.

The electric field produced by a dipole can be calculated at any point from Coulomb's law. The calculation is particularly simple for the special case of a point $P$ on the axis of the dipole (Fig. 16.15). Let $q_1$ and $q_2 = -q_1$ be the dipole charges, and suppose that there is a test charge $q$ on the axis of the dipole a distance $r$ from the

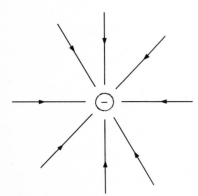

**FIGURE 16.13**
The lines of force of a negative
point charge.

dipole's center. The total force $\mathbf{F}$ on $q$ is the sum of the forces $\mathbf{F}_1$ and $\mathbf{F}_2$ due to $q_1$ and $q_2$. Since these forces are parallel, the magnitude of $\mathbf{F}$ is

$$F = F_1 + F_2$$

$$= K\frac{qq_1}{(r + \frac{1}{2}d)^2} + K\frac{qq_2}{(r - \frac{1}{2}d)^2}$$

$$= Kqq_1\left[\frac{1}{(r + \frac{1}{2}d)^2} - \frac{1}{(r - \frac{1}{2}d)^2}\right]$$

$$= Kqq_1\frac{-2rd}{(r + \frac{1}{2}d)^2(r - \frac{1}{2}d)^2}$$

When $r$ is very large compared to $d$, the terms $\frac{1}{2}d$ can be neglected in the denominator. The force then is given by

$$F = -Kqq_1\frac{2rd}{r^4} = -Kqq_1\frac{2d}{r^3}$$

FIGURE 16.14
The lines of force of a dipole. The electric field at any point $P$ is tangent to the line of force at that point.

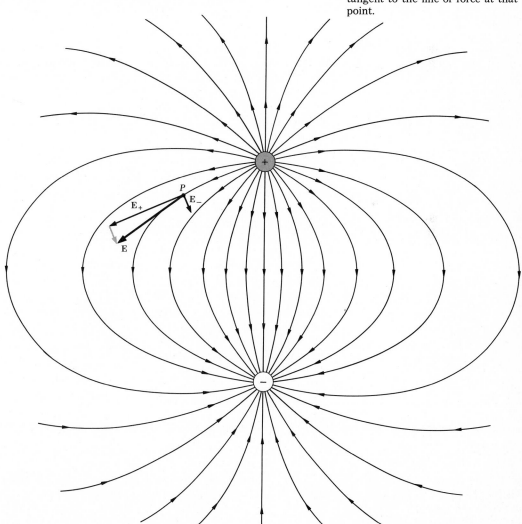

This shows that the force exerted by a dipole decreases inversely as the cube of the distance from the dipole, in contrast to the force exerted by a single charge, which decreases as the square of the distance. At large distances, the dipole force is much weaker than the force produced by either $q_1$ or $q_2$ alone because the individual forces cancel each other to a large extent (although not completely). The magnitude of the electric field produced by the dipole at $P$ is

$$E = \frac{F}{q} = -K\frac{2dq_1}{r^3}$$

## 16.4   ELECTRIC POTENTIAL

Consider an object of mass $m$ and charge $q$ in the presence of a fixed charge $q_1$ (Fig. 16.16). If both charges are positive, $q_1$ exerts a repulsive force

$$F = K\frac{qq_1}{r^2}$$

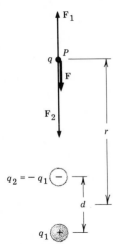

on the object. If the object is initially at rest at point $A$, it will accelerate outward along a line of force. At point $B$ it will have a speed $v_B$ and a kinetic energy $K_B = \frac{1}{2}mv_B{}^2$. According to the work-energy theorem (Sec. 5.2), the work $W_{AB}$ done on $q$ by the force $F$ in moving the object from $A$ to $B$ is equal to the change in its kinetic energy,

$$W_{AB} = K_B - K_A \qquad\qquad 16.5$$

Since the object starts from rest, $K_A$ is zero in this case.

The work $W_{AB}$ done by the electric force, like the work done by gravity (Sec. 5.4), can be written as the difference of the *potential energy* $U$ of the object at $A$ and $B$,

$$W_{AB} = U_A - U_B \qquad\qquad 16.6$$

That is, a number $U$ can be assigned to every point such that the work done by the electric force in moving the object between any two points is given by Eq. 16.6. Clearly, if the same number is added to the value of $U$ at every point, Eq. 16.6 will not be changed. This freedom in the definition of $U$ is used to set $U$ equal to zero at some convenient point in a given situation.

From Eqs. 16.5 and 16.6 we get the result

$$U_A - U_B = K_B - K_A$$

or
$$U_A + K_A = U_B + K_B \qquad\qquad 16.7$$

This says that the sum of the kinetic and potential energies of the object at $A$ is equal to the sum of the energies at $B$. Since $A$ and $B$ are arbitrary points, the sum of these energies is the same at all points; i.e., it is a constant.

It can be shown that the potential energy of a point charge $q$ due to a point charge $q_1$ is given by

$$U = K\frac{qq_1}{r} \qquad\qquad 16.8$$

**FIGURE 16.15**
Force on a test charge $q$ located on the axis of a dipole.

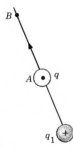

**FIGURE 16.16**
A test charge $q$ moving from $A$ to $B$ in the presence of a fixed charge $q_1$.

This is very similar to the gravitational potential energy (Eq. 5.18), except for a difference in sign, which just reflects the difference in sign of the force laws (Eqs. 16.1 and 16.2). If $q$ and $q_1$ have the same sign, they repel each other and $U$ is positive. If $q$ and $q_1$ have opposite sign, they attract each other and $U$ is negative. Since the gravitational force is always attractive, Eq. 5.18 must have a minus sign. The potential energy given by Eq. 16.8 is never zero, but it gets smaller and smaller as $r$ gets larger and larger. Thus we can say that the potential energy is zero when the charges are infinitely far apart.

As an example, consider an electron at a point $A$, a distance $r_A = 0.53 \times 10^{-10}$ m from a proton.† Using Table 16.1 and Eq. 16.8, we find the potential energy of the electron at $A$ to be

$$U_A = K \frac{q_e q_p}{r_A}$$

$$= \frac{(9.0 \times 10^9 \text{ N-m}^2/\text{C}^2)(-1.6 \times 10^{-19} \text{ C})(1.6 \times 10^{-19} \text{ C})}{5.3 \times 10^{-11} \text{ m}}$$

$$= -4.3 \times 10^{-18} \text{ N-m} = -4.3 \times 10^{-18} \text{ J}$$

The energy is negative because the electron is attracted to the proton. Equation 16.7 can be used to calculate the minimum speed $v_A$ that the electron must have at $A$ in order to escape completely from the proton. An electron with the minimum escape speed can move infinitely far from the proton, but its speed at infinity will be zero. Thus when the electron is infinitely far from the proton, both its potential energy $U_\infty$ and its kinetic energy $K_\infty$ are zero. Therefore, from Eq. 16.7 we have

$$K_A + U_A = K_\infty + U_\infty = 0$$

or

$$K_A = \tfrac{1}{2}mv_A{}^2 = -U_A = 4.3 \times 10^{-18} \text{ J}$$

where $m$ is the mass of the electron. Solving this for $v_A$, we get

$$v_A{}^2 = \frac{4.3 \times 10^{-18} \text{ J}}{\tfrac{1}{2}m}$$

$$= \frac{4.3 \times 10^{-18} \text{ J}}{0.5(9.1 \times 10^{-31} \text{ kg})}$$

$$= 9.4 \times 10^{12} \text{ J/kg}$$

so that

$$v_A = 3.1 \times 10^6 \text{ m/s}$$

**REMARK** This calculation should be compared to the calculation in Sec. 5.4 of the speed needed for an object to escape completely from the earth. The principles are the same in both calculations. However, the escape speed in the gravitational case does not depend on the mass of the object, whereas it does in the electrical case. Why?

The potential energy of a test charge $q$ due to a configuration of source charges $(q_1, q_2, \ldots, q_n)$ is just the sum of the potential energies due to each charge individually. The potential energy depends jointly on the test charge $q$ and on the source charges $q_1, q_2, \ldots, q_n$. It is useful to express the potential energy, like the electric force, as

† This is the average distance between an electron and a proton in a hydrogen atom (Sec. 19.2).

the product of a factor that depends only on the test charge and a factor that depends only on the source charges. Because $U$ is proportional to $q$, the ratio

$$V = \frac{U}{q} \qquad 16.9$$

is independent of $q$. The quantity $V$, called the *electric potential*, depends only on the source charges. The potential energy $U$ of a test charge is simply expressed in terms of the potential $V$,

$$U = qV \qquad 16.10$$

The unit of potential is joules per coulomb (J/C), which is called a *volt* (V):

$$1 \text{ V} = 1 \text{ J/C}$$

From Eqs. 16.8 and 16.9, the potential due to a point charge $q_1$ is found to be

$$V = \frac{U}{q} = K\frac{q_1}{r} \qquad 16.11$$

For example, the potential at a distance $r = 5.3 \times 10^{-11}$ m from a proton is

$$V = \frac{(9.0 \times 10^9 \text{ N-m}^2/\text{C}^2)(1.6 \times 10^{-19} \text{ C})}{5.3 \times 10^{-11} \text{ m}}$$

$$= 27.2 \text{ N-m/C} = 27.2 \text{ J/C} = 27.2 \text{ V}$$

The potential at all points of a charge configuration is graphically displayed by use of *equipotentials*, i.e., lines along which the potential is constant. Figure 16.17 shows the equipotentials (dashed lines) and the lines of force (solid lines) of a dipole (two equal and opposite charges). The electric field does no work on a test charge that is moved along an equipotential since the potential energy of the charge does not change. This means that the lines of force must be perpendicular to the equipotentials, as shown in Fig. 16.17, for if this were not the case, the electric force would have a component parallel to an equipotential and some work would be done on a test charge moved along it. The force on a positive test charge is always directed from a region of high potential to a region of low potential, perpendicular to the equipotentials. The force on a negative test charge is the reverse; it is directed from a region of low potential to a region of high potential.

When a metal conductor is placed in a static electric field, different parts of the conductor may be momentarily at different potentials. If this is the case, the negatively charged electrons, which are free to move in the metal, will flow from the regions of low potential to the regions of high potential. The electrons redistribute themselves until, in less than a millionth of a second, the potential is the same everywhere in the metal. Thus, when there is no flow of charge in it, a metal or other conducting object is an equipotential region. Furthermore, since any electric field in a conductor would cause the free electrons to move, the electric field in a conductor must be zero when there is no flow of charge.

A conductor that is in good contact with the earth is said to be *grounded*. The earth's surface is itself a moderately good conductor, so that the earth and all grounded conductors together form a single larger conductor, all at the same potential, namely, that of the earth. In practical applications, the potential of the earth is taken to be zero.

For example, the electric outlet in a house is generally at 120 V. This means that one side of the outlet is maintained at a potential of 120 V relative to the other side, which is grounded. When an appliance is plugged into the outlet, positive charge in effect flows from the high-potential side, through the appliance, and back to ground. The work done on the charge is

$$W = U_{120} - U_0$$
$$= q \times 120 \text{ V} - 0 = q \times 120 \text{ V}$$

For instance, the work done on a charge of 10 C is

$$W = 10 \text{ C} \times 120 \text{ V} = 1200 \text{ C-V} = 1200 \text{ J}$$

This work is converted into heat, light, or mechanical energy, depending on the appliance connected to the outlet.

FIGURE 16.17
Equipotentials and lines of force of a dipole. The lines of force are everywhere perpendicular to the equipotentials.

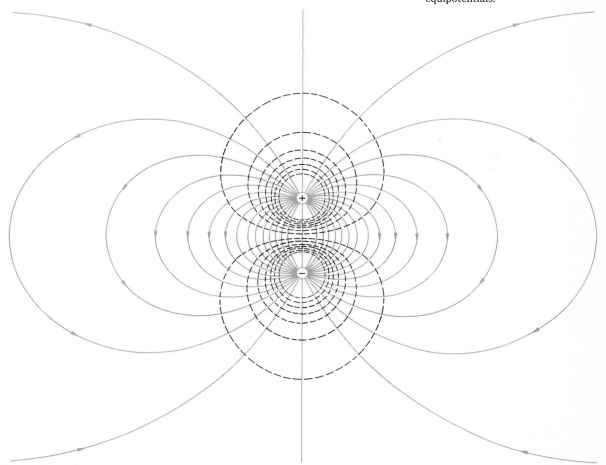

# 16.5   ELECTRON BEAMS

A number of important devices, such as oscilloscopes, television sets, x-ray machines, electron microscopes, and electronic vacuum tubes, use a beam of electrons accelerated by an electric field.  In all cases the beam is contained in an evacuated glass tube, as shown in Fig. 16.18.  Metal plates, or *electrodes*, are mounted inside the tube, and wires attached to the electrodes pass through the wall of the tube.  One electrode, called the *cathode*, is heated by a wire filament in which there is a current (Sec. 17.1).  When the temperature of the cathode is sufficiently high, some of its free electrons have enough energy to escape from the metal, just like molecules evaporating from a liquid (Sec. 9.1).  These evaporated electrons form an electron cloud about the cathode; when it becomes dense enough, the cloud prevents further evaporation.

If the other electrode, called the *anode*, is maintained at a high positive potential relative to the cathode, the electrons in the cloud will be attracted to it.  Because there is no air in the tube, these electrons move freely to the anode without colliding with air molecules.  As they move away from the electron cloud, they are replaced by more electrons evaporated from the cathode.

Let $V_A$ be the potential of the anode, and let $V_C$ be the potential of the cathode.  The potential energies of an electron of charge $q = -e$ at the anode and cathode are

$$U_A = -eV_A \qquad \text{and} \qquad U_C = -eV_C$$

The electric force is the only force on the electrons, so as an electron moves from the cathode to the anode, the sum of its kinetic and potential energies remains constant:

$$K_C + U_C = K_A + U_A$$

Solving this for the kinetic energy $K_A$ at the anode, we get

$$K_A = K_C + (U_C - U_A)$$
$$= K_C + e(V_A - V_C)$$

The kinetic energy $K_C$ of the electrons as they emerge from the cathode is nearly zero, so to a good approximation we have

$$K_A = \tfrac{1}{2}mv_A{}^2 = e(V_A - V_C) \qquad 16.12$$

or

$$v_A{}^2 = \frac{2e(V_A - V_C)}{m} \qquad 16.13$$

where $m$ is the mass of an electron.

For example, if the anode is at a potential of 5000 V relative to the cathode, the speed of the electron when it reaches the anode is

$$v_A{}^2 = \frac{2(1.60 \times 10^{-19} \text{ C})(5000 \text{ V})}{9.11 \times 10^{-31} \text{ kg}}$$

$$= 17.6 \times 10^{14} \text{ J/kg}$$

or

$$v_A = 4.2 \times 10^7 \text{ m/s}$$

**REMARK**   The electron speed calculated from Eq. 16.13 exceeds the speed of

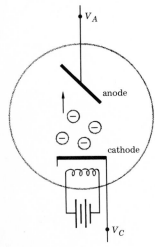

$V_A$

anode

cathode

$V_C$

**FIGURE 16.18**
A vacuum tube. Electrons that evaporate from the heated cathode are accelerated toward the anode, which is maintained at a positive potential with respect to the cathode.

light ($c = 3.0 \times 10^8$ m/s) when $V_A - V_C$ is greater than 256,000 V. But according to the theory of relativity (Sec. 14.6), an object cannot move faster than $c$. Therefore, Eq. 16.13 cannot be completely valid. It gives the correct speed when the speed is much less than $c$, but it becomes increasingly inaccurate for values of $v$ greater than $\frac{1}{2}c$. The error is caused by using the expression $\frac{1}{2}mv^2$ for the kinetic energy in Eq. 16.12. For speeds near the speed of light, the kinetic energy is given by the expression

$$K = mc^2 \left[ \frac{1}{\sqrt{1 - (v/c)^2}} - 1 \right] \qquad \text{16.14}$$

This expression is equal to $\frac{1}{2}mv^2$ when $v$ is very much smaller than $c$, but it differs greatly from $\frac{1}{2}mv^2$ when $v$ is close to $c$. (According to Eq. 16.14, the kinetic energy becomes larger and larger, as $v$ gets closer and closer to $c$. When $v = c$, the kinetic energy is infinite.) Fortunately, it will not be necessary to use Eq. 16.14, because it is the kinetic energy of the electron, rather than its speed, that is most important, and the relation between $K$ and $V$ given by Eq. 16.12 is valid at all energies.

### X-ray tube

In an x-ray tube the electrons collide with the anode, as shown in Fig. 16.18. The sudden deceleration of the electrons generates the x-rays, which are just electromagnetic waves of very short wavelength (Sec. 14.1). The wavelength of the x-rays produced becomes shorter as the potential difference $V_A - V_C$ becomes larger (Sec. 19.1). Because shorter-wavelength x-rays are more penetrating than longer-wavelength x-rays, x-ray machines with very large potential differences are now used.

With a potential difference of 8000 V, the kinetic energy of the electrons as they strike the anode is

$$K_A = e(V_A - V_C)$$
$$= 1.60 \times 10^{-19} \text{ C} \times 8000 \text{ V} = 12.8 \times 10^{-16} \text{ J}$$

It is usually more convenient to express the energy of an elementary particle in *electron volts* (eV). An electron volt is the energy gained by an electron in going through a potential difference of 1 V. That is, the electron volt is related to the joule by

$$1 \text{ eV} = e \times 1\text{V} = 1.60 \times 10^{-19} \text{ J}$$

and

$$1 \text{ J} = 6.25 \times 10^{18} \text{ eV}$$

For instance, the kinetic energy of the electron in electron volts is

$$K_A = (12.8 \times 10^{-16} \text{ J})(6.25 \times 10^{18} \text{ eV/J})$$
$$= 8000 \text{ eV}$$

Thus the energy in electron volts is equal to the potential difference.

### Cathode-ray tube

The cathode-ray tube (Fig. 16.19) is used both in oscilloscopes and television sets to provide an electrically controlled picture. Electrons evaporated from the cathode are accelerated to the anode, as in an x-ray tube. However, there is a hole in the anode of a cathode-ray tube through which some of the electrons pass. These electrons then go on to strike the inside surface of the flattened end of the tube. This surface is coated with a fluorescent material that produces a bright spot where the beam hits it.

The position of the spot is controlled by two pairs of deflecting plates, oriented at right angles to each other. As the electrons pass between a pair of plates, they are deflected toward the plate at the higher potential. The amount of deflection is controlled by varying the potential difference between the plates. One pair of plates produces horizontal deflection, and the other pair produces vertical deflection. The two pairs together can move the spot anywhere on the screen.

An oscilloscope is used to display a time-varying potential, e.g., that produced by impulses moving along the axon of a nerve cell. By means of electrodes implanted in the cell, the potential of interest is applied to the vertical plates of an oscilloscope, causing vertical deflections of the spot. To display the time variation of the pulses, the beam is simultaneously swept horizontally from left to right at constant speed. At the end of its sweep, it is rapidly returned to the left and swept across again. This sweep requires that the potential applied to the horizontal plates have the sawtooth waveform shown in Fig. 16.20. This sweep potential is provided by electronic circuits in the oscilloscope, and there are controls for varying the frequency of the sweep.

In operation, the sweep frequency is adjusted to equal the frequency at which the pulses arrive on the vertical plates. Then, with each horizontal sweep, a new pulse is displayed in the same position as the previous pulse, giving a stationary image of a single pulse. Figure 16.21 shows the action potential (Sec. 17.5) of the giant axon of a squid, as it appears on an oscilloscope screen.

### Television tube

A television picture tube is very similar to a cathode-ray tube, except that the beam is deflected magnetically rather than electrically. The picture is composed of 525 individual horizontal lines, and it is changed 30 times a second. This means that the beam is swept horizontally across the screen $525 \times 30 = 15{,}750$ times a second. As the beam is swept horizontally, it is also swept vertically, at a rate of 60 times a second. It takes two vertical sweeps to form one picture

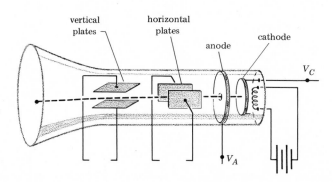

**FIGURE 16.19**
Cathode-ray tube. Some electrons pass through the hole in the anode and go on to strike the fluorescent screen. The position of the beam on the screen is controlled by the potentials applied to the deflection plates.

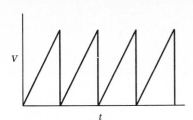

FIGURE 16.20
Time variation of the sweep
potential applied to the horizontal
deflection plates of an oscilloscope.

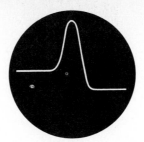

FIGURE 16.21
Action potential of the giant axon
of a squid as displayed on an
oscilloscope.

because only half the lines are displayed on each sweep.† The horizontal and vertical sweeps are produced by electronic circuits in the
set, but they are synchronized to the signal transmitted by the station. This signal also controls the intensity of the electron beam
and hence the brightness of the spot produced as the beam is swept
across the screen. It is the variation of the beam intensity as the
beam sweeps back and forth across the screen that produces a particular picture.

† The eye sees a single image because the eye retains the image of a line for about $\frac{1}{20}$ s
after the line vanishes (persistence of vision). Thus, the eye still sees the first line of
an individual picture when the last line is formed $\frac{1}{30}$ s later.

## PROBLEMS

*1* (*a*) What is the magnitude of the force exerted
on a $+7$-C charge by a $-3$-C charge 2 m
away? (*b*) Is the force attractive or repulsive?
*Ans.* (*a*) $4.72 \times 10^{10}$ N; (*b*) attractive

*2* Suppose that the charge on the metal spheres
of an induction machine (Fig. 16.9) are $+1.4 \times 10^{-8}$ and $-1.4 \times 10^{-8}$ C, respectively. What is
the force that one sphere exerts on the other
when the spheres are 5 cm apart?

*3* (*a*) What is the mass of a group of protons
with a total charge of 1 C? (*b*) What is the
total charge of 1 kg of protons?
*Ans.* (*a*) $1.04 \times 10^{-8}$ kg; (*b*) $0.96 \times 10^{8}$ C

*4* (*a*) Find the electric force between a kilogram
of protons and a kilogram of electrons separated by $6 \times 10^{6}$ m (the radius of the earth).
(*b*) What is the gravitational force between
these same objects?

*5* A glass rod rubbed with silk acquires a charge
of $+3 \times 10^{-10}$ C. How many electrons were
transferred from the glass to the silk?
*Ans.* $1.9 \times 10^{9}$

*6* Find the force on a charge of $5 \times 10^{-8}$ C exerted by a charge of $3 \times 10^{-9}$ C for the following values of $r$: 0.5, 1.0, 2.0, 2.5, and 3.0 m. Plot
the force against $r$, and connect the points
with a smooth curve.

*7* Find the force (magnitude and direction) that
the dipole in Fig. 16.22 exerts on a test charge
$q = +10^{-10}$ C at point $P$.　　*Ans.* $-0.176$ N

*8* Find the force (magnitude and direction) that
the dipole in Fig. 16.22 exerts on a test charge
$q = +10^{-10}$ C at point $Q$.

*9* Find the force (magnitude and direction) that
the dipole in Fig. 16.22 exerts on a test charge
$q = +10^{-10}$ C at point $R$. (Vector addition is
required.)　　*Ans.* 1.8 N

*10* Find the force (magnitude and direction) that
the dipole in Fig. 16.22 exerts on a test charge
$q = +10^{-10}$ C at point $S$.

*11* A charge $q_1$ exerts a force of 100 N on a test
charge $q_2 = 2 \times 10^{-5}$ C located at a point
0.20 m from $q_1$. (*a*) What is the electric field
due to $q_1$ at the point? (*b*) What is the magnitude of $q_1$?
*Ans.* (*a*) $5 \times 10^{6}$ N/C; (*b*) $2.2 \times 10^{-5}$ C

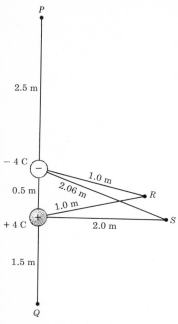

FIGURE 16.22   Problems 7 to 10, 21, and 22.

12 Find the magnitude of the electric field at 0.2, 0.5, and 0.8 m from a charge of $2 \times 10^{-10}$ C. Make a scale drawing similar to Fig. 16.11 by drawing arrows to represent the field at these points.

13 (*a*) What are the magnitude and direction of the total force on charge $q_3 = +5$ C exerted by charges $q_1$ and $q_2$ in Fig. 16.23? (*b*) What is the electric field at point *P* due to $q_1$ and $q_2$?
    *Ans.* (*a*) $1.06 \times 10^{11}$ N; (*b*) $2.12 \times 10^{10}$ N/C

14 (*a*) What are the magnitude and direction of the total force on charge $q_2 = 10$ C exerted by charges $q_1$ and $q_3$ in Fig. 16.23? (*b*) What is the electric field at point *Q* due to $q_1$ and $q_3$?

15 (*a*) What are the magnitude and direction of the total force on charge $q_3 = +3$ C exerted by charges $q_1$ and $q_2$ in Fig. 16.24? (Vector

FIGURE 16.23   Problems 13, 14, and 23.

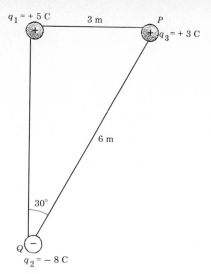

FIGURE 16.24   Problems 15, 16, and 24.

addition is required.) (*b*) What is the electric field at point *P* due to $q_1$ and $q_2$?
    *Ans.* (*a*) $1.3 \times 10^{10}$ N; (*b*) $4.3 \times 10^9$ N/C

16 (*a*) What are the magnitude and direction of the total force on charge $q_2 = -8$ C exerted by charges $q_1$ and $q_3$ in Fig. 16.24? (*b*) What is the electric field at point *Q* due to $q_1$ and $q_3$?

17 (*a*) What is the potential at a distance of 3 m from a charge $q_1 = 15 \times 10^{-6}$ C? (*b*) A charge $q = +3$ C is originally 3 m from $q_1$. How much work is done on $q$ by the electric field when $q$ is moved to a point 5 m from $q_1$?
    *Ans.* (*a*) $4.5 \times 10^4$ V; (*b*) $5.4 \times 10^4$ J

18 (*a*) What is the potential energy of an electron that is 20 cm from a charge of $6 \times 10^{-8}$ C? (*b*) How much work is required to move the electron very far from the charge?

19 The potential energy $U$ of a proton is $3 \times 10^{-18}$ J at a particular point. What is the electric potential $V$ at this point.
    *Ans.* 18.75 V

20 Show that the unit of electric field strength is equal to volts per meter.

21 (*a*) Find the potentials at points *P*, *Q*, *R*, and *S* in Fig. 16.22. (*b*) How much work is required to move a charge of $7.5 \times 10^{-7}$ C from point *P* to point *Q*?
    *Ans.* (*a*) $-2.4 \times 10^9$, $+6.0 \times 10^9$, 0, and $+0.52 \times 10^9$ V; (*b*) 6300 J

22 Find several points in Fig. 16.22 where the potential is $-4.0 \times 10^9$ V.

23 (a) What is the potential energy of the charge $q_3 = +5$ C in Fig. 16.23? (b) What is the potential at point $P$ due to $q_1$ and $q_2$?

        *Ans.* (a) $3.75 \times 10^{11}$ J; (b) $7.5 \times 10^{10}$ V

24 (a) What is the potential energy of the charge $q_3 = +3$ C in Fig. 16.24? (b) What is the potential at point $P$ due to $q_1$ and $q_2$?

25 The nucleus of uranium has a radius of $8 \times 10^{-15}$ m and contains 92 protons. (a) What is the potential at a point $P$ just outside the nucleus? (b) A proton originally at rest at $P$ is repelled by the positive charge of the nucleus. What is the speed of the proton when it is very far from the nucleus?

        *Ans.* (a) $1.65 \times 10^7$ V; (b) $5.62 \times 10^7$ m/s

26 The anode of an x-ray tube is maintained at a potential of 12,000 V with respect to the cathode. What is the speed of the electrons when they hit the anode?

# CHAPTER 17 current

A current is a flow of charge. As positive charge moves from a region of high potential to a region of low potential, its potential energy is transformed into other forms of energy. For example, in a heating coil the potential energy of the moving charge is transformed into heat, in a light bulb it is transformed into light (and heat), and in a motor it is transformed into mechanical energy (the kinetic energy of the rotor). By pulsing and modulating a current, it can be used for communication, as in telegraphy and television, and for control, as in computers. In fact, all electric and electronic devices use currents in one way or another.

Currents are also used by biological systems. The electric eel creates a large current for defense, and certain other fish navigate by means of the small currents they create in the water surrounding them. More important, currents are involved in the transportation of nerve impulses along a nerve fiber.

## 17.1 OHM'S LAW

Figure 17.1 shows two metal spheres with equal and opposite charge supported on insulated stands. When the spheres are connected by a conducting wire, electrons flow from the negatively charged sphere through the wire to the positively charged sphere. During this process there is a flow of charge, or *current*, in the wire. Specifically, the current $I$ in the wire is the charge per second that passes through the wire. The unit of current is coulombs per second (C/s), called an *ampere* (A):

$$1\,A = 1\,C/s$$

The current exists in the wire until the spheres are completely discharged, a period typically of about 1 microsecond ($1\,\mu s = 10^{-6}\,s$). Thus, if the initial charge on the spheres is $q = 10^{-8}\,C$ and the current persists for the time $t = 10^{-6}\,s$, the average current $I$ during this time is

$$I = \frac{q}{t} = \frac{10^{-8}\,C}{10^{-6}\,s}$$
$$= 10^{-2}\,C/s = 10^{-2}\,A$$

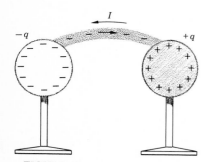

**FIGURE 17.1**
Two metal spheres with equal and opposite charge. Electrons flow from the negative sphere to the positive sphere when the spheres are connected by a wire.

The wire itself remains uncharged, even with a current in it, because as many electrons flow into it from the negative sphere as flow out from it to the positive sphere. There is no accumulation of charge in the wire.

**REMARK** By convention, the direction of a current is defined as the direction in which positive charge would flow. Since the moving charges in a metal happen to be negatively charged electrons, the conventional current in a wire is directed opposite to the real flow of charge. This causes no difficulty, however, because a flow of negative charge in one direction is equivalent to a flow of positive charge in the other direction. Thus we can avoid unnecessary minus signs by considering a current to consist of positive charge flowing in one direction, though in fact it consists of negative charge flowing in the opposite direction.

The current in Fig. 17.1 is directed from the positive (high-potential) sphere to the negative (low-potential) sphere and persists only until the potential difference between the spheres vanishes. However, a steady current can be maintained between the spheres if there is a mechanism for transporting the positive charge back to the positive sphere as rapidly as it flows to the negative sphere. Any device that does this is called a *seat of emf*.† For example, the induction machine shown in Fig. 16.9 (Sec. 16.2) is a seat of emf that uses mechanical energy to separate negative and positive charge. This mechanical energy is supplied by the work done in rotating the wheel and is transformed into the potential energy of the charges on the metal spheres. When the spheres are connected by a wire, charge flows through the wire and the potential energy is transformed into heat and other forms of energy. A steady current can be maintained in the wire by steady rotation of the wheel.

**REMARK** When the potential difference between the spheres becomes sufficiently large (20,000 V or more), the air between them becomes ionized and able to conduct a current. Such a current is visible as a spark. A steady spark can be maintained between the spheres by continuously rotating the wheel of the induction machine. This dramatically demonstrates the conversion from mechanical energy to potential energy and from potential energy to heat and light. A power-generating station similarly converts the mechanical energy of the generator's turbine into electric energy. The process, however, uses the principle of magnetic induction (Sec. 18.5) rather than electric induction.

A battery is a seat of emf that uses chemical energy to maintain a potential difference between its two terminals. When the terminals are not connected, a potential difference $\mathcal{E}$, called the emf of the battery, is maintained between them. The emf of a flashlight battery is 1.5 V. That is, the positive terminal is 1.5 V above the negative terminal. No energy is expended when there is no current, but when the terminals are connected by a conducting wire, charge flows through the wire to equalize the potential difference. The battery then expends chemical energy to separate the charges as rapidly as they combine. This can be seen from Fig. 17.2, which shows that inside the battery positive charge is moved toward the positive terminal. Chemical energy is required to move the charge against the repulsive electric force. (If the current is large, the potential difference $V$ between the terminals may be less than $\mathcal{E}$. However, we shall neglect this possibility here.)

The energy expended by the battery is released in the external circuit. Thus suppose a current $I$ is maintained in the circuit in Fig.

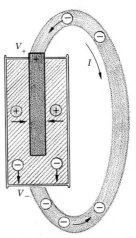

**FIGURE 17.2**
Inside a battery, chemical energy is used to move positive charge to the positive terminal and negative charge to the negative terminal.

---

† Emf is an abbreviation for *electromotive force*, but the term itself is inaccurate and is seldom used anymore.

17.2. In a time $t$, the amount of positive charge transferred from the positive to the negative terminal is

$$q = It \qquad 17.1$$

The work done on this charge in moving it through the wire is equal to the change in its potential energy. From Eqs. 16.6 and 16.10 we find

$$W = U_+ - U_- = qV_+ - qV_- = q(V_+ - V_-) = qV$$

where $V = V_+ - V_-$ is the potential difference between the terminals. The work done per second, or power $P$, is

$$P = \frac{W}{t} = \frac{qV}{t} = IV \qquad 17.2$$

For example, if $I = 0.2$ A and $V$ is equal to the emf of the battery, the power output is

$$P = I\mathcal{E} = 0.2 \text{ A} \times 1.5 \text{ V} = 0.3 \text{ A-V}$$

But 1 A = 1 C/s and 1 V = 1 J/C, so

$$1 \text{ A-V} = (1 \text{ C/s})(1 \text{ J/C}) = 1 \text{ J/s} = 1 \text{ W}$$

That is, the product of the potential difference in volts and the current in amperes is equal to the power in watts. Therefore, the power delivered to the external circuit is

$$P = 0.3 \text{ A-V} = 0.3 \text{ W}$$

This power is supplied at the expense of the internal chemical energy of the battery.

**REMARK**  The mks system has been adopted primarily because it gives a simple relation between the common electrical units (volts and amperes) and the mechanical units of energy and power (joules and watts).

The current $I$ in a wire depends on the potential difference $V$ between its ends. For most metals, $I$ is proportional to $V$, and so the relation between $I$ and $V$ can be written

$$I = \frac{V}{R} \qquad 17.3$$

or

$$V = RI \qquad 17.4$$

This is *Ohm's law*. The constant of proportionality $R$ is called the *resistance*, and its unit is volts per ampere (V/A), called the *ohm* ($\Omega$†):

$$1 \, \Omega = 1 \text{ V/A}$$

Ohm's law is true only for certain materials, notably metals. It is of great importance, nevertheless, because it applies to the materials commonly used in electric circuits.

The resistance of a conductor to a flow of charge is caused by the frequent collisions which the moving electrons make with the stationary atoms. When a potential difference is applied across a wire, an electric field is established which exerts a force on each electron

---

†$\Omega$ is the Greek letter capital omega.

in the wire. This force accelerates a free electron and increases its kinetic energy until it collides with a stationary atom. In the collision, the excess kinetic energy of the electron is transformed into vibrational energy of the atom. After the collision, the electron is again accelerated by the field until it again collides with an atom. In each collision, the electron loses the kinetic energy it acquired since the previous collision. Thus, as the electron moves through the wire in starts and stops, its electric energy is transformed into thermal energy.

**REMARK** Equation 17.3 is analogous to Eq. 7.16 for the flow of a fluid through a pipe. The fluid flow $Q$ is the analog of the current $I$, the pressure difference $p_2 - p_1$ is the analog of the potential difference $V$, and the fluid resistance of a pipe (Eq. 7.17) is the analog of the electric resistance of a wire. The electric resistance of a wire is related to the length $L$ and radius $r$ of the wire by

$$R = \frac{\rho L}{\pi r^2} \qquad 17.5a$$

where $\rho$ is a constant, called the *resistivity*, that is characteristic of the metal of which the wire is made. This equation is similar to Eq. 7.17 for fluid resistance, except that electric resistance is proportional to $1/r^2$, whereas fluid resistance is proportional to $1/r^4$. But, like a pipe, the resistance of a wire increases with increasing length and decreases with increasing radius.

The analogy between electric currents and fluid flow can be carried quite far. For instance, a pump is the fluid analog of a seat of emf. A pump uses an outside source of energy to create a pressure difference between its intake and output openings. When these openings are connected by a pipe, the fluid flows from the high-pressure side through the pipe to the low-pressure side. The pump expends energy by raising the pressure of the fluid that flows through it.

Figure 17.3 shows a battery connected to a light bulb. The resistance of the wires that make the connections is very small and can be neglected. The entire resistance $R$ of the circuit comes from the extremely thin filament in the bulb. This circuit is drawn schematically in Fig. 17.4. The symbol ⊣║⊢ represents a battery, the long vertical line being the positive (high-potential) terminal. The symbol –⌇⌇⌇– represents a *resistor*, i.e., a circuit element, such as the bulb, that has a finite resistance $R$. The solid lines represent resistanceless wires. The current $I$ is directed from the positive terminal to the negative terminal in the external circuit because this is the direction that positive charge would flow.

Equation 17.4 can be used to calculate the potential difference between various points in the circuit. The potential difference between points $a$ and $b$ is

$$V_b - V_a = 0$$

because the resistance between $a$ and $b$ is zero. Likewise, the potential difference between $c$ and $d$ is

$$V_d - V_c = 0$$

The potential difference between $b$ and $c$, however, is

$$V_c - V_b = RI$$

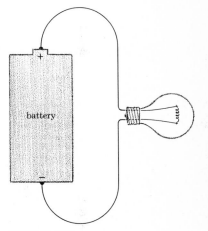

FIGURE 17.3
A battery connected to a light bulb.

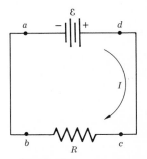

FIGURE 17.4
Schematic diagram of a circuit consisting of a battery connected to a resistor (light bulb).

If we add these equations together, we get

$$(V_b - V_a) + (V_d - V_c) + (V_c - V_b) = RI$$

or

$$V_d - V_a = RI$$

But $V_d - V_a$, the potential difference between $d$ and $a$, is the emf $\mathcal{E}$ of the battery. Therefore we have

$$\mathcal{E} = RI \qquad\qquad 17.5b$$

In other words, the potential difference across the resistor is equal to the emf of the battery.

According to Eq. 17.2, energy is dissipated in the resistor at the rate

$$P = VI \qquad\qquad 17.6a$$

If $V$ is replaced by $RI$ (Eq. 17.4), this can be written

$$P = RI^2 \qquad\qquad 17.6b$$

On the other hand, if $I$ is replaced by $V/R$ (Eq. 17.3), the power can also be written

$$P = \frac{V^2}{R} \qquad\qquad 17.6c$$

All three expressions for the power are useful. It is necessary to remember only one expression, however, because the other two are easily derived from it using Ohm's law.

As an example, consider a 1.5-V battery connected to a light bulb. The brightness of the bulb depends on the power dissipation. Which will be brighter, a 2-$\Omega$ bulb or a 5-$\Omega$ bulb? Using Eq. 17.6$b$, you might say that more power is dissipated in the bulb with the larger resistance. This is not correct, however, because the current is smaller with the larger resistance. From Eq. 17.3 we see that the current in the 5-$\Omega$ bulb is

$$I = \frac{V}{R} = \frac{1.5 \text{ V}}{5 \ \Omega} = 0.3 \text{ A}$$

and so, by Eq. 17.6$b$, the power is

$$P = RI^2 = 5 \ \Omega \times (0.3 \text{ A})^2 = 0.45 \text{ W}$$

The current in the 2-$\Omega$ bulb is

$$I = \frac{1.5 \text{ V}}{2 \ \Omega} = 0.75 \text{ A}$$

and the power is

$$P = 2 \ \Omega \times (0.75 \text{ A})^2 = 1.125 \text{ W}$$

Since the potential is the same in both cases, the same results are obtained more directly using Eq. 17.6$c$. For the 5-$\Omega$ bulb we have

$$P = \frac{V^2}{R} = \frac{(1.5 \text{ V})^2}{5 \ \Omega} = 0.45 \text{ W}$$

and for the 2-$\Omega$ bulb we have

$$P = \frac{(1.5 \text{ V})^2}{2 \ \Omega} = 1.125 \text{ W}$$

Circuits often consist of a network of interconnected resistors, like the one shown in Fig. 17.5. The basic problem of circuit theory is to find the current in each branch of the network, given the values of the resistors. The analysis of this or any other network uses only two principles, known as *Kirchhoff's laws.*

**Kirchhoff's first law**  *The total current entering any point of the circuit is equal to the total current leaving the point.* This is a consequence of the fact that no charge accumulates at a point in a circuit, so that as much charge must flow out from a point as flows into it.

**Kirchhoff's second law**  *The potential difference between any two points in a circuit is the same along any path connecting the points.* This means, for instance, that the potential difference between the points $a$ and $e$ in Fig. 17.6 is the same along the path $abcde$ as it is along the path through the battery.

These principles can be used to find the current in the circuit in Fig. 17.6, which has two resistors connected in *series*. From Kirchhoff's first law, the current $I$ that enters point $c$ from $R_1$ is equal to the current that leaves point $c$ and passes through $R_2$. In fact the current $I$ is the same everywhere in the circuit because there is only one path through every point. The potential difference between points $a$ and $e$ along the path $abcde$ can be calculated from Ohm's law (Eq. 17.4). The potential difference between points $a$ and $b$ is

$$V_b - V_a = 0$$

because the resistance between $a$ and $b$ is zero. Likewise, the potential difference between $d$ and $e$ is

$$V_e - V_d = 0$$

The potential across resistor $R_2$ is

$$V_c - V_b = R_2 I$$

and the potential across $R_1$ is

$$V_d - V_c = R_1 I$$

If we add these four equations together, we get

$$(V_b - V_a) + (V_e - V_d) + (V_c - V_b) + (V_d - V_c) = R_1 I + R_2 I$$

or
$$V_e - V_a = (R_1 + R_2)I$$

which is the potential difference between $e$ and $a$ along the path $abcde$. From Kirchhoff's second law, this is equal to the potential difference between $e$ and $a$ along the path through the battery, which is the emf of the battery. Therefore we have

$$\varepsilon = (R_1 + R_2)I$$

and so the current in the circuit is

$$I = \frac{\varepsilon}{R_1 + R_2}$$

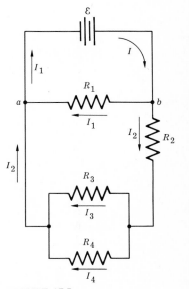

FIGURE 17.5
A complex circuit network.

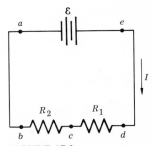

FIGURE 17.6
Two resistors connected in series to a battery.

Comparing this to Eq. 17.5, we see that two resistors $R_1$ and $R_2$ connected in series act as a single resistor of magnitude

$$R = R_1 + R_2 \quad \text{series} \quad\quad 17.7$$

For example, consider a battery with $\mathcal{E} = 1.5$ V connected in series to two bulbs with resistances $R_1 = 2\,\Omega$ and $R_2 = 5\,\Omega$. What is the current in the circuit and the power dissipation in each bulb? From Eq. 17.7 the total resistance of the circuit is

$$R = R_1 + R_2 = 2\,\Omega + 5\,\Omega = 7\,\Omega$$

From Eq. 17.5 the current is

$$I = \frac{\mathcal{E}}{R} = \frac{1.5\text{ V}}{7\,\Omega} = 0.214\text{ A}$$

The potential across $R_1$ is

$$V_1 = R_1 I = 2\,\Omega \times 0.214\text{ A} = 0.43\text{ V}$$

and the potential across $R_2$ is

$$V_2 = R_2 I = 5\,\Omega \times 0.214\text{ A} = 1.07\text{ V}$$

The sum of the potentials across the two resistors is equal to the emf of the battery. The power dissipation in the resistors is found using Eq. 17.6b:

$$P_1 = R_1 I^2 = (2\,\Omega)(0.214\text{ A})^2 = 0.092\text{ W}$$
$$P_2 = R_2 I^2 = (5\,\Omega)(0.214\text{ A})^2 = 0.229\text{ W}$$

In this example, more power is dissipated in the larger resistor because the current is the same in both resistors. In the example in Sec. 17.1, more power was dissipated in the smaller resistor because the potential across the resistors was the same.

Figure 17.7 shows two resistors connected in *parallel* to a battery. Compare this with the series circuit in Fig. 17.6. In the series circuit all the current passes through both $R_1$ and $R_2$, while in the parallel circuit the current divides, some passing through $R_1$ and some through $R_2$. Since the parallel connection provides two paths between points $a$ and $b$, the total resistance between these points is less than it would be if only one resistor were in the circuit. Let $I$ be the current through the battery, and let $I_1$ and $I_2$ be the currents in $R_1$ and $R_2$, respectively. The current entering point $b$ is $I$, and the currents leaving point $b$ are $I_1$ and $I_2$. Therefore, from Kirchhoff's first law we have

$$I = I_1 + I_2 \quad\quad 17.8$$

The potentials across $R_1$ and $R_2$ are equal because each resistor is a path between points $a$ and $b$. This potential difference is also equal to the emf of the battery because the battery is another path between these points. Consequently we have

$$V_b - V_a = R_1 I_1 = R_2 I_2 = \mathcal{E}$$

and so the currents in the resistors are

$$I_1 = \frac{\mathcal{E}}{R_1} \quad\text{and}\quad I_2 = \frac{\mathcal{E}}{R_2}$$

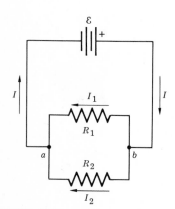

FIGURE 17.7
Two resistors connected in parallel to a battery.

$$I = I_1 + I_2 = \frac{\mathcal{E}}{R_1} + \frac{\mathcal{E}}{R_2}$$

$$= \mathcal{E}\left(\frac{1}{R_1} + \frac{1}{R_2}\right)$$

$$= \frac{\mathcal{E}}{R}$$

That is, two resistors connected in parallel act as a single resistor $R$, given by

$$\frac{1}{R} = \frac{1}{R_1} + \frac{1}{R_2} \qquad \text{parallel} \qquad\qquad 17.9$$

The magnitude of $R$ is always less than either $R_1$ or $R_2$.

For example, suppose two bulbs with resistances $R_1 = 2\,\Omega$ and $R_2 = 5\,\Omega$ are connected in parallel to a 1.5-V battery. The total resistance of the circuit is given by

$$\frac{1}{R} = \frac{1}{2\,\Omega} + \frac{1}{5\,\Omega} = 0.7\,\Omega^{-1}$$

or

$$R = \frac{1}{0.7\,\Omega^{-1}} = 1.43\,\Omega$$

and so the current through the battery is

$$I = \frac{\mathcal{E}}{R} = \frac{1.5\text{ V}}{1.43\,\Omega} = 1.05\text{ A}$$

This is to be compared to the value of 0.214 A that was obtained when the resistors were connected in series. The currents in the individual resistors are

$$I_1 = \frac{\mathcal{E}}{R_1} = \frac{1.5\text{ V}}{2\,\Omega} = 0.75\text{ A}$$

and

$$I_2 = \frac{\mathcal{E}}{R_2} = \frac{1.5\text{ V}}{5\,\Omega} = 0.30\text{ A}$$

The total current of 1.05 A is divided between the two resistors, the smaller resistor having the greater current.

**REMARK**   The rules for adding resistors in series and parallel also apply to fluid resistance. The circulatory system of the body is a complex network of vessels in series and parallel connection, and the methods used to analyze electric circuits are applicable to problems of fluid flow in the body. For example, Fig. 17.8 shows two blood vessels with resistances $R_1$ and $R_2$ connected in parallel. Let $p = p_b - p_a$ be the pressure difference between points $a$ and $b$. The fluid flow $Q$ through the two vessels is given by Eq. 7.17,

$$Q = \frac{p}{R}$$

where

$$\frac{1}{R} = \frac{1}{R_1} + \frac{1}{R_2} = \frac{R_1 + R_2}{R_1 R_2}$$

Thus we can write

$$Q = p\,\frac{R_1 + R_2}{R_1 R_2} \qquad\qquad 17.10a$$

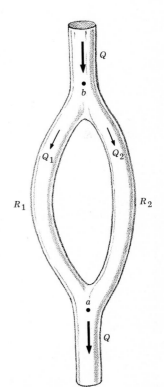

FIGURE 17.8
Two blood vessels connected in parallel.

Suppose vessel 2 becomes clogged, so that no blood flows through it. (This means that $R_2$ is infinitely large or that $1/R_2$ is zero.) All the blood will then flow through vessel 1, so that

$$Q' = \frac{p'}{R_1} \qquad\qquad 17.10b$$

To maintain the same fluid flow, the pressure $p'$ across the vessel must increase. That is, the body tries to make $Q' = Q$ by increasing the pressure. The required increase in pressure is found by setting $Q'$ equal to $Q$ in Eq. 17.10b and solving for $p'$:

$$p' = R_1 Q$$

Substituting the value of $Q$ given by Eq. 17.10a, we get

$$p' = p\frac{R_1 + R_2}{R_2}$$

For instance, if $R_2$ originally was equal to $R_1$, the pressure $p'$ must equal $2p$ to maintain the same fluid flow once vessel 2 becomes clogged. What is the equivalent of the seat of emf of the circulatory system?

The current in all parts of a complex network is found by repeated use of Eqs. 17.7 and 17.9. For example, let the resistances in Fig. 17.5 be

$$R_1 = 15\,\Omega \qquad R_2 = 8\,\Omega \qquad R_3 = 3\,\Omega \qquad R_4 = 6\,\Omega$$

The resistors $R_3$ and $R_4$ are connected in parallel, so they are equivalent to a single resistance $R'$ given by

$$\frac{1}{R'} = \frac{1}{3\,\Omega} + \frac{1}{6\,\Omega}$$

or $$R' = 2\,\Omega$$

The resistance $R'$ is in series with $R_2$, so the effective resistance of $R'$ and $R_2$ together is

$$R'' = R_2 + R' = 8\,\Omega + 2\,\Omega = 10\,\Omega$$

The circuit is redrawn in Fig. 17.9, with $R_2$, $R_3$, and $R_4$ replaced by $R''$. This shows that $R_1$ and $R''$ are in parallel, and so the total resistance $R$ of the entire circuit is

$$\frac{1}{R} = \frac{1}{R_1} + \frac{1}{R''} = \frac{1}{15\,\Omega} + \frac{1}{10\,\Omega}$$

or $$R = 6\,\Omega$$

If the emf of the battery is 3.0 V, the current $I$ is

$$I = \frac{\mathcal{E}}{R} = 0.5\ \text{A}$$

The potential across $R_1$ and $R''$ is also 3.0 V, so the currents $I_1$ and $I_2$ are

$$I_1 = \frac{3.0\ \text{V}}{R_1} = \frac{3.0\ \text{V}}{15\,\Omega} = 0.2\ \text{A}$$

and $$I_2 = \frac{3.0\ \text{V}}{R''} = \frac{3.0\ \text{V}}{10\,\Omega} = 0.3\ \text{A}$$

Note that $I_1 + I_2 = I$, in accordance with Kirchhoff's first law.

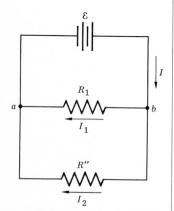

**FIGURE 17.9**
The circuit in Fig. 17.5 redrawn with $R_2$, $R_3$, and $R_4$ replaced by the equivalent resistor $R''$.

The potential difference between points $a$ and $b$ can be written (see Fig. 17.5)

$$V_b - V_a = 3.0 \text{ V} = I_2 R_2 + I_3 R_3$$
$$= 0.3 \text{ A} \times 8 \, \Omega + I_3 \times 3 \, \Omega$$

Solving this for $I_3$, we get

$$I_3 \times 3 \, \Omega = 3.0 \text{ V} - 2.4 \text{ V} = 0.6 \text{ V}$$

so

$$I_3 = \frac{0.6 \text{ V}}{3 \, \Omega} = 0.2 \text{ A}$$

Finally, from the condition that $I_2 = I_3 + I_4$, the current in $R_4$ is found to be

$$I_4 = I_2 - I_3 = 0.3 \text{ A} - 0.2 \text{ A} = 0.1 \text{ A}$$

This shows how the currents in all the resistors of a circuit can be found by a systematic analysis of the circuit.

## 17.3 ALTERNATING CURRENT

A battery maintains a constant potential difference between its terminals. When these terminals are connected to a resistor, a steady current is produced, directed from the high-potential terminal through the resistor to the low-potential terminal. Because the direction of the current is constant, it is called a *direct current* (dc).

A power generator produces an oscillating potential across its terminals, as shown in Fig. 17.10. The potential goes through one complete cycle of oscillation in the time $\tau$, called the *period*. The potential curve has the shape of a sine wave (Sec. 12.4) and is represented by the equation

$$V = V_p \sin\left(360° \frac{t}{\tau}\right) \qquad 17.11$$

where $V_p$ is the peak potential. The current $I$ in a resistor placed across the terminals of a generator is given by Ohm's law

$$I = \frac{V}{R} = I_p \sin\left(360° \frac{t}{\tau}\right) \qquad 17.12$$

where

$$I_p = \frac{V_p}{R} \qquad 17.13$$

Because the current oscillates with time, it is called *alternating current* (ac).

In the United States and Canada, the period of all commercial alternating current is $\tau = \frac{1}{60}$ s. The *frequency f*, which is defined as $1/\tau$, is

$$f = \frac{1}{\tau} = 60 \text{ s}^{-1}$$

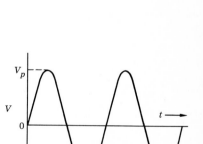

FIGURE 17.10
Time variation of the potential across the terminals of an ac generator.

The unit of frequency is the hertz, so

$$f = 60 \text{ s}^{-1} = 60 \text{ Hz}$$

In Europe the frequency of the alternating current is 50 Hz.

If a 63-$\Omega$ resistor is placed across the terminals of an ac generator whose peak potential is 170 V, the current in the resistor is given by Eqs. 17.12 and 17.13, with

$$I_p = \frac{V_p}{R} = \frac{170 \text{ V}}{63 \, \Omega} = 2.7 \text{ A}$$

This current is plotted in Fig. 17.11, assuming $f = 60$ Hz. The current alternates in sign, as does the potential. When the current is positive, charge flows in one direction, and when the current is negative, charge flows in the other direction. Thus, there is no net flow of charge around the circuit; instead the charge surges back and forth in the resistor.

The power dissipation in a resistor is still given by Eqs. 17.6. Using Eqs. 17.6b and 17.12, we find

$$P = RI^2 = R[I_p \sin (360° \, ft)]^2$$
$$= RI_p^2 \sin^2 (360° \, ft)$$

This is plotted in Fig. 17.12. Thus, although the current oscillates in sign, the power, which is proportional to the square of the current, is always positive. The power varies between zero and $RI_p^2$ twice each cycle, or 120 times a second, but for most purposes only the average power $\bar{P}$ is of interest. It can be shown that

$$\bar{P} = R\bar{I^2} = \tfrac{1}{2}RI_p^2$$

where $\bar{I^2}$ is the average of the square of the current.

### Rms current and potential

It is useful to define a direct current $I_{rms}$ that gives the same power dissipation as the alternating current. That is, $I_{rms}$ is defined by the condition†

$$RI_{rms}^2 = \bar{P} = R\bar{I^2} = \tfrac{1}{2}RI_p^2$$

so

$$I_{rms}^2 = \bar{I^2} = \tfrac{1}{2}I_p^2$$

and

$$I_{rms} = \sqrt{\bar{I^2}} = \frac{I_p}{\sqrt{2}} = \frac{I_p}{1.41}$$

The direct or rms potential $V_{rms}$ is defined in terms of $I_{rms}$ by Ohm's law:

$$V_{rms} = RI_{rms}$$

But $I_{rms} = I_p/\sqrt{2}$, so

$$V_{rms} = \frac{RI_p}{\sqrt{2}} = \frac{V_p}{\sqrt{2}}$$

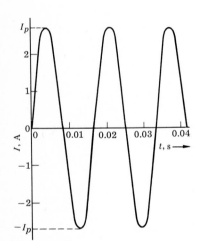

**FIGURE 17.11**
Time variation of a 60-Hz alternating current in a resistor.

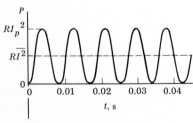

**FIGURE 17.12**
Time variation of the power dissipation in a resistor due to a 60-Hz alternating current.

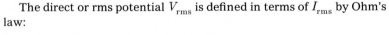

† The subscript rms stands for *root mean square*. The current $I_{rms}$ is the square *root* of the average (*mean*) of the *square* of the current $I$: $I_{rms} = \sqrt{\bar{I^2}}$.

since $RI_p = V_p$. Thus an alternating current can be treated as a direct current with potential

$$V_{rms} = \frac{V_p}{\sqrt{2}} \qquad\qquad 17.14a$$

and current

$$I_{rms} = \frac{I_p}{\sqrt{2}} \qquad\qquad 17.14b$$

and so the methods of Sec. 17.2 can be used to analyze ac circuits. The significance of $I_{rms}$ and $V_{rms}$ is that they can be used in Eqs. 17.6 to find the average power.

In the United States and Canada, the rms potential across an electric outlet is $V_{rms} = 120$ V. The peak potential across the outlet is

$$V_p = \sqrt{2} \times 120 \text{ V} = 170 \text{ V}$$

In Europe, the rms potential is 240 V, which corresponds to a peak potential of 340 V.

For example, suppose an American toaster is rated at 1320 W. What is the resistance of its heating element? The power rating gives the average power dissipation at 120 V. Therefore, using Eq. 17.6c, we find

$$R = \frac{V_{rms}^2}{\bar{P}}$$

$$= \frac{14,400 \text{ V}^2}{1320 \text{ W}} = 10.9 \text{ }\Omega$$

If this toaster were plugged into an outlet in England, the power dissipation would be

$$\bar{P} = \frac{(240 \text{ V})^2}{10.9 \text{ }\Omega} = 5280 \text{ W}$$

At this power, the heating element in the toaster would melt. In general, an appliance designed to operate at one potential will not operate properly at a significantly different potential.

### Household wiring

Electric power is distributed from the generator to individual houses by a pair of power lines. Each house is connected in parallel to these lines, and the power company keeps a constant rms potential between the lines. This is diagrammed in Fig. 17.13, where the symbol ─◯─ designates an ac generator and the symbol ─╫╷ designates a connection to ground. The grounded power line is connected to the earth by means of wires buried in the ground, which keeps the line at zero potential. The potential of the other ("live") line oscillates relative to the grounded line.

A fuse $R_f$ and a meter $M$ are placed in series with the line entering a house, but all other appliances are connected in parallel. Line $a$ shows the connection to a wall outlet. The full 120-V potential is

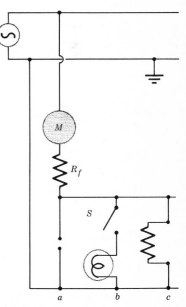

FIGURE 17.13
Electric connections in a house.

maintained between the terminals of the outlet, but when no appliance is connected to the outlet, there is no current in this part of the circuit. Line $b$ shows the connection to a ceiling light controlled by a wall switch $S$. Line $c$ shows another wall outlet with an appliance connected to it.

As more appliances are connected in parallel to the house line, the total resistance across the line decreases and the total current $I_{rms}$ entering the house increases. According to Eq. 17.6a, the total power to the house is

$$\bar{P} = V_{rms} I_{rms}$$

For instance, if $I_{rms} = 35$ A, the power is

$$\bar{P} = 120 \text{ V} \times 35 \text{ A} = 4200 \text{ W} = 4.2 \text{ kW}$$

where 1 kW = $10^3$ W. If the power of 4.2 kW is maintained for 5 h, the energy consumed is

$$E = 4.2 \text{ kW} \times 5 \text{ h} = 21 \text{ kWh}$$

where the *kilowatthour* (kWh) is the unit of energy used by the power industry. It is related to the joule by

$$1 \text{ kWh} = 10^3 \text{ W} \times 3600 \text{ s} = 3.6 \times 10^6 \text{ J}$$

In 1970 the total generating capacity of the United States was $3 \times 10^5$ MW (1 megawatt = 1 MW = $10^6$ W), and a total of $1.5 \times 10^{12}$ kWh of electric energy was produced. It is energy that the power industry sells to the public; i.e., the power industry charges for kilowatthours not kilowatts.

The meter $M$ in Fig. 17.13 is a small rotor that turns at a speed proportional to the current and records, for billing purposes, the total energy consumed. The fuse $R_f$ consists of a special metal resistor that melts when the current exceeds a certain value, thus breaking the connection between the house and the power line. An excessively large current occurs either when too many appliances are connected to the line (*overloaded circuit*) or when a very low resistance is accidentally placed across the terminals of an outlet (*short circuit*). A short circuit reduces the total resistance of the house circuit to nearly zero, resulting in an extremely large current. If the fuse did not immediately break the circuit, the wires in the walls could become hot enough to start a fire.

### Electrical safety

Since one side of an electric outlet is grounded, a circuit is completed when the other (live) side is connected directly to the ground. This happens, for instance, if the live side is connected to a water pipe or sewer pipe in good electric contact with the earth. Thus a person can be electrocuted by simultaneously touching a water pipe and the live side of an electric outlet. He may also get a shock by just touching the live outlet if he is standing outside on wet ground or on a damp basement floor, because water is a fairly good conductor.

If a person is in good contact with the ground when he touches the electric outlet, the only resistance in the circuit is that of his body. This varies from 500 $\Omega$ when the body is wet to 1500 $\Omega$ when the body is dry. Taking an average value of 1000 Ω, we see that with good contact to a 120-V outlet, the current in the body is

$$I_{\text{rms}} = \frac{120 \text{ V}}{1000 \text{ }\Omega} = 0.12 \text{ A} = 120 \text{ mA}$$

where 1 milliampere = 1 mA = $10^{-3}$ A.

A 60-Hz current of 1.0 mA is perceptible as a slight tingle by 75 percent of the population. A larger current begins to disrupt the body's own electrical system, interfering with the transmission of nerve impulses. A current of 10 mA is large enough to paralyze the hand holding the wire, preventing the person from letting go. A current of 120 mA (at 60 Hz), if it persists for more than a few seconds, is sufficient to cause ventricular fibrillation (i.e., a twitching of part of the heart muscle that prevents coordinated contractions of the heart).

A person can be electrocuted by touching an electric appliance in which a live wire has accidently become loose because there is now a complete circuit from the live side of the outlet through the appliance and the person's body back to ground. To prevent this, all modern appliances have a third wire that connects the body of the appliance to the water pipes of the house. If a wire becomes loose, there is an immediate short circuit through the third wire and the fuse will break the circuit.

## 17.4   ELECTRONICS

A resistor is only one of a number of elements, both active and passive, that are used in electric circuits. Active elements, such as transistors and integrated circuits, are of special importance because their behavior in one circuit can be controlled by another circuit. Electronic circuits, consisting of active and passive elements in various combinations, have a variety of functions, e.g., amplifying signals and performing logical operations. Amplifying circuits are used in radios, televisions, and high-fidelity sound equipment; logical circuits are used in calculators, computers, and control devices. This section introduces the most important circuit elements and describes some electronic circuits in which they are used.

### Capacitors

A *capacitor* is a passive circuit element that consists of two conducting surfaces separated by a thin insulating sheet (Fig. 17.14). Wires attached to the surfaces permit the capacitor to be connected in an electric circuit. Figure 17.15 is a diagram of a circuit in which a capacitor, symbolized by ——│ │——, is connected in series to a resistor $R$ and a battery $\mathcal{E}$. Because of the insulation between the plates of the capacitor, charge cannot flow through the capacitor, and so there can be no direct current through a capacitor. However, when the switch $S$ is closed, there will be a *transient current* through the resistor, as electrons flow from one plate of the capacitor to the other (Fig. 17.16). Consequently, positive charge $q$ will accumulate on one

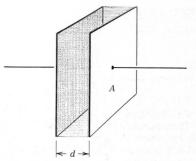

FIGURE 17.14
A capacitor.

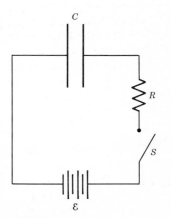

FIGURE 17.15
A capacitor connected in series to a battery and a resistor.

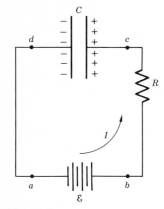

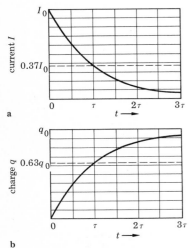

FIGURE 17.16
The transient current in a capacitor circuit exists until the potential across the plates of the capacitor is equal to the emf of the battery.

**a**

**b**

FIGURE 17.17
(*a*) Time variation of the transient current in a resistor-capacitor circuit. (*b*) Time variation of the charge on either plate of a capacitor in a resistor-capacitor circuit. The unit of time on the horizontal axis is $\tau = RC$, the characteristic time of the circuit.

plate, while an equal amount of negative charge $q$ accumulates on the other, until the potential $V = V_c - V_d$ across the capacitor is equal to the emf of the battery. The transient current $I$ is large at the instant the switch is closed, but it rapidly decreases to zero as the capacitor becomes charged (Fig. 17.17*a*). At the same time, the magnitude of the charge on each plate of the capacitor increases from zero to its final value $q_0$, as shown in Fig. 17.17*b*.

At any instant the charge $q$ on the capacitor is proportional to the potential $V$ across it. This can be written

$$q = CV \qquad 17.15$$

where $C$ is a constant, called the *capacitance*. The unit of capacitance is coulombs per volt (C/V), called a *farad* (F).

**REMARK** The magnitude of $C$ is related to the area $A$ of the capacitor plates and the distance $d$ between them by

$$C = \frac{\epsilon A}{4\pi K d} \qquad 17.16$$

where $K = 9.0 \times 10^9$ N-m²/C² is the electric constant (Sec. 16.2) and $\epsilon$ is the *dielectric constant*, a dimensionless number characteristic of the insulating material. It is 1 for air and varies between 2 and 6 for most insulating materials. Because $K$ is so large, the capacitance (in farads) of a normal-sized capacitor is very small. Capacitors used in circuits vary between 0.1 F and 1 pF (1 picoFarad = $10^{-12}$ F).

The product $RC$ of resistance and capacitance has the dimension of time. This can be seen by noting that $RC$ has the unit

$$\frac{V}{A}\frac{C}{V} = \frac{C}{A} = \frac{C}{C/s} = s$$

In the time $\tau = RC$, the transient current in Fig. 17.17*a* decays to 37 percent of its initial value while the charge on the capacitor (Fig. 17.17*b*) reaches 63 percent of its maximum value. From Eq. 17.15, the potential across the capacitor is just $V = q/C$, so $V$ also reaches 63 percent of its maximum value in the time $\tau$. Thus, $\tau = RC$ is the *characteristic time* required for a circuit with a capacitor to change its condition. Such circuits can be used to produce electric pulses with a definite frequency.

Once the transient current decays to zero, there is no direct current in the circuit in Fig. 17.16 because a capacitor is a circuit element with infinite resistance for a direct potential. On the other hand, suppose an alternating potential is applied to a capacitor (Fig. 17.18). The charge on each plate alternates between positive and negative as the potential across the capacitor alternates, and so there is an alternating current in the circuit. Thus a capacitor in effect is an element that offers little resistance to an alternating potential and infinite resistance to a direct potential. In many electronic circuits in which the potential is a mixture of a direct and an alternating component, capacitors are used to filter the direct component from the alternating component.

### Diodes

A *diode* is a circuit element that conducts a current in only one direction. The characteristics of a diode are best displayed by a graph, as in Fig. 17.19, which plots the current $I$ in the diode against the potential $V$ across it. A similar curve for a resistor is shown for comparison. With a resistor, the current is proportional to $V$ (Ohm's law), so the $IV$ curve is a straight line. Furthermore, $I$ is negative when $V$ is negative, which means that the current reverses direction when the potential across the resistor is reversed. The $IV$ curve of a diode is not a straight line; this is equivalent to saying that the effective resistance of the diode depends on $V$. More important, no (or very little) current passes through the diode when the potential is reversed; this is equivalent to saying that the resistance of the diode is infinite when $V$ is negative. For most purposes it is sufficient to remember that the effective resistance of a diode is very small (essentially zero) when $V$ is positive and very large (essentially infinite) when $V$ is negative.

A diode can be either a vacuum tube or a solid-state device. Solid-state circuit elements, e.g., diodes and transistors, have replaced tubes in most electronic circuits because of their small size, reliability, and small power consumption. In fact, the solid-state diode is a passive element, requiring no outside power supply, whereas the vacuum-tube diode is an active element that must be powered. Nevertheless, the physics of a vacuum tube is easier to understand, so it is described here.

A vacuum-tube diode, shown in Fig. 17.20, is similar to the electron-beam tubes discussed in Sec. 16.5. A small filament heats one electrode (the cathode), evaporating electrons from it. A second electrode (the anode) is maintained at a potential $V$ relative to the cathode. When $V$ is positive, the evaporated electrons flow to the anode, producing a current $I$ directed from the anode to the cathode. When $V$ is negative, the anode repels the electrons, and there is no current in the diode.

Solid-state diodes are similar to vacuum tubes except that they have no filaments requiring a separate power supply. The symbols used to designate a vacuum-tube diode and a solid-state diode are shown in Fig. 17.21. The filament circuit is omitted from the symbol for the vacuum-tube diode. The vacuum-tube diode conducts a current from the anode to the cathode, and the solid-state diode conducts a current in the direction of the arrow.

When an ac generator is connected to a diode (Fig. 17.22), the potential across the diode reverses sign twice during each full oscillation of the potential. When the anode is positive, the diode conducts and there is a current $I$ in the resistor $R$ and a potential $RI$ across it. When the anode is negative, there is no current. That is, the diode conducts only during the positive half of each cycle, so the potential across the resistor has the time variation shown in Fig. 17.23. By attaching wires to either side of the resistor, this nonreversing potential can be supplied to another circuit. The process of converting an alternating potential to a direct (nonreversing) potential is called *rectification*. It is used in all radio receivers.

Computers represent numbers by a sequence of electric pulses that can be processed by logic circuits. Suppose, for instance, that the number 1 is represented by a potential pulse of 3 V and width 1 $\mu$s and that the number 0 is represented by the absence of a pulse (0 V).

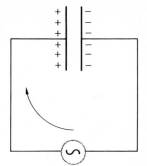

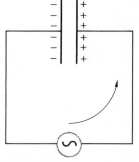

FIGURE 17.18
Charge flows from one plate of a capacitor to the other each time the potential of an ac generator changes sign.

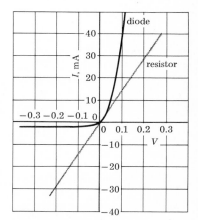

FIGURE 17.19
Comparison of the $IV$ characteristics of a resistor and a solid-state diode.

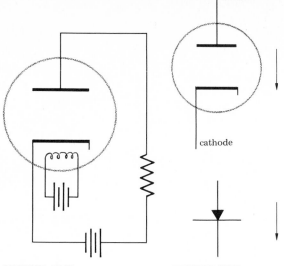

FIGURE 17.20
A vacuum-tube diode.

FIGURE 17.21
Symbols used in circuit diagrams to represent a vacuum-tube diode and a solid-state diode.

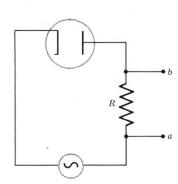

FIGURE 17.22
A diode connected in series to a resistor and an ac generator.

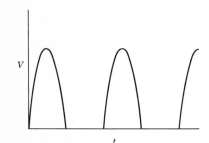

FIGURE 17.23
Time variation of the potential across the resistor in Fig. 17.22.

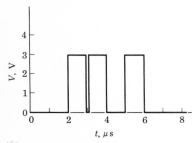

FIGURE 17.24
A sequence of electric pulses representing a binary number.

Then the sequence of pulses in Fig. 17.24 represents the sequence of digits 1101, which is the binary form of the number

$$1 \times 2^3 + 1 \times 2^2 + 0 \times 2^1 + 1 \times 2^0 = 13$$

Electronic circuits exist that can add and multiply binary numbers represented by pulses. As an example, Fig. 17.25 shows a logical AND circuit that can multiply two one-digit binary numbers. The potential $V_a$ at point $a$ is taken to be zero; i.e., all other potentials are measured relative to it. The terminals $V_1$ and $V_2$ are each either at zero potential (representing the digit 0) or at 3 V (representing the digit 1). If both are at zero, both diodes are conducting and the potential $V_b$ at point $b$ is just the potential across the diodes, which is small compared to 3 V. This small positive potential represents 0 on the output side of the circuit. If $V_1$ is raised to 3 V while $V_2$ remains zero, diode 1 becomes nonconducting but diode 2 remains conducting. The potential at point $b$ is still just the potential drop across the conducting diode, and so $V_b$ remains small. The situation is the same if $V_1 = 0$ and $V_2 = 3$ V. However, if both $V_1$ and $V_2$ are 3 V, both diodes are nonconducting and there is no current through $R$. The potential at $b$ is then the potential of the battery, which represents the digit 1 on the output side of the circuit.

The table in Fig. 17.26$a$ summarizes the output potential $V_b$ for the four possible combinations of inputs. In the table in Fig. 17.26$b$ these potentials are translated into the digits they represent. Since this table is the multiplication table for 0 and 1, the circuit is seen to perform this basic operation. This illustrates in a very simple case how an electronic circuit can perform an arithmetic operation.

**REMARK**   In logic, propositions are assigned a *truth value* which is either

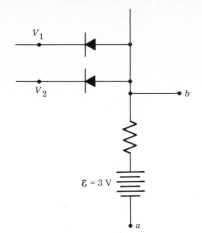

FIGURE 17.25
A logical AND circuit.

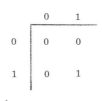

FIGURE 17.26
(*a*) The output potential $V_b$ of the AND circuit in Fig. 17.25, for all possible combinations of $V_1$ and $V_2$. (*b*) Same as (*a*) with the potentials replaced by the numbers they represent.

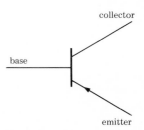

FIGURE 17.27
Schematic diagram of a three-terminal transistor.

*true* or *false*. From two propositions, $V_1$ and $V_2$, complex propositions such as "$V_1$ *or* $V_2$" and "$V_1$ *and* $V_2$" can be formed. The truth value of a complex proposition follows from the truth values of $V_1$ and $V_2$, just as the numerical value of the algebraic expressions "$V_1 + V_2$" and "$V_1 \times V_2$" follow from the numerical values of $V_1$ and $V_2$. For example, the proposition "$V_1$ *and* $V_2$" is true if and only if both $V_1$ and $V_2$ are true. Letting 0 represent *false* and 1 represent *true*, the table in Fig. 17.26*b* gives the truth value of "$V_1$ *and* $V_2$" for all possible values of $V_1$ and $V_2$. Thus the logical *and* operation is identical to the arithmetic *multiply* operation.

The AND circuit in Fig. 17.25 is used in automatic control systems. For instance, the output $V_b$ might control a welding operation that is to start only when the two pieces to be welded are both in position. The inputs $V_1$ and $V_2$ are supplied by sensors that detect the presence of the pieces. The condition "weld if and only if both pieces are present" is logically equivalent to an AND circuit.

### Transistors

A *transistor* is an active three-terminal solid-state circuit element used to control and amplify current. It is a specially processed piece of germanium or silicon to which three wire terminals are attached, as shown schematically in Fig. 17.27. In normal operation, a potential difference $V_{CE}$ is maintained between the two end terminals (*emitter* and *collector*) by means of a battery $\mathcal{E}_C$, as shown in Fig. 17.28. When the middle terminal (*base*) is disconnected, the transistor is nonconducting; i.e., there is no current between the emitter and the conductor. However, if a small current $I_B$ is drawn from the base by means of a second battery $\mathcal{E}_B$, the transistor is turned on and there is a large

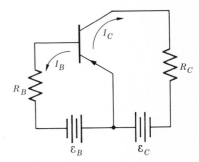

FIGURE 17.28
Typical transistor circuit. A large current between the emitter and the collector is controlled by a small current in the base.

current $I_C$ between the emitter and the collector.† Thus, the current between the emitter and the collector is controlled by the base current $I_B$.

The behavior of a transistor is displayed in Fig. 17.29. This graph plots, for different values of $I_B$, the current $I_C$ in the collector against the potential difference $V_{CE} = V_E - V_C$ between the emitter and the collector. With $I_B = 0$, the collector current $I_C$ is essentially zero for all values of $V_{CE}$; that is, the transistor is nonconducting. For nonzero values of $I_B$, there is a large collector current that increases as $V_{CE}$ increases. For a fixed value of $V_{CE}$, the collector current increases as the base current increases. For example, with $V_{CE} = 7.5$ V and $I_B = 0.15$ mA, the collector current is $I_C = 14.5$ mA. If $V_{CE}$ is kept at 7.5 V while the base current is increased to 0.20 mA, the collector current increases to 21 mA. Thus a change in base current of only 0.05 mA produces a 6.5-mA change in the collector current. It is this property that enables a transistor to operate as an amplifying and controlling element.

Figure 17.30 shows a circuit designed to amplify the output of an ac generator. To understand how this circuit works, we first consider the circuit in Fig. 17.28, which is similar to Fig. 17.30 but without the generator. The resistor $R_B$ is chosen in conjunction with the battery $\mathcal{E}_B$ to give a reasonable base current, say $I_B = 0.25$ mA. The current $I_C$ and the potential $V_{CE}$ are then related by the $I_B = 0.25$ mA curve in Fig. 17.29. For instance, if $V_{CE} = 10$ V, the current is 29 mA.

The potential $V_{CE}$ across the transistor depends on the battery $\mathcal{E}_C$ and the potential drop across the resistor $R_C$. From Kirchhoff's second law we have

$$V_{CE} = \mathcal{E}_C - R_C I_C \qquad 17.17$$

This is a second relation between $V_{CE}$ and $I_C$. The values of $I_C$ and $V_{CE}$ must simultaneously satisfy this equation and lie on the $I_B = 0.25$ mA curve in Fig. 17.29.

For example, suppose $\mathcal{E}_C = 16$ V and $R_C = 400\ \Omega$. Then Eq. 17.17 becomes

$$V_{CE} = 16\ \text{V} - (400\ \Omega)I_C \qquad 17.18$$

which is the equation of a straight line. To determine $I_C$ and $V_{CE}$, this equation is plotted on a graph together with the $I_C V_{CE}$ curves of the transistor (Fig. 17.31). The line is easily found by observing that when $I_C$ is zero, $V_{CE}$ equals 16 V and when $V_{CE}$ is zero, $I_C$ equals $16\ \text{V}/400\ \Omega = 40$ mA. Thus the straight line representing Eq. 17.18, which is called the *load line*, connects the 16-V point on the horizontal axis and the 40-mA point on the vertical axis. All points on this line satisfy Eq. 17.18, and all points on the $I_B = 0.25$ mA curve are possible values of $I_C$ and $V_{CE}$ when the base current is 0.25 mA. The intersection of the line and the curve at $Q$ gives the values of $I_C$ and $V_{CE}$ that simultaneously satisfy both these conditions. From Fig. 17.31 the values of $I_C$ and $V_{CE}$ are found to be 25.5 mA and 5.5 V, respectively. These are the *quiescent* operating values of the circuit.

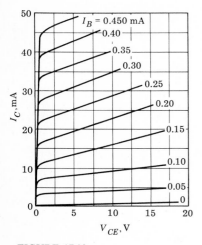

**FIGURE 17.29**
Characteristics of a transistor. The variation of the collector current $I_C$ with the collector potential $V_{CE}$ depends on the value of the base current $I_B$.

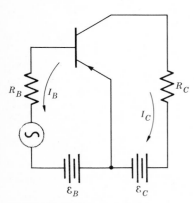

**FIGURE 17.30**
Amplifying circuit. The alternation of the base current produces a much larger alternation of the collector current.

† The symbol for a transistor (Fig. 17.27) has an arrow on the emitter pointing in the direction of the current under these circumstances.

Now consider the circuit in Fig. 17.30 which has, in addition to the battery $\mathcal{E}_B$, an ac generator connected to the base. The battery and the generator together produce a current of the form

$$i_B = I_B + I_p \sin (360° ft)$$

which is a mixture of the direct current $I_B$ produced by the battery and an alternating current of amplitude $I_p$ produced by the generator. This current is plotted in Fig. 17.32 for $I_B = 0.25$ mA and $I_p = 0.05$ mA. As the base current oscillates between 0.20 and 0.30 mA, the operating point $Q$ moves back and forth between $Q_1$ and $Q_2$ along the load line, showing that the collector current oscillates between 21 and 30 mA. This current is a mixture of a direct current of 25.5 mA and an alternating current with a peak of 4.5 mA:

$$i_C = 25.5 \text{ mA} + 4.5 \text{ mA} \sin (360° ft)$$

Since an alternating current in the base of 0.05 mA produces an alternating current in the resistor $R_C$ of 4.5 mA, this circuit has amplified the base current by a factor of 4.5 mA/0.05 mA = 90. If the amplified current is connected to the base of a second transistor, it can be further amplified. Amplification factors of $10^5$ are obtained using several transistors this way.

Amplifiers are essential for the study of the electrical effects produced by living organisms. For instance, a nerve impulse consists of a time-varying potential that propagates along an axon (Sec. 17.5). By using microelectrodes implanted inside a nerve cell, it is possible to study these pulses. Figure 17.33 shows an arrangement for displaying such pulses on an oscilloscope. The time-varying current from the electrodes is fed into the base of an amplifier in order to produce a large time-varying potential across the vertical deflection plates of the oscilloscope. The capacitor $C$ prevents the direct current produced by the battery $\mathcal{E}_C$ from passing through the resistor $R_L$, so the potential across the plates is proportional to the time-varying input signal. A special sweep circuit puts a sawtoothed potential across the horizontal plates. As the electron beam passes through these plates, it is simultaneously swept horizontally at constant speed and deflected vertically in proportion to the input signal, producing a picture on the screen of the time-varying potential in the axon.

### Integrated circuits

An *integrated circuit* is a specially processed silicon chip that operates as an entire electronic circuit. A chip only a few millimeters across can be the equivalent of several capacitors, resistors, and transistors all wired together. The increased availability of these devices in recent years is having as revolutionary effects on electronics as the introduction of the transistor did in the early 1960s. Already low-priced pocket calculators have appeared on the market, and many other important devices can be expected in the near future. Of particular interest will be microminiature transmitters that can be implanted in a patient to monitor a vital body function or attached to a bird or insect to monitor its migratory habits.

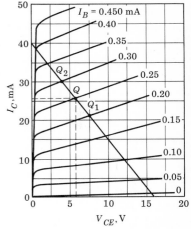

**FIGURE 17.31**
The load line for the circuit in Fig. 17.30 with $R_C = 400 \ \Omega$ and $\mathcal{E}_C = 16$ V.

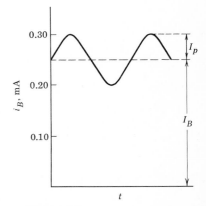

**FIGURE 17.32**
The base current is a mixture of a direct current $I_B$ and an alternating current with a peak current $I_p$.

In most animal cells there is a potential difference between the interior and the exterior of the cell. This is most pronounced in nerve and muscle cells, where in the resting state the interior potential is about $-85$ mV relative to the exterior potential. A nerve impulse is a change in this potential that propagates along a nerve fiber, or axon. Nerve impulses carry information signals from sensory cells to the brain and command signals back from the brain to muscle cells. A nerve impulse, upon reaching a muscle fiber, invokes similar potential changes that propagate along the fiber, initiating the contraction of the fiber. Electricity thus plays a fundamental role in the neuromuscular organization of animals. The muscles of some fish have lost their ability to contract and are used exclusively to generate an electric potential. The well-known electric eel generates a large potential that is used for defense, but many fish generate a small potential used to detect objects in the surrounding water.

### The Nernst potential

The resting potential of a cell is caused by differences in the concentration of ions inside and outside the cell and by differences in the permeability of the cell wall to different ions. To understand this, let us first consider the simple case of two potassium chloride, KCl, solutions of different concentrations separated by a semipermeable membrane (Fig. 17.34). In solution, KCl dissociates into positively charged potassium ions, $K^+$, and negatively charged chlorine ions, $Cl^-$. These ions move about independently of each other, but any small region of fluid has on the average equal numbers of $K^+$ and $Cl^-$ ions and so is electrically neutral. If the membrane is permeable

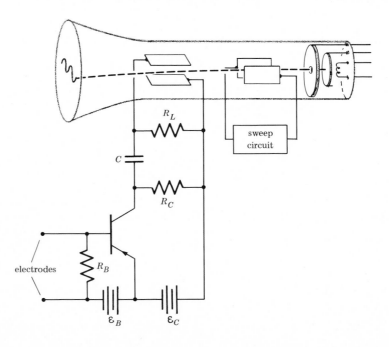

**FIGURE 17.33**
Circuit for displaying a pulse on an oscilloscope. The small time-varying potential difference that the pulse puts across the electrodes is amplified by the transistor circuit, and a large potential difference, with the same time variation as the pulse, is put across the vertical deflection plates of the oscilloscope.

to $K^+$ ions but not to $Cl^-$ ions, the $K^+$ ions will diffuse freely back and forth across the membrane. However, because there are more ions in compartment 1 than in compartment 2, more $K^+$ ions initially will flow from 1 to 2 than from 2 to 1. Since the $Cl^-$ cannot diffuse across the membrane, there is soon an excess of positive charge in compartment 2 and an excess of negative charge in compartment 1 (Fig. 17.34$b$). These excess charges are all concentrated along the membrane, whereas the bulk fluid in each compartment remains neutral. The layers of positive and negative charge on each side of the membrane produce a potential difference $V = V_1 - V_2$ across the membrane and an electric field $\mathbf{E}$ that retards the flow of positive ions from compartment 1 to compartment 2 and accelerates their flow from compartment 2 to compartment 1.

A dynamic equilibrium is soon established in which as many $K^+$ ions diffuse from 1 to 2 as diffuse from 2 to 1. This equilibrium results from the balance of two opposite tendencies, the concentration difference, which favors diffusion from 1 to 2, and the potential difference, which favors diffusion from 2 to 1. At equilibrium the potential difference $V$ is given in terms of the concentrations† $c_1$ and $c_2$ of $K^+$ ions in the two compartments by the equation

$$V = V_1 - V_2 = \pm 2.3 \frac{kT}{e} \log \frac{c_1}{c_2} \qquad 17.19$$

This is the *Nernst equilibrium potential*. It is negative when the membrane is permeable to positive ions and positive when the membrane is permeable to negative ions. Here $k$ is the Boltzmann constant and $T$ is the absolute temperature. The quality $kT$ is proportional to the average kinetic energy of the ions in solution (Sec. 8.4), and the quantity $kT \log(c_1/c_2)$ is proportional to the net flow of ions due to the concentration difference. The quantity $eV$ is proportional to the net flow of ions due to the potential difference. Equation 17.19 is the condition that these two flows be equal and opposite.

At body temperature 37°C the quantity $kT/e$ is

$$\frac{kT}{e} = \frac{(1.38 \times 10^{-23} \text{ J/K})(310 \text{ K})}{1.60 \times 10^{-19} \text{ C}}$$

$$= 0.0267 \text{ V} = 26.7 \text{ mV}$$

so the Nernst potential is

$$V = V_1 - V_2 = \pm(61.4 \text{ mV}) \log \frac{c_1}{c_2} \qquad 17.20$$

The intracellular fluid of a nerve cell has a $K^+$ concentration of 0.141 mol/l, whereas the extracellular fluid has a $K^+$ concentration of only 0.005 mol/l (Table 17.1). If these concentrations are in equilibrium, the potential across the membrane will be

$$V = -61.4 \text{ mV} \log \frac{0.141}{0.005}$$

$$= -61.4 \text{ mV} \times 1.45 = -89.2 \text{ mV}$$

The potential difference is found by measurement to be $-85$ mV, so

†Concentration is measured in moles per unit volume. It is the number density $\eta$ (Sec. 8.1) divided by Avogadro's number.

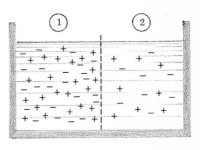

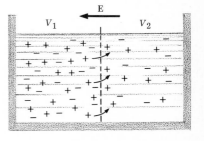

b

FIGURE 17.34
($a$) Two KCl solutions separated by a membrane permeable only to $K^+$ ions. ($b$) The diffusion of $K^+$ ions from the more concentrated solution in compartment 1 to the less concentrated solution in compartment 2 results in an excess of negative charge in 1 and an excess of positive charge in 2.

the K$^+$ concentrations are nearly in equilibrium. However, because the interior of the cell is slightly less negative than is required to balance the concentration difference, there is a small net diffusion of K$^+$ ions out of the cell. These ions are actively pumped back into the cell by an as yet unknown metabolic process in the cell wall. The term "pumped" is used to emphasize that it is not a diffusion process but an active mechanism that requires metabolic energy. On the other hand, the number of K$^+$ ions actively transported this way is very small.

**REMARK** It is very important to realize that only a very small fraction of the K$^+$ ions in a cell are involved in establishing the Nernst potential. A typical cell has a volume of about $10^{-9}$ cm$^3$ and a K$^+$ concentration of about 0.14 mol/l, or about $10^{20}$ K$^+$ ions per cubic centimeter. Thus there are about $(10^{20}$ ions/cm$^3)(10^{-9}$ cm$^3) = 10^{11}$ K$^+$ ions in a cell. As K$^+$ ions diffuse out of the cell, the excess charge inside and outside the cell accumulates along the cell wall (Fig. 17.35). The cell wall acts as a capacitor with an area $A$ of about $10^{-6}$ cm$^2$ and a thickness of about $10^{-6}$ cm. Its capacitance $C$ is estimated from Eq. 17.16 to be

$$C = \frac{\epsilon \times 10^{-10} \text{ m}^2}{4\pi \times (9 \times 10^9 \text{ N-m}^2/\text{C}^2)(10^{-8} \text{ m})} = 2.6 \times 10^{-13} \text{ F}$$

where $\epsilon$ has been assumed to be 3. With a potential difference of 0.085 V across the cell wall, the net charge on either side of the wall is found from Eq. 17.15 to be

$$q = CV = (2.6 \times 10^{-13} \text{ F})(0.085 \text{ V}) = 2.2 \times 10^{-14} \text{ C}$$

The number of K$^+$ ions required to produce this charge is

$$\frac{q}{e} = \frac{2.2 \times 10^{-14} \text{ C}}{1.6 \times 10^{-19} \text{ C}} = 1.4 \times 10^5$$

which is only about a millionth of the K$^+$ ions inside the cell. Thus, while the Nernst potential requires some K$^+$ ions to leave the cell, the fraction of ions involved is too small to affect the concentration of ions on either side of the cell wall to any significant degree.

Many kinds of ions are dissolved in the intracellular and extracellular fluid, but only ions that can diffuse through the cell wall contribute to the Nernst potential. In the resting state the cell wall

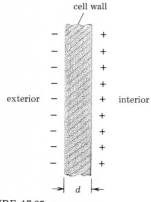

cell wall

exterior — interior

$d$

FIGURE 17.35
A cell wall acts as a capacitor with positive charge on the inside and negative charge on the outside.

TABLE 17.1 **Concentrations of K$^+$, Na$^+$, and Cl$^-$ ions in intracellular and extracellular fluid**
All other ions are denoted by A$^-$.

| Ion | Concentration, mol/l | |
| --- | --- | --- |
| | Extracellular $c_E$ | Intracellular $c_I$ |
| K$^+$ | 0.005 | 0.141 |
| Na$^+$ | 0.142 | 0.010 |
| | 0.147 | 0.151 |
| Cl$^-$ | 0.103 | 0.004 |
| A$^-$ | 0.044 | 0.147 |
| | 0.147 | 0.151 |

is permeable only to $K^+$ and $Cl^-$ ions. The $Cl^-$ ions are distributed between the intracellular and extracellular compartments in accordance with Eq. 17.20. That is, the ratio $c_I/c_E$ of the intracellular to extracellular $Cl^-$ concentrations is given in terms of the cellular potential $V = V_I - V_E$ by

$$\log \frac{c_I}{c_E} = \frac{+V}{61.4 \text{ mV}}$$

where the plus sign is used because the ions are negatively charged. Since the potential is $-85$ mV, this gives

$$\log \frac{c_I}{c_E} = -1.38$$

or

$$\frac{c_I}{c_E} = 0.0417$$

Measurements show that the extracellular $Cl^-$ concentration is $c_E = 0.104$ mol/l and that the intracellular concentration $c_I$ is

$$c_I = 0.0417 \times 0.104 \text{ mol/l} = 0.0043 \text{ mol/l}$$

The negative potential of the cellular interior favors a high intracellular $K^+$ concentration and a low intracellular $Cl^-$ concentration. The overall neutrality of the fluid in both compartments is achieved by the presence of ions to which the cell wall is impermeable. The most important of these are sodium ions, $Na^+$, and various negative organic ions, $A^-$. Table 17.1 gives the concentrations of these ions in the intracellular and extracellular fluids, showing that the total charge in both compartments is zero. (As noted in the preceding remark, the excess charge in each compartment is concentrated along the cell wall, and the number of ions involved is too small to affect the value of the concentrations significantly.)

The Nernst potential is determined both by the difference in concentration of the ions in the intracellular and extracellular fluids and by the selective permeability of the cell wall to ions of different kinds. For instance, the difference in the concentration of $Na^+$ ions in the two compartments would have a Nernst potential of

$$V = -(61.4 \text{ mV}) \log \frac{0.010}{0.142}$$

$$= +70.6 \text{ mV}$$

if the cell wall were permeable to $Na^+$ ions. This potential would cancel and even reverse the $K^+$ and $Cl^-$ potentials. The situation is analogous to the electric circuit shown in Fig. 17.36. The batteries $\mathcal{E}_K$, $\mathcal{E}_{Cl}$, and $\mathcal{E}_{Na}$ represent the Nernst potentials resulting from the concentration differences of $K^+$, $Cl^-$, and $Na^+$ ions, respectively. These batteries are connected in parallel to wires that represent the interior and exterior of the cell. In the resting state the circuit containing the $Na^+$ battery is open, and only the $K^+$ and $Cl^-$ batteries contribute to the potential difference. (There is a small current $I$ through the resistor $R_K$ because the total potential between the wires is 85 mV $= \mathcal{E}_K - R_K I$.)

The wall of nerve and muscle cells has the ability to change its relative permeability to $K^+$ and $Na^+$ ions. When a nerve cell is stimulated, whether electrically, chemically, or mechanically, the cell wall suddenly becomes permeable to Na⁺. This is analogous to closing the switch $S$ in the circuit in Fig. 17.36. Na⁺ ions now flow through the cell wall into the cell, neutralizing the negative charge inside the cell. For a period of about 0.2 ms the cell wall is about 100 times more permeable to Na⁺ ions than to K⁺ ions. (This is represented in Fig. 17.36 by making $R_{Na}$ less than $R_K$ and $R_{Cl}$.) During this period the Na⁺ battery dominates, and the interior potential changes from $-85$ to $+60$ mV. (The resistors in Fig. 17.36 are chosen so that the potential across the wires changes from $-85$ to $+60$ mV when the switch is closed.) After this period the cell wall again becomes impermeable to Na⁺ ions, so that the K⁺ battery again dominates. K⁺ ions diffuse out of the cell until the equilibrium potential of $-85$ mV is reestablished. These changes in the cellular potential are shown in Fig. 17.37. The sudden rise and fall of the cellular potential is called the *action potential*.

During the following resting period, the excess Na⁺ ions that had entered the cell during the action potential are actively pumped out of the cell. The mechanism of the Na⁺ pump is unknown, but it is believed to be similar to the K⁺ pump. It must be emphasized again, however, that the total number of ions that move across the cell wall during these potential changes is very small compared to the number of ions in the cell. Therefore a single action potential does not produce a significant change in the ion concentration of the cell. The cell can fire many thousands of times before the Na⁺ concentration is appreciably affected. The Na⁺ pump is thus a mechanism for maintaining the proper Na⁺ concentration in the long run, and it is not significant for any one firing of the action potential.

**FIGURE 17.36**
Circuit analog of the Nernst potentials in a cell. (*a*) In the resting state the cell wall is impermeable to Na⁺ ions, which corresponds to switch $S$ being open. The Cl⁻ and K⁺ batteries maintain a potential difference of $-85$ mV between the inside and the outside of the cell. (*b*) During an action potential, the cell wall suddenly becomes permeable to Na⁺, which corresponds to switch $S$ being closed. The Na⁺ battery then makes the interior of the cell $+60$ mV with respect to the exterior.

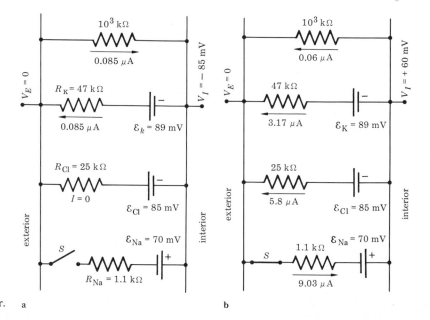

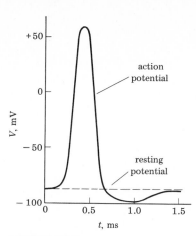

FIGURE 17.37
Time variation of the action potential.

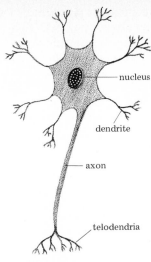

FIGURE 17.38
A nerve cell.

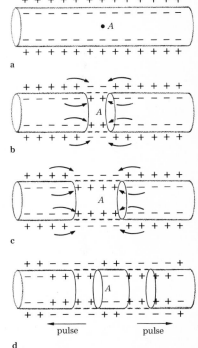

a

b

c

d

FIGURE 17.39
Transmission of a nerve impulse along an axon. (*a*) Resting state. (*b*) Stimulation at point *A* produces a local action potential there, so the interior of the cell becomes positive. (*c*) As charge flows toward *A* from adjacent regions, action potentials are stimulated further along the axon. (*d*) The axon returns to its resting state at *A*, while the pulse continues to propagate along the axon.

## Nerve impulses

A nerve cell, shown in Fig. 17.38, consists of a cell body and a single long extension called an *axon*. The fluid in an axon is similar in composition to the fluid in the cell body, and in the resting state the interior of the axon is at a potential of $-85$ mV relative to the extracellular fluid. The axons act as cables that transmit nerve impulses from one nerve cell to another nerve or muscle cell. A nerve impulse is an action potential that propagates along the axon.

Figure 17.39*a* shows a section of an axon in the resting state. When the axon is stimulated at a point *A*, only the region of the cell wall around *A* becomes permeable to $Na^+$ ions. An action potential is established in this region, which momentarily causes the outside of the cell to become negative and the inside positive (Fig. 17.39*b*). On the outside of the axon, positive charge is attracted to this negative region, and so the adjacent regions become less positive. Similarly, on the inside, negative charge is attracted to the positive region, and the adjacent regions become less negative. This causes a small change in the potential across the cell wall in these adjacent regions, which in turn causes the cell wall in these regions to suddenly become permeable to $Na^+$ ions, thus triggering an action potential. $Na^+$ ions flow into the cell, reversing the potential in these regions (Fig. 17.39*c*). Charge from areas further along the axon are attracted to these regions, stimulating action potentials in these areas. Thus a potential pulse, with the shape of the action potential, is propagated along the axon in both directions. Shortly after the pulse moves away from *A*, the potential at *A* returns to its resting value and is ready to fire again.

A nerve impulse moves along an axon at a speed of 30 m/s, which is small compared to the speed of electric impulses in electronic circuits. The speed of a nerve impulse is determined by the time required for ions to flow back and forth across the cell wall. A nerve impulse is analogous to a wave on a string in that the flow of charge is mainly perpendicular to the direction of propagation of the pulse.

The basic mechanism by which the cell wall changes its permeability to Na$^+$ ions is not yet understood. The cell wall is a functionally complex structure that controls the passage of molecules between the intracellular and extracellular compartments. It is currently the subject of intensive research by biologists and biophysicists throughout the world, and we can look forward to major advances in our understanding of this fundamental structure in the near future.

When a nerve impulse reaches a muscle cell, it generates an action potential in the muscle cell. This potential propagates along the muscle fiber, in the same way it propagates along an axon. It is the movement of this impulse along the fiber that initiates the contraction of the fiber. A very large action potential spreads through the heart prior to each heart beat. This potential produces currents in the surrounding tissue, some of which reach the skin, where they can be detected by electrodes placed on the chest. The signals from the electrodes are amplified and recorded on a moving chart. The resulting record, called an *electrocardiogram* (EKG), is of great value in the diagnosis of heart disease. Figure 17.40 shows a normal electrocardiogram for two heart beats.

### Electric fish

Some fish have developed modified muscle cells that are used primarily for generating an action potential. These cells are not long thin fibers, like a normal muscle cell, but short flat plates, called *electroplaques*. They are arranged on top of each other, like batteries in a flashlight. When stimulated by nerve impulses, an action potential is initiated simultaneously on one side of each cell. While each cell generates an action potential of only 0.1 V, a stack of hundreds and thousands of these cells generates a momentary potential pulse of many volts. The electric eel (*Electrophorus*) generates a momentary potential of 300 V, sufficient to stun prey or potential enemies.

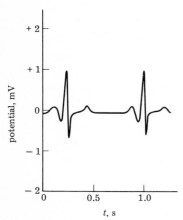

**FIGURE 17.40**
A normal electrocardiogram.

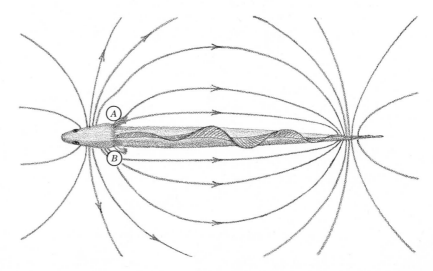

**FIGURE 17.41**
Lines of force around an electric fish.

More interesting are the weak electric fish that maintain a continuous potential differences of a few volts between their head and tail. This potential difference is produced in pulses, at a rate of several hundred pulses per second. As a consequence, a pulsating electric field is established around the fish and electric charge flows along the lines of force of this field (Fig. 17.41). As the fish swims around, the electric field and current are modified by the presence of objects of different electric resistance in the water. Figure 17.41 shows the electric field around the knife fish. When an object with a resistance less than that of the surrounding water is present at a point $B$ near the fish, the local current is larger than normal at this point. Similarly, when an object with a resistance greater than that of the surrounding water is present at a point $A$ near the fish, the local current is smaller than normal at this point.

Electric fish have sensory cells in their skin that can detect these currents. Furthermore, these fish can determine the size and position of objects in their vicinity from the effect these objects have on the current. Experiments have shown that the Nile fish (*Gymnarchus*) can detect the presence of a thin glass rod concealed inside a porous pot from the effect of the rod's electric resistance on the surrounding current.

Electrically sensitive cells may have first evolved in some fish in order to detect the normal action potentials in other fish. Sharks and rays, for instance, can detect a plaice at a distance of 5 to 10 cm from the action potential of the plaice's respiratory muscles. But two orders of fish (Mormyriformes and Gymnotoidei) are characterized by adaptations which make it possible for them to generate and control large action potentials; these enable them to detect animate and inanimate objects in the dark or murky waters where they live. In addition, some of these fish appear to use their electric fields to signal aggressive and mating intentions to each other.

The electrical sense of fish has been known only for the last 30 years. We have seen that other animals are able to detect physical properties that man cannot. Bats, for example, detect ultrasonic sound, and bees detect polarized light. Over the long course of evolution, other physical properties may also have been adopted as a specialized sense by some animals.† We can reasonably expect that new and interesting discoveries are still to be made in the field of animal senses.

† See Sec. 18.5 for a discussion of the possibility of a magnetic sense in birds.

## PROBLEMS

1 A charge of 75 C flows through a wire in 120 s. (*a*) What is the current in the wire during this time? (*b*) How many electrons pass through the wire during this time?
   *Ans.* (*a*) 0.625 A; (*b*) 4.69 × 10²⁰

2 (*a*) In 30 s, how much charge passes through a wire in which there is a current of 4.5 A? (*b*) How many electrons pass through the wire in this time?

3 The *faraday* (F) is the charge of 1 mol of protons. (*a*) What is the value of the faraday in coulombs? (*b*) How long would it take for 1 F of charge to pass through a wire in which the current is 15 A?
   *Ans.* (*a*) 9.6 × 10⁴ C; (*b*) 106.7 min

4 A battery can deliver a total of 0.40 F (see Prob. 3). (*a*) How long can the battery maintain a current of 0.7 A in a circuit? (*b*) If the emf of the battery is 1.5 V, what is the total electric energy that the battery can deliver?

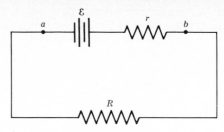

FIGURE 17.42   Problems 13 and 14.

5 A current of 0.3 A is maintained by a battery with an emf of 1.5 V. (*a*) What is the power output of the battery? (*b*) How much energy does the battery expend in 5 min?

Ans. (a) 0.45 W; (b) 135 J

6 A battery with an emf of 3.0 V dissipates energy at the rate of 0.80 W in an external circuit. (*a*) What is the current in the circuit? (*b*) If the total electric energy that the battery can deliver is $3.5 \times 10^4$ J, how long can the battery maintain this current?

7 Suppose the battery in Fig. 17.4 has an emf of 5.0 V and the resistance is 10 $\Omega$. What is (*a*) the current in the circuit and (*b*) the power dissipation in the resistor?

Ans. (a) 0.5 A; (b) 2.5 W

8 Suppose the current in Fig. 17.4 is 0.3 A and the resistance is 20 $\Omega$. What is (*a*) the emf of the battery and (*b*) the power dissipation in the resistor?

9 Suppose the emf of the battery in Fig. 17.4 is 4.5 V and the current in the circuit is 0.2 A. What is (*a*) the resistance in the circuit and (*b*) the power dissipation in it?

Ans. (a) 22.5 $\Omega$; (b) 0.9 W

10 Suppose the current in Fig. 17.4 is 0.4 A and the power dissipation in the resistor is 1.2 W. What is (*a*) the emf of the battery and (*b*) the resistance of the circuit?

11 Suppose the emf of the battery in Fig. 17.4 is 3.0 V and the power dissipation in the resistor is 2.0 W. What is (*a*) the current in the circuit and (*b*) the resistance?

Ans. (a) 0.67 A; (b) 4.5 $\Omega$

12 Suppose the resistance in Fig. 17.4 is 27 $\Omega$ and the power dissipation in it is 3.0 W. What is (*a*) the emf of the battery and (*b*) the current in the circuit?

13 A battery usually has a small internal resistance of its own. This is indicated by the resistor $r$ in Fig. 17.42. If the emf of the battery is 3.0 V, $r = 0.5\ \Omega$, and $R = 5\ \Omega$, what is the potential difference between the terminals $a$ and $b$ of the battery?          Ans. 2.73 V

**REMARK**   Because of the internal resistance of a battery, the potential across its terminals can be less than the emf of the battery.

14 The circuit in Fig. 17.42 has a current of 0.5 A when $R$ is 10 $\Omega$ and a current of 0.27 A when

$R$ is 20 $\Omega$. What is (*a*) the internal resistance and (*b*) the emf of the battery?

15 Suppose the emf of the battery in Fig. 17.6 is 7.5 V and the resistors are $R_1 = 8\ \Omega$ and $R_2 = 12\ \Omega$. What is (*a*) the current in the circuit and (*b*) the power dissipation in each resistor?

Ans. (a) 0.375 A; (b) 1.125 and 1.69 W

16 Suppose the emf of the battery in Fig. 17.6 is 9.0 V and the resistor $R_1$ is 100 $\Omega$. What must the resistor $R_2$ be for the current to be 20 mA?

17 Figure 17.43 shows a potential difference of 120 V placed across a circuit that has a lamp with resistance $R_1 = 144\ \Omega$ connected in series to a variable resistor $R_2$. The brightness of the lamp is controlled by changing the magnitude of $R_2$. What is the power dissipation in the lamp (*a*) when $R_2$ is zero and (*b*) when $R_2 = 144\ \Omega$? (*c*) What must $R_2$ be for the power dissipation in the lamp to be 50 W?

Ans. (a) 100 W; (b) 25 W; (c) 59.7 $\Omega$

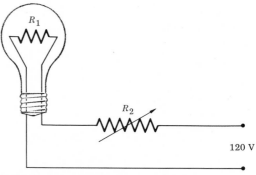

FIGURE 17.43   Problem 17.

18 A three-way light bulb has two filaments which are connected to three wires, as shown in Fig. 17.44. By turning the socket switch, 120 V is put across either $ab$, $bc$, or $ac$. (*a*) If $R_1 = 144\ \Omega$ and $R_2 = 216\ \Omega$, what are the three

FIGURE 17.44   Problem 18.

possible power dissipations of the light bulb? (b) A different three-way light bulb can operate at 300, 100, and 75 W. What are the resistances of its two filaments?

19 Suppose the emf of the battery in Fig. 17.7 is 7.5 V and the resistors are $R_1 = 8\ \Omega$ and $R_2 = 12\ \Omega$. What are the currents $I$, $I_1$, $I_2$?

   *Ans. 1.56, 0.94, and 0.62 A*

20 Suppose the emf of the battery in Fig. 17.7 is 9.0 V and the resistor $R_1$ is 100 $\Omega$. What must the resistor $R_2$ be in order for the current $I$ to be 150 mA?

21 Six Christmas tree lights are arranged in a parallel circuit, as shown in Fig. 17.45. Each bulb dissipates 10 W when operated at

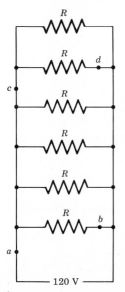

FIGURE 17.45   Problem 21.

120 V. (a) What is the resistance $R$ of each bulb? (b) What is the resistance of the entire array of bulbs? (c) What is the total power consumption of the array? (d) What are the currents at points $a$, $b$, $c$, and $d$?

   *Ans. (a) 1440 $\Omega$; (b) 240 $\Omega$; (c) 60 W; (d) 0.50, 0.083, 0.167, and 0.083 A*

22 Six bulbs are arranged in a series circuit, as shown in Fig. 17.46. (Old-fashioned Christmas tree lights were connected this way.) Each bulb has the same resistance $R$, and the entire array is designed to dissipate 60 W when operated at 120 V. (a) What is the resistance of each bulb? (b) What is the current in the circuit when it is operated at 120 V? (c) What would be the power dissipated in a single bulb operated at 120 V? (d) Compare this circuit to the circuit in Prob. 21 and discuss the disadvantages of arranging Christmas tree lights in series.

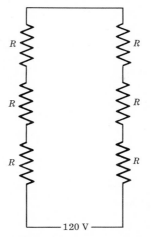

FIGURE 17.46   Problem 22.

23 Figure 17.47 shows the three resistors $R_1 = 5\ \Omega$, $R_2 = 15\ \Omega$, and $R_3 = 25\ \Omega$ in four different circuits. For each circuit find the currents $I_1$, $I_2$, and $I_3$ in each resistor, and the current $I$ in the battery.

   *Ans. (a) $I_1 = I_2 = I_3 = I = 0.067$ A; (b) 0.6, 0.2, 0.12, and 0.92 A; (c) 0.15, 0.15, 0.12, and 0.27 A; (d) 0.209, 0.130, 0.079, and 0.209 A*

24 For each circuit in Fig. 17.47, find the power

ε = 3.0 V

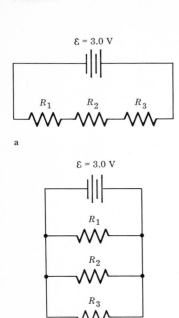

a

b

c

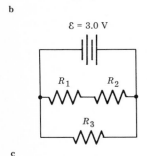

d

FIGURE 17.47   Problems 23 and 24.

dissipation in each resistor and the total power output of the battery (see Prob. 23).

25 Find all the resistances that can be made with three 10 Ω resistors in various combinations.

Not every combination need use all three resistors. *Ans.* 3.33, 5, 6.66, 10, 15, 20, and 30 Ω

26 Arrange a 5-, a 10-, and a 20-Ω resistor in a combination that has a total resistance of 14 Ω.

27 Find all the resistances that can be made with a 6-, a 9-, and a 15-Ω resistor in various combinations. Not every combination need use all three resistors.

*Ans.* 2.90, 3.6, 4.28, 4.8, 5.625, 6, 6.3, 7.5, 9, 11.625, 13.28, 15, 18.6, 21, 24, and 33 Ω

28 Arrange an 8-, a 12-, and a 16-Ω resistor in a combination that has a total resistance of 8.89 Ω.

29 An alternating potential with a peak potential of 75 V is placed across a 15-Ω resistor. What is (a) the rms current and (b) the average power dissipation in the resistor?

*Ans.* (a) 3.53 A; (b) 187.5 W

30 An alternating current with an rms current of 2.4 A passes through a 25-Ω resistor. What is (a) the peak potential across the resistor and (b) the average power dissipation in it?

31 From the data in Sec. 18.3, estimate the total electric energy required to kill a person by electrocution. Compare this to the energy required to kill a person with a bullet (see Prob. 7 of Chap. 5). *Ans.* 14.4 J

32 Why aren't birds electrocuted when they sit on an overhead power line? (The line does not have an insulating cover.)

33 What is the magnitude of the maximum charge on a plate of a 0.25-μF capacitor that is connected to a 3.0-V battery?

*Ans.* $7.5 \times 10^{-7}$ C

34 A 200-μF capacitor has a charge of 0.015 C on each plate. What is the potential difference between the plates?

35 A 50-μF capacitor initially uncharged is connected through a 300-Ω resistor to a 12-V battery. (a) What is the magnitude of the final charge $q_0$ on the capacitor? (b) How long after the capacitor is connected to the battery will it be charged to $\frac{1}{2}q_0$? (c) How long will it take for the capacitor to be charged to $0.90q_0$?

*Ans.* (a) $6 \times 10^{-4}$ C; (b) 10.5 ms; (c) 37.5 ms

36 A 400-μF capacitor is connected through a resistor to a battery. Find (a) the resistance R and (b) the emf of the battery ε if the characteristic time of the circuit is 0.5 s and the maximum charge on the capacitor is 0.024 C.

37 What is the number represented by the sequence of pulses in Fig. 17.48? *Ans.* 89

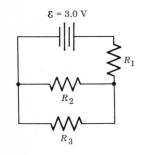

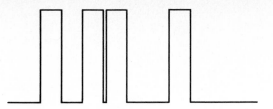

FIGURE 17.48   Problem 37.

*38* Represent the number 37 by a sequence of electric pulses.

*39* What is the binary representation of 100?

$Ans.$ 1100100

*40* Write a table giving the truth value of "$V_1$ or $V_2$" for all combinations of the truth values of $V_1$ and $V_2$. (*Hint:* In logic "$V_2$ or $V_2$" is taken to mean "either $V_1$ or $V_2$ or both.")

*41* What is the collector current in a transistor when the collector potential $V_{CE}$ is 7.5 V and the base current is 0.40 mA (use Fig. 17.29)?

$Ans.$ 45 mA

*42* What must the base current be in a transistor to give a collector current of 25 mA when the collector potential $V_{CE}$ is 10 V?

*43* What is (*a*) the maximum and (*b*) the minimum potential difference across the 400-$\Omega$ resistor $R_C$ in the circuit in Fig. 17.30 when the circuit is operating between the points $Q_1$ and $Q_2$ in Fig. 17.31?

$Ans.$ (*a*) 8.4 V; (*b*) 12 V

*44* On Fig. 17.31 draw the load line for the amplifying circuit in Fig. 17.30 with $\mathcal{E}_C = 18$ V and $R_C = 400\ \Omega$. Find the quiescent point when $I_B = 0.30$ mA. What is the amplification of an alternating current whose peak current $I_p$ is 0.05 mA?

*45* If the cell wall were permeable to the negatively charged organic ions, A$^-$, in the cellular fluids (Table 17.1), what would the Nernst potential due to these ions be?

$Ans.$ +32.1 mV

*46* Suppose the intracellular concentration of Cl$^-$ were 0.025 mol/l. What would the extracellular concentration be if the Nernst potential due to Cl$^-$ were $-72$ mV?

## BIBLIOGRAPHY

ALEXANDER, R. M.: "Functional Design in Fishes," Hutchinson University Library, London, 1967. The electrical sense of fish is described in detail in chap. 6.

BULLOCK, THEODORE HOLMES: Seeing the World through a New Sense: Electroreception in Fish, *American Scientist*, **61:** 316 (1973). Survey of the types of electroreception—active and passive, high- and low-frequency—found in fish and their function in object detection and social communication.

BURTON, ROBERT: "Animal Senses," Taplinger Publishing Co., Inc., New York, 1970. A popular account of our present knowledge of animal senses, including the electrical sense of fish.

DALZIEL, CHARLES F.: Electric Shock Hazard, *IEEE Spectrum*, **9:** 41 (February 1972). Authoritative account of the known effects of electric current on the human body. A section of the article deals with the special problems of electrical safety in hospitals.

GRUNDFEST, HARRY: Electric Fish, *Scientific American*, **203:** 115 (October 1960). Basic physiology of the electroplaques in electric fish. Some discussion of the variety of forms of electric fish that have developed by convergent evolution.

GUYTON, ARTHUR C.: "Textbook of Medical Physiology," W. B. Saunders Company, Philadelphia, 1971. The basic physiology of the resting and action potential is given in chaps. 4 and 5.

KATZ, BERNARD: "Nerve, Muscle, and Synapse," McGraw-Hill Book Company, New York, 1966. Elementary introduction to the physics and physiology of nerve-impulse transmission. The experimental techniques used in neurophysiology are briefly discussed.

LISSMANN, H. W.: Electric Location by Fishes, *Scientific American*, **208:** 50 (March 1963). A fascinating account of the author's discovery of the electrical sense in fish.

# CHAPTER 18 magnetism

Magnetism is a fundamental force of nature, closely related to electricity. Because of the existence of naturally occurring magnetic materials, some simple magnetic effects have been known since antiquity. However, the important electromagnetic phenomena that established the connections between electricity and magnetism were discovered only in the nineteenth century. In fact, all the devices used in the commercial generation and distribution of electricity, such as generators, transformers, and motors, are based on electromagnetic principles discovered between 1820 and 1831. Moreover, in 1873 Maxwell showed mathematically that these principles imply the existence of self-propagating waves of electric and magnetic fields that travel with the speed of light. Thus the principles of electromagnetism are the basis of our technology, and our understanding of the nature of light and other forms of electromagnetic radiation.

## 18.1   MAGNETS

The most familiar example of magnetism is the attraction of small pieces of iron by the ends, or *poles*, of a magnet. This phenomenon is similar in some respects to the attraction of a small piece of cork by an electrified rod (Sec. 16.2). However, iron is one of the few substances attracted by a magnet, whereas any substance is attracted by an electrified rod. Furthermore, a rod remains electrified for only a short time, whereas a magnet retains its magnetism indefinitely.

A compass needle is a long thin magnet suspended at its center of gravity in such a manner that it is free to rotate in the horizontal plane (Fig. 18.1). If no other magnetic material is nearby, the needle will align itself in a nearly north-south direction. The end of the needle that always points north is called the north (N) pole of the magnet; the other end is called the south (S) pole. *North* and *south* are used to distinguish opposite magnetic poles, just as *positive* and *negative* are used to distinguish opposite electric charge.

Furthermore, the force between two magnetic poles depends on the poles in the same way that the force between two electric charges depends on the charges: opposite poles attract each other, and like poles repel each other. This is easily demonstrated with two bar magnets whose poles have been determined from their north-seeking (or south-seeking) tendencies. When the north pole of one magnet is brought near the south pole of the other magnet, the two poles pull together, whereas when the north pole of one magnet is brought near the north pole of the other magnet, the two poles push apart.

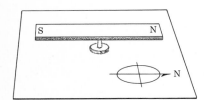

FIGURE 18.1
A bar magnet suspended at its center of gravity.

Because of the attraction of unlike poles, a compass needle brought near a bar magnet will align itself with its north pole directed toward the south pole of the magnet (Fig. 18.2). The earth itself is a giant magnet whose south pole is located near the earth's north geographical pole. Therefore, in the absence of other magnets, the north pole of a compass needle points north toward the south magnetic pole of the earth. However, the earth's magnetism is so weak that its effect on a compass needle is negligible compared to that of a nearby magnet. Consequently we shall neglect the earth's magnetism when considering the effects of a magnet on a nearby compass.

Unlike electric charge, isolated magnetic poles do not exist. For instance, if we try to separate the north and south poles of a long magnet by breaking it in half (Fig. 18.3), we find that north and south poles immediately appear on the broken ends, so that each fragment remains a complete magnet with both a north and south pole. In fact, no matter how many pieces the magnet is broken into, each piece has a north and south pole of equal strength. Furthermore, even elementary particles, such as electrons, protons, and neutrons, act as complete magnets with both north and south poles. Because elementary particles with only one magnetic pole have never been found, magnetic poles lack the fundamental significance that electric charges have.

Nevertheless the *magnetic field* **B** can be defined in the same way as we defined the electric field **E** in Sec. 16.3. Recall that the electric field at any point in space is the force that a system of charges would exert on one unit of positive electric charge placed at that point. Likewise, the magnetic field at any point in space is the force that a magnetic system (a bar magnet, for instance) would exert on one unit of north magnetic pole placed at that point. Since isolated poles do not exist, we have to use little compass needles to measure the field.

For example, consider a small compass needle located at a point $P$ in the magnetic field produced by a bar magnet (Fig. 18.4). The two poles of the magnet exert a net force $\mathbf{F}_1$ on the north pole of the compass needle and a net force $\mathbf{F}_2$ on the south pole. If the compass needle is not already aligned with the magnetic field at $P$, Fig. 18.4$a$ shows that these forces exert a torque that will rotate the needle about the pivot until it is aligned with the field. Because the torque on a compass needle is zero only when it is aligned with the magnetic field, at equilibrium a compass needle points in the direction of the magnetic field in its vicinity. Consequently, the direction of

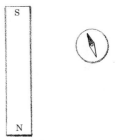

**FIGURE 18.2**
A compass needle near a bar magnet. The dark end is the north pole of the needle.

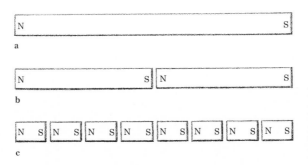

**FIGURE 18.3**
A bar magnet broken in half and then into smaller fragments. Each fragment is a complete magnet with both a north and a south pole.

the magnetic field at any point is given by the direction of a compass needle placed at that point.

Figure 18.5 shows the magnetic field around a bar magnet. As with the electric field, the lines of force are drawn closer together wherever the magnitude of the magnetic field is stronger. A comparison of Figs. 18.5 and 16.14 shows that the magnetic field around a bar magnet is identical to the electric field around an electric dipole. This is not surprising when you realize that a magnet is a magnetic dipole. That is, it is a system composed of two opposite poles separated by a distance $d$.

**REMARK** Although isolated magnetic poles do not exist in nature, magnetic dipoles do. Elementary particles are the ultimate carriers of magnet dipoles, just as they are the ultimate carriers of charge. Normally the magnetic dipoles of the electrons in matter are oriented in all directions, so that they cancel each other (Fig. 18.6$a$). In a permanent magnet, however, many of these dipoles are aligned in the same direction, so that the substance as a whole acts as a single large magnetic dipole that is the vector sum of the elementary dipoles (Fig. 18.6$b$). This is discussed more fully in Sec. 18.3.

Figure 18.7 shows the magnetic field of a C magnet, which is a magnet bent so that its two poles face each other. This configuration has many important applications because it produces a nearly uniform field in the gap between the two poles.

## 18.2 CURRENTS AND MAGNETISM

Most of the basic facts about static electricity, currents, and permanent magnets were known by the end of the eighteenth century. Because of the similarities between electric and magnetic phenomena, many scientists suspected a connection between them, but none had yet been discovered. Then in 1820 the Danish physicist Hans Christian Oersted (1777–1851) made an epochal discovery that changed civilization. While performing some demonstrations for a small group of students, Oersted noticed that a compass needle was deflected whenever there was a current in a nearby wire. Experimenting further, he discovered that a current-carrying wire has the same effect on a compass needle as a magnet. This was one of the long-sought connections between electricity and magnetism: magnetism is produced by a current.

**REMARK** Until Oersted's discovery of electromagnetism, electricity had been a laboratory curiosity, with no practical application. The discovery of electromagnetism was the breakthrough that led to the practical utilization of electricity. For example, the telegraph, which initiated the modern era of high-speed communication, was developed within 20 years of Oersted's discovery.

André Marie Ampère (1775–1836) investigated Oersted's discovery further and soon developed a complete mathematical description of the relationship between currents and magnetism. He found that the magnetic lines of force near a current-carrying wire are in the form of concentric circles about the wire, as shown in Fig. 18.8. This can be demonstrated by placing a small compass in a plane perpendicular to the wire. The compass needle, which points in the direction of the field where it is located, is always found to be perpendicular to

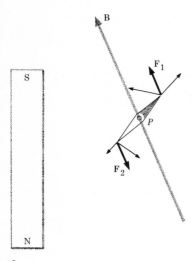

a

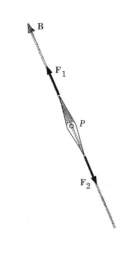

b

**FIGURE 18.4**
A compass needle near a bar magnet. ($a$) When the needle is not aligned with the magnetic field, the forces $\mathbf{F}_1$ and $\mathbf{F}_2$ exerted by the magnet on the needle cause the needle to rotate. ($b$) Only when the needle is aligned with the field do $\mathbf{F}_1$ and $\mathbf{F}_2$ exert no torque about the pivot.

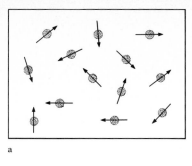

a

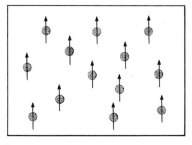

b

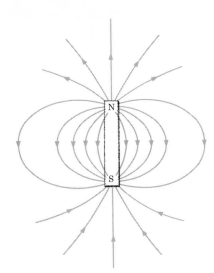

FIGURE 18.5
Magnetic field around a bar
magnet.

FIGURE 18.6
(*a*) The magnetic dipoles of the
electrons are randomly arranged in
ordinary matter. (*b*) The magnetic
dipoles of the electrons are aligned
in magnetized material.

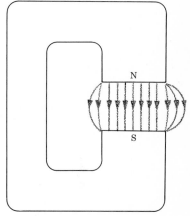

FIGURE 18.7
Magnetic field in the gap of a C
magnet.

the radius $r$ drawn from the wire to the needle, which indicates that
the field forms a circle about the wire. The relation between the
direction of the field and the direction of the current is given by the
*right-hand rule: When a wire is grasped with the right hand in such
a way that the thumb points in the direction of the current, the fingers
encircle the wire in the same sense as the magnetic field.*

The magnitude $B$ of the magnetic field at a point near a very long
current-carrying wire is proportional to the current $I$ and inversely
proportional to the perpendicular distance $r$ from the point to the
wire. In symbols this relation is

$$B = k\frac{I}{r} \qquad 18.1$$

where $k$ is a constant of proportionality. In the mks system of units,
where the unit of current is the ampere and the unit of distance is
the meter, the unit of magnetic field strength is the *tesla* (T). It is
defined by taking $k$ to be exactly $2 \times 10^{-7}$ T-m/A. That is, by definition the magnitude of the magnetic field one meter from a long wire
that has a current of one ampere is

$$B = k\frac{I}{r} = 2 \times 10^{-7}\ \text{T-m/A}\frac{1\ \text{A}}{1\ \text{m}}$$
$$= 2 \times 10^{-7}\ \text{T}$$

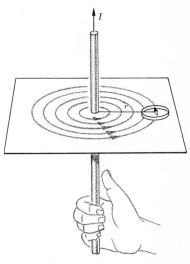

FIGURE 18.8
Magnetic field around a
current-carrying wire. The relation
between the current and the field
is the same as that between the
thumb and the fingers of the right
hand.

The constant $k$ is usually written

$$k = \frac{\mu_0}{2\pi}$$

where

$$\mu_0 = 4\pi \times 10^{-7} \text{ T-m/A}$$

is called the *magnetic permeability*. In terms of $\mu_0$, Eq. 18.1 becomes

$$B = \frac{\mu_0 I}{2\pi r} \qquad\qquad 18.2$$

**REMARKS** (1) The tesla is not a fundamental unit. As shown in Sec. 18.4, it is related to the newton, the ampere, and the meter by

$$1 \text{ T} = 1 \text{ N/A-m}$$

As is well known, the newton and the ampere are in turn related to the fundamental units of length, mass, time, and charge. (2) The gauss (G) is another common unit of magnetic field strength. It is related to the tesla by

$$1 \text{ T} = 10^4 \text{ G}$$

or

$$1 \text{ G} = 10^{-4} \text{ T}$$

As an example, let us calculate the field at a point 5 cm from a wire that has a current of 3 A. From Eq. 18.2 the field is

$$B = \frac{(4\pi \times 10^{-7} \text{ T-m/A})(3 \text{ A})}{2\pi \times 0.05 \text{ m}}$$

$$= 1.20 \times 10^{-5} \text{ T} = 0.12 \text{ G}$$

In the middle latitudes, the magnitude of the earth's magnetic field is between 0.1 and 0.4 G, so that the effect of the field of the wire on a compass needle is small compared to the effect of the earth's field.

The magnetic field produced by a current-carrying wire is greatly increased if the wire is formed into a circular coil with many turns. This can be understood by considering the coil with two turns shown in Fig. 18.9. Each turn produces a field proportional to the current $I$ in the wire, so that the two turns together produce a field proportional to $2I$. Similarly, a coil with $n$ turns produces a field proportional to $nI$. Since the center of a coil of radius $a$ is the distance $a$ from the wire, we might expect from Eq. 18.2 that the field at the center is

$$B = \frac{\mu_0 n I}{2\pi a}$$

This is not quite correct, however, because Eq. 18.2 is valid only for a very long straight wire, whereas here we are dealing with a circular wire. The correct expression for a circular coil is

$$B = \frac{\mu_0 n I}{2a} \qquad\qquad 18.3$$

which differs from the previous expression only by a factor of $\pi$, due to the difference in geometry between a straight wire and a circular wire. The field at the center of the coil is perpendicular to the plane of the coil.

For example, consider a coil of radius 5 cm that has 100 turns of

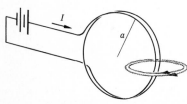

FIGURE 18.9
The magnetic field produced by a coil with two turns is the sum of the fields produced by each turn separately.

wire. When the current in the wire is 3 A, the magnetic field at the center of the coil is

$$B = \frac{(4\pi \times 10^{-7} \text{ T-m/A})(100)(3 \text{ A})}{2 \times 0.05 \text{ m}}$$

$$= 3.77 \times 10^{-3} \text{ T} = 37.7 \text{ G}$$

which is 100 times the earth's magnetic field.

Figure 18.10 shows a compass at the center of a coil containing many turns. When the switch $S$ is closed, the compass aligns itself parallel to the magnetic field produced by the current. When the switch is opened, the field vanishes and the needle swings back to a north-south orientation. Thus the orientation of the needle indicates whether there is a current in the wire.

**REMARK**  The first telegraph line, built by William Fothergill Cooke (1806-1879) and Charles Wheatstone (1802-1875) in 1838, used a coil and compass device to receive messages from a distant transmitter. A message was transmitted by an operator opening and closing a switch in accordance with a predetermined code and was received by another operator noting the corresponding deflections of the compass needle (Fig. 18.10).

The line, built in 1838, ran 13 mi from Paddington Station (London) to West Drayton. It was extended farther in the following years, but the general public first became aware of the importance of the new device in 1845. A man suspected of murder was seen to board a train in Slough for Paddington Station. His description was telegraphed to Paddington, where he was apprehended as he got off the train.

The first telegraph line in the United States was built by Samuel Morse (1791-1872) in 1844. It ran 40 mi from Washington to Baltimore. The Morse receiver is discussed in the next section.

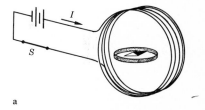

a

## 18.3  FERROMAGNETISM

The extensive use made of electricity in the modern world would not be possible if it were not for a peculiar property of iron, called *ferromagnetism*. Other transition elements (Table 19.2), such as nickel and cobalt, also exhibit this property, but iron is the only abundant ferromagnetic material.

Ferromagnetism is the tendency of the magnetic dipoles of the outer electrons of an atom to align themselves parallel to the magnetic dipoles of the corresponding electrons in a neighboring atom. In a permanent magnet, the dipoles of the outer electrons of all the atoms in a large volume of the material are aligned, so that their magnetic fields add together to produce a large magnetic field outside the magnet. In unmagnetized iron, the dipoles are aligned within a small volume, or *domain*, but the direction of alignment is different in different domains (Fig. 18.11). Consequently, the magnetic fields of these domains cancel each other, and there is no field outside the iron. Each domain, which is typically a fraction of a millimeter in width, acts as a tiny permanent magnet, so that unmagnetized iron can be thought of as a random arrangement of millions of these magnets.

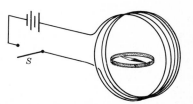

b

FIGURE 18.10
(*a*) When the switch is closed, the compass needle aligns itself with the magnetic field at the center of the circular coil. (*b*) When the switch is opened, the compass needle swings back to a north-south orientation.

When unmagnetized iron is placed in an external magnetic field, domains aligned parallel to the field increase in size at the expense of the other domains. Figure 18.12 shows this for a piece of iron near one pole of a magnet. Because of the increase in the size of the domains of iron parallel to the external field, the iron develops a magnetization in the direction of the field. As seen from Fig. 18.12, the end of the iron nearest the magnet's north pole develops a south pole, so that the iron is attracted to the magnet. The situation is analogous to that of a piece of cork near an electrified rod (Fig. 16.5).

An *electromagnet* consists of a wire coil wrapped around an iron cylinder (Fig. 18.13). The magnetic field produced by a current in the

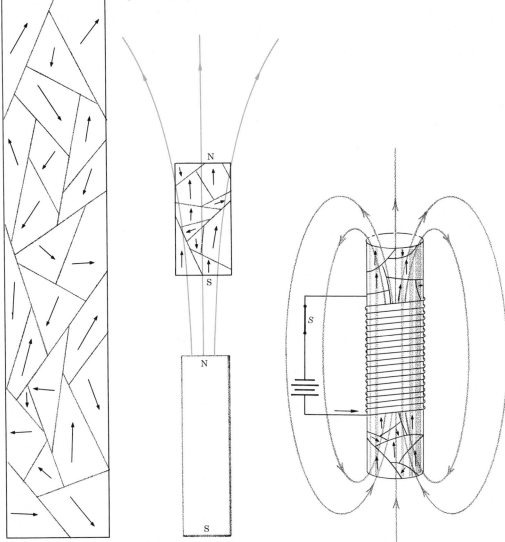

FIGURE 18.11
The domains in a bar of unmagnetized iron. In unmagnetized irons the magnetizations of the domains are randomly oriented.

FIGURE 18.12
A piece of iron in the magnetic field of a magnet. The domains of the iron that are parallel to the external field increase in size at the expense of the other domains.

FIGURE 18.13
An electromagnet. The magnetic field of the coil magnetizes the iron, which greatly increases the magnetic field around the iron.

coil increases the size of the domains in the iron that are magnetized in the direction of the field. The iron thus develops its own magnetic field, which adds to the field of the current. The field of the iron can easily be thousands of times larger than the field of the current alone, so that the iron in effect magnifies the magnetic field of the current. Furthermore, if the current is not too large, the domains instantly return to their original size when the current stops, and so the magnetic field vanishes. That is, the magnetic field of an electro-magnet exists only as long as there is a current in the coil.

**REMARK** The domains of a piece of iron placed in a very large magnetic field become permanently aligned in the direction of the field, so that the iron retains its magnetization after the external field is removed. Permanent mag-nets are made simply by placing a bar of iron in the magnetic field of a coil in which there is a very large current. In most applications, however, elec-tromagnets are operated at values of the current too small to magnetize the iron permanently.

Many important applications of electricity are based on the ability to turn a large magnetic field on and off with an electromagnet. We illustrate this by considering the following devices.

### The Morse telegraph

The receiver of the Morse telegraph (Fig. 18.14) consists of an electro-magnet mounted just below an iron bar, called the *clapper*. The clapper is held above one pole of the electromagnet by a spring. The circuit connecting the electromagnet to a battery is closed whenever the key at the transmitting end of the line is depressed. This causes a momentary current in the coil of the electromagnetic, which then attracts the clapper. When the key is released, the current stops and the spring returns the clapper to its resting position. Thus a coded message tapped on the key at one end of the line causes the clapper to make a corresponding sequence of clicking sounds at the other end of the line.

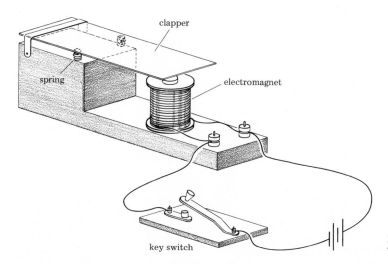

FIGURE 18.14
The Morse telegraph.

### Electric bell

An electric bell (Fig. 18.15) is similar to a telegraph receiver in that it consists of a clapper and an electromagnet that attracts the clapper when the circuit is closed. Each time the clapper moves toward the magnet, it strikes a metal bell, which makes a ringing sound. Moreover, the clapper is made part of the circuit in such a way that the movement of the clapper toward the magnet opens the circuit. Thus, just as the clapper hits the bell, the circuit opens and the spring returns the clapper to its resting position. Then, just as the clapper reaches its resting position, the circuit is closed and the clapper is again attracted to the magnet. Therefore, as long as the main switch $S$ is closed, the clapper will move back and forth, repeatedly striking the bell. This arrangement is one of the simplest mechanisms for producing continuous mechanical motion from electricity.

### Electric motors

A simple *direct-current* (dc) *motor*, like that used in a mechanical toy, consists of an electromagnet on a shaft (the *armature*) that rotates between the poles of a C magnet (Fig. 18.16). The ends of the wire of the electromagnet are connected to two metal contacts (*commutators*) on the shaft. The circuit is completed whenever the commutators touch the external contacts (*brushes*) that are connected to a battery. As the armature rotates, the commutators repeatedly reverse the connection between the battery and the electromagnet, so that the direction of the current in the electromagnet is repeatedly reversed.

When the armature is in the position shown in Fig. 18.16$a$, the current is directed in such a way that pole $A$ of the electromagnet is south and pole $B$ is north. Consequently, the attraction of the C magnet causes the armature to rotate counterclockwise. When the armature reaches the position shown in Fig. 18.16$b$, the brushes no longer make contact with the commutators, so that the electromagnet is momentarily turned off and thus is able to rotate past the poles of the C magnet. When the armature reaches the position shown in Fig. 18.16$c$, the brushes again contact the commutators, but with the connection to the battery reversed. Thus pole $A$ of the electromagnet is now north, and pole $B$ is south. Consequently, the attraction of the C magnet continues to rotate the armature in a counterclockwise direction.

**REMARK** The battery converts chemical energy into electric energy, which the electric motor converts into the mechanical energy of the rotating armature.

Most of the small electric motors used in clocks and other small household appliances are *synchronous alternating-current* (ac) *motors* (Fig. 18.17). These are similar to a dc motor except that it is the normal alternation of the current, and not the commutators, that reverses the polarity of the electromagnet. Figure 18.17 shows that the only difference between a synchronous motor and a dc motor is the form of the commutators. Since the armature of a synchronous motor must make one complete revolution during one complete alternation of the current, the speed of a synchronous motor is determined solely by the frequency of the current.

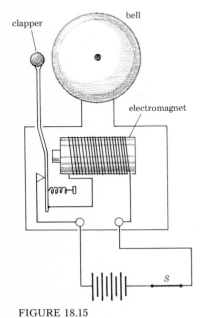

FIGURE 18.15
An electric bell.

### Forces on a current

Oersted's discovery showed that an electric current exerts a force, through its magnetic field, on a magnet. Therefore, according to Newton's third law (Property 3, Sec. 2.1), a magnet must exert a force, through its magnetic field, on a current. More generally, any external magnetic field exerts a force on a current.

Consider, for instance, a section of wire of length $l$ in which there is a current $I$ (Fig. 18.18). (This section of wire is part of a larger circuit that is not shown.) If the wire makes an angle $\theta$ with a uniform magnetic field $B$, the magntidue of the magnetic force $F_m$ on the wire is

$$F_m = BIl \sin\theta \qquad 18.4$$

When the field is parallel to the wire, the force is zero because $\sin 0° = 0$. When the field is perpendicular to the wire, the magnitude of the force is

$$F_m = BIl \qquad 18.5$$

because $\sin 90° = 1$. These are the only cases we shall have to consider.

**REMARK**   From Eq. 18.5 we see that the tesla, the unit of magnetic field strength, is related to the newton, ampere, and meter by

$$1\ T = \frac{1\ N}{1\ A \times 1\ m} = 1\ N/A\text{-}m \qquad 18.6$$

The direction of the magnetic force on a current is perpendicular to both the magnetic field and the current; it is found by using the right hand, as shown in Fig. 18.18. The index finger of the right hand is pointed in the direction of the current while the middle finger is pointed in the direction of the field. The thumb then points in the direction of the force.

Of course, a current-carrying piece of wire cannot exist by itself; it must be part of a larger circuit. Let us consider, therefore, the

**FIGURE 18.16**
A dc motor. (*a*) In this position, the commutators connect the coil to the battery in such a way that side *A* of the electromagnet is a south pole and side *B* is a north pole. As a consequence, the armature is rotated counterclockwise by the forces exerted on it by the C magnet. (*b*) In this position, the commutators disconnect the battery, so that the electromagnet rotates freely between the poles of the C magnet. (*c*) In this position, the commutators connect the coil to the battery in such a way that side *A* is a north pole and side *B* is a south pole. As a consequence, the armature continues to be rotated counterclockwise.

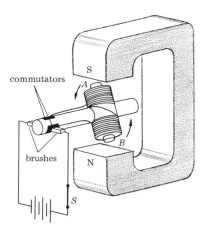

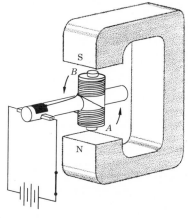

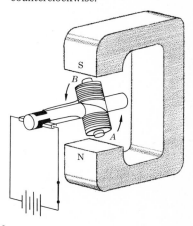

a                                    b                                    c

forces on the complete circuit shown in Fig. 18.19. Part of the circuit consists of a rectangular loop of wire of length $l$ and width $d$ placed between the poles of a C magnet. The loop is connected to a source of emf through wires that lie outside the magnet. Consequently, there is no magnetic force on these external wires, and so only the forces on the loop itself have to be considered.

Furthermore, two of the four straight sections of the loop are parallel to the magnetic field, so that the magnetic force on them is zero also. The other two sections are perpendicular to the field, so the magnitude of the magnetic force on each of them is $BIl$. The forces on these sections have equal magnitude, but opposite direction, so the total force on the loop is zero.

There is, however, a total torque on the loop, which tends to rotate it around the axis. The magnitude of this torque about any point on the axis is

$$\tau_m = \tfrac{1}{2}dF_m + \tfrac{1}{2}dF_m = dBIl$$
$$= BIA$$

where $A = ld$ is the area of the loop. If the loop contains $n$ turns of wire, the torque is

$$\tau_m = nBIA \qquad\qquad 18.7$$

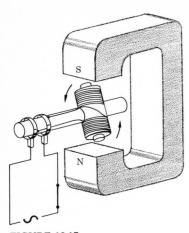

FIGURE 18.17
A synchronous ac motor. The commutators are circular rings that make continuous contact with the brushes. It is the alternation of the source current that reverses the polarity of the electromagnet.

For example, suppose a loop 6 cm high and 2 cm wide is placed in a magnetic field of 0.02 T. If the loop contains 200 turns and carries a current of 50 mA, the torque on it is

$$\tau_m = nBIA$$
$$= (200)(0.02 \text{ T})(50 \times 10^{-3} \text{ A})(12 \times 10^{-4} \text{ m}^2)$$
$$= 2.4 \times 10^{-4} \text{ T-A-m}^2$$
$$= 2.4 \times 10^{-4} \text{ N-m}$$

where the last equality is obtained by using Eq. 18.6 to convert from electrical to mechanical units.

Even though the magnetic torque on a coil is rather small, it is sufficient to rotate a delicately suspended coil. This fact is used in a number of instruments that measure current and potential difference.

A *galvanometer*, which is an instrument for measuring very small currents, consists of a many-turn coil suspended between the poles of a C magnet by a fine wire. The suspension is arranged in such a way that when there is no current in the coil, the coil is in the position shown in Fig. 18.19. When there is a small current in the coil, the magnetic torque $\tau_m$ on the coil causes it to rotate, thus twisting the suspension wire. As the wire is twisted, it exerts a countertorque $\tau_w$ proportional to the angle of rotation $\theta$. This torque can be written

$$\tau_w = k\theta$$

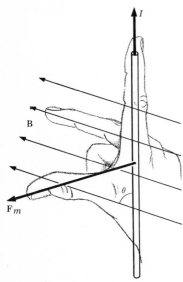

FIGURE 18.18
Force on a section of current-carrying wire in a magnetic field. The directions of the force, current, and field are related to each other as the thumb, index finger, and middle finger of the right hand.

where $k$ is a constant characteristic of the wire. At equilibrium the coil comes to rest at the angle at which $\tau_w$ equals $\tau_m$. Therefore, from Eq. 18.7 the condition for equilibrium is

$$k\theta = nBIA$$

or

$$I = \frac{k\theta}{nBA}$$

Thus the current in the galvanometer is determined by measuring the angle of rotation of the coil. With the use of very fine suspension wire, such a device will give measurable deflections with a current of $10^{-6}$ A.

An *ammeter* is a portable instrument for measuring current. It operates on the same principle as a galvanometer except that the coil is mounted on a pivot about which it rotates and the countertorque is applied to the coil by means of a spiral spring, as shown in Fig. 18.20. When there is a current in the coil, the coil rotates until the magnetic torque equals the countertorque exerted by the spring. A needle attached to the coil indicates the degree of rotation.

The current in a circuit is measured by connecting an ammeter in series with the circuit (Fig. 18.21). To avoid altering the circuit, the resistance of the ammeter must be small.

A *voltmeter* is an instrument for measuring the potential difference between two points. It consists of an ammeter in series with a large resistor $R_v$, as shown in Fig. 18.21. The potential $V$ across the resistor $R_1$ in Fig. 18.21 is measured by connecting the voltmeter in parallel with $R_1$. If $R_v$ is very much larger than $R_1$, the voltmeter will not significantly change the current in $R_1$. There will, however, be a small current

$$I' = \frac{V}{R_v}$$

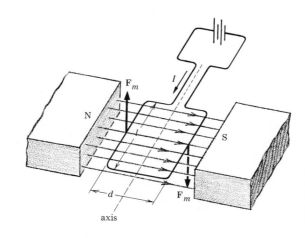

FIGURE 18.19
A rectangular loop in the magnetic field of a C magnet. In a galvanometer the loop is suspended by a fine wire.

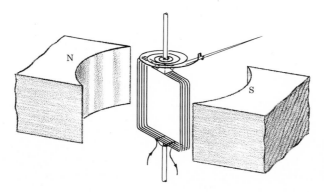

FIGURE 18.20
An ammeter. At equilibrium, the spiral spring balances the magnetic torque on the coil.

through the voltmeter, which causes the meter to deflect. Since $I'$ is proportional to $V$, the deflection of the meter is proportional to $V$.

### Forces on a moving charge

A magnetic force is exerted on a charged particle moving in a magnetic field. In fact, the magnetic force on a current-carrying wire is just the sum of the magnetic forces on the charges flowing in the wire. The magnetic force on a single charged particle can be found from Eq. 18.5 by considering the average current of a particle of charge $q$ moving with speed $v$ at right angles to a uniform magnetic field **B** (Fig. 18.22). The particle travels the distance $l$ from point $A$ to point $B$ in the time

$$t = \frac{l}{v}$$

so that the average current between $A$ and $B$ during this time is

$$I = \frac{q}{t} = \frac{q}{l/v} = \frac{qv}{l}$$

But from Eq. 18.5 the magnetic force on this current during this time is

$$F_m = BIl = B\frac{qv}{l}l$$

so the force that a magnetic field exerts on a moving charge is

$$F_m = Bqv \qquad 18.8$$

Like the magnetic force on a current, this force is perpendicular to both the magnetic field and the direction of motion of the particle.

It was found in Sec. 5.1 that the work done by a force **F** on a moving object is $F_x d$, where $F_x$ is the magnitude of the component of **F** parallel to the direction of motion of the object and $d$ is the distance the object moves. Since the magnetic force is always perpendicular to the direction of motion of a charged particle, its component parallel to the direction of motion is zero and consequently the magnetic force does no work on the particle. Therefore, according to the work-energy theorem (Sec. 5.2), a magnetic field cannot change the particle's speed, although it can change the particle's direction.

Magnetic fields are used in many devices to change and control the direction of a beam of charged particles. For instance, the electron beam in the picture tube of a television set is moved across the screen by magnetic fields. A television tube differs in this respect from a cathode-ray tube, which uses electric fields to deflect the electrons (Sec. 16.5). Similarly, the electrons scattered by a specimen in an electron microscope are focused by magnetic fields onto a fluorescent screen. The optics of an electron microscope is similar to that of a light microscope except that magnetic lenses, in the form of current-carrying coils, are used instead of glass lenses.

The *mass spectrometer*, an instrument that measures the mass of individual atoms and molecules, also uses a magnetic field to bend a beam of charged particles. A sample of the material to be analyzed

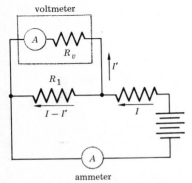

**FIGURE 18.21**
A voltmeter and an ammeter connected to a circuit to measure the potential difference $V$ across resistor $R_1$ and the current $I$ in the circuit. The voltmeter is just an ammeter in series with a large resistor $R_v$.

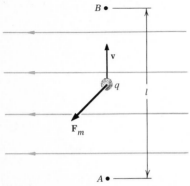

**FIGURE 18.22**
Force on a charged particle moving in a magnetic field.

is placed in a chamber $C$ of the spectrometer (Fig. 18.23), where an electric current forms ions by stripping electrons away from some of the atoms of the sample. These positively charged ions are then accelerated to the plate $P$, which is maintained at a large negative potential $-V = V_P - V_C$ with respect to $C$. Suppose that each atom loses only one electron, so the charge of an ion is $q = +e$. Then, if the kinetic energy $K_C$ of an ion is nearly zero when it leaves the chamber, its kinetic energy $K_P$ when it arrives at $P$ is†

$$K_P = K_C + e(V_C - V_P)$$
$$= eV$$

But, by definition, $K_P = \frac{1}{2}mv^2$, so that we have

$$\tfrac{1}{2}mv^2 = eV$$

where $m$ is the mass of the ion and $v$ is the speed of the ion when it reaches $P$. Thus the square of the speed of the ion at $P$ is

$$v^2 = \frac{2eV}{m} \qquad\qquad 18.9$$

Some of the ions reaching $P$ pass through an aperture into a region in which there is a constant magnetic field $B$ perpendicular to the direction of motion of the ions. This field, which is directed out of the plane of the diagram in Fig. 18.23, exerts a magnetic force of magnitude

$$F_m = Bqv = Bev \qquad\qquad 18.10$$

on each ion. Since this force is always perpendicular to the direction of motion of an ion, the ion moves in a circular path of radius $r$ with

† This result differs by a minus sign from Eq. 16.12 in Sec. 16.5 because here we are considering positively charged particles, whereas in Sec. 16.5 the particles are negatively charged.

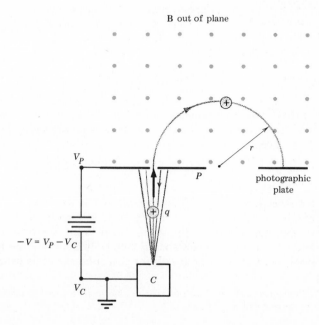

FIGURE 18.23
Mass spectrometer. Ions accelerated by the potential difference $-V$ move in circles of radius $r$ in the magnetic field **B**. The radius is determined from the position of the spot the ions make on the photographic plate.

a constant speed $v$. As a consequence, the ion has a centripetal acceleration

$$a = \frac{v^2}{r} \qquad\qquad 18.11$$

From Newton's second law (Property 7, Sec. 4.3)

$$F = ma$$

we see that Eqs. 18.10 and 18.11 are related by

$$Bev = m\frac{v^2}{r}$$

Therefore the ion's mass is related to its speed and radius by

$$m = \frac{Ber}{v}$$

Squaring both sides of this expression gives

$$m^2 = \frac{B^2 e^2 r^2}{v^2}$$

Then using Eq. 18.9 to replace $v^2$ by $2eV/m$, we get

$$m^2 = \frac{B^2 e^2 r^2}{2eV/m} = \frac{B^2 r^2 em}{2V}$$

or

$$m = \frac{B^2 r^2 e}{2V} \qquad\qquad 18.12$$

Equation 18.12 gives the mass $m$ of an ion in terms of the potential $V$, the magnetic field $B$, and the radius $r$ of the ion's orbit. The radius is obtained by measuring the position of the spot made by the ion when it strikes a photographic plate. Even if $V$ and $B$ are not known, Eq. 18.12 allows the mass of one ion to be measured relative to another.

For example, suppose that in a mass spectrometer carbon ions move in a circle of radius $r_C = 9.0$ cm, and oxygen ions move in a circle of radius $r_O = 10.4$ cm. Then if $V$ and $B$ are the same in both cases, we have

$$\frac{m_O}{m_C} = \frac{r_O{}^2}{r_C{}^2} = \frac{10.4^2}{9.0^2} = 1.33$$

so that the mass of an oxygen ion is 1.33 times the mass of a carbon ion. Since the mass of carbon is defined to be exactly 12 u, the mass of oxygen is†

$$m_O = 1.33 m_C = 1.33 \times 12 \text{ u}$$
$$= 16 \text{ u}$$

Knowledge of $V$ and $B$ is needed only if we want to relate atomic mass units to kilograms.

The masses of all the atoms have been determined with great accuracy with the mass spectrometer. In addition, the instrument is used to detect minute quantities of a substance in a sample material.

---

† We are neglecting the small (0.00055-u) difference between the mass of an atom and the mass of its ion.

The spectrometer produces a sequence of spots on a photographic plate, each of which corresponds to an atom or molecule of a specific mass. Since only a few atoms are needed to produce a detectable spot, the presence of trace amounts of a substance of known mass is readily detected.

## 18.5  MAGNETIC INDUCTION

Oersted's discovery of electromagnetism not only initiated the first practical use of electricity (telegraphy) but also stimulated a great deal of scientific research into electrical and magnetic phenomena. Michael Faraday (1791–1867), one of the nineteenth century's greatest scientists, began his electromagnetic investigations soon after learning of Oersted's work. Faraday reasoned that if an electric current produces a magnetic field, then a magnetic field should somehow be able to produce an electric current. He set about systematically trying one arrangement of wires and magnets after another until, in 1831, he found the effect for which he had been searching.

Faraday discovered *magnetic induction,* the creation of an electric field by a magnetic field. Magnetic induction ranks with electromagnetism as one of the fundamental principles on which our technology is based. For whereas electromagnetism provides the means whereby electric energy can be converted into mechanical work, magnetic induction provides the means whereby mechanical work can be converted into electric energy. All our electric power comes from generators that operate on the principle of magnetic induction.

Magnetic induction can best be understood by a demonstration of one of Faraday's first experiments with induction. A coil of wire is connected to a sensitive galvanometer, as shown in Fig. 18.24. As a magnet is moved toward the coil, the galvanometer is deflected, which indicates the presence of a current in the coil. The magnitude of the current depends on the speed with which the magnet is moved, and the current stops whenever the magnet is at rest. That is, a magnet at rest near the coil does not generate a current. As the magnet is moved away from the coil, the galvanometer is deflected in the opposite direction, which indicates that the direction of the current is reversed.

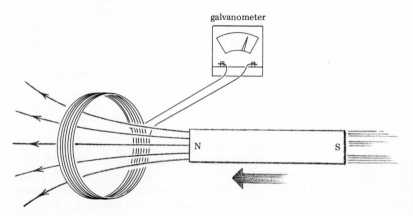

FIGURE 18.24
Magnetic induction. A current is generated in the coil as the magnet is moved toward it.

The current in the coil is the result of an electric field created by the moving magnetic field of the magnet. A current is also produced when the coil is moved toward the magnet, instead of the magnet moving toward the coil, since it is the relative motion of one with respect to the other that is responsible for the electric field. An electric field is always created by a moving magnetic field, even if no coil is present in which it can produce a current.

A detailed account of the theory of magnetic induction is beyond the scope of this book, but we shall discuss a few of its most significant consequences.

### Electric generators

In Sec. 17.1 we discussed how an electric current can be generated by means of electric induction (Sec. 16.2). All commercial electricity, however, is generated by means of magnetic induction. A magnetic induction generator, or *electric generator*, is identical to the ac motor shown in Fig. 18.17. When this device is used as a motor, an external source of alternating current is connected to the coil by the brushes. The alternation of the current causes the polarity of the electromagnet to reverse in such a way that the forces exerted on the electromagnet by the C magnet keep the armature rotating at constant speed.

When the device in Fig. 18.17 is used as a generator, the armature is rotated by an external source of power, usually a steam turbine. The motion of the coil in and out of the magnetic field of the C magnet generates a current in the external circuit. Each time the armature makes half a revolution, the coil reverses its orientation with respect to the magnetic field. This reverses the direction of the current, so that, as the armature is rotated, an alternating current is generated. Furthermore, if the armature is rotated at constant speed, the current will vary sinusoidally with time (Fig. 17.11).

**REMARK** A dc motor (Fig. 18.16) generates a direct current when its armature is rotated by an external power source because the split commutator reverses the connection to the external circuit every time the current is reversed.

An electric-power plant consists of a steam generator, a steam turbine, and an electric generator (Fig. 20.6). The steam is produced either from burning fossil fuel (coal and oil) or from a nuclear reactor (Sec. 20.3). The steam is the means whereby the energy of the fuel, whether fossil or nuclear, is converted into the mechanical energy of the turbine. The turbine drives the armature of the electric generator, which converts the mechanical energy of the turbine into electric energy. Thus, except for the fuels used to produce the steam, conventional and nuclear power plants are identical in the way they generate electricity.

### Electromagnetic waves

The concept of the electric and magnetic field was developed by Faraday as a pictorial way of viewing electromagnetic phenomena. The field concept was not generally used by other physicists, however, until Maxwell showed how all the basic laws of electricity and magnetism could be described by four equations involving the electric and magnetic fields. Known as *Maxwell's equations*, they are the starting point for all modern discussions of electromagnetism.

As originally formulated, each of Maxwell's equations is a quantitative description of one of the following four laws:

*1* Electric charges produce electric fields (*Coulomb's law*, Sec. 16.2).

*2* Isolated magnetic poles cannot exist (Sec. 18.1).

*3* Currents produce magnetic fields (*electromagnetism*, Sec. 18.3).

*4* A changing magnetic field produces an electric field (*magnetic induction*, Sec. 18.5).

Maxwell soon realized, however, that these equations were inconsistent with another fundamental law:

*5* The total charge of an isolated system cannot change (*conservation of charge*, Sec. 16.2).

Maxwell could eliminate this inconsistency only by assuming still another law:

*6* A changing electric field produces a magnetic field.

This last law, which is the counterpart of magnetic induction, modifies one of Maxwell's equations in such a way that they are now consistent with the conservation of charge.

Although the effect of a changing electric field is too small to be observed directly, it has profound consequences. For if a changing electric field produces a magnetic field and a changing magnetic field produces an electric field, then the possibility exists that a self-perpetuating disturbance of electric and magnetic fields might be produced. Maxwell showed that such an electromagnetic disturbance would act as a wave that propagates through space with the speed

$$v = \sqrt{\frac{4\pi K}{\mu_0}}$$

where $K = 9 \times 10^9$ N-m$^2$/C$^2$ = electric constant

$\mu_0 = 4\pi \times 10^{-7}$ N-s$^2$/C$^2$ = magnetic permeability

With these values of the constants, the speed is

$$v = \sqrt{\frac{4\pi(9 \times 10^9 \text{ N-m}^2/\text{C}^2)}{4\pi \times 10^{-7} \text{ N-s}^2/\text{C}^2}}$$
$$= \sqrt{9 \times 10^{16} \text{ m}^2/\text{s}^2} = 3 \times 10^8 \text{ m/s}$$

which is identical to the speed of light.

In an electromagnetic wave, it is the fields themselves, rather than the displacement of a medium, that propagate through space. Figure 18.25 shows how the electric and magnetic fields vary in a sinusoidal electromagnetic wave. Both fields are perpendicular to the direction of propagation, so the wave is transverse.

As a result of Maxwell's work, it soon became established that light is just a form of electromagnetic radiation of a certain wavelength. Thus the centuries of research into electric and magnetic phenomena led unexpectedly to the discovery of the true nature of light. This discovery is one of the major triumphs of nineteenth-century physics.

### Biomagnetism

We have seen throughout this book that some animals are able to detect a variety of unusual physical stimuli, e.g., ultrasonic sound, electric current, and the polarization of light. It seems reasonable to expect, therefore, that some animals can detect a magnetic field. This becomes even more likely when we consider that the earth's magnetic field has existed throughout evolutionary history and that it provides a reliable means of determining direction. Thus it would be of great benefit to migratory birds to be able to detect this field.

Some experiments on birds do seem to indicate that they have the ability to orient themselves with respect to a magnetic field. In one of these experiments [Wiltschko (1972)], migrating birds were captured and placed in cages. In the cages the birds still tended on the average to orient themselves in the direction of their migratory flight, but when large coils were placed on either side of the cage to change the direction of the magnetic field in the cage, the birds changed their average direction of orientation.

In spite of experiments such as this, however, the existence of a magnetic sense cannot be established until a physiological basis for the sense is found. As far as is known, birds contain no ferromagnetic material that could be influenced by the earth's magnetic field. Therefore magnetic induction seems to be the only reasonable mechanism for a magnetic sense. One consequence of induction is that a potential difference $V$ is established between any two points on an object moving through a magnetic field. In the special case in which the field **B**, the velocity of the object **v**, and the distance $l$ between the points are all perpendicular to each other (Fig. 18.26), this potential difference is

$$V = Bvl$$

For example, consider a bird flying at a speed of 10 m/s perpendicular to a magnetic field of 0.4 G. If the bird's head is 2 cm wide, the potential across it is

$$V = (0.4 \times 10^{-4} \text{ T})(10 \text{ m/s})(2 \times 10^{-2} \text{ m})$$
$$= 8 \times 10^{-6} \text{ V}$$

Although this is an extremely small potential difference, weak electric fish (Sec. 17.5) are able to detect even smaller potentials. In these

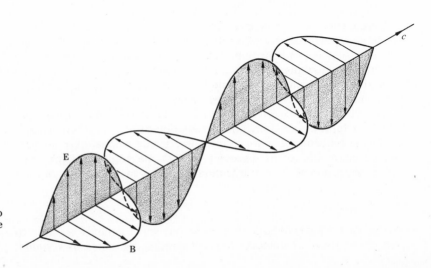

**FIGURE 18.25**
An electromagnetic wave. The electric and magnetic fields are perpendicular to each other and to the direction of propagation of the wave. Both **E** and **B** vary sinusoidally.

fish, however, cells specialized for this function have been identified, whereas no similar cells have yet been found in birds.

Furthermore, even if a bird in flight can detect a magnetic field, it does not explain experiments with caged birds, who presumably cannot move fast enough to generate as large a potential difference.

Needless to say, the subject of biomagnetism still has more questions than answers. Much more research is needed before the nature or even the existence of a magnetic sense can be definitely established.

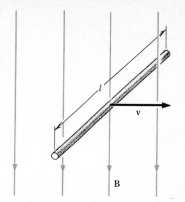

FIGURE 18.26
Potential difference induced across a wire moving in a magnetic field.

## PROBLEMS

*1* What is the magnitude of the magnetic field at a point 2 cm from a long wire carrying a current of 7 A?           *Ans.* $0.7 \times 10^{-4}$ T

*2* The magnitude of the magnetic field at a point 8 cm from a long wire is $0.3 \times 10^{-4}$ T. (*a*) What is the current in the wire? (*b*) What is the magnitude of the field at a point 24 cm from the wire?

*3* A long wire carrying a current of 15 A produces a magnetic field of $0.35 \times 10^{-4}$ T at a nearby point. What would the magnitude of the field at this point be if the current were 20 A?           *Ans.* $0.467 \times 10^{-4}$ T

*4* (*a*) Show that 1 T = 1 kg/C-s. (*b*) Show that $\mu_0$ has the units $N/A^2 = kg\text{-}m/C^2$.

*5* Figure 18.27 shows two long parallel wires separated by a distance of 15 cm. There is a current of 5 A in wire 1 and an opposite current of 10 A in wire 2. (*a*) Find the total magnetic field at point *A* on the line joining the wires. Point *A* is 9 cm from wire 1 and 24 cm from wire 2. (*b*) At what point on the line joining the wires is the magnetic field zero?
*Ans.* (*a*) $0.278 \times 10^{-5}$ T; (*b*) 15 cm from wire 1

*6* Find the total magnetic field at point *C* produced by the two currents in Fig. 18.27. Point *C* is 8 cm from wire 1 and 20 cm from wire 2. (*Hint:* The total field at a point is the vector sum of the fields due to each wire separately. Make a scale drawing to determine the direction of these fields.)

*7* Figure 18.28 shows two long parallel wires

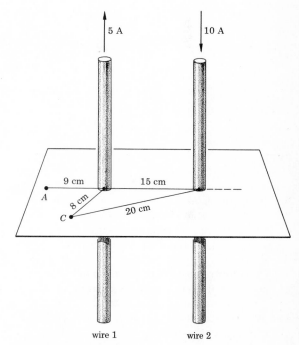

FIGURE 18.27   Problems 5, 6, and 15.

separated by a distance of 18 cm. There is a current of 8 A in wire 1 and a current of 12 A in wire 2. (*a*) Find the total magnetic field at the point *A*, which is on the line joining the wires and 3 cm from wire 1 and 15 cm from wire 2. (*b*) At what point on the line joining the wires is the magnetic field zero?
*Ans.* (*a*) $3.73 \times 10^{-5}$ T; (*b*) 7.2 cm from wire 1

*8* Find the total magnetic field at point *C* produced by the two currents in Fig. 18.28. Point

411

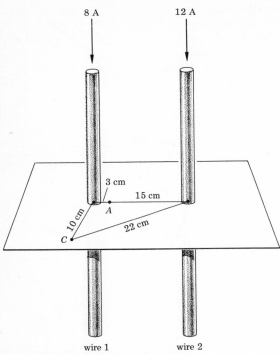

FIGURE 18.28   Problems 7, 8, and 11.

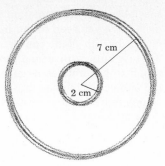

FIGURE 18.29   Problems 11 and 12.

C is 10 cm from wire 1 and 22 cm from wire 2 (see Prob. 6).

9 A circular coil of radius 4 cm consists of 250 turns of wire in which the current is 20 mA. What is the magnetic field in the center of the coil?                                    *Ans.* 0.785 G

10 A coil of radius 20 cm is to produce a field of 0.4 G in its center with a current of 0.25 A. How many turns must there be in the coil?

11 Figure 18.29 shows a coil of radius 2 cm concentric with a coil of radius 7 cm. Each coil has 100 turns. With a current of 5 A in the larger coil, find the currents needed in the smaller coil to give the following values for the total magnetic field at the center: (a) $9.0 \times 10^{-3}$ T, (b) $2.0 \times 10^{-3}$ T, (c) zero. In each case, determine whether the direction of the current in the smaller coil is the same as the current in the larger coil or opposite.

*Ans.* (a) 1.44 A same; (b) 0.793 A opposite; (c) 1.43 A opposite

12 Figure 18.29 shows a coil of radius 2 cm concentric with a coil of radius 7 cm. Each coil has 100 turns. With a current of 3 A in the small coil, find the total magnetic field at the center for each of the following currents in the larger coil: (a) $+7$ A, (b) $-5$ A, (c) $-12$ A. The minus sign indicates that the direction of the current in the larger coil is opposite in direction to the current in the smaller coil.

13 A wire loop is suspended between the poles of a magnet, as shown in Fig. 18.30. The horizontal section of the loop is 8 cm long, and the magnetic field is 3000 G. (a) What is the force on the loop when there is a current of 5 A in it? (b) What current is required to produce a force of 2.5 N on the loop?

*Ans.* (a) 0.12 N; (b) 104 A

14 A wire loop is suspended between the poles of a magnet, as shown in Fig. 18.30. When the

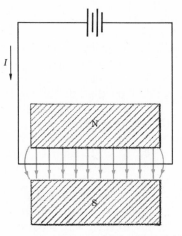

FIGURE 18.30   Problems 13 and 14.

current in the loop is 3 A, there is a magnetic force of 0.25 N on the 8-cm-long horizontal section of the loop. (*a*) What is the magnitude of the magnetic field? (*b*) What is the force on the loop when the current is 35 A?

15 (*a*) Calculate the force that wire 1 in Fig. 18.27 exerts on a 1-m-long section of wire 2. That is, calculate the force per unit length on wire 2. (*b*) Is this force attractive or repulsive? (*Hint:* Use Eq. 18.2 to find the magnetic field at wire 2 due to wire 1. Then use Eq. 18.5 to find the force on wire 2.)

    *Ans.* (*a*) $6.67 \times 10^{-5}$ N; (*b*) repulsive

16 Repeat Prob. 15 for the wires in Fig. 18.28.

17 (*a*) What is the force on a singly charged carbon ion moving with a speed of $3 \times 10^5$ m/s at right angles to a magnetic field of 7500 G? (*b*) What is the centripetal acceleration of the ion? (*c*) What is the radius of the circle in which the ion moves?

*Ans.* (*a*) $3.6 \times 10^{-14}$ N; (*b*) $1.81 \times 10^{12}$ m/s$^2$; (*c*) 4.97 cm

18 A singly charged ion of unknown mass moves in a circle of radius 12.5 cm in a magnetic field of 12,000 G. The ion was accelerated through a potential difference of 7000 V. What is the mass of the ion?

19 Natural carbon consists of two types of atoms, which have the same chemical properties but different masses (such atoms are called *isotopes;* see Sec. 20.1). The mass of the more abundant carbon isotope is taken to be exactly 12.0000 u. When carbon is placed in a mass spectrometer, two lines are formed on the photographic plate, corresponding to these isotopes. Suppose the lines indicate that the more abundant isotope moved in a circle of radius 15.0 cm, while the rarer isotope moved in a circle of radius 15.6 cm. What is the mass of the rarer isotope?     *Ans.* 13.0 u

20 After being accelerated through a potential difference of 5000 V, a singly charged carbon ion moves in a circle of radius 21 cm in the magnetic field of a mass spectrometer. What is the magnitude of the field?

21 A sample containing carbon (12 u), oxygen (16 u), and an unknown element is placed in a mass spectrometer. The carbon and oxygen lines are separated by 2.250 cm on the photo-

graphic plate, and the unknown element makes a line between them that is 1.160 cm from the carbon line. What is the mass of the unknown element?     *Ans.* 14 u

22 A sample containing sulfur (32 u), manganese (55 u), and an unknown element is placed in a mass spectrometer. The sulfur and manganese lines are separated by 3.20 cm, and the unknown element makes a line between them that is 1.07 cm from the sulfur line. What is the mass of the unknown element? What is the name of the element?

## BIBLIOGRAPHY

BECK, A. H. W.: "Words and Waves," McGraw-Hill Book Company, New York, 1967. History, theory, and operation of the modern forms of communication: telegraph, telephone, radio, and television.

GARRATT, G. R. M.: Telegraphy, in Charles Singer, E. J. Holmyard, A. R. Hall, and Trevor I. Williams (eds.), "A History of Technology," vol. IV, Oxford University Press, London, 1958. A brief history of the development of telegraphy. Most interesting is the account of attempts, as early as 1750, to build telegraphs using static electricity.

SHAMOS, MORRIS H.: "Great Experiments in Physics," Holt, Rinehart and Winston, New York, 1959. Includes excerpts from original articles by Oersted, Faraday, and Maxwell on their important discoveries in electromagnetism.

WILTSCHKO, WOLFGANG, and ROSWITHA WILTSCHKO: Magnetic Compass of European Robins, *Science,* **176:** 62 (1972). One of the few recent articles in English that describes experiments on the orientation of birds in magnetic fields. Although the data presented suggest that birds have a magnetic sense, more evidence is required before a definite conclusion can be reached.

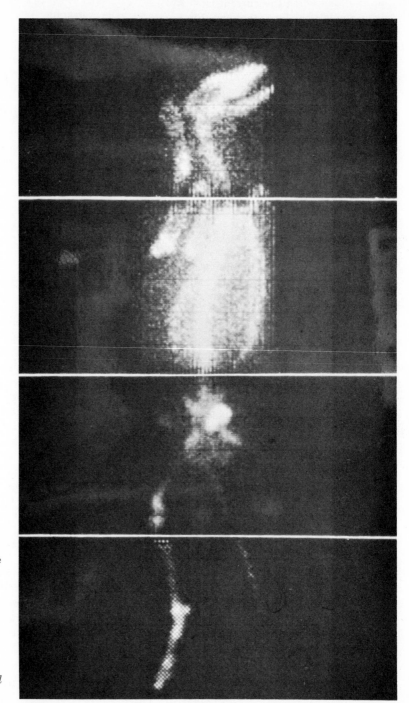

Positron-camera image of the bone structure of a dog. The dog was injected with $^{18}$F, a positron-emitting isotope of fluorine, which is rapidly absorbed by the bones. The image is obtained using two arrays of scintillation counters that detect the annihilation radiation (Sec. 20.5). [*Physics Research Laboratory, Massachusetts General Hospital, Boston.*]

# v modern physics

By the year 1875, the development of physics seemed to be complete. The laws of mechanics (Part I), thermodynamics (Chap. 11), and electromagnetism (Part IV) formed a coherent body of fundamental knowledge that appeared to include all the basic laws of the universe. This knowledge, now called *classical physics*, was so confidently felt to explain all physical phenomena that when Max Planck entered the University of Munich in 1875, he was discouraged from pursuing a career in physics because, as he was told, "all the important discoveries in physics have been made."

Contrary to this appraisal, however, the next 50 years proved to be the most exciting and fruitful period in the history of physics. Extraordinary discoveries were made in the fields of atomic and nuclear physics that could not be understood in terms of classical physics. It was found that fundamentally new principles and concepts were needed to reconcile the behavior of electrons and atoms with the behavior of ordinary-sized objects. The theories that emerged during this period—relativity and quantum mechanics—constitute *modern physics*, a coherent body of fundamental knowledge that replaces classical physics in discussions of atomic and nuclear phenomena but which is equivalent to classical physics in discussions of macroscopic phenomena.

# CHAPTER 19 atoms

In 1875 almost nothing was known about the structure of atoms. Indeed, some physicists of the period still questioned the existence of atoms, while others considered it impossible to ever investigate the inner structure of an atom. But so rapidly did physics develop in the next 50 years that by 1925 atomic structure was known in complete and exhaustive detail.

An important breakthrough came in 1913, when Niels Bohr developed a model of the atom that explained the characteristic spectrum of the radiation emitted by a gas when an electron beam passes through it. Bohr succeeded where others had failed because he was bold enough to introduce several new physical principles, even though they contradicted principles of classical physics. In the period from 1913 to 1925 these new principles were developed into quantum mechanics, a comprehensive theory of the motion of atomic particles that provides the necessary framework for the complete understanding of atomic phenomena.

This chapter introduces the basic principles of quantum mechanics and uses them to study atomic structure, the chemical properties of the elements, and the nature of the chemical bond.

## 19.1 WAVE-PARTICLE DUALITY

Classical physics makes a sharp distinction between waves and particles. A particle is localized in space and is characterized by a definite mass $m$. It can move with any speed $v$, and its kinetic energy is related to its mass and speed by

$$K = \tfrac{1}{2}mv^2$$

A wave, on the other hand, is extended in space and is characterized by a definite speed. It can have any amplitude $A$ and frequency $f$. According to classical physics, an electron is a particle and electromagnetic radiation (light) is a wave.

Quantum mechanics blurs the distinction between waves and particles by stating that all entities, such as electrons and light, have both a wavelike and a particlelike nature. This paradoxical concept has no counterpart in classical physics. It is impossible to form a mental picture of an entity that is simultaneously a wave and a particle because our mental pictures are based on our experiences in the macroscopic world of classical physics. Such nonclassical entities must be handled formally with the aid of mathematics.

The first hint that light had particle properties was given by Max

Planck (1858–1947) in 1900. On the basis of his study of the radiation emitted by heated objects (*blackbody radiation*), he showed that electromagnetic radiation is absorbed or emitted by an object only in discrete bundles of energy. For radiation of a given frequency $f$ this bundle, or *quantum*, of energy is

$$E = hf \qquad\qquad 19.1$$

where $h = 6.63 \times 10^{-34}$ J-s is a universal constant, called *Planck's constant.*

For example, a quantum of yellow light, which has a frequency of $6.0 \times 10^{14}$ Hz, is

$$\begin{aligned} E = hf &= (6.63 \times 10^{-34} \text{ J-s})(6.0 \times 10^{14} \text{ s}^{-1}) \\ &= 4.0 \times 10^{-19} \text{ J} \end{aligned}$$

This is the smallest amount of (yellow) light that can be absorbed (or emitted) in any process.

In particular, it is the smallest unit of light that can be absorbed by a rod cell in the retina of the eye. (A rod cell is so sensitive that it can send an impulse to the brain upon absorbing a single quantum of light. That is, a rod cell is sensitive enough to detect the smallest possible light signal.)

In 1905 Einstein carried the idea of light quanta a step further by proposing that light itself is composed of corpuscular units, called *photons.* A photon travels with the speed of light, and its energy is related to the frequency of the radiation by Eq. 19.1. Thus Newton's suggestion that light is corpuscular in nature was reintroduced into physics less than 100 years after the wave nature of light had been established by Young and Fresnel (Sec. 14.1). Only now the corpuscular nature of light somehow coexists with its wave nature. This wave-particle duality is built into Eq. 19.1, since the frequency $f$ is a wave property, whereas the quantum of energy $E$ is a corpuscular property.

Such a paradoxical concept would be considered completely absurd if it did not rest on incontrovertible experimental evidence. The wave nature of light is firmly established by the interference experiment of Young (Sec. 14.2). The particle nature of light was first clearly established by experiments on the photoelectric effect by Philipp Lenard (1862–1947) in 1902.

When radiation of certain frequencies is incident on a metal, electrons are ejected from the surface. This is the *photoelectric effect.* Lenard discovered that for a given metal there is a *critical frequency* $f_0$ for the ejection of electrons. Radiation with a frequency less than $f_0$ never ejects electrons, regardless of its intensity. Radiation with a frequency greater than $f_0$ always ejects some electrons, and the number of electrons ejected per second from a particular surface is proportional to the intensity of radiation incident on the surface. However, the energy of an individual electron is not affected by the intensity. That is, as the intensity of the radiation is increased, the number of electrons increases but the energy of each electron does not. Instead, the energy of an individual electron is proportional to the frequency of the radiation. That is, as the frequency of the radiation is increased, the energy of each electron increases.

The wave theory of light is unable to explain why an increase in intensity increases the number of electrons but not their energy, or why the energy of an electron is proportional to frequency. However, Einstein showed that these facts are easily explained if light is assumed to consist of photons, because then an individual electron can receive only the energy $hf$ of a single photon. Increasing the intensity of the radiation increases the number of photons hitting the metals, and hence the number of electrons ejected, but it does not increase the energy received by an individual electron. On the other hand, increasing the frequency of the radiation does increase the energy received by an individual electron. Because there is a minimum energy $E_{min}$ required to eject an electron from a metal, radiation with a frequency less than $f_0 = E_{min}/h$ cannot eject electrons.

**REMARK** Einstein wrote his paper on the photoelectric effect in the same year (1905) that he published his first paper on relativity. It was for his explanation of the photoelectric effect, rather than his work on relativity, that he received the Nobel prize in 1921.

In the photoelectric effect, a photon is destroyed and its energy is converted into the kinetic energy of an electron. The reverse process, in which the kinetic energy of an electron is converted into a photon, is the basis for the production of x-rays. X-rays are short-wavelength electromagnetic radiation produced when high-speed electrons, which have been accelerated through an electric potential $V$, suddenly collide in a metal electrode (Sec. 16.5). The kinetic energy of an electron just before collision is

$$K = eV$$

where $e$ is the electron's charge. During the collision this energy may be converted into many low-energy photons or a few high-energy photons. The highest-frequency (shortest-wavelength) radiation is produced when all the energy of an individual electron is converted into a single photon of the same energy. From Eq. 19.1 the frequency of this photon is

$$f = \frac{K}{h} = \frac{eV}{h}$$

and its wavelength is

$$\lambda = \frac{c}{f} = \frac{hc}{eV}$$

For example, the shortest-wavelength x-ray produced by a 10,000-V x-ray machine is

$$\lambda = \frac{hc}{eV} = \frac{(6.63 \times 10^{-34} \text{ J-s})(3.0 \times 10^8 \text{ m/s})}{(1.6 \times 10^{-19} \text{ C})(10^4 \text{ V})}$$

$$= 1.2 \times 10^{-10} \text{ m}$$

Because of their great penetrating power, x-rays of very short wavelength are desirable for many types of radiation therapy. To obtain such x-rays, machines are used that accelerate electrons through potentials of a million volts and more.

The wave-particle duality of light was accepted, although not fully understood, by the scientific community in the decade following Einstein's work on the photoelectric effect. This led Louis de Broglie (1892-    ) in 1923 to speculate that matter, and in particular electrons, may have wavelike properties and thus have a dual nature similar to that of light. In analogy with Planck's formula (Eq. 19.1), he proposed that the wavelength $\lambda$ of an electron is related to its mass $m$ and speed $v$ by the equation

$$mv = \frac{h}{\lambda} \qquad 19.2$$

This equation is an expression of the dual nature of the electron, because the mass on the left is a particle property, whereas the wavelength on the right is a wave property. This equation was verified in 1925, when George Thomson (1892-     ) in England and Clinton Davisson (1881-1958) in the United States independently obtained interference patterns using a beam of electrons instead of light. The experiments were similar in principle to Young's first interference experiment using a double slit except that Thomson and Davisson used the spacing between the atoms of a crystal instead of slits.

Modern quantum mechanics is based on the principle that all fundamental physical entities, such as electrons, protons, and photons, have a dual (wave-particle) nature. This radical break with classical physics has been the key to unlocking the secrets of the atom. Unfortunately, duality is an abstract concept which, like four-dimensional space, cannot be visualized. However, the implications of duality can be deduced mathematically. This can be done even when it is not completely clear what the symbols in the equations represent. For instance, Eq. 19.2 gives the wavelength of an electron without saying anything about the nature of the wave itself. Nevertheless, Eq. 19.2 is sufficient, as we shall see in the next section, to determine the basic structure of the atom. In Sec. 19.3 we shall say more about the nature of the electron wave.

## 19.2  BOHR'S MODEL OF THE ATOM

### Rutherford's scattering experiment

By the beginning of the twentieth century, it was known that an atom has a diameter of about $10^{-10}$ m and contains a number of negatively charged electrons. Since the atom as a whole is neutral, it was inferred that it must contain an equal quantity of positive charge. Furthermore, it was known that the mass of an atom is thousands of times greater than the mass of an electron, so that the positive charge must constitute almost all the mass of an atom. One early model of the atom assumed that the positive charge formed a continuous fluid filling the volume of the atom in which the electrons floated, like "plums in a pudding" (Fig. 19.1).

In 1911 Ernest Rutherford (1871-1937) tested this plum-pudding model by bombarding a thin sheet of gold foil with alpha particles, the high-speed positively charged particles emitted by radioactive elements such as radium (Sec. 20.2). The alpha particles, most of which passed through the foil without appreciable deflection, were detected by the light produced when they struck a fluorescent screen (Fig. 19.2). Occasionally, however, a particle was scattered through

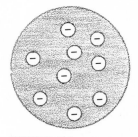

FIGURE 19.1
Pre-1911 plum-pudding model of the atom.

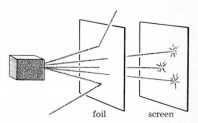

FIGURE 19.2
Rutherford's scattering experiment. Alpha particles from a radioactive source are aimed at a thin gold foil. The scattered particles are detected by the light emitted when they hit a fluorescent screen.

a large angle, and about 1 particle in 20,000 was scattered by more than 90°. This was very unexpected, because alpha particles are too heavy and move too fast to be appreciably deflected by an atom composed of electrons and a diffuse positively charged fluid.

Rutherford could account for the observed number of large-angle deflections only by assuming that almost all the mass of the atom is concentrated in a central core of radius $10^{-15}$ m. This core, or *nucleus*, is positively charged, and so exerts a repulsive electric force on an alpha particle that passes near it. Since the radius of the nucleus is only 1/100,000 the radius of the atom, almost all the alpha particles pass through the atom without getting close enough to the nucleus to experience an appreciable force. However, the few that do get close to the nucleus experience a very strong force and consequently are deflected through a large angle (Fig. 19.3).

In the model of the atom that emerged from Rutherford's work, the electrons were thought to revolve in orbits about the nucleus, just as the planets revolve about the sun. Most of the volume of an atom is filled with orbiting electrons, which are too light to deflect an incident alpha particle. The electrons are maintained in orbits about the nucleus by the attractive electric force exerted on them by the nucleus, just as the planets are maintained in orbits about the sun by the attractive gravitational force exerted on them by the sun.

The simplest atom, hydrogen, has only one electron in orbit about the nucleus. It is tempting to treat this case just as we treated the case of a satellite in orbit about the earth (Sec. 4.5). Thus suppose that the electron is revolving about the nucleus in a circular orbit of radius $r$ (Fig. 19.4). The force on the electron is the electric force (Eq. 16.2)

$$F_e = K \frac{q_e q_n}{r^2}$$

where $q_e$ and $q_n$ are the charges of the electron and nucleus, respectively. The electrons's charge is denoted $-e$, and the nucleus has the opposite charge $+e$, so

$$F_e = -K \frac{e^2}{r^2}$$

The minus sign indicates that this is an attractive force, directed toward the nucleus, as shown in Fig. 19.4.

An electron moving at constant speed $v$ in a circular orbit has the centripetal acceleration (Eq. 4.5)

$$a = \frac{v^2}{r}$$

directed toward the nucleus. From Newton's second law (Eq. 4.7) the force on the electron is equal to its mass times its acceleration, so

$$K \frac{e^2}{r^2} = m \frac{v^2}{r}$$

or

$$mv^2 = K \frac{e^2}{r} \qquad\qquad 19.3$$

**FIGURE 19.3**
Alpha particles traveling through an atom. Only particles that get close to the small massive nucleus are deflected through large angles.

alpha particles

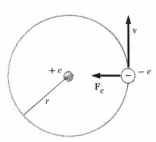

**FIGURE 19.4**
An electron moving in a circular orbit about a hydrogen nucleus.

This equation gives the speed $v$ of the electron for any orbital radius $r$.

The energy of the electron is the sum of its kinetic energy and its potential energy

$$E = K + U$$

The potential energy is

$$U = q_e V = -eV$$

where $V$ is the electric potential at a distance $r$ from the nucleus. From Eq. 16.11 this potential is

$$V = K\frac{q_n}{r} = K\frac{e}{r}$$

so the potential energy can be written

$$U = -eV = -K\frac{e^2}{r}$$

The minus sign indicates that the potential energy of an electron a distance $r$ from the nucleus is less than its potential energy at infinity.

The kinetic energy $K$ of the electron is $\frac{1}{2}mv^2$. From Eq. 19.3 this can be written

$$K = \tfrac{1}{2}mv^2 = \tfrac{1}{2}K\frac{e^2}{r}$$

so the total energy of the electron is

$$E = K + U = \tfrac{1}{2}K\frac{e^2}{r} - K\frac{e^2}{r}$$

$$= -\tfrac{1}{2}K\frac{e^2}{r} \qquad\qquad 19.4$$

Again the minus sign indicates that the energy of an electron in orbit about the nucleus is less than the energy it would have if it were at rest and infinitely far from the nucleus. Thus $-E$ is the energy required to separate the electron from the nucleus completely, i.e., to *ionize* the atom.

### Bohr's quantum condition

This was the situation in 1912, when Niels Bohr (1885–1962) was a student of Rutherford's at the University of Manchester. Bohr understood that this orbital model, although very appealing, could not be entirely correct, because it did not explain, for instance, why all atoms of hydrogen have identical chemical properties. According to classical physics, the electron can be in an orbit of any radius, and so there is a continuous range of possible energies for the electron. Yet hydrogen behaves as though all its atoms have the same energy. Furthermore, even if the electron of each atom was originally in a specially preferred orbit, in time this orbit would change because of collisions between atoms.

An even more serious objection to this model comes from electromagnetic theory, which says that a charged particle emits electromagnetic radiation whenever it is accelerated. In particular, an orbiting electron is expected to radiate electromagnetic radiation at the expense of its own energy. Consequently the orbital motion of an electron would be unstable, because the electron would spiral closer and closer to the nucleus as it radiated away its energy.

Bohr worked on these problems after he returned to Copenhagen in 1913, trying to introduce the new quantum ideas of Planck and Einstein into the model. What was needed was some restriction on Eq. 19.3 that would limit the possible orbits. He succeeded in finding this restriction and so obtained, for the first time, a quantitative description of the atom.

The Bohr condition is best understood today in terms of the wave nature of the electron, though this was discovered 10 years after Bohr's original paper. Recall (Sec. 12.5) that a standing wave on a string fixed at both ends can have only certain wavelengths, determined by the condition that there must be an integral number of half-wavelengths on the string (Fig. 19.5). Suppose, similarly, that only standing electron waves exist in an atom. Figure 19.6 shows that only an integral number of whole wavelengths can exist on a circular string because a wave with an odd number of half-wavelengths does not join smoothly together (Fig. 19.6b). Thus the condition for a standing wave on a circle of radius $r$ is

$$n\lambda = 2\pi r$$

where $n$ is an integer ($n = 1, 2, 3, \ldots$) and $2\pi r$ is the circumference of the circle. This condition, in conjunction with Eq. 19.2 for the wavelength of an electron wave, gives

$$mv = \frac{h}{\lambda} = \frac{h}{2\pi r/n} = \frac{nh}{2\pi r}$$

which is a nonclassical equation relating the speed of the electron and the radius of the orbit. Squaring both sides of this equation, we get

$$m^2v^2 = \frac{n^2h^2}{4\pi^2r^2}$$

or $\qquad m v^2 = \dfrac{n^2h^2}{4\pi^2mr^2}$ $\qquad$ Bohr's quantum condition $\qquad$ 19.5

This is the additional restriction needed to limit the possible orbits given by Eq. 19.3. Combining Eqs. 19.3 and 19.5, we get

$$K\frac{e^2}{r} = \frac{n^2h^2}{4\pi^2mr^2}$$

or $\qquad r = \dfrac{n^2h^2}{4\pi^2Kme^2}$ $\qquad\qquad$ 19.6

According to Bohr, the only allowed orbits of an electron are those obtained from Eq. 19.6 with integer values of $n$.

For example, the smallest orbit occurs with $n = 1$. Substituting the known values of $h$, $m$, $e$, and $K$ into Eq. 19.6, we get the radius

$$r = \frac{(1^2)(6.63 \times 10^{-34} \text{ J-s})^2}{4\pi^2(9.0 \times 10^9 \text{ N-m}^2/\text{C}^2)(9.11 \times 10^{-31} \text{ kg})(1.60 \times 10^{-19} \text{ C})^2}$$

$$= 0.0053 \times 10^{-8} \frac{\text{J}^2\text{-s}^2}{\text{N-m}^2\text{-kg}}$$

$$= 0.53 \times 10^{-10} \text{ m}$$

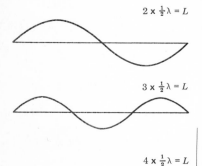

$1 \times \frac{1}{2}\lambda = L$

$2 \times \frac{1}{2}\lambda = L$

$3 \times \frac{1}{2}\lambda = L$

$4 \times \frac{1}{2}\lambda = L$

FIGURE 19.5
Possible standing waves on a string fixed at both ends.

This quantity, known as the *Bohr radius*, is consistent with what was known in 1913 about the size of a hydrogen atom. The next orbit occurs when $n = 2$, and its radius is 4 ($2^2$) times the Bohr radius; the radius of the $n = 3$ orbit is 9 ($3^2$) times the Bohr radius, and so on.

The allowed energies $E_n$ of an electron in a hydrogen atom are obtained by substituting the allowed radii from Eq. 19.6 into Eq. 19.4:

$$E_n = -\tfrac{1}{2}K\frac{e^2}{r} = -\frac{2\pi^2K^2me^4}{n^2h^2} \qquad 19.7$$

The lowest energy $E_1$ is given by Eq. 19.7 with $n = 1$:

$$E_1 = -\frac{2\pi^2(9.0\times10^9 \text{ N-m}^2/\text{C}^2)^2(9.11\times10^{-31} \text{ kg})(1.60\times10^{-19} \text{ C})^4}{(1^2)(6.63\times10^{-34} \text{ J-s})^2}$$

$$= -2170\times10^{-21} \text{ m}^2\text{-kg/s}^2 = -2.17\times10^{-18} \text{ J}$$

From Eq. 19.7 we see that the $n$th energy $E_n$ is

$$E_n = \frac{E_1}{n^2} = -\frac{2.17\times10^{-18} \text{ J}}{n^2}$$

For instance, with $n = 2$, the energy is

$$E_2 = -\frac{2.17\times10^{-18} \text{ J}}{2^2} = -0.542\times10^{-18} \text{ J}$$

which is greater than $E_1$ since it is less negative.

In atomic physics it is convenient to measure energies in electron volts. An electron volt is the energy gained by an electron in passing through a potential difference of 1 V (Sec. 16.5), so that the conversion between electron volts and joules is

$$1 \text{ eV} = e \times 1 \text{ V} = 1.60\times10^{-19} \text{ J}$$

or
$$1 \text{ J} = 6.25\times10^{18} \text{ eV}$$

For instance, the lowest energy $E_1$ of hydrogen is

$$E_1 = (-2.17\times10^{-18} \text{ J})(6.25\times10^{18} \text{ eV/J})$$
$$= -13.6 \text{ eV}$$

and the $n$th energy is

$$E_n = -\frac{13.6 \text{ eV}}{n^2} \qquad 19.8$$

**REMARK** It is useful to express Planck's constant in electron volts rather than joules. The conversion is

$$h = (6.63\times10^{-34} \text{ J-s})(6.25\times10^{18} \text{ eV/J})$$
$$= 4.14\times10^{-15} \text{ eV-s}$$

The allowed energies of the electron in a hydrogen atom form an increasing series, starting with the lowest energy ($E_1 = -13.6$ eV) and ending with the highest energy ($E_\infty = 0$). This is shown in Fig. 19.7, where each allowed energy is represented by a horizontal line. An electron can only be in an orbit, or *state*, corresponding to an allowed energy. Normally the electron is in the lowest, or *ground* state, from which it cannot radiate away energy since it already has its lowest possible energy.

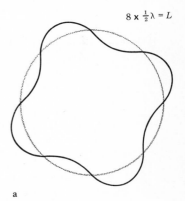

$8 \times \tfrac{1}{2}\lambda = L$

a

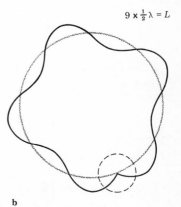

$9 \times \tfrac{1}{2}\lambda = L$

b

FIGURE 19.6
(*a*) A possible standing wave on a circular string. (*b*) A wave with an odd number of half-wavelengths does not fit smoothly on a circular string.

Furthermore, since there is a 10.2-eV difference between the ground state of hydrogen and the next higher state, a hydrogen atom in its ground state must absorb at least 10.2 eV in order to change its state. The kinetic energy of a molecule at room temperature is only about 0.02 eV, so that a hydrogen atom is unaffected by collisions with other gas molecules. Thus an atom in its ground state is completely stable against radiating away energy and is relatively stable against absorbing energy. All hydrogen atoms normally have identical properties because they are all in their ground state.

### Atomic spectra

Hydrogen atoms can be excited into higher energy states by bombarding them with a beam of energetic electrons. This is most conveniently done in a gas-discharge tube (Fig. 19.8), which is similar to the vacuum tubes discussed in Sec. 16.5 except that it contains hydrogen (or some other gas) at very low pressure. Metal plates, or electrodes, are mounted inside the sealed tube, and wires attached to the electrodes pass through the wall of the tube. One electrode (the cathode) is heated to drive electrons out of it, and these electrons are attracted to the other electrode (the anode), which is maintained at a positive potential relative to the cathode. The electrons gain kinetic energy as they move toward the anode, and occasionally they collide with a hydrogen atom. In the collision process some of the electron's kinetic energy may be transferred to the atom by promoting the atom's electron to a higher energy state. At least 10.2 eV must be absorbed to excite an atom to the $n = 2$ state, and an atom that absorbs more than 13.6 eV is ionized, its electron being permanently separated from its nucleus. As a result of collisions between electrons and atoms, the atoms of the gas in a discharge tube are in all possible states.

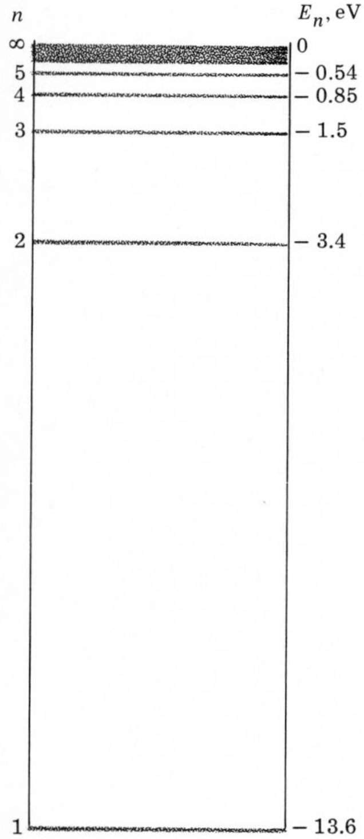

An atom in an excited state soon radiates away some of its excess energy and falls into a lower energy state. Bohr showed that this radiation consists of a single photon whose frequency $f$ is related to the energy change by Eq. 19.1. That is, if the atom makes a transition directly from a state of energy $E_n$ to a state of energy $E_{n'}$, a single photon of frequency

$$f = \frac{E_n - E_{n'}}{h} \qquad 19.9$$

is emitted. For instance, if the atom makes a transition from the $n = 3$ state to the ground state ($n = 1$), the energy change is

$$E_n - E_{n'} = -\frac{13.6\ \text{eV}}{n^2} - \left(-\frac{13.6\ \text{eV}}{n'^2}\right)$$

$$= -\frac{13.6\ \text{eV}}{3^2} - \left(-\frac{13.6\ \text{eV}}{1^2}\right)$$

$$= -1.51\ \text{eV} - (-13.6\ \text{eV})$$

$$= 12.1\ \text{eV}$$

and the frequency of the emitted photon is

$$f = \frac{E_n - E_{n'}}{h} = \frac{12.1\ \text{eV}}{4.14 \times 10^{-15}\ \text{eV-s}}$$

$$= 2.92 \times 10^{15}\ \text{Hz}$$

| $n$ | $E_n$, eV |
|---|---|
| $\infty$ | 0 |
| 5 | − 0.54 |
| 4 | − 0.85 |
| 3 | − 1.5 |
| 2 | − 3.4 |
| 1 | − 13.6 |

**FIGURE 19.7**
Energy levels of a hydrogen atom.

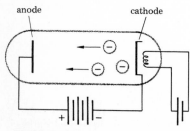

**FIGURE 19.8**
A gas-discharge tube.

which is ultraviolet radiation (see Fig. 14.1). This transition is represented in Fig. 19.9 by the arrow drawn from the $n = 3$ state to the $n = 1$ state.

On the other hand, an atom in the $n = 3$ state may first make a transition to the $n = 2$ state, emitting a photon of frequency

$$f = \frac{-1.51 \text{ eV} - (-3.40 \text{ eV})}{h}$$

$$= \frac{1.89 \text{ eV}}{4.14 \times 10^{-15} \text{ eV-s}}$$

$$= 4.56 \times 10^{14} \text{ Hz}$$

and then make a transition from the $n = 2$ state to the $n = 1$ state, emitting a photon of frequency

$$f = \frac{-3.40 \text{ eV} - (-13.6 \text{ eV})}{4.14 \times 10^{-15} \text{ eV-s}}$$

$$= 2.46 \times 10^{15} \text{ Hz}$$

In this two-step process, two photons are emitted, one in the visible region and one in the ultraviolet region. These transitions are represented in Fig. 19.9 by the arrows from the $n = 3$ to the $n = 2$ state and from the $n = 2$ to the $n = 1$ state.

The spectrum of the radiation emanating from a gas-discharge tube contains all the frequencies that can be obtained from transitions between any two energy states. Hundreds of the frequencies in the spectrum of hydrogen have been measured, and they all correspond to frequencies calculated from Eqs. 19.8 and 19.9. Thus the spectrum of radiation emitted by a gas in a discharge tube gives direct information about the energy levels of an atom. Hydrogen is the simplest atom to analyze, and its spectrum was used to verify Bohr's model of atomic structure. The energy levels of complex atoms cannot be calculated as easily as those of hydrogen, so they are determined experimentally from their spectra.

## 19.3  QUANTUM MECHANICS

Although it successfully describes the atom and correctly predicts the spectrum of radiation emitted by hydrogen, Bohr's model is based on intrinsically contradictory concepts. Specifically, newtonian mechanics, used to derive Eq. 19.3, allows a particle to be in an orbit of any radius, whereas Eq. 19.5 restricts the allowed radii with a condition foreign to newtonian mechanics. This contradiction was resolved only with the development of quantum mechanics, which is a comprehensive theory of the motion of particles of atomic size. For particles of ordinary size, quantum mechanics is equivalent to newtonian mechanics; i.e., quantum mechanics is more general than newtonian mechanics and reduces to it when the mass of the particle is large.

One form of quantum mechanics was developed in 1925 by Erwin Schrödinger (1887–1961) in terms of an equation for the electron wave de Broglie had proposed the year before (Sec. 19.1). This equation

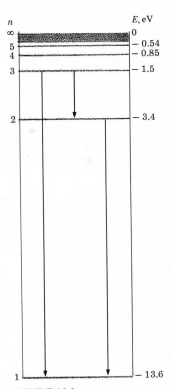

FIGURE 19.9
A hydrogen atom in the $n = 3$ state can get to the ground state either in a single transition or in a two-step transition.

is similar to the classical equations for waves on a string and for electromagnetic waves, but it differs in some essential details. Like these classical equations, it supposes that the particle wave obeys the superposition principle (Sec. 12.3). Unlike these other equations, however, the wave itself represents neither the displacement of a medium nor the magnitude of a physical quantity. Instead, the wave has to be interpreted as a *probability amplitude.*

To understand this, consider a particle of mass $m$ moving in a box with perfectly rigid walls, as shown in Fig. 19.10. According to new-tonian mechanics, the particle moves in a straight line at constant speed $v$ until it hits a wall of the box. It then bounces off the wall and moves in the opposite direction with the same speed $v$ until it hits the other wall. At any instant of time the particle has a definite position in the box, and in the course of time it moves back and forth in the box. The particle can have any speed $v$ and any kinetic energy $K = \frac{1}{2}mv^2$.

The quantum mechanics of Schrödinger represents the particle by a wave $\psi$.† Since the wave is confined in a finite space, only standing waves with certain wavelengths are possible. Like a wave on a string fixed at both ends, the particle wave must be zero at each wall of the box, and so a standing wave must contain an integral number of half-wavelengths (Fig. 19.11). If the box is of length $L$, the wavelength satisfies the condition

$$n\frac{\lambda}{2} = L$$

or

$$\lambda = \frac{2L}{n} \qquad\qquad 19.10$$

where $n$ is an integer. This equation is identical to Eq. 12.7. However, the wave $\psi$ has a very different interpretation.

At any point in the box the quantity $\psi^2$, the square of the wave, gives the probability of finding the particle at that point. Figure 19.12 shows the square of the wave in Fig. 19.11. From Fig. 19.12 we see that the particle is most likely to be found at points $A$, $B$, and $C$ and will never be found at points $W$, $X$, $Y$, and $Z$. The particle therefore cannot be thought of as having a definite position at each instant of time. All that can be said about the particle is that, at each instant of time, there is a definite probability of its being found at any point in the box.

**REMARK** This is perhaps the strangest assertion of quantum mechanics, and if you find it hard to believe, you are in good company because Einstein never accepted it. In a famous series of discussions he argued with Bohr that quantum mechanics is not a complete description of nature because it does not give the position of an electron at all times. Bohr argued that quantum mechanics is a complete description because it describes everything that is physically measurable.

To measure the position of an electron, it is necessary to reflect some radiation from it. Since radiation is quantized, at least one photon must strike the electron and bounce into the measuring instrument. In the process, however, the electron's energy is changed and the electron is no longer in its

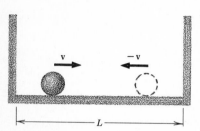

FIGURE 19.10
Classical description of a particle in a box.

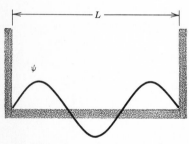

FIGURE 19.11
Quantum-mechanical description of a particle in a box.

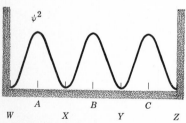

FIGURE 19.12
The square of the quantum-mechanical wave gives the probability of finding the particle at various points in the box.

†$\psi$ is the Greek letter *psi.*

original state. Thus the position of an electron in a particular state can be measured only once because as a consequence of the measurement itself the electron is knocked into another state.

In order for a particle to be said to have a definite position at every instant of time, it must be possible to monitor the particle's position continuously without otherwise disturbing it. For example, a marble rolling on a table has a definite position at every instant because the light used to observe it exerts a negligible force on it. But the photon nature of light limits the ultimate gentleness with which a particle can be observed. The motion of a very light particle, such as an electron, cannot be observed because a single photon bouncing off it is sufficient to alter the particle's motion radically.

It might be thought that the electron can possess a definite position at every instant even though it is not possible to observe it. Bohr, however, argued vigorously against this idea, showing that it is inconsistent with experiments on the interference of electron beams. Besides, a quantity that in principle cannot be measured cannot be physically meaningful.[†] Einstein had maintained much the same point of view regarding the nature of simultaneity (Sec. 14.6), but he never accepted Bohr's viewpoint of quantum mechanics. Today almost all physicists accept Bohr's contention that quantum mechanics provides a complete description of atomic processes. This viewpoint is sometimes referred to as the *Copenhagen interpretation* because it was developed by Bohr and his collaborators at the University of Copenhagen.

A particle moving in a box of length $L$ exists only in the states with wavelengths given by Eq. 19.10. From the de Broglie relation (Eq. 19.2), a particle with wavelength $\lambda$ and mass $m$ has a speed

$$v = \frac{h}{m\lambda}$$

so that the possible speeds of a particle in a box are

$$v = \frac{h}{2Lm/n} = \frac{nh}{2Lm} \qquad 19.11$$

and the possible kinetic energies are

$$K = \tfrac{1}{2}mv^2 = \frac{n^2h^2}{8L^2m} \qquad 19.12$$

Thus the energies of a particle in a box are quantized: only those energies are allowed which satisfy Eq. 19.12 with an integer value of $n$.

The lowest allowed energy occurs with $n = 1$. That is, a particle confined to a box of length $L$ must have a kinetic energy of at least $h^2/8L^2m$. This is the so-called *zero-point energy*. For example, an atom has a diameter of about $10^{-10}$ m, which means the electrons are effectively confined to a region of length $L = 10^{-10}$ m. Inserting this into Eq. 19.12, we obtain (with $n = 1$)

$$K = \frac{n^2h^2}{8L^2m} = \frac{(1^2)(6.63 \times 10^{-34}\text{ J-s})^2}{(8)(10^{-10}\text{ m})^2(9.11 \times 10^{-31}\text{ kg})}$$

$$= 6.03 \times 10^{-18}\text{ J} = 37.7\text{ eV}$$

which is the correct order of magnitude of the kinetic energy of an

[†] This is the operational viewpoint introduced in Sec. 1.2. It has had a major influence on the development of modern scientific philosophy.

electron in the ground state of hydrogen.† Equation 19.12 is not expected to give a more exact result, since it applies to a particle in a box and not to a particle attracted by a point charge. Nevertheless, Eq. 19.12 correctly shows that the smaller the region of confinement, the larger the kinetic energy.‡

For example, a nucleus consists of protons and neutrons confined to a region with a diameter $10^{-5}$ times the diameter of an atom. The mass of a proton is $2 \times 10^3$ times the mass of an electron, so from Eq. 19.12 we expect the smallest kinetic energy of a proton in a nucleus to be about

$$\frac{1}{(10^{-5})^2 (2 \times 10^3)} = 5 \times 10^6$$

times the kinetic energy of an electron in the ground state of an atom. This is in fact the case. The kinetic energy of a particle in a nucleus is more than a million times the energy of an electron in an atom. This is why nuclear reactions involve so much more energy than chemical reactions.

The existence of a zero-point energy seems to contradict the fact that an ordinary-sized particle can have zero energy. A marble, for instance, can sit inside a bowl with no apparent motion, whereas, according to Eq. 19.11, the marble has a zero-point speed. For a marble of mass $10^{-2}$ kg inside a bowl of diameter $6 \times 10^{-3}$ m, this speed is

$$v = \frac{h}{2Lm} = \frac{6.63 \times 10^{-34} \text{ J-s}}{(2)(6 \times 10^{-3} \text{ m})(10^{-2} \text{ kg})}$$

$$= 5.5 \times 10^{-30} \text{ m/s}$$

At this speed it would take the marble a thousand billion years to move a distance of one atomic diameter. A speed this small is indistinguishable from zero. It is the small size of $h$ that ensures that special quantum-mechanical effects, such as the zero-point energy, are significant only for particles of atomic size. For ordinary-sized objects, newtonian mechanics still applies to very good approximation.

## 19.4 COMPLEX ATOMS AND THE PERIODIC TABLE

### Quantum states of an atom

When the Schrödinger wave equation is solved for an electron bound to a positively charged nucleus, it is found (as in the case of a particle in a box) that only certain states are possible. The energies of these states are identical to the energies that Bohr found earlier using his simpler model. However, quantum mechanics abandons the picture of the electrons revolving in orbits about the nucleus and replaces

---

† Equation 19.4 shows that the kinetic energy $K$ of an electron in an atom is equal to $-E$. Therefore the kinetic energy of an electron in the ground state of hydrogen is 13.6 eV.

‡ This is a special case of a more general quantum principle, known as the *uncertainty principle*, which says that the greater the certainty of a particle's position, i.e., the smaller the region of confinement, the greater the uncertainty in the particle's velocity, i.e., the larger the speed.

it by distributions of the probability of finding the electrons at every point around the nucleus.

For example, Fig. 19.13 shows the probability distribution of an electron in the lowest $n = 1$ state, with the old Bohr orbit superimposed for comparison. This is a spherical distribution, and the probability of finding the electron is greatest at the nucleus and decreases rapidly with distance from it. Although this distribution bears no resemblance to the Bohr model of an orbiting electron, the figure shows that the Bohr orbit gives the correct size of the atom. **REMARK** An electron in an atom is sometimes said to form a cloud about the nucleus. The word is meant to describe the fuzzy probability distribution shown in Fig. 19.13 and not to imply that the electron itself is smeared out. Any measurement of an electron in an atom always reveals one whole electron with the full charge $-e$, and never a fraction of an electron or a continuous distribution of charge.

The Schrödinger equation gives four different waves for $n = 2$ and nine different waves for $n = 3$. This means there are four states with the energy $E_2$ and nine states with the energy $E_3$. The states of the hydrogen atom are designated by the three *quantum numbers n, l, m*. The *principal quantum number n* corresponds to the Bohr orbits and determines the energy of the state; the *orbital quantum number l* determines the general symmetry of the wave; and the *magnetic quantum number m* determines the spatial orientation of the wave. For instance, $l = 0$ is a spherically symmetric wave which has a probability distribution like the one shown in Fig. 19.13. Thus $l$ is zero in the single $n = 1$ state. Furthermore, since orientation is meaningless for a spherical distribution, $m$ is zero also. Therefore the quantum numbers of the $n = 1$ state are:

| $n$ | $l$ | $m$ |
|-----|-----|-----|
| 1 | 0 | 0 |

One of the $n = 2$ states also has $l = 0$, $m = 0$, and the other three have $l = 1$. The $l = 1$ waves are doughnut-shaped rings in two orientations ($m = \pm 1$) and form two isolated blobs in the third orientation ($m = 0$). These distributions are shown in Fig. 19.14, with the $n = 1$ Bohr orbit superimposed to show that the $n = 2$ distributions extend much further than the $n = 1$ distribution. The distributions with $m = +1$ and $m = -1$ are identical, but the waves corresponding to them differ in sign. Essentially $m = +1$ describes an electron rotating clockwise around the nucleus, and $m = -1$ describes an electron rotating counterclockwise around the nucleus. An $l = 1$ distribution is mostly outside the Bohr orbit, whereas an $l = 0$ distribution has a significant portion inside the orbit. These facts are important for understanding atomic structure. The quantum numbers of the four $n = 2$ states are:

| $n$ | $l$ | $m$ |
|-----|-----|-----|
| 2 | 0 | 0 |
|   | 1 | $-1, 0, 1$ |

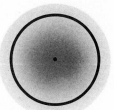

FIGURE 19.13
Probability distribution of an electron in the ground ($n = 1$) state of hydrogen. The Bohr orbit is shown for comparison.

In general, for a given value of the principal quantum number $n$, states exist with orbital quantum numbers $l$ varying in integer steps from 0 to $n - 1$, and for each value of $l$, states exist with magnetic quantum numbers $m$ varying in integer steps from $-l$ to $+l$. Thus the quantum numbers of the nine $n = 3$ states are:

| $n$ | $l$ | $m$ |
|---|---|---|
| 3 | 0 | 0 |
|   | 1 | $-1, 0, 1$ |
|   | 2 | $-2, -1, 0, 1, 2$ |

A group of states with the same value of $n$ is a *shell*, and a group of states with the same values of $n$ and $l$ is a *subshell*. Table 19.1 lists all the shells and subshells with principal quantum numbers 1 to 6, together with the number of states in each subshell. (Note that there are $2l + 1$ states in a subshell with orbital quantum number $l$.). A subshell is denoted by its principal quantum number $n$ and a code letter for its orbital quantum number $l$. The code letters for $l = 0, 1, 2, 3, 4, 5$ are $s, p, d, f, g, h$, respectively, as indicated in Table 19.1. Thus the $n = 3$, $l = 2$ subshell is denoted $3d$.

The properties of complex (many-electron) atoms and the structure of the periodic table are determined by the quantum numbers $n$, $l$, $m$, in conjunction with two other discoveries made in the critical year of 1925.

**Spin**  From analysis of atomic spectra, it was found that atoms have twice the number of quantum states given in Table 19.1. S. A. Goudsmit and G. E. Uhlenbeck showed that this was because the electron itself has two intrinsic states. The electron can be thought

$n=2,\ l=0$

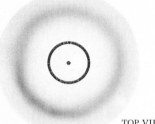

TOP VIEW

$n=2,\ l=1,\ m=\pm1$

SIDE VIEW

$n=2,\ l=1,\ m=0$

**FIGURE 19.14**
Probability distributions of an electron in the first excited ($n = 2$) states of hydrogen. The Bohr orbit is shown for comparison. (*a*) The $n = 2$, $l = 0$ state is spherically symmetric. It has two regions with appreciable probability, one of which is inside the Bohr orbit. (*b*) The $n = 2$, $l = 1$, $m = \pm1$ states are doughnut-shaped and correspond closely to Bohr's original picture of orbiting electrons. (*c*) The $n = 2$, $l = 1$, $m = 0$ state consists of two isolated blobs on either side of the nucleus.

**TABLE 19.1  States of an electron bound to a positively charged nucleus**
The notation used to denote each subshell is given, together with the number of states in the subshell. This number must be doubled to include the two possible spin orientations of the electron.

| $n$ | $l$ | Subshell symbol | $m$ | Number of states |
|---|---|---|---|---|
| 1 | 0 | $1s$ | 0 | 1 |
| 2 | 0 | $2s$ | 0 | 1 |
|   | 1 | $2p$ | $-1,0,1$ | 3 |
| 3 | 0 | $3s$ | 0 | 1 |
|   | 1 | $3p$ | $-1,0,1$ | 3 |
|   | 2 | $3d$ | $-2,-1,0,1,2$ | 5 |
| 4 | 0 | $4s$ | 0 | 1 |
|   | 1 | $4p$ | $-1,0,1$ | 3 |
|   | 2 | $4d$ | $-2,-1,0,1,2$ | 5 |
|   | 3 | $4f$ | $-3,-2,-1,0,1,2,3$ | 7 |
| 5 | 0 | $5s$ | 0 | 1 |
|   | 1 | $5p$ | $-1,0,1$ | 3 |
|   | 2 | $5d$ | $-2,-1,0,1,2$ | 5 |
|   | 3 | $5f$ | $-3,-2,-1,0,1,2,3$ | 7 |
|   | 4 | $5g$ | $-4,-3,-2,-1,0,1,2,3,4$ | 9 |

of as spinning on an axis,† and the two states correspond to a spin axis up or down with respect to an arbitrary reference line. These states are designated by the *spin quantum number* $m_s$, which has only two possible values: $m_s = +1$ for spin up and $m_s = -1$ for spin down. Thus two states exist for every set of quantum numbers $n$, $l$, $m$, one with $m_s = +1$, and one with $m_s = -1$.

**Exclusion principle** This principle, formulated by Wolfgang Pauli (1900–1958), states that two or more electrons in an atom cannot occupy the same state at the same time. Consequently, all the electrons in a many-electron atom are in different states. Two states differ if any of their four quantum numbers $(n, l, m, m_s)$ are different.

### Periodic table

An atom consists of $Z$ electrons bound to a nucleus of charge $+Ze$, where $Z$ is an integer, called the *atomic number*. The atoms of each chemical element have a specific value of $Z$. For instance, hydrogen atoms have $Z = 1$, carbon atoms have $Z = 6$, and uranium atoms have $Z = 92$. The periodic table of the elements (Table 19.2) is obtained by arranging the elements in order of increasing $Z$, with elements with similar chemical properties placed under each other. Dmitri Mendeleev (1834–1907) discovered empirically in 1869 that in such an arrangement elements with similar properties reoccur at regular (periodic) intervals. Note that in Table 19.2 the atomic number $Z$ increases across each row (period) and elements in the same column (group) have similar properties. For instance, group I contains the alkali halides, elements (such as sodium and potassium) that easily form singly charged positive ions, while group VII contains the halogens, elements (such as chlorine and fluorine) that easily form singly charged negative ions.

We shall now derive the periodic table from the principles of quantum mechanics. In this construction we assume that the electrons in the ground state of an atom occupy the lowest energy levels consistent with the exclusion principle.

**Period 1** The periodic table starts with hydrogen ($Z = 1$), whose one electron occupies the lowest $1s$ subshell. Hydrogen is placed in group I because it easily forms an ion of charge $+e$. The exclusion principle allows the two electrons of helium ($Z = 2$) to occupy the $1s$ subshell if their spins have opposite orientations. This arrangement has the lowest energy, and so it is the ground-state configuration of helium. Since the helium nucleus has twice the charge of the hydrogen nucleus, each helium electron is more tightly bound than the electron in hydrogen. As a consequence, helium atoms have no tendency to gain or lose electrons and do not react with other atoms to form molecules. Helium is an inert gas, belonging to group 0 of the periodic table. An inert gas is formed whenever the electrons of an atom completely fill the last occupied shell.

**Period 2** The exclusion principle allows only two of the three lithium ($Z = 3$) electrons to occupy the $1s$ state. The third electron must go

---

† The spin axis of an electron coincides with the axis of its magnetic dipole (Sec. 18.3).

TABLE 19.2 **Periodic table of the elements**

As shown in the key, each entry gives the atomic number $Z$, the chemical symbol, the atomic mass, and the electron configuration. A number in parentheses is the mass number of the most stable isotope of that element. The electron configurations are the number of electrons in the last three or four subshells. See Table 19.3 for the order in which the subshells are filled.

**Key**

| | |
|---|---|
| 88.91 | —— Atomic mass |
| **Y** | |
| 39 | —— Atomic number |
| 6,2,1 | |

Symbol → **Y**
Electron configuration → 6,2,1

*Transition elements*

Main table (mass / symbol / atomic number / electron configuration):

| Period | Group I | Group II | Transition elements | | | | | | | | | | Group III | Group IV | Group V | Group VI | Group VII | Group 0 |
|---|---|---|---|---|---|---|---|---|---|---|---|---|---|---|---|---|---|---|
| 1 | 1.008 **H** 1 / 1 | | | | | | | | | | | | | | | | | 4.00 **He** 2 / 2 |
| 2 | 6.94 **Li** 3 / 2,1 | 9.01 **Be** 4 / 2,2 | | | | | | | | | | | 10.81 **B** 5 / 2,2,1 | 12.01 **C** 6 / 2,2,2 | 14.01 **N** 7 / 2,2,3 | 16.00 **O** 8 / 2,2,4 | 19.00 **F** 9 / 2,2,5 | 20.18 **Ne** 10 / 2,2,6 |
| 3 | 22.99 **Na** 11 / 2,6,1 | 24.31 **Mg** 12 / 2,6,2 | | | | | | | | | | | 26.98 **Al** 13 / 6,2,1 | 28.09 **Si** 14 / 6,2,2 | 30.97 **P** 15 / 6,2,3 | 32.06 **S** 16 / 6,2,4 | 35.45 **Cl** 17 / 6,2,5 | 39.95 **Ar** 18 / 6,2,6 |
| 4 | 39.10 **K** 19 / 2,6,1 | 40.08 **Ca** 20 / 2,6,2 | 44.96 **Sc** 21 / 6,2,1 | 47.90 **Ti** 22 / 6,2,2 | 50.94 **V** 23 / 6,2,3 | 52.00 **Cr** 24 / 6,1,5 | 54.94 **Mn** 25 / 6,2,5 | 55.85 **Fe** 26 / 6,2,6 | 58.93 **Co** 27 / 6,2,7 | 58.71 **Ni** 28 / 6,2,8 | 63.54 **Cu** 29 / 6,1,10 | 65.37 **Zn** 30 / 6,2,10 | 69.72 **Ga** 31 / 2,10,1 | 72.59 **Ge** 32 / 2,10,2 | 74.92 **As** 33 / 2,10,3 | 78.96 **Se** 34 / 2,10,4 | 79.90 **Br** 35 / 2,10,5 | 83.80 **Kr** 36 / 2,10,6 |
| 5 | 85.47 **Rb** 37 / 10,6,1 | 87.62 **Sr** 38 / 10,6,2 | 88.91 **Y** 39 / 6,2,1 | 91.22 **Zr** 40 / 6,2,2 | 92.91 **Nb** 41 / 6,1,4 | 95.94 **Mo** 42 / 6,1,5 | (97) **Tc** 43 / 6,2,5 | 101.1 **Ru** 44 / 6,1,7 | 102.91 **Rh** 45 / 6,1,8 | 106.4 **Pd** 46 / 6,0,10 | 107.87 **Ag** 47 / 6,1,10 | 112.40 **Cd** 48 / 6,2,10 | 114.82 **In** 49 / 2,10,1 | 118.69 **Sn** 50 / 2,10,2 | 121.75 **Sb** 51 / 2,10,3 | 127.60 **Te** 52 / 2,10,4 | 126.90 **I** 53 / 2,10,5 | 131.30 **Xe** 54 / 2,10,6 |
| 6 | 132.91 **Cs** 55 / 10,6,1 | 137.34 **Ba** 56 / 10,6,2 | 57–71 **†** | 178.49 **Hf** 72 / 2,14,2 | 180.95 **Ta** 73 / 2,14,3 | 183.85 **W** 74 / 2,14,4 | 186.2 **Re** 75 / 2,14,5 | 190.2 **Os** 76 / 2,14,6 | 192.2 **Ir** 77 / 2,14,7 | 195.09 **Pt** 78 / 1,14,9 | 196.97 **Au** 79 / 1,14,10 | 200.59 **Hg** 80 / 2,14,10 | 204.37 **Tl** 81 / 14,10,1 | 207.19 **Pb** 82 / 14,10,2 | 208.98 **Bi** 83 / 14,10,3 | (209) **Po** 84 / 14,10,4 | (210) **At** 85 / 14,10,5 | 222 **Rn** 86 / 14,10,6 |
| 7 | (223) **Fr** 87 / 10,6,1 | (226) **Ra** 88 / 10,6,2 | 89–103 **‡** | | | | | | | | | | | | | | | |

**† Lanthanides**

| | | | | | | | | | | | | | | |
|---|---|---|---|---|---|---|---|---|---|---|---|---|---|---|
| 138.91 **La** 57 / 6,2,0,1 | 140.12 **Ce** 58 / 6,2,0,2 | 140.91 **Pr** 59 / 6,2,3 | 144.24 **Nd** 60 / 6,2,4 | (145) **Pm** 61 / 6,2,5 | 150.35 **Sm** 62 / 6,2,6 | 151.96 **Eu** 63 / 6,2,7 | 157.25 **Gd** 64 / 6,2,7,1 | 158.92 **Tb** 65 / 6,2,9 | 162.50 **Dy** 66 / 6,2,10 | 164.93 **Ho** 67 / 6,2,11 | 167.26 **Er** 68 / 6,2,12 | 168.93 **Tm** 69 / 6,2,13 | 173.04 **Yb** 70 / 6,2,14 | 174.97 **Lu** 71 / 2,14,1 |

**‡ Actinides**

| | | | | | | | | | | | | | | |
|---|---|---|---|---|---|---|---|---|---|---|---|---|---|---|
| (227) **Ac** 89 / 6,2,0,1 | (232) **Th** 90 / 6,2,0,2 | (231) **Pa** 91 / 6,2,2,1 | (238) **U** 92 / 6,2,3,1 | (237) **Np** 93 / 6,2,4,1 | (244) **Pu** 94 / 6,2,5,1 | (243) **Am** 95 / 6,2,6,1 | (247) **Cm** 96 / 6,2,7,1 | (247) **Bk** 97 / 6,2,8,1 | (251) **Cf** 98 / 6,2,10 | (254) **Es** 99 / 6,2,11 | (257) **Fm** 100 / 6,2,12 | (256) **Md** 101 / 6,2,13 | (254) **No** 102 / 6,2,14 | (−) **Lw** 103 / 2,14,1 |

into the next higher ($n = 2$) shell. In a one-electron atom all the $n = 2$ states have the same energy, but in an atom in which the $1s$ subshell is filled, the $2s$ states have slightly lower energy than the $2p$ states because the electrons in the $1s$ subshell partially neutralize the charge of the nucleus for electrons outside the $1s$ cloud. A $2p$ electron, which is entirely outside the $1s$ distribution (Fig. 19.14), is attracted by an effective charge of only $+e$, whereas a $2s$ electron, which is partially inside the $1s$ distribution, is sometimes attracted by the full $+3e$ charge of the nucleus. Consequently the $2s$ state has lower energy (is more tightly bound) than the $2p$ state, and so the third lithium electron is in the $2s$ state. Lithium is placed in group I because it can easily lose its $2s$ electron to form an ion of charge $+e$.

The four electrons of beryllium ($Z = 4$) fill the $1s$ and $2s$ subshells. Beryllium is in group II because it can easily lose the two $2s$ electrons.

The next six elements ($Z = 5$ to $Z = 10$) are formed by adding electrons to the $2p$ subshell. Table 19.3 shows how the $2p$ subshell can hold up to six electrons with properly oriented spins. Fluorine ($Z = 9$), which is one electron short of filling the $2p$ shell, has a strong tendency to fill the last position by picking up an additional electron to form an ion of charge $-e$. Therefore fluorine is in group VII. The inert gas neon ($Z = 10$) fills the $2p$ shell, completing period 2.

**Period 3** The next eight elements ($Z = 11$ to $Z = 18$) are formed similar to the preceding eight by adding electrons in turn to the $3s$ and $3p$ subshells (Table 19.3). Thus sodium ($Z = 11$), which has one electron in the $3s$ state, has properties similar to lithium, whereas chlorine ($Z = 17$), which is one electron short of filling the $3p$ subshell, has properties similar to fluorine. The inert gas argon ($Z = 18$) fills the $3p$ shell, completing period 3.

The order in which succeeding subshells are filled is governed by two competing factors. On the one hand, a subshell with a smaller principal quantum number $n$ tends to have lower energy than a subshell with larger $n$. On the other hand, a subshell with a smaller orbital quantum number $l$ tends to have lower energy than a subshell with larger $l$. Thus the $2s$ subshell is lower than the $3s$ subshell and the $3s$ subshell is lower than the $3p$ subshell. However, there exist competing cases, such as the $2p$ and the $3s$ subshells, in which $n$ is smaller for one subshell while $l$ is smaller for the other. In such an event the subshell with the smaller $n$ has the lower energy unless the difference in the $l$ values is greater than the difference in the $n$ values. Thus the $2p$ subshell is lower than the $3s$ subshell, but the $4s$ subshell is lower than the $3d$ subshell. The $4s$ subshell has lower energy, in spite of its larger value of $n$, because its value of $l$ is 2 units lower than the $l$ value of the $3d$ subshell. This peculiar inversion of the ordering of the subshells is shown in Table 19.3. It results in the existence of a group of transition elements which have similar properties but do not belong to any of the standard groups in the periodic table. These elements occur for the first time in period 4.

**Period 4** The first two elements ($Z = 19$ and $Z = 20$) of period 4 are formed by adding electrons to the $4s$ subshell and so belong to groups I and II. Thereafter, electrons are added to the $3d$ subshell, forming a series of 10 transition elements ($Z = 21$ to $Z = 30$) between groups II and III. These elements have similar properties because they all have two electrons in the 4s subshell.† After filling the 3d subshell, electrons go into the 4p subshell, and the elements formed resemble the 2p and 3p elements in the preceding periods. The inert gas krypton ($Z = 36$) fills the 4p subshell, completing period 4.

**Period 5** This period is identical in form to period 4. Electrons fill the 5s, 4d, and 5p subshells in period 5, just as they fill the 4s, 3d, and 4p shells in period 4.

**Period 6** The first two elements ($Z = 55$ and $Z = 56$) of period 6 are formed by filling the 6s subshell. Thereafter the 4f subshell is filled because it has lower energy than the 5d subshell. Because the 4f electrons are inside the 6s electrons, they have little effect on the

† There are some irregularities in the filling of the 3d subshell. Some transition elements have only one 4s electron, the other electron being in the 3d subshell. These and other minor irregularities can be seen by studying the electron configurations shown in Table 19.2.

TABLE 19.3 **Order of subshell filling in atoms**

Each arrow represents a quantum state, with the point representing the electron's spin orientation. The number under each arrow is the atomic number of the element that fills all the preceding states. An asterisk beside a number indicates an irregularity in subshell filling. In these cases the actual order can be found from Table 19.2. For example, Table 19.2 shows that niobium ($Z = 41$) has four electrons in the 4d subshell and one electron in the 5s subshell, in contrast to the expected arrangement (three in the 4d and two in the 5s) shown in this table.

| Period | Sequence | Subshell | Quantum states |
|---|---|---|---|
| 7 | Transition | 6d | ↑103 ↑ ↑ ↑ ↑ ↓ ↓ ↓ ↓ ↓ |
|  | Actinides | 5f | ↑89* ↑90* ↑91* ↑92* ↑93* ↑94* ↑95* ↓96* ↓97* ↓98 ↓99 ↓100 ↓101 ↓102 |
|  |  | 7s | ↑87 ↓88 |
| 6 |  | 6p | ↑81 ↑82 ↑83 ↓84 ↓85 ↓86 |
|  | Transition | 5d | ↑71 ↑72 ↑73 ↑74 ↑75 ↑76 ↓77 ↓78* ↓79* ↓80 |
|  | Lanthanides | 4f | ↑57* ↑58 ↑59 ↑60 ↑61 ↑62 ↑63 ↓64 ↓65 ↓66 ↓67 ↓68 ↓69 ↓70 |
|  |  | 6s | ↑55 ↓56 |
| 5 |  | 5p | ↑49 ↑50 ↑51 ↓52 ↓53 ↓54 |
|  | Transition | 4d | ↑39 ↑40 ↑41* ↑42* ↑43 ↓44* ↓45* ↓46* ↓47* ↓48 |
|  |  | 5s | ↑37 ↓38 |
| 4 |  | 4p | ↑31 ↑32 ↑33 ↓34 ↓35 ↓36 |
|  | Transition | 3d | ↑21 ↑22 ↑23 ↑24* ↑25 ↓26 ↓27 ↓28 ↓29* ↓30 |
|  |  | 4s | ↑19 ↓20 |
| 3 |  | 3p | ↑13 ↑14 ↑15 ↓16 ↓17 ↓18 |
|  |  | 3s | ↑11 ↓12 |
| 2 |  | 2p | ↑5 ↑6 ↑7 ↓8 ↓9 ↓10 |
|  |  | 2s | ↑3 ↓4 |
| 1 |  | 1s | ↑1 ↓2 |

chemical properties of these elements. Consequently the next 14 elements ($Z = 57$ to $Z = 70$) form a sequence with nearly identical chemical properties. This sequence, called the *lanthanides* (or rare earths), does not correspond to any previous sequence of elements, so it is given a special place in the periodic table. The $4f$ shell is filled at $Z = 70$. After this, electrons go into the $5d$ subshell, forming a sequence of 10 transition elements analogous to the ones in periods 4 and 5. The first member of the transition sequence ($Z = 71$) is also considered to be the last member of the lanthanide sequence. The inert gas radon fills the $6p$ subshell, completing period 6.

**Period 7** This period begins the same as the preceding period, with the filling of the $7s$ and the $5f$ subshells. The filling of the $5f$ subshell forms a new sequence of elements ($Z = 89$ through $Z = 102$), called the *actinides*. The last naturally occurring element is uranium ($Z = 92$), but elements with $Z$ as high as 103 have been produced artificially in nuclear reactions. This last element has one electron in the $6d$ subshell, and so it is the first member of the period 7 transition sequence and the last member of the actinides.

**REMARK** Elements with $Z$ greater than 103 have been reported but not yet confirmed. The existence of high-$Z$ elements is limited by the ability of their nuclei to hold together. If higher-$Z$ elements are produced, they are expected to fill period 7 in regular sequence.

The construction of the periodic table from fundamental physical principles is one of the great achievements of quantum mechanics. From this construction it is seen that the properties of the elements are not arbitrary whims of nature but follow in an orderly and understandable way from fundamental laws.

## 19.5   CHEMICAL BONDS

A molecule consists of two or more atoms bound together by attractive forces between the atoms. These interatomic bonding forces, or *chemical bonds*, are electric in origin, but they have different characteristics in different molecules. We shall discuss the extreme types of chemical bond, *ionic* (or heteropolar) and *covalent* (or homopolar), because they are the easiest to understand. In most molecules, however, the type of bonding is intermediate between these two extremes.

Pure ionic bonding occurs in simple molecules, such as NaCl and KF, that are composed of one atom from group I and one atom from group VII. Group I atoms have only one electron in their outer $s$ subshell and $Z - 1$ inner electrons. These inner electrons tend to neutralize most of the charge of the nucleus, so that the outer electron is bound to the atom by an effective charge much less than the full charge ($+Ze$) of the nucleus. Consequently the outer electron is easily removed from a group I atom, forming a positively charged ion. Group VII atoms, on the other hand, are one electron short of filling their outer $p$ subshell, so they tend to attract an extra electron, forming a negatively charged ion.

An ionic bond is formed when the outer electron leaves a group I atom and attaches itself to a group VII atom, forming two ions of opposite charge. When these ions are far apart, the electric force

FIGURE 19.15
Probability distributions of the electrons in two hydrogen atoms before and after the atoms are bound together to form a hydrogen molecule, $H_2$.

between them is attractive and they move toward each other. When they get close, however, their electron clouds start to repel each other, and at a certain separation distance the attractive force just balances the repulsive force. This is the equilibrium separation distance for the ions in an ionic bond. At this distance the ions are prevented from moving closer together by the repulsion of their electron clouds and from moving farther apart by the attraction of their opposite charges. Thus a stable molecule is formed, with a fixed distance between the ions.

A covalent bond is formed by two atoms sharing electrons, rather than exchanging them. Figure 19.15 shows how the probability distributions of the electrons in two separated hydrogen atoms rearrange themselves to form a covalently bonded hydrogen molecule, $H_2$, in which the two electrons are shared equally with each nucleus. The probability distribution is denser between the nuclei than outside, which means that an electron is more likely to be between the nuclei than outside. It is this concentration of negative charge between the two nuclei that holds the molecule together. It can be proved from the Schrödinger equation that this distribution forms a stable molecule.

## PROBLEMS

**REMARK**  Express all energies in electron volts.

1 What is the energy of a photon of light of wavelength $4.0 \times 10^{-7}$ m?          *Ans.* 3.1 eV

2 Find the wavelength and frequency of a 100-eV photon.

3 The critical frequency $f_0$ of silver is $1.04 \times 10^{15}$ Hz. What is the minimum energy $E_{min}$ required to remove an electron from silver?
          *Ans.* 4.35 eV

4 The minimum energy required to remove an electron from cesium is 2.6 eV. What is the longest-wavelength photon that will eject an electron from cesium?

5 What is the shortest-wavelength x-ray produced by a 200-kV x-ray machine?
          *Ans.* $6.21 \times 10^{-12}$ m

6 What potential difference is required to produce x-rays with a minimum wavelength of $4.5 \times 10^{-11}$ m?

7 (a) What is the wavelength of an electron with a speed of $5 \times 10^7$ m/s? (b) What is the wavelength of a proton with the same speed?
          *Ans.* (a) $1.46 \times 10^{-11}$ m; (b) $7.94 \times 10^{-15}$ m

8 What is the kinetic energy of an electron with a wavelength of $5 \times 10^{-10}$ m?

9 The kinetic energy of an electron is 100 eV. (a) What is its wavelength? (b) What is the wavelength of a 100-eV proton?
          *Ans.* (a) $1.23 \times 10^{-10}$ m; (b) $2.87 \times 10^{-12}$ m

10 What are the wavelengths of a 1-eV photon and a 1-eV electron, respectively?

11 An electron microscope can resolve structures that are 10 or more times the electron's wavelength in size. What is the smallest structure that can be resolved in an electron microscope using 10,000-eV electrons? Compare this to the resolving power of an optical microscope.
          *Ans.* 0.123 nm

12 Equations 19.1 and 19.2 give the frequency $f$ and wavelength $\lambda$ of a particle of energy $E = \frac{1}{2}mv^2$. Show that $f\lambda$ is *not* equal to the speed $v$ of the particle.

13 What is the speed of an electron in the lowest Bohr orbit of hydrogen?  *Ans.* $2.18 \times 10^6$ m/s

14 What is the radius of the $n = 20$ Bohr orbit of hydrogen?

15 Find the wavelength of the radiation emitted when a hydrogen atom makes a transition from the $n = 6$ state to the $n = 3$ state.
          *Ans.* $1.096 \times 10^{-6}$ m

16 How much energy is required to ionize a hydrogen atom in the $n = 2$ state?

17 The spectrum of hydrogen contains four wavelengths in the visible region. Find the transitions that emit these wavelengths and determine the wavelengths.
          *Ans.* 411, 435, 487, and 657 nm

18 All transitions that go directly to the ground state of hydrogen produce a sequence of wavelengths called the *Lyman series*. (a) What are

the longest and shortest wavelengths in the Lyman series? (*b*) To what region of the electromagnetic spectrum does this radiation belong?

19 All transitions that go directly to the $n = 2$ state of hydrogen produce a sequence of wavelengths called the *Balmer series*. What is (*a*) the longest and (*b*) the shortest wavelength in the Balmer series? (*c*) To what regions of the electromagnetic spectrum do these radiations belong?

*Ans.* (*a*) 657 nm; (*b*) 365 nm; (*c*) visible and ultraviolet

20 What is the minimum kinetic energy of an electron confined to a region the size of a nucleus ($10^{-15}$ m)?

21 A 10-g marble moves back and forth in a box 10 cm long at a speed of 10 cm/s. (*a*) What is the number $n$ of the quantum state of the marble? (*b*) Why is it possible to vary the energy of the marble continuously even though its energy can change only by discrete amounts?　　　　　　*Ans.* (*a*) $3 \times 10^{29}$

22 List all the subshells in the $n = 7$ shell. How many states are there in this shell?

23 List the subshells with $n \leq 6$ that are not occupied in any atom.

24 (*a*) What is the atomic number of the atom that just fills the $6d$ subshell? (*b*) After the $6d$ subshell is filled, is the next subshell expected to be $5g$ or $7p$? (*c*) What is the atomic number of the last element in period 7?

## BIBLIOGRAPHY

BEISER, ARTHUR: "Concepts of Modern Physics," McGraw-Hill Book Company, New York, 1963. For students with some calculus background, this book provides an excellent introduction to the mathematical treatment of quantum mechanics.

BOHR, NIELS: "Atomic Physics and Human Knowledge," John Wiley & Sons, Inc., New York, 1958. A collection of essays on the significance of quantum mechanics for biology, human culture, and epistemology. Of particular importance is an essay that recounts Bohr's discussions with Einstein about quantum mechanics.

CLINE, BARBARA LOVETT: "Men Who Made a New Physics," The New American Library, New York, 1969. A simply written account of the development of modern physics, with special emphasis on the ideas and personalities of the men involved.

GAMOW, GEORGE: "Mr. Tomkins in Wonderland: Stories of *c*, *G*, and *h*," The Macmillan Company, New York, 1940. Mr. Tomkins visits a world in which *h* is so large that quantum effects are everyday occurrences.

REICHENBACH, HANS: "The Rise of Scientific Philosophy," University of California Press, Berkeley, 1951. A clear and scholarly account of the implications of quantum mechanics for modern philosophy.

SHAMOS, MORRIS H. (ed.): "Great Experiments in Physics," Holt, Rinehart and Winston, New York, 1959. Contains excerpts from Einstein's 1905 paper on the photoelectric effect and from Bohr's 1913 paper on the structure of the atom.

# CHAPTER 20 nuclei

The only properties of the nucleus used in atomic theory are its small size (compared to the size of an atom), its large mass (compared to the mass of an electron), and its positive charge. Although the nucleus has a complex internal structure of its own, this structure has almost no effect on the electronic orbits because the nucleus is so much smaller than the orbits. For this reason all the properties of the atom, including the chemical properties of the elements, are largely independent of the internal structure of the nucleus. This is fortunate, since otherwise nuclear physics would have had to be developed before atomic physics.

Much was known empirically about the nucleus before 1930, but it was only after the discovery of the neutron by James Chadwick (1891- ) in 1932 that a systematic understanding of the internal structure of the nucleus could be obtained. While atomic physics has provided a deeper understanding of the known properties of atoms and molecules, nuclear physics has discovered new and unexpected properties of matter. These discoveries have had the most profound consequences, resulting in major advances in medical diagnosis, radiation therapy, and power generation, as well as in military weapons of great destructive capabilities.

## 20.1 THE STRUCTURE OF THE NUCLEUS

A nucleus is composed of two kinds of particles, *protons* and *neutrons*, bound together by nuclear forces. The proton is an elementary particle (Sec. 16.1) with charge $+e$; the *atomic number $Z$* of the element is equal to the number of protons in the nucleus. For instance, a carbon nucleus ($Z = 6$) contains six protons, and a uranium nucleus ($Z = 92$) contains 92 protons. The neutron is an elementary particle with no charge and a mass slightly greater than the mass of a proton. The *mass number $A$* of a nucleus is the sum of the proton number $Z$ and the neutron number $N$:

$$A = Z + N$$

The mass number is the total number of *nucleons* (neutrons and protons) in the nucleus.

All nuclei with the same value of $Z$ are denoted by the chemical symbol of the corresponding element. All nuclei with the same values of $Z$ and $A$ constitute a particular nuclear species, or *nuclide*. A particular nuclide is denoted by affixing the mass number $A$ as a superscript to the chemical symbol. For example, the nuclide composed of six protons and six neutrons is written $^{12}C$, the nuclide composed of six protons and eight neutrons is written $^{14}$C, and the nuclide composed of seven protons and seven neutrons is written $^{14}$N. This notation completely specifies the nuclide if the atomic numbers and symbols of the elements are known. To avoid confusion, however, the atomic number $Z$ is sometimes affixed as a subscript to the symbol, for example, $^{12}_{6}$C, $^{14}_{6}$C, $^{14}_{7}$N.

Nucleons are bound together in a nucleus by a fundamental force (Sec. 16.1) that differs from either the electric or gravitational force. At small distances the nuclear force is much stronger than the electric force, but it decreases rapidly as the distance between two nucleons increases. The nuclear force is best illustrated by plotting the nuclear potential energy $U_n$ between two nucleons against the distance $r$ between them. Figure 20.1 shows the neutron-proton ($n$-$p$) and the neutron-neutron ($n$-$n$) potentials. Both potentials rapidly approach zero as $r$ increases, and they dip around $1.0 \times 10^{-15}$ m. The negative potential energy indicates an attractive force between the nucleons because positive work must be done to separate them. The $n$-$p$ force is more attractive than the $n$-$n$ because its potential is more negative. The sharp rise in the potential energy at distances less than $1.0 \times 10^{-15}$ m indicates a strong repulsive force at small distances. Nucleons behave as though they had a hard core preventing them from getting closer than $0.5 \times 10^{-15}$ m.

The nuclear force between two protons is the same as the $n$-$n$ force, but in addition there is the electrical repulsion between the protons. Figure 20.2 shows the nuclear and electric potential energies between two protons and the combined potential energy.

Two factors govern the ratio of neutrons to protons in a nucleus: (1) The $n$-$p$ force, being the most attractive, favors an equal number of neutrons and protons. (2) The repulsive electric force between the

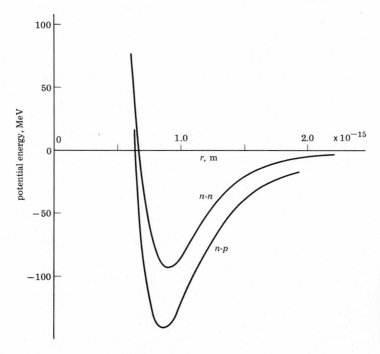

FIGURE 20.1
Nuclear potential energy $U_n$ between two neutrons ($n$-$n$) and between a neutron and a proton ($n$-$p$) plotted against the distance $r$ between them.

protons, which increases as the number of protons increase, favors an excess of neutrons over protons. In light nuclides ($A \leq 40$) the first factor dominates, and stable nuclides have approximately equal numbers of protons and neutrons. In heavier nuclides, however, the second factor dominates, and $N$ exceeds $Z$. This is shown in Fig. 20.3, which is a plot of $N$ against $Z$ for the stable nuclides. The $Z = N$ line is shown for comparison. Since no stable nuclide exists with $Z > 84$, the points with $Z \geq 85$ represent the longer-lived radioactive nuclides (Sec. 20.2).

Nuclides with the same $Z$ but different $A$ are called *isotopes*. They form atoms with identical chemical properties but very different nuclear properties. There are 10 to 20 isotopes for every $Z$, only a few of which are stable and naturally occurring. The others are radioactive and must usually be produced artificially by a nuclear reaction. Table 20.1 is a selected list of some important isotopes. A complete table of the properties of all the known isotopes can be found in the "Handbook of Chemistry and Physics."

The mass of an atom is measured in atomic mass units (u). An atomic mass unit is defined as one-twelfth the mass of a $^{12}$C atom:

$$1 \text{ u} = \tfrac{1}{12} m_{^{12}\text{C}}$$

or

$$m_{^{12}\text{C}} = 12.00000 \text{ u}$$

The masses of some common nuclides are given in Table 20.1. These are the masses of the entire atom, including the electrons. For instance, the mass of $^1$H is the combined mass of a proton and an electron. Table 20.1 also gives the proton, neutron, and electron masses in atomic mass units.

The mass of an atom is approximately equal to the mass of $Z$ hydrogen atoms and $N$ neutrons. Since the masses of $^1$H and $n$ are both approximately 1.0 u, all atomic masses are approximately equal to the mass number $A$. For example, the mass of eight $^1$H atoms and eight neutrons is

$$8 \times 1.007825 \text{ u} + 8 \times 1.008665 \text{ u} = 16.1319 \text{ u}$$

which is 0.137 u more than the mass of $^{16}$O given in Table 20.1. In fact, the mass of every atom is slightly less than the sum of the masses

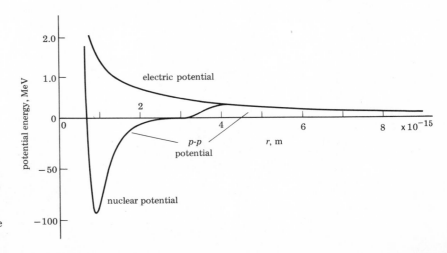

**FIGURE 20.2**
Potential energy between two protons plotted against the distance $r$ between them. The potential energy is the sum of the negative nuclear potential energy and the positive electric potential energy. Note that the positive scale is different from the negative scale.

of its constituent parts. The *mass defect* $\Delta m$ of an atom is defined as the difference between the masses of its parts and the mass of the atom. For instance, the mass defect of $^{16}$O is $\Delta^{16}$O $= 0.137$ u, and the mass defect of $^{32}$S is

$$\Delta^{32}\text{S} = 16 \times 1.007825 \text{ u} + 16 \times 1.008665 \text{ u} - 31.97207 \text{ u}$$
$$= 0.292 \text{ u}$$

The energy of any bound system is less than the energy of its separated parts because work has to be done to separate the parts. For instance, a hydrogen atom has 13.6 eV less energy than a sepa-

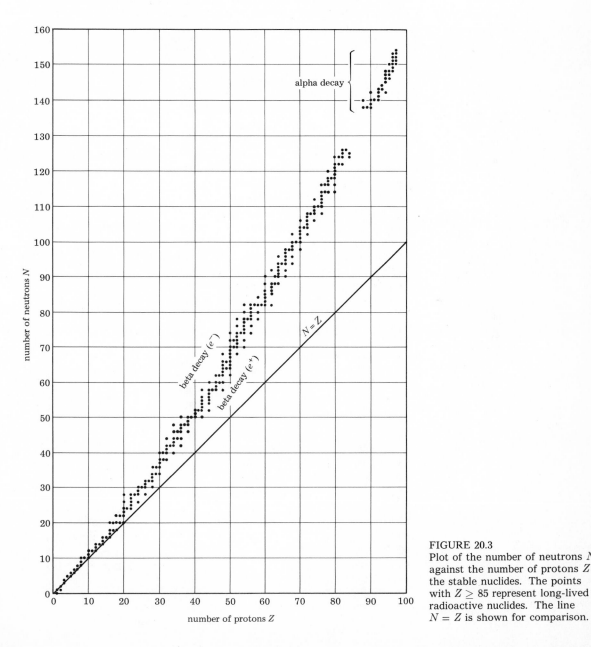

FIGURE 20.3
Plot of the number of neutrons $N$ against the number of protons $Z$ in the stable nuclides. The points with $Z \geq 85$ represent long-lived radioactive nuclides. The line $N = Z$ is shown for comparison.

rated proton and electron pair, which means it requires 13.6 eV to ionize hydrogen. Similarly, a carbon nucleus has 127.6 MeV (1 MeV = $10^6$ eV) less energy than its separated parts, which means it requires 127.6 MeV to break an $^{16}$O nucleus into 16 separated nucleons. The scale of nuclear binding is a million times greater than that of atomic binding, but the principle is the same.

TABLE 20.1 **Properties of selected nuclides**
Besides the atomic number, mass number, and atomic mass of selected nuclides, the table gives the percentage of abundance of the naturally occurring nuclides and the half-life and decay mode of the radioactive nuclides.

| Z | A | Symbol | Atomic mass | Percent abundance | Half-life | Decay mode |
|---|---|---|---|---|---|---|
| | | $e$ | 0.0005486 | | | |
| 0 | 1 | $n$ | 1.008665 | | 12 min | $e^-$ |
| 1 | 1 | $p$ | 1.007277 | | | |
| 1 | 1 | $^1_1$H | 1.007825 | 99.985 | | |
| 1 | 2 | $^2_1$H | 2.01402 | 0.015 | | |
| 1 | 3 | $^3_1$H | 3.01605 | † | 12.26 y | $e^-$ |
| 2 | 4 | $^4_2$He | 4.00260 | 100 | | |
| 6 | 12 | $^{12}_6$C | 12.00000 | 98.89 | | |
| 6 | 13 | $^{13}_6$C | 13.00335 | 1.11 | | |
| 6 | 14 | $^{14}_6$C | 14.00324 | † | 5730 y | $e^-$ |
| 7 | 13 | $^{13}_7$N | 13.00574 | | 10 min | $e^+$ |
| 7 | 14 | $^{14}_7$N | 14.00307 | 99.63 | | |
| 7 | 15 | $^{15}_7$N | 15.00011 | 0.37 | | |
| 8 | 16 | $^{16}_8$O | 15.99491 | 99.759 | | |
| 8 | 17 | $^{17}_8$O | 16.99913 | 0.037 | | |
| 8 | 18 | $^{18}_8$O | 17.99916 | 0.204 | | |
| 15 | 31 | $^{31}_{15}$P | 30.97376 | 100 | | |
| 15 | 32 | $^{32}_{15}$P | 31.97391 | | 14.3 d | $e^-$ |
| 16 | 32 | $^{32}_{16}$S | 31.97207 | 95.0 | | |
| 19 | 39 | $^{39}_{19}$K | 38.96371 | 93.70 | | |
| 19 | 40 | $^{40}_{19}$K | 39.97400 | 0.00118 | $1.28 \times 10^9$ y | $e^-, e^+$ |
| 19 | 41 | $^{41}_{19}$K | 40.96183 | 6.88 | | |
| 27 | 59 | $^{59}_{27}$Co | 58.93319 | 100 | | |
| 27 | 60 | $^{60}_{27}$Co | 59.93381 | | 5.26 y | $e^-$ |
| 38 | 88 | $^{88}_{38}$Sr | 87.90564 | 82.56 | | |
| 38 | 90 | $^{90}_{38}$Sr | 89.90775 | | 28.1 y | $e^-$ |
| 43 | 99 | $^{99}_{43}$Tc$^m$ | 98.90625 | | 6.0 h | $\gamma$ |
| 43 | 99 | $^{99}_{43}$Tc | 98.90625 | | $2.12 \times 10^5$ y | $e^-$ |
| 83 | 209 | $^{209}_{83}$Bi | 208.98039 | 100 | | |
| 83 | 214 | $^{214}_{83}$Bi | 213.99869 | | 19.7 min | $e^-, \alpha$ |
| 92 | 235 | $^{235}_{92}$U | 239.04392 | 0.7 | $7.1 \times 10^8$ y | $\alpha$ |
| 92 | 238 | $^{238}_{92}$U | 238.05077 | 99.3 | $4.51 \times 10^9$ y | $\alpha$ |
| 92 | 239 | $^{239}_{92}$U | 239.05430 | | 23.5 min | $e^-$ |
| 93 | 239 | $^{239}_{93}$Np | 239.05292 | | 2.35 d | $e^-$ |
| 94 | 239 | $^{239}_{94}$Pu | 239.05215 | | $2.44 \times 10^4$ y | $\alpha$ |

†Occurs naturally in trace amounts.

The mass defect is a manifestation of the lower energy of the bound system. According to the theory of relativity, mass is a form of energy. This means that an energy change $\Delta E$ is equivalent to a mass change $\Delta m$. These quantities are related by

$$\Delta m = \frac{\Delta E}{c^2} \qquad 20.1$$

where $c$ is the speed of light. To use this equation $m$, $E$, and $c$ must be in mks units. In nuclear physics, however, mass is usually given in atomic mass units and energy is given in million electron volts (MeV). The conversions between these units and the mks units are

$$1 \text{ u} = 1.66053 \times 10^{-27} \text{ kg}$$
$$1 \text{ MeV} = 1.60219 \times 10^{-13} \text{ J}$$

so if $\Delta m$ and $\Delta E$ are expressed in atomic mass units and MeV, respectively, Eq. 20.1 becomes

$$\Delta m(1.66053 \times 10^{-27} \text{ kg/u}) = \frac{\Delta E(1.60219 \times 10^{-13} \text{ J/MeV})}{(2.99792 \times 10^8 \text{ m/s})^2}$$

or
$$\Delta m = 1.07356 \times 10^{-3} \text{ u/MeV } \Delta E \qquad 20.2$$
$$\Delta E = 931.48 \text{ MeV/u } \Delta m$$

For example, the mass defect of $^{16}$O corresponds to a total binding energy of

$$\Delta E = 931.48 \text{ MeV/u} \times 0.137 \text{ u}$$
$$= 127.6 \text{ MeV}$$

as previously stated. The average binding energy per nucleon in $^{16}$O is

$$\frac{\Delta E}{A} = \frac{127.6 \text{ MeV}}{16} = 7.98 \text{ MeV}$$

which is typical of most nuclei. This is the average energy required to remove one nucleon from the nucleus.

Very accurate measurements of the masses of the nuclides have been made with mass spectrometers (Sec. 18.4). From these data the mass defect $\Delta m$ and the average binding energy per nucleon $\Delta E/A$ can be determined. The results of many such measurements are summarized in Fig. 20.4, which is a plot of $\Delta E/A$ against $A$. This shows that although $\Delta E/A$ is approximately 8 MeV for almost all nuclei, it increases gradually with $A$ to a maximum of 8.8 MeV around $A = 58$ and then decreases gradually to 7.6 MeV at $A = 238$. This behavior is of critical importance for the production of energy from nuclear processes (Sec. 20.3).

**REMARK** According to Eq. 20.2 the mass of a hydrogen atom is less than the mass of a proton and an electron by

$$\Delta m = (1.073 \times 10^{-3} \text{ u/MeV})(13.6 \times 10^{-6} \text{ MeV})$$
$$= 1.46 \times 10^{-9} \text{ u}$$

Since this mass difference is less than the uncertainties in the measured masses, it is not detectable. In general, the energy changes in chemical reactions are too small to result in measurable mass changes.

Only a few hundred of the thousands of known nuclides are stable. An unstable, or *radioactive*, nuclide is one that spontaneously transforms itself (decays) into another nuclide. If the daughter nuclide is also radioactive, it decays further until a stable (nonradioactive) nuclide is formed. Although the decay process itself is instantaneous, a radioactive nucleus can survive hours, days, or years before suddenly decaying. Radioactive nuclides with sufficiently long lifetimes have found important applications in technology, medicine, and biological research.

### Modes of decay

The two principal modes of decay are called *alpha* and *beta decay*. Each radioactive nuclide has a characteristic decay mode, as listed in Table 20.1, although a few nuclides, such as $^{214}$Bi, decay through either mode. Alpha decay is confined primarily to isotopes of the heavier elements ($Z \geq 78$), whereas beta decay occurs in isotopes of all elements. As a rule, alpha decay occurs for nuclides near the line marked "radioactive (alpha)" in Fig. 20.3, and beta decay occurs for nuclides above and below the line of stable nuclides.

*Beta decay* is a process in which a neutron inside a nucleus spontaneously converts into a proton by emitting an electron $e^-$ and a chargeless, massless particle called a *neutrino* $\nu$. For instance, the decay of $^{14}$C is

$$^{14}_{6}\text{C} \longrightarrow {}^{14}_{7}\text{N} + e^- + \nu$$

Electrons produced this way are called *beta rays*. Notice that the total charge ($+6e$) and the total number of nucleons (14) is unchanged in this reaction. It is a general result, true for any nuclear reaction

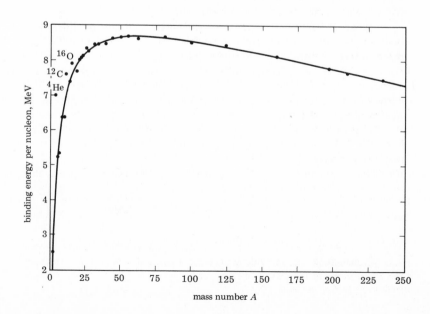

FIGURE 20.4
The binding energy per nucleon $\Delta E/A$ plotted against the mass number $A$ for the stable nuclides.

that *the total charge and the total nucleon number of a system never change.* This is a statement of the laws of *conservation of charge* (Sec. 16.2) and *conservation of nucleon number.* These laws have been found to hold without exception in all physical processes.

From Table 20.1 it is seen that the mass of $^{14}$C is greater than the mass of $^{14}$N by 0.00017 u. In the decay process this excess mass is converted into 0.156 MeV of energy, which appears as the kinetic energy of the electron and neutrino.† In any particular decay, the electron has a kinetic energy between 0 and 0.156 MeV, and the neutrino has the remainder. The neutrino seldom interacts with other matter once it is produced, so that it has no physiological effect. It is important in astronomy, however, because it is one of the ways a star loses energy.

Beta decay is the process by which a nuclide that lies above the stability line, i.e., a nuclide with an excess of neutrons, is converted into a stable nuclide. A nuclide that lies below the stability line, i.e., a nuclide with an excess of protons, undergoes an analogous decay process in which a proton converts into a neutron by emitting a positron $e^+$ and a neutrino. A *positron* is a particle identical to an electron in all respects except that it is positively charged. For example, the decay of $^{13}$N is

$$^{13}_{7}\text{N} \longrightarrow {}^{13}_{6}\text{C} + e^+ + \nu$$

Note again that charge and nucleon number are conserved. Positrons are discussed more fully in Sec. 20.5, where their use in nuclear medicine is considered.

*Alpha decay* is a process in which a $^4$He nucleus is spontaneously emitted by a heavy nucleus. For example, the decay of $^{238}$U is

$$^{238}_{92}\text{U} \longrightarrow {}^{234}_{90}\text{Th} + {}^4_2\text{He}$$

Helium nuclei produced in this way are called *alpha particles*‡. The alpha particles used by Rutherford in his famous scattering experiment (Sec. 19.2) were obtained from the decay of $^{214}$Bi. The mass difference between the parent nuclide and the sum of the masses of $^4$He and the daughter nuclide appears as the kinetic energy of the alpha particle. As the alpha particle moves through matter, it quickly transfers this energy to the surrounding atoms, increasing their thermal energy.

*Gamma rays* are short-wavelength photons that sometimes accompany an alpha or beta decay. The motion of the nucleons inside a nucleus is governed by the same quantum-mechanical laws that govern the motion of electrons in an atom, so that a nucleus, like an atom, can exist only in certain discrete quantum states. When a radioactive nucleus decays, the daughter nucleus is not necessarily produced in the state of lowest energy (*ground state*). But, like an atom, a nucleus in an excited state rapidly decays to the ground state by the emission of photons. The photons produced this way are called gamma rays. Because the energy difference between quantum states

---

† The mass of the beta ray is already included in the mass of $^{14}_{7}$N because atomic masses include the mass of the $Z$ orbital electrons in addition to the mass of the nucleus. That is, the mass of $^{14}_{7}$N includes the mass of one more electron than the mass of $^{14}_{6}$C does.
‡ The use of the word "ray" in beta ray and "particle" in alpha particle has only historic significance. Alphas and betas are just helium nuclei and electrons, respectively.

of a nucleus is often millions of electron volts, gamma rays have very high energy and very short wavelengths. Gamma rays are physically the same as short-wavelength x-rays, and they have the same use in radiation therapy.

### Half-life

The transition from an excited state to the ground state of a nucleus by the emission of a gamma ray is usually instantaneous. However, a radioactive nucleus can exist for a long time before undergoing alpha or beta decay. According to the laws of quantum mechanics, it is completely uncertain when a particular nucleus will decay. All that can be predicted is the probability that a nucleus will decay within a given period. This is usually expressed by giving the period $\tau$ within which a nucleus has a 50 percent chance of decaying. This period, called the *half-life*, is characteristic of each nuclide.

For example, from Table 20.1 we see that $^{13}$N has a half-life of 10 min. This means that in a sample of 1000 $^{13}$N nuclei, 500 will decay in the first 10 min; of the 500 remaining, 250 will decay in the next 10 min; and of the 250 nuclei remaining after 20 min, 125 will decay in the next 10 min. A $^{13}$N nucleus that survives 30 min has the same fifty-fifty chance of decaying in the next 10-min period as it did in the first 10-min period.

The number of nuclei that decay in any given time period (the *rate of decay R*) is proportional to the number $N$ of nuclei present. For periods short compared to a half-life the rate of decay is

$$R = \frac{0.693}{\tau} N \qquad\qquad 20.3$$

For example, with a half-life of 10 min, the rate of decay is

$$R = \frac{0.693}{10 \text{ min}} N = 0.0693N \text{ min}^{-1}$$

If there are initially 1000 nuclei in the sample, then

$$R = 69.3 \text{ min}^{-1}$$

which means there are 69.3 decays per minute. However, after 10 min there are only 500 nuclei in the sample, so the decay rate is only 34.6 min$^{-1}$. The decay rate is of special importance because it can be directly measured with a Geiger counter (Sec. 20.4), whereas the number of nuclei in the sample is not observable. The half-life is determined by measuring the decrease in the decay rate. The details are given in Appendix VII.

The half-lives of the nuclides vary from a fraction of a second to billions of years. For instance, $^{238}$U has a half-life of 4.5 billion years, which is about the age of the solar system. This means that of all the $^{238}$U present when the solar system was formed, about half still exists. The isotope $^{235}$U, on the other hand, has a half-life of only 0.7 billion years, so that over 6 half-lives have elapsed since the formation of the solar system. Consequently only the fraction

$$\frac{1}{2^6} = \frac{1}{64} = 0.0156$$

of the original $^{235}$U still exists. At present $^{235}$U constitutes only 0.7 percent of naturally occurring uranium, while $^{238}$U constitutes the other 99.3 percent.

Nuclides with half-lives less than 0.1 billion years will not have survived their original formation in any detectable amounts. Short-lived nuclides that do occur naturally have been recently formed by one of two processes.

**Decay of a long-lived parent**  For example, $^{238}$U is the longest-lived member of a long chain of decays, which ends with the stable isotope $^{206}$Pb. The nuclides in this chain, along with their half-lives and decay modes, are displayed on a plot of $A$ versus $Z$ in Fig. 20.5. In a sample of naturally occurring uranium, all the decay products are present in proportion to their half-lives. Note that only nuclides with $A = 238 - 4n$ can be formed from $^{238}$U by alpha and beta decay. Other chains exist starting with $^{235}$U, $^{232}$Th, and $^{237}$Np.

**Production by cosmic rays**  Cosmic rays are high-energy particles, mostly electrons and protons, that bombard the earth from outer space. The collisions of cosmic-ray protons with nuclei in the upper atmosphere, if violent enough, produce a number of secondary protons, neutrons, and alphas by breaking the nucleus apart. The secondary neutrons in turn create radioactive nuclides through collisions

**FIGURE 20.5**
The $^{238}$U decay chain. The mass number $A$ and atomic number $Z$ of each nuclide in the chain are represented by a dot, and the half-life of each decay is shown.

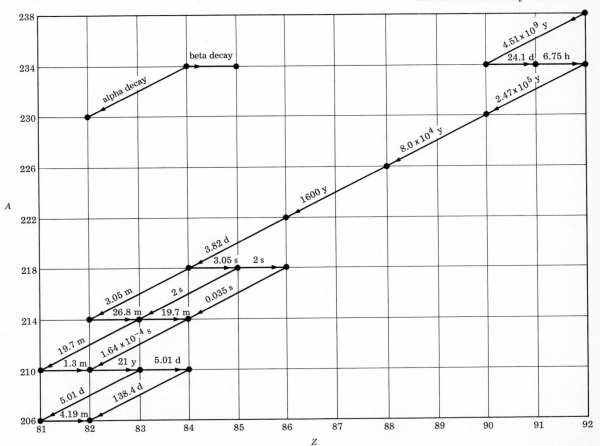

with other nuclei. One of the most important reactions of this type is

$$^{14}_{7}\text{N} + ^{1}_{0}n \longrightarrow ^{14}_{6}\text{C} + ^{1}_{1}\text{H}$$

in which a neutron converts $^{14}\text{N}$ to $^{14}\text{C}$ by changing places with a proton. This process occurs continuously in the atmosphere, and a dynamic equilibrium exists between the rate of $^{14}\text{C}$ production and the rate of $^{14}\text{C}$ decay.

### Carbon 14 dating

Soon after a $^{14}\text{C}$ atom is formed in the atmosphere, it reacts with oxygen to form a $CO_2$ molecule, and in this form $^{14}\text{C}$ enters the biosphere. Since living organisms continuously exchange $CO_2$ with the atmosphere, the ratio of $^{14}\text{C}$ to $^{12}\text{C}$ in an organism is equal to the equilibrium ratio of $^{14}\text{C}$ to $^{12}\text{C}$ that exists in the atmosphere. After an organism dies, however, the $^{14}\text{C}$ supply can no longer be replenished, while the $^{14}\text{C}$ already in the organism continues to decay with a half-life of 5730 y. Consequently, the date of an archeological site can be determined by a measurement of the current rate of $^{14}\text{C}$ decay in a sample of organic material found at the site.

For example, suppose the rate of $^{14}\text{C}$ decay in a 12-g sample of carbon taken from a bone found at an excavation is

$$R = 45.9 \text{ min}^{-1}$$

The half-life of $^{14}\text{C}$ is

$$\tau = 5730 \text{ y} = 3.0 \times 10^9 \text{ min}$$

so from Eq. 20.3 we find there are

$$N = \frac{R\tau}{0.693} = \frac{(45.9 \text{ min}^{-1})(3.0 \times 10^9 \text{ min})}{0.693}$$

$$= 1.99 \times 10^{11}$$

$^{14}\text{C}$ nuclei in the sample. Since the sample is 1 mol of $^{12}\text{C}$, it contains $6.02 \times 10^{23}$ normal $^{12}\text{C}$ atoms. Thus the ratio of $^{14}\text{C}$ to $^{12}\text{C}$ in the sample is

$$\frac{^{14}\text{C}}{^{12}\text{C}} = \frac{1.99 \times 10^{11}}{6.02 \times 10^{23}} = 3.3 \times 10^{-13}$$

whereas other measurements have found that the equilibrium ratio of $^{14}\text{C}$ to $^{12}\text{C}$ in the atmosphere is $1.3 \times 10^{-12}$. If we assume that the ratio was the same when the bone was formed, the amount of $^{14}\text{C}$ in the sample must have decreased by a factor of

$$\frac{3.3 \times 10^{-13}}{1.3 \times 10^{-12}} = 0.254$$

or about one-fourth. From this it is concluded that the bone was formed 2 half-lives ago, or that the site is $2 \times 5730 \text{ y} = 11,460 \text{ y}$ old.

This technique is valuable for dating samples up to 20,000 y old. For samples much older than this, the amount of $^{14}\text{C}$ remaining is too small to be detected with accuracy.

## Fission

Because of the large energy changes involved in nuclear reactions, it has been realized since the early 1900s that a self-sustaining nuclear reaction would liberate millions of times more energy than an ordinary chemical reaction. However, no nuclear reaction capable of being made self-sustaining was known until German scientists discovered in 1938 that when the uranium isotope $^{235}U$ absorbs a neutron, the newly formed $^{236}U$ nucleus, instead of undergoing alpha or beta decay, immediately breaks apart into two nearly equal fragments. This type of nuclear decay is called *fission*.

The fission fragments vary in mass number between 75 and 160, but a typical fission reaction is

$$^{235}_{92}U + {}^{1}_{0}n \longrightarrow {}^{236}_{92}U \longrightarrow {}^{95}_{38}Sr + {}^{139}_{54}Xe + 2\,{}^{1}_{0}n$$

There are a number of important features to note about this reaction.

1 The $^{236}U$ nucleus exists for less than $10^{-12}$ s, so that the fission process can be considered to be instantaneous.

2 Figure 20.4 shows that the average binding energy per nucleon of a stable nuclide with $A$ between 75 and 160 is about 8.5 MeV, whereas the binding energy per nucleon of $^{235}U$ is 7.6 MeV. This means that the 234 nucleons are more tightly bound in the fission fragments than in $^{235}U$. The binding energy per nucleon is the energy required to remove a nucleon from a nucleus, or, alternatively, it is the energy released when a nucleon is bound to a nucleus. Therefore, increasing the binding energy from 7.6 to 8.5 MeV releases 0.9 MeV per nucleon, or

$$234 \times 0.9 \text{ MeV} = 211 \text{ MeV}$$

per fission event. About 86 percent of this energy is released immediately in the form of kinetic energy of the fragments and the neutrons.

3 Because the neutron-to-proton ratio of $^{235}U$ is greater than that of stable nuclides in the $A = 125$ mass region, the fission fragments have an excess of neutrons and lie above the stability curve in Fig. 20.3. These fragments, therefore, beta-decay in a series of steps until stable nuclides with the same mass numbers are formed. The $^{236}U$ nucleus breaks apart in many different ways, and each fragment forms a series of three or four radioactive nuclides, so that over 200 different radioactive nuclides are produced by fission. Most of these do not occur naturally on earth.

**REMARK** These radioactive nuclides are a dangerous and undesirable by-product of fission. During the 1950s, when both the United States and the Soviet Union exploded many atomic bombs, a large amount of radioactive material was released in the atmosphere from the fission process. The most notorious of these was $^{90}Sr$, which is chemically similar to calcium (Table 19.2) and has a relatively long half-life (Table 20.1). As a consequence, it lived long enough to enter the food chain, where it was deposited along with calcium in the bones of growing children. The worldwide radiation hazard that this presented led eventually to the ban on atomic tests in the atmosphere.

*4* The fission process, which requires a neutron to initiate, releases (on an average) 2.5 neutrons per fission. It is this fact that immediately attracted the attention of scientists, because it meant that a chain of self-sustaining nuclear reactions was possible if the extra neutrons released in one fission event could be made to initiate additional fission events.

News of the discovery of fission reached the United States in January 1939. In July Einstein wrote to President Roosevelt about fission, urging that research into the military potential of this discovery be started at once. In September 1939 Germany invaded Poland, plunging Europe into war. The United States, which was then in the process of general rearmament, embarked on an ambitious program to build a fission bomb.

The first chain of self-sustaining fission reactions (a *chain reaction*) was achieved by Enrico Fermi (1901–1954) and his colleagues at the University of Chicago on December 2, 1942. This date is usually taken to mark the beginning of the Atomic Age,† since it is only through a chain reaction that usable amounts of energy can be obtained from a nuclear reaction.

The average number of neutrons released in a fission event that go on to initiate other fission events is called the *multiplication factor f*. The maintenance of a chain reaction requires that $f$ be greater than or equal to 1. Since on average 2.5 neutrons are released in each fission event, it is not necessary that every neutron initiate another fission. This is fortunate, because there are competing processes that remove neutrons from the reaction region. Of these, the most important are:

*1* Escape of neutrons from the reactor

*2* Capture of neutrons by $^{238}U$ in the reaction

$$^{238}U + {}^{1}n \longrightarrow {}^{239}U \qquad\qquad 20.4$$

*3* Capture of neutrons by other nuclides

Process 1 is minimized by making the reactor so large that a neutron is very likely to be captured before it escapes. Process 3 is minimized by the proper choice of the materials placed in the reactor.

Figure 20.6 is a schematic diagram of the type of nuclear reactor now being used in electric-power generating plants. There are four essential components to any reactor:

*1* The fuel, usually natural uranium in the form of long rods of uranium dioxide, $UO_2$. Only $^{235}U$, which constitutes 0.7 percent of natural uranium, is fissionable.‡ (Modern reactors sometimes use enriched uranium, which contains up to 90 percent $^{235}U$.)

*2* The moderator, usually graphite in the form of rods or blocks. The graphite and uranium rods are stacked together in a large array, or *pile*, which is the reactor. The purpose of the moderator is to slow down the neutrons released in the fission process.

---

† In this book *atomic* usually refers to the electronic properties of an atom, and *nuclear* refers to the nuclear properties. In common usage, however, atomic is synonymous with nuclear.

‡ A nuclide is fissionable only if it can sustain a chain reaction. There are nuclides, such as $^{238}U$, which can be made to fission under certain circumstances but which cannot sustain a chain reaction. $^{235}U$ is the only naturally occurring fissionable nuclide. Two other fissionable nuclides, $^{233}U$ and $^{239}Pu$, are produced artificially.

3 Control rods, usually made of cadmium. Cadmium readily absorbs neutrons, so that the multiplication factor can be increased or decreased by removing or inserting these rods.

4 Coolant, usually water, which circulates through the reactor to remove the heat generated. The coolant exchanges its heat in a boiler that produces steam to drive a steam turbine.

To start a reactor, the control rods are gradually withdrawn. This increases the multiplication factor by decreasing the opportunity for neutrons to be captured by nonfissionable cadmium. When the multiplication factor exceeds 1, the reactor is *critical*. At this stage a single fission event, initiated by a stray cosmic-ray neutron, can rapidly multiply until billions of fissions per second are occurring. For instance, starting with a single fission event and a multiplication factor $f = 1.007$, there will be $f^n = 1.007^n$ fissions after $n$ generations, which amounts to $10^3$ fissions by the thousandth generation. Since the time interval between generations is only $10^{-8}$ s, the fission rate builds up very rapidly once the reactor becomes critical.

When the reactor reaches the desired level of activity, the control rods are adjusted to reduce $f$ to 1.00000. Thereafter the reactor maintains a constant rate of activity. Through collisions with other nuclei, the kinetic energy of the fission fragments and neutrons released by the fission process is converted into thermal energy, which raises the temperature of the interior of the reactor. This heat is removed by the coolant that circulates through the reactor and is transferred to a boiler, where it is used to make the steam to turn the turbine to run the generator to produce electricity. In this way the energy of the nucleus is converted into usable electric energy. The only difference between a nuclear power plant and a conventional power plant is the fuel used to produce the steam. Once the steam is produced, both types of plants are identical in their operation.

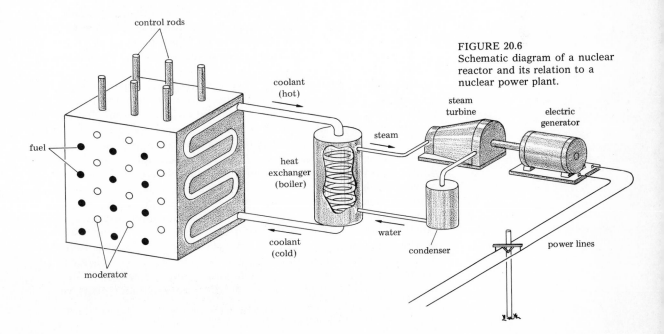

FIGURE 20.6
Schematic diagram of a nuclear reactor and its relation to a nuclear power plant.

Nuclear power plants are certain to become an increasingly important part of our power network as our supply of coal and oil diminishes. Nuclear power has the advantage of not releasing pollutants into the atmosphere, but the radioactive waste products that accumulate inside a reactor have to be removed every few years. The safe disposal of this hazardous material is a serious problem that will become more critical as the nuclear power industry expands.

The only naturally occurring fissionable material, $^{235}$U, constitutes only 0.7 percent of natural uranium. Thus the world supply of fissionable material is very limited and is probably not sufficient to sustain the fully developed nuclear power industry expected in the next century. Fortunately two other fissionable nuclides, $^{233}$U and $^{239}$Pu, can be produced in a nuclear reactor. Plutonium 239 is a natural decay product of $^{239}$U, which is produced from nonfissionable $^{238}$U by the capture of a neutron (Eq. 20.4). The beta decay of $^{239}$U produces $^{239}$Np, a new element with atomic number 93. The reaction is

$$^{239}_{92}\text{U} \longrightarrow {}^{239}_{93}\text{Np} + e^- + \nu$$

and the half-life of $^{239}$U is 23.5 min. Neptunium 239 in turn beta-decays with a half-life of 2.35 d into $^{239}$Pu, a new element with atomic number 94. The reaction is

$$^{239}_{93}\text{Np} \longrightarrow {}^{239}_{94}\text{Pu} + e^- + \nu$$

Thus the $^{239}$U produced in a reactor decays within a few days to $^{239}$Pu. Since $^{239}$Pu has a half-life of 24,400 y, it is stable enough for industrial purposes.

In conventional nuclear reactors, the amount of $^{239}$Pu produced is less than the amount of $^{235}$U consumed. However, *breeder reactors* are now being developed that produce more fissionable material than they consume. In these reactors there is no moderator to slow the neutrons, which are captured either by $^{235}$U (causing fission) or by $^{239}$U (producing $^{239}$Pu). Of the 2.5 neutrons released in each fission event, 1 is required to initiate another fission and 1.5 are available to produce $^{239}$Pu. Since $^{238}$U (from which $^{239}$Pu is made) is 140 times more plentiful than $^{235}$U, breeders open up the possibility of a virtually unlimited fuel supply.

An atomic bomb, unlike a reactor, requires fissionable material in pure form. The major activities in the wartime development of the atomic bomb were centered on the separation of $^{235}$U from $^{238}$U and on the production of $^{239}$Pu in a reactor. Since $^{235}$U is chemically identical to $^{238}$U, the separation requires a physical process that depends on their mass difference. The most successful process is gas diffusion, in which gaseous uranium, in the form of uranium hexafluoride, $UF_6$, is allowed to diffuse through a porous membrane. A molecule composed of $^{235}$U has (on the average) a slightly greater speed than a molecule composed of $^{238}$U (Sec. 8.4) and diffuses more rapidly. Thus the diffusion of $UF_6$ through a porous membrane results in a minute enrichment of $^{235}$U. In 1943 an entire city, Oak Ridge, Tennessee, was built for the purpose of separating $^{235}$U by repeatedly passing $UF_6$ through a porous membrane. Simultaneously, in Hanford, Washington, a giant reactor† and chemical separation plant was

† The first Hanford production reactor was started on June 7, 1943, only 18 months after Fermi had achieved the first nuclear chain reaction.

built to produce and purify $^{239}$Pu. These were parallel developments, since the success of either one would have resulted in a bomb. In fact, both projects succeeded in producing usable quantities of purified material before the end of the war.

When a stray cosmic-ray neutron initiates a fission event in a block of pure $^{235}$U (or $^{239}$Pu), the neutrons released can either escape from the block or initiate another fission event. The multiplication factor of the block thus depends entirely on its mass. If the mass is too small, too many neutrons escape before initiating fission, so $f$ is less than 1. Such a mass is said to be *subcritical*. If the mass is large enough, however, enough neutrons initiate fission events before they reach the surface, so $f$ is greater than 1. Such a mass is said to be *critical*. An uncontrolled chain reaction will spontaneously erupt in a block of critical mass.

In an atomic bomb, two subcritical masses of $^{235}$U (or $^{239}$Pu) are suddenly forced together by conventional explosives, so that a critical mass is formed. Before this mass can disintegrate, enough fission takes place to generate an enormous amount of energy in a very small volume, which results in an immense explosion.

The energy released in the fission of 1 mol (235 g) of $^{235}$U is about

$$6 \times 10^{23} \times 200 \text{ MeV} = 1.2 \times 10^{26} \text{ MeV}$$
$$= 2 \times 10^{13} \text{ J}$$

In the bombs dropped on Japan approximately 1 kg (4.25 mol) of $^{235}$U (or $^{239}$Pu) underwent fission. The energy released, therefore, was

$$(4.25)(2 \times 10^{13} \text{ J}) = 8.5 \times 10^{13} \text{ J}$$

which is equivalent to the energy released in the explosion of 20,000 tons of TNT.

**Fusion**

The energy of the sun, and hence ultimately all nonnuclear energy on earth, comes from the fusion of four hydrogen nuclei (protons) to form helium. The overall reaction is

$$4\,^1_1\text{H} \longrightarrow \,^4_2\text{He} + 2e^+ + 2\nu \qquad\qquad 20.5$$

From the mass change

$$\begin{aligned} \Delta m &= 4\, m_{^1\text{H}} - m_{^4\text{He}} \\ &= 4 \times 1.007825 \text{ u} - 4.0026 \text{ u} \\ &= 0.0287 \text{ u} \end{aligned}$$

the energy released in each fusion event is found to be 26.7 MeV. The masses of the positrons are not included in the calculation of $\Delta m$ because these masses are soon converted into energy in the reaction

$$e^+ + e^- \longrightarrow 2\gamma \qquad\qquad 20.6$$

(This annihilation process is discussed more fully in Sec. 20.5.) The energy released by the fusion of 1 g of hydrogen is equal to the energy released by burning 16 tons of oil.

Equation 20.5 is just the overall reaction, but in reality four protons can never all come together at once. Instead the helium is built up in a series of reactions, each one of which involves the fusion of only two nuclei. In 1938 Hans Bethe (1906–    ) suggested two reaction sequences, the *carbon cycle* and the *proton-proton chain,* which are believed to be the primary processes taking place in the sun and similar stars. These sequences are shown in Table 20.2.

In the carbon cycle, hydrogen nuclei, instead of combining directly with each other, combine with a $^{12}$C nucleus. After four hydrogen nuclei have attached themselves to the $^{12}$C nucleus, it breaks apart into $^4$He and $^{12}$C. Thus $^{12}$C is just a catalyst for this reaction, since it is not ultimately consumed. The overall reaction is given by Eq. 20.5.

In the proton-proton chain, two hydrogen nuclei first combine to form deuterium, $^2$H. This then reacts with normal hydrogen to form $^3$He, and finally two $^3$He nuclei react with each other to produce a $^4$He nucleus and two protons, $^1$H. The overall reaction is again given by Eq. 20.5.

These fusion reactions do not occur at normal temperature because the electrical repulsion between the nuclei keep them from getting close enough to react. Figure 20.2 shows how the repulsive electric force between two protons dominates the potential energy at large distances, while the attractive nuclear force dominates at small distances. Only at temperatures of 20 million degrees or more, such as exist in the interior of stars, do nuclei have sufficient kinetic energy to overcome the electrical repulsion. The situation is similar to any combustion process, which occurs only above a certain temperature.

Because of its great energy potential, efforts are being made to produce a controlled fusion reaction on earth. The primary reactions being studied in these experiments are

TABLE 20.2 **Carbon cycle and proton-proton chain**

| Reaction | MeV |
|---|---|
| *Carbon cycle* | |
| $^{12}_{6}C + ^1_1H \longrightarrow \ ^{13}_{7}N$ | + 2.0 |
| $\qquad\qquad ^{13}_{7}N \longrightarrow \ ^{13}_{6}C + e^+ + \nu$ | + 2.2† |
| $\qquad\qquad \tau = 10$ min | |
| $^{13}_{6}C + ^1_1H \longrightarrow \ ^{14}_{7}N$ | + 7.5 |
| $^{14}_{7}N + ^1_1H \longrightarrow \ ^{15}_{8}O$ | + 7.3 |
| $\qquad\qquad ^{15}_{8}O \longrightarrow \ ^{15}_{7}N + e^+ + \nu$ | + 2.7† |
| $\qquad\qquad \tau = 2$ min | |
| $^{15}_{7}N + ^1_1H \longrightarrow \ ^{12}_{6}C + ^4_2He$ | + 5.0 |
| | 26.7 |
| *Proton-proton chain* | |
| $^1_1H + ^1_1H \longrightarrow \ ^2_1H + e^+ + \nu \quad + 1.5†$ (twice) = | 3.0 |
| $^2_1H + ^1_1H \longrightarrow \ ^3_2He \qquad\qquad + 5.4$ (twice) = | 10.8 |
| $^3_2He + ^3_2He \longrightarrow \ ^4_2He + 2\ ^1_1H \ +12.9$ (once) = | 12.9 |
| | 26.7 |

†Includes the energy obtained from the ultimate annihilation of the positron (Eq. 20.6).

$$\begin{array}{c}{}^{2}_{1}\text{H} + {}^{2}_{1}\text{H} \begin{cases} {}^{3}_{2}\text{He} + {}^{1}_{0}n + 3.2 \text{ MeV} \\ {}^{3}_{1}\text{H} + {}^{1}_{1}\text{H} + 4.0 \text{ MeV} \end{cases}\end{array} \qquad 20.7$$

and $$\qquad {}^{3}_{1}\text{H} + {}^{2}_{1}\text{H} \longrightarrow {}^{4}_{2}\text{He} + {}^{1}_{0}n + 17.6 \text{ MeV}$$

where ³H is a radioactive isotope of hydrogen (tritium).†

These are called *thermonuclear reactions* because they only occur at temperatures of millions of degrees. Since no material can contain a gas at these temperatures, it is necessary to contain the hydrogen nuclei in a strong magnetic field. So far it has not been possible to contain enough nuclei long enough to achieve fusion. If and when controlled fusion is achieved, the world will have an unlimited, pollution-free source of energy which produces fewer radioactive by-products than fission.

An uncontrolled fusion reaction (hydrogen bomb) is produced by placing a mixture of ²H and ³H around an ordinary fission bomb. When the fission bomb explodes, temperatures of millions of degrees initiate fusion in the hydrogen. The result is an explosion of astro-nomical proportions.

## 20.4  NUCLEAR RADIATION

### Properties

Nuclear radiation refers collectively to the high-energy particles (alpha particles, beta rays, protons, neutrons) and the electromagnetic radiation (gamma rays, x-rays) that accompany nuclear decay and nuclear reactions. All these radiations penetrate through matter, ionizing atoms and breaking apart molecules in their path. As a consequence they have very deleterious effects on living cells and are a health hazard to people who work with radioactive materials or near nuclear reactors. At the same time, the ability of nuclear radiation to destroy cell function makes it very effective in the treat-ment of cancer and related diseases.

Although all nuclear radiations have the same general effect on living matter, each radiation has its own characteristic properties that effect its potency and usefulness in different situations. Thus it is necessary to consider the individual properties of each type of radiation.

*Alpha particles*, because they are so much heavier than electrons, travel through matter in a straight line, knocking aside the electrons in their path.‡ In each alpha-electron collision, the alpha loses about 33 eV by knocking an electron out of an atom. Thus an alpha particle with an initial kinetic energy of 5 MeV makes about

$$\frac{5 \text{ MeV}}{33 \times 10^{-6} \text{ MeV}} = 151,000$$

† Tritium is produced by placing ⁶Li inside a reactor, where it absorbs a neutron in the reaction

$$\qquad {}^{6}_{3}\text{Li} + {}^{1}_{0}n \longrightarrow {}^{3}_{1}\text{H} + {}^{4}_{2}\text{He}$$

‡ The Rutherford scattering experiment (Sec. 19.2) studied the rare instances in which an alpha gets close enough to a nucleus to suffer a large-angle deflection.

collisions before it comes to rest. Since each collision ionizes an atom or breaks a molecule apart, an alpha particle does considerable damage before it stops.

The distance a particle travels before stopping is called its *range*. In a particular material, all alpha particles with a given energy have the same range. This range increases with the energy of the alphas and decreases with the density of the material in which they travel. This is seen from Table 20.3, which gives the range of alphas of various energies in air, body tissue, and aluminum. This table also shows that the range of alphas is very small. A 5-MeV alpha, for instance, travels only 0.21 mm through tissue and is completely stopped by a thin sheet of aluminum foil. Consequently it is very easy to shield against alphas, and even without shielding they cannot penetrate through the skin. However, a radioactive alpha emitter, if swallowed, may accumulate in certain parts of the body, where it can deliver very harmful doses of radiation.

*Beta rays*, being simply high-speed electrons, are more easily deflected by collisions with atomic electrons than alphas are. Nevertheless, like alphas, betas continuously lose energy by ionizing the atoms they pass and come to rest after they have traveled a definite distance through a particular material. Table 20.3 gives the range of beta rays of various energies. Note that the range increases rapidly with energy and that a 3-MeV beta ray travels 150 times farther through tissue than a 3-MeV alpha particle.

In addition to losing energy through collisions with electrons, beta rays produce electromagnetic radiation (x-rays) whenever they suffer a sudden deceleration. This braking radiation, or *bremsstrahlung*, is the source of the x-rays produced when an electron beam collides with the anode of an x-ray tube (Sec. 16.5). X-rays penetrate much farther through matter than beta rays, so that even after all the beta rays have stopped in a substance, the x-rays they have generated continue to travel through it.

Figure 20.7 shows the tracks of beta rays incident on a lead plate thick enough to stop them. However, the x-rays generated by the

TABLE 20.3 **Range of alpha particles and beta rays of various energies in air, body tissue, and aluminum**

| Energy, MeV | Range, cm | | |
| --- | --- | --- | --- |
| | Air | Body tissue | Aluminum |
| *Alpha particles* | | | |
| 1.0 | 0.55 | $0.33 \times 10^{-2}$ | $0.32 \times 10^{-3}$ |
| 2.0 | 1.04 | $0.63 \times 10^{-2}$ | $0.61 \times 10^{-3}$ |
| 3.0 | 1.67 | $1.00 \times 10^{-2}$ | $0.98 \times 10^{-3}$ |
| 4.0 | 2.58 | $1.55 \times 10^{-2}$ | $1.50 \times 10^{-3}$ |
| 5.0 | 3.50 | $2.10 \times 10^{-2}$ | $2.06 \times 10^{-3}$ |
| *Beta rays* | | | |
| 0.01 | 0.23 | 0.00027 | |
| 0.10 | 12.0 | 0.0151 | 0.0043 |
| 0.50 | 150 | 0.18 | 0.059 |
| 1.0 | 420 | 0.50 | 0.15 |
| 2.0 | 840 | 1.00 | 0.34 |
| 3.0 | 1260 | 1.50 | 0.56 |

beta rays in the lead pass through to the other side, where they may still present a radiation hazard. Thus, to provide adequate protective shielding from beta rays, the barrier must be much thicker than the range of the beta rays.

*Gamma rays*, unlike charged particles, lose all their energy in a single interaction, rather than continuously in a series of collisions. For instance, when a gamma-ray photon interacts with an atomic electron, all the photon's energy is transferred to the electron, which consequently is ejected from the atom with a large kinetic energy. This process is identical to the photoelectric effect (Sec. 19.1) and results in the destruction of the photon. The high-energy electron produced in the process behaves like a beta ray, ionizing other atoms as it travels through matter. (These electrons can in turn generate lower-energy x-rays, as shown in Fig. 20.7.)

The distance a photon travels before it interacts with an electron cannot be predicted. All that can be predicted is the distance in which a photon has a 50 percent chance of interacting. This is called the *half-value layer* and is analogous to the half-life of a radioactive nuclide. Table 20.4 gives the half-value layer for gamma rays of various energies in tissue and lead.

For example, the half-value layer for a 0.1-MeV gamma ray in tissue is 4.05 cm. This means that after it has traveled through 4.05 cm of tissue, the intensity of a gamma-ray beam is half its original value. After it has traveled through another 4.05 cm, the intensity of the remaining beam is reduced by one-half again. Thus, after the beam has traveled through 8.10 cm of tissue its intensity is only a quarter of its original value. To reduce a beam to less than 1 percent of its original value, it has to travel through at least seven half-value layers, or 28.35 cm in this case. Gamma rays and x-rays are thus far more penetrating than charged particles.

*Neutrons*, like gamma rays, lose all their energy in a single collision, although neutrons collide with nuclei rather than with electrons. Neutrons are very penetrating, and it requires massive amounts of lead and concrete shielding to protect workers from the neutrons produced by a reactor.

## Detection

Nuclear radiation is always detected by the ionization it produces. For instance, the track of a charged particle through a *cloud* or *bubble*

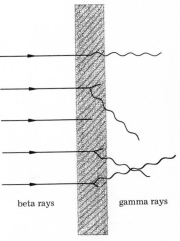

**FIGURE 20.7**
Beta rays incident on a lead plate thick enough to stop them. As the betas stop, they generate x-rays that pass through the plate.

TABLE 20.4 **Half-value layer for gamma rays of various energies in body tissue and lead**

| Energy, MeV | Half-value layer, cm | |
| | Body tissue | Lead |
|---|---|---|
| 0.01 | 0.131 | 0.00076 |
| 0.05 | 3.12 | 0.012 |
| 0.10 | 4.05 | 0.012 |
| 0.50 | 7.20 | 0.42 |
| 1.0 | 9.80 | 0.89 |
| 5.0 | 23.0 | 1.52 |

*chamber* is made visible by the bubbles that condense along the ionized path produced by the particle. Similarly, as a charged particle moves through the photosensitive emulsion on photographic film, the ionization it produces leaves a record of its path that is revealed when the film is developed. Radiation workers wear film badges to monitor the total radiation to which they are exposed. Cloud chambers, bubble chambers, and photographic film are valuable detectors in nuclear research because they give a complete record of the fate of individual particles.

Electronic devices are used for the instantaneous detection of radiation. The oldest and simplest device is the Geiger-Müller counter (*Geiger tube*). This is just a metallic gas-discharge tube with a wire running through its center (Fig. 20.8). The wire is maintained at a large positive potential (about +500 V) relative to the wall of the tube, and the tube contains a gas at low pressure (about 0.1 atm). When a single charged particle passes through the gas in the tube, it ionizes one or more atoms in the gas, releasing several free electrons. These electrons are then accelerated toward the wire, but before they reach it, they usually ionize other atoms in their path, thereby releasing more electrons. These additional electrons in turn ionize still more atoms as they accelerate toward the wire. A chain reaction (or *avalanche*) develops, which rapidly ionizes most of the gas in the tube. As a consequence, the gas becomes electrically conducting, and there is a momentary surge of current through the tube. This current pulse goes to an amplifier, and the output of the amplifier operates a counter which records the discharge produced each time a charged particle passes through the tube.

Geiger tubes do not detect photons and neutrons directly. However, if the tube is surrounded by the proper material, some of the incident gamma rays and neutrons will knock electrons and protons out of the atoms of the material, and these charged particles will be detected. Of course, only a fraction of the incident photons and neutrons are detected this way, since only a fraction of the incident particles interact with the surrounding material.

Geiger tubes are relatively inexpensive and easy to operate. Their main disadvantages are their inability to measure the energy of the incident particle and their rather long *dead time*, or time between the passage of one charged particle through a detector and the moment when the detector is able to detect a second particle. In a Geiger tube this is the time required for all the ions and electrons produced by the first particle to recombine into neutral atoms, so that the tube is back in its resting state. The dead time of a Geiger tube is of the order of milliseconds, which means that the tube is limited to a few hundred counts per second. If particles are incident on a Geiger tube at a faster rate, it will fail to count them all since many will arrive during a dead period.

In nuclear medicine (Sec. 20.5) a more sophisticated detector, called a *scintillation counter*, is used. This is an electronic version of the method used by Rutherford to detect alpha particles in 1911 (Sec. 19.2). It is based on the fact that certain materials convert some of the energy of an ionizing particle into light and ultraviolet radiation. In Rutherford's time this light was detected by eye, but in a modern scintillation counter it is detected by a *phototube*.

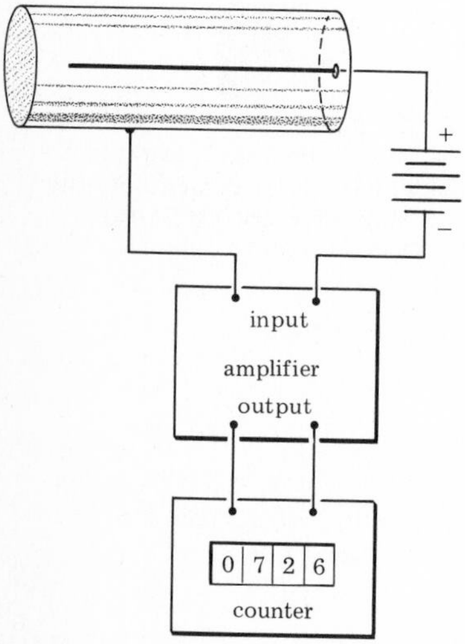

**FIGURE 20.8**
Schematic diagram of a Geiger counter.

The scintillation material most often used today is a transparent crystal of sodium iodide, NaI. When a charged particle stops in the crystal, a quantity of light proportional to the energy of the particle is generated. Some of this light falls on the first electrode of a phototube (Fig. 20.9), from which it knocks out electrons by the photoelectric effect (Sec. 19.1). These electrons are accelerated to the second electrode, which is maintained at a potential of several hundred volts above the first electrode. Upon hitting this electrode, each electron knocks out several electrons, which in turn accelerate toward a third electrode maintained at a potential several hundred volts above the second electrode. At each stage the number of electrons released is 2 to 10 times the number incident, so the total number of electrons rapidly increases. For instance, in a phototube with 10 electrodes, 256 electrons will reach the last electrode for each electron leaving the first electrode if the multiplication factor is 2. Thus only a few photons of light have to arrive at the first electrode to produce a measurable signal at the last electrode.

Since the amount of light generated when a charged particle stops in the NaI crystal is proportional to the energy of the particle, and since the current at the last electrode is proportional to the number of photons incident on the first electrode, the output of the phototube is proportional to the energy of the particle. Thus a scintillation counter not only detects charged particles, but, once calibrated, it can measure their energy as well. Furthermore, the dead time of a scintillation counter is the order of microseconds, so it is capable of much higher counting rates than a Geiger counter.

## Dosimetry

Radiation medicine involves the protection of radiation workers and the general population from the harmful effects of radiation, as well as the use of radiation for the diagnosis and treatment of disease. All these applications require a quantitative measure of the unit of absorbed radiation dose. The unit currently adopted is the *rad* (rd), defined as the absorption of $10^{-2}$ J of ionizing radiation per kilogram of absorbing material. Only the energy actually absorbed is counted; a gamma ray that passes through the body without interaction does not contribute to the dose. Since dense material, such as bone, has a smaller half-value layer for gamma rays than soft tissue, such as tissue, the same intensity of radiation results in a greater dose in bone than in tissue.

The maximum permissible whole-body exposure† dose of x-rays and gamma rays is 5 rd/year (0.1 rd/week). A whole-body dose assumes that each part of the body has received about the same exposure, so that a 60-kg person should absorb no more than

$$(0.1 \text{ rd})(10^{-2} \text{ J/kg})(60 \text{ kg}) = 0.06 \text{ J}$$

of ionizing radiation per week throughout his entire body. However, a small volume of the body, say a few grams on the skin, can receive as much as 5000 rd during cancer therapy without serious conse-

†This is the occupational standard set for radiation workers. The suggested maximum exposure for the general population is usually taken to be one-fiftieth to one-hundredth the occupational exposure.

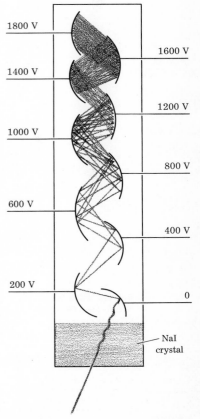

FIGURE 20.9
Schematic diagram of a scintillation counter.

TABLE 20.5 **Effects of acute doses of whole-body gamma radiation**

| Dose, rd | Effect |
|---|---|
| 0–25 | No observable effect |
| 25–100 | Slight blood change |
| 100–200 | Moderate blood change; vomiting in 5–50% of cases within 3 h; complete recovery within a few weeks (except for blood forming system) |
| 200–600 | Severe blood changes; vomiting in 50–100% of cases within 3 h; loss of hair within 2 weeks; hemorrhaging and infection; death in 0–80% of cases within 2 months |
| 600–1000 | Severe blood changes; vomiting within 1 h; loss of hair; hemorrhaging and infection; death in 80–100% of cases within 2 months |

quences, even though 5000 rd delivered to a 0.1-kg mass of skin amounts to the absorption of

$$(5000 \text{ rd})(10^{-2} \text{ J/kg})(0.1 \text{ kg}) = 5 \text{ J}$$

of ionizing radiation.

The body has the ability to recover from small amounts of radiation exposure, so that the time factor is important. For example, an exposure of 5 rd/year amounts to a lifetime exposure of 350 rd, whereas a single exposure of a whole-body dose of 350 rd is often fatal. Table 20.5 shows the effects of a single exposure to various dosages of gamma radiation.

The rad is a physical unit. However, equal doses (in rads) of different types of radiation have different biological effects. The *relative biological effectiveness* (RBE) of a particular type of radiation is the ratio of the dose (in rads) of x-rays or gamma rays to the dose (in rads) of the radiation that produces the same biological effect. From Table 20.6, which gives the RBE of various radiations, it is seen that beta rays have the same effectiveness as gamma rays but that alpha particles are 10 to 20 times more effective.

The biological unit of dose is the *rad equivalent man* (rem), which is defined in terms of the RBE by

$$\text{rem} = \text{RBE} \times \text{rad}$$

For instance, a dose of 50 rd of alpha particles is equal to 500 to 1000 rem. Thus a 50-rd dose of whole-body alpha particles would be as lethal as a 500-rd (= 500-rem) dose of gamma rays. Of course, because of the short range of alpha particles, whole-body exposure is possible only by ingesting an alpha emitter that diffuses throughout the entire body.

TABLE 20.6 **Relative biological effectiveness (RBE) of different types of radiation**

| Radiation | RBE |
|---|---|
| Gamma rays, x-rays, beta rays | 1.0 |
| Fast neutrons and protons | 10 |
| Slow neutrons | 4–5 |
| Alpha particles | 10–20 |

## 20.5 NUCLEAR MEDICINE

Nuclear medicine is the application of the techniques of nuclear physics to medicine. Most applications involve the use of radioactive isotopes, which have become readily available as a by-product of the nuclear power industry. These isotopes are finding increasing use in biological and medical research and in the treatment and diagnosis of disease.

461

20.5   NUCLEAR MEDICINE

## Research

In research, radioactive isotopes are used as *tracers* to label individual atoms in a molecule. For example, the metabolism of iron in the body has been studied using radioactive $^{59}$Fe. Most of the iron in the body occurs in the hemoglobin in the red blood cells. By first feeding an animal a diet containing $^{59}$Fe and then measuring the subsequent radioactivity of the animal's red cells, it has been found that very little iron is absorbed from the diet by an animal under normal conditions. That is, unlike most other elements in the body, the iron in the body is not constantly being replaced by new iron from the diet. An animal is found to incorporate $^{59}$Fe into its red blood cells only when it has lost body iron through blood loss.

Once the $^{59}$Fe is incorporated into a red blood cell, it remains there throughout the life of the cell. Furthermore, even after the cell is destroyed, its iron is not eliminated from the body but is reused to make the hemoglobin for a new cell.

The stability of the iron in the red blood cells has resulted in an interesting technique for measuring the total number of red cells in an animal. A known number of red cells containing $^{59}$Fe is removed from an animal that was fed a $^{59}$Fe diet after blood loss and is injected into a test animal fed a normal diet. The number of radioactive cells in a sample of blood taken from the test animal is determined by measuring the radioactivity of the sample. The total number of red blood cells in the animal can then be calculated from the number of cells in the sample and the total number of radioactive cells injected.

## Therapy

In the treatment of disease, radioactive $^{60}$Co is used directly as a source of ionizing radiation. $^{60}$Co beta-decays to an excited state of $^{60}$Ni, which then immediately decays to its ground state by emitting in succession a 1.17-MeV gamma ray and a 1.33-MeV gamma ray. The energies of these gamma rays are much higher than the energies of the x-rays produced by all but the largest x-ray machines, so that $^{60}$Co is a relatively inexpensive and convenient source of very penetrating radiation. The required amount of $^{60}$Co (about 1 g) is kept inside a heavy lead box, mounted above a table. A patient lying on the table is exposed to radiation by opening a small hole in the box. The room itself is heavily shielded from the rest of the hospital. Since no personnel can be in the room during the treatment, the procedure is monitored on closed-circuit television.

As an example, let us calculate the dosage involved in a $^{60}$Co treatment, assuming a 1-g $^{60}$Co source. The half-life of $^{60}$Co is

$$\tau = 5.26 \text{ y} = 1.65 \times 10^8 \text{ s}$$

so from Eq. 20.3 the decay rate of 1 g ($\frac{1}{60}$ mol) of $^{60}$Co is

$$R = \frac{(0.693)(6.02 \times 10^{23})}{(1.65 \times 10^8 \text{ s})(60)} = 4.2 \times 10^{13} \text{ s}^{-1}$$

Since each $^{60}$Co decay produces two gamma rays, there are $8.4 \times 10^{13}$ gamma rays produced per second. These gamma rays are emitted uniformly in all directions, and only about 1 percent, or $8.4 \times 10^{11}$, fall on the patient per second. From Table 20.4 we see that the half-layer value of a 1-MeV gamma ray in tissue is about 10 cm, which is about the front-to-back thickness of the body. This means that approximately half the gamma rays incident on the patient pass through without interaction and the other half are absorbed by the body. Thus about $4.2 \times 10^{11}$ gamma rays per second are absorbed by the body.

To get the dose per second (rd/s) we must know the mass of the patient. For instance, if the patient's mass is 60 kg, the rate of absorption of gamma rays per kilogram of body mass is

$$\frac{4.2 \times 10^{11} \text{ s}^{-1}}{60 \text{ kg}} = 7 \times 10^9 \text{ kg}^{-1} \text{ s}^{-1}$$

The average energy of a $^{60}$Co gamma ray is 1.25 MeV, so the energy absorbed per kilogram per second is

$$(7 \times 10^9 \text{ kg}^{-1} \text{ s}^{-1})(1.25 \text{ MeV}) = 8.75 \times 10^9 \text{ MeV/kg-s}$$
$$= 14.0 \times 10^{-4} \text{ J/kg-s} = 0.14 \text{ rd/s}$$

where the rad is defined to be $10^{-2}$ J/kg of absorbed radiation (Sec. 20.4). Thus a 3-min exposure to the $^{60}$Co results in a whole-body dose of 25 rd.

Radiopharmaceuticals are also used for radiation therapy because certain elements, e.g., iodine, are preferentially absorbed by certain organs in the body, such as the thyroid. Thus some diseases of the thyroid are treated by administering pharmaceuticals containing radioactive $^{131}$I. Since the iodine concentrates in the thyroid, the beta rays that accompany the decay of $^{131}$I are delivered directly to the diseased organ without too much radiation being absorbed by healthy tissue.

## Diagnosis

Conventional x-ray pictures of the body are only able to distinguish organs of different density. In particular, dense bone is clearly seen in an x-ray picture in contrast to the soft surrounding tissue because less radiation passes through the bone than through the tissue. However, x-ray pictures are unable to detect an abnormal mass of soft tissue, such as a tumor, embedded in normal tissue. For this reason radioactive pharmaceuticals have been developed that are preferentially absorbed by the tumor. Special devices detect the radiation coming from a patient who has been injected with such a radiopharmaceutical to determine whether a tumor is present.

The radioactive nuclide usually used in these studies is $^{99}_{43}\text{Tc}^m$ (the superscript $m$ stands for *metastable* state). This particular technetium isotope is obtained from $^{99}$Mo, which beta-decays to an excited state of $^{99}$Tc rather than to the ground state. This is common in many decays. However, unlike most other excited states, the excited state of $^{99}$Tc is semistable (metastable). The half-life for the decay of $^{99}$Tc$^m$ to the ground state is 6.0 h, rather than $10^{-8}$ s, which is the usual

半half-life for such a decay. Thus $^{99}$Tc$^m$ lives long enough to be used in a medical procedure. Furthermore, since it decays to its ground state by emitting a gamma ray, no charged particles are involved. (The ground state beta-decays to $^{99}$Ru with a half-life of 212,000 y, so the technetium is eliminated from the body long before any beta rays are produced.) The absence of charged particles minimizes the radiation hazard involved in using $^{99}$Tc$^m$, because many of the gamma rays pass out of the body before interacting.

**REMARK** Technetium ($Z = 43$) and promethium ($Z = 61$) are the only elements with $Z < 84$ that have no stable isotope. Neither element occurs naturally on earth; like plutonium, both must be produced artificially from nuclear reactions.

The $^{99}$Tc$^m$ is obtained in the form of the pertechnetate ($^{99}$Tc$^m$O$_4^-$) by washing a salt of $^{99}$MoO$_4^-$ with saline solution. The $^{99}$MoO$_4^-$ continuously decays into $^{99}$Tc$^m$O$_4^-$, which remains mixed with the permolybdenate until it is dissolved in the saline. The molybdenum salt is referred to as the generator since it can be stored for several days and washed with saline solution whenever $^{99}$Tc$^m$O$_4^-$ is needed. Because the half-life of $^{99}$Tc$^m$ is only 6 h, the pertechnetate must be injected into the patient immediately and the patient examined within a few hours after injection.

The blood-brain barrier, which normally prevents foreign chemicals from diffusing into the brain, does not exist in a brain tumor, so that the pertechnetate diffuses into a brain tumor but not into the surrounding tissue. Consequently, more gamma rays are given off by a brain tumor than by the surrounding tissue, making possible the detection of the tumor. However, because gamma rays cannot be focused, special techniques are required to get a meaningful image.

A *rectilinear scanning device* consists of a single scintillation counter with a lead collimator covering the front of the crystal (Fig. 20.10). The collimator is perforated by a number of holes angled in such a way that only gamma rays coming from a small volume of space can pass through them all. Thus the counter "sees" only this small volume.† A complete picture of the head is obtained point by point by passing the counter back and forth across the head. At each point the number of counts recorded by the counter measures the amount of $^{99}$Tc$^m$ at that point.

A picture of the head, or *brain scan*, is obtained indirectly by having a small light move back and forth across a photographic plate in synchronization with the counter. The intensity of the light at each point is adjusted to be proportional to the number of counts recorded at the corresponding point on the patient's head, so that a visual record of the radioactivity detected by the counter is obtained. A tumor appears as a light spot on the plate (see Fig. 20.13). By taking scans from different angles, the tumor can be located in three dimensions.

Because the rectilinear scanning device constructs an image point by point, it takes about 15 min to get a single picture. Faster results

†Each hole detects gamma rays coming from anywhere along the line of sight of the hole. However, since only gamma rays coming from the point of intersection of all the lines of sight can pass through all the holes, more gamma rays will be detected from this point than from any other.

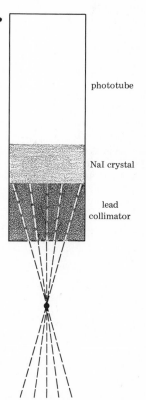

FIGURE 20.10
Schematic diagram of the radiation detector in a rectilinear scanning device.

phototube

NaI crystal

lead collimator

are possible with the Anger gamma-scintillation camera (Fig. 20.11). This device consists of a single large scintillation crystal with 19 phototubes mounted in back of it (Fig. 20.12). The front of the crystal is covered by a large lead collimator containing thousands of parallel collimating holes. The device is called a camera because it receives gamma rays simultaneously from all parts of the field of view. Each collimating hole views a different point in the field because a gamma ray coming from a particular point can only pass through the collimator directly above it.

The light given off when a gamma ray interacts in the crystal is detected simultaneously by each of the 19 phototubes but with an intensity that depends on the distance of the gamma ray from the phototube. The position of the gamma ray in two dimensions is obtained from an analysis of the relative intensities detected by the phototubes. This analysis is performed electronically in less than a

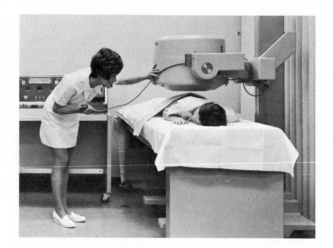

FIGURE 20.11
An Anger camera in operation.
[*Nuclear-Chicago.*]

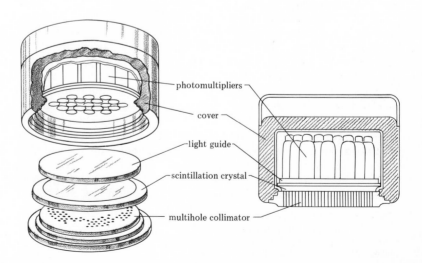

photomultipliers

cover

light guide

scintillation crystal

multihole collimator

FIGURE 20.12
The arrangement of phototubes in an Anger camera.

microsecond, and a spot corresponding to the position of each gamma ray is displayed on an oscilloscope. A camera mounted in front of the oscilloscope screen records each spot as it is displayed until a complete image is formed on the film. Figure 20.13 is a typical picture of the oscilloscope display of a brain scan taken with an Anger camera. A satisfactory image can be taken in only a few minutes with an Anger camera.

One of the newest imaging techniques, still in the experimental stage, uses the annihilation properties of positrons. A positron is an antielectron. Every elementary particle has an antiparticle counterpart. The antiproton and the antineutron have been produced in high-energy-particle accelerators, but the antielectron, or positron, is given off naturally by certain radioactive nuclides. A particle and its antiparticle have the same mass but opposite charge. Furthermore, a particle and its antiparticle annihilate each other when they collide.

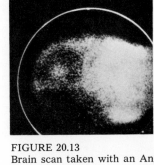

FIGURE 20.13
Brain scan taken with an Anger camera. The white spot indicates a tumor. [*Nuclear-Chicago.*]

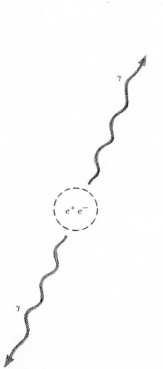

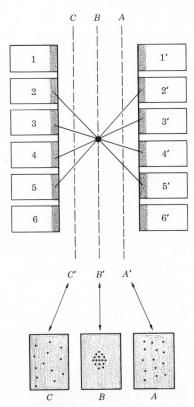

FIGURE 20.14
Electron-positron annihilation into two gamma rays.

FIGURE 20.15
Schematic diagram of an electron-positron camera. The images at the bottom are the intersections of the lines with various imaginary planes. The sharpest image is obtained when a plane passes through the position of the radioactive source.

When a positron and an electron annihilate each other, their mass is converted into two gamma rays (Eq. 20.6). From Eq. 20.2 and Table 20.1 the destruction of one electron is found to liberate the energy

$$E = 931.5 \text{ MeV/u} \times 0.0005486 \text{ u}$$
$$= 0.511 \text{ MeV}$$

Since two electron masses are destroyed in the annihilation process, a total of $2 \times 0.511$ MeV is liberated. This appears in the form of two 0.511-MeV gamma rays that move in opposite directions from the point of annihilation (Fig. 20.14).

A positron camera is designed to detect this annihilation radiation and to use the fact that the two gammas are produced back to back. The experimental camera developed at the Massachusetts General Hospital consists of two arrays of scintillation counters, each composed of 127 counters (Fig. 20.15). The patient, who has been administered a positron-emitting radioactive isotope, is placed between the two arrays. Table 20.3 shows that low-energy electrons travel only a fraction of a millimeter in tissue. A positron behaves similarly, so that it comes to rest very close to the nucleus from which it was emitted. It then annihilates with an electron, giving off two gamma rays. The simultaneous detection of a 0.511-MeV gamma ray by a counter in each array signifies that the two gamma rays came from the same annihilation event.

For example, if counters 2 and 5′ in Fig. 20.15 both detect a gamma ray at the same instant, the annihilation event must have taken place at some point along the line connecting these counters. Thus the annihilation radiation detected by the two arrays determines a large number of straight lines. This information is stored in a computer. Now imagine a plane $AA'$ parallel to the counter arrays, positioned at some point between the arrays. The intersection of the lines and the plane is a two-dimensional array of points that can be determined by the computer and displayed on an oscilloscope. As the position of the plane is changed, the array of points changes. An operator adjusts the position of the plane until the points form the sharpest image on the oscilloscope screen. In this way an image of the source is obtained and the perpendicular distance of the source from the counters is determined.

The positron camera is able to get a satisfactory image in a fraction of a second, making dynamic imaging possible. Figure 20.16 shows

**FIGURE 20.16**
Lung images taken with an electron-positron camera. These pictures, taken at 0.1-s intervals, shows air entering the passageways of the lung. [*Physics Research Laboratory, Massachusetts General Hospital.*]

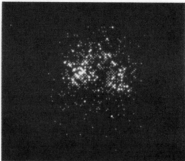
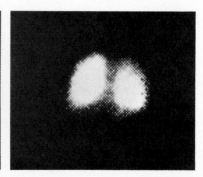

three images taken at 0.1-s intervals of a normal subject inhaling air containing a small quantity of $^{13}$N. The first picture shows the air just entering the primary and secondary bronchi; the second picture shows the air perfusing the smaller bronchi; and the third picture shows the lung fully inflated. The positron camera offers exciting possibilities for detecting obstructions and other pathologies of the lung.

## PROBLEMS

**REMARK** Refer to Table 20.1 for data concerning nuclides mentioned in these problems.

1 How many neutrons are there in a $^{197}$Au nucleus? *Ans.* 118

2 What is the symbol for the nuclide with 13 protons and 15 neutrons?

3 What is the mass in kilograms of a $^{56}$Fe nucleus? *Ans.* $9.3 \times 10^{-26}$ kg

4 Which of the following unidentified nuclides are isotopes of each other? $^{175}_{71}$X, $^{71}_{32}$X, $^{175}_{74}$X, $^{167}_{71}$X, $^{71}_{30}$X, $^{180}_{74}$X

5 The radius of a nucleus with mass number $A$ is $r = 1.3 \times 10^{-15}A^{1/3}$ m. (*a*) What is the radius of a $^{125}$Te nucleus? (*b*) What is the volume of the nucleus? (*c*) What is the density of the nucleus?
*Ans.* (*a*) $6.5 \times 10^{-15}$ m; (*b*) $1.15 \times 10^{-42}$ m$^3$; (*c*) $1.8 \times 10^{17}$ kg/m$^3$

6 Show that the nuclear density is the same for all nuclei, independent of $A$ (see Prob. 5).

7 In a neutron star all the matter is condensed to the density of a nucleus. That is, the density of matter in a neutron star is $1.8 \times 10^{17}$ kg/m$^3$. What is the radius of a neutron star with the same mass as the sun? *Ans.* 13.8 km

8 What is the mass defect and the binding energy of $^2$H (deuterium)?

9 What is the binding energy per particle of $^{88}$Sr? *Ans.* 8.73 MeV

10 Show that the decay $^{16}$O $\longrightarrow$ $^{12}$C + $^4$He is not possible. (*Hint:* Consider the masses of the nuclei involved.)

11 (*a*) How many decays per second are there in 1 mol of $^{32}$P? (*b*) One *curie* is a quantity of radioactive material that produces $3.70 \times 10^{10}$ decays per second. How many curies is 1 g of $^{32}$P?
*Ans.* (*a*) $3.38 \times 10^{17}$ s$^{-1}$; (*b*) $2.85 \times 10^5$ curies

**REMARK** One curie of radioactive material is very "hot" and requires special handling procedures. A milli-curie of material can be handled using routine precautions.

12 (*a*) How many decays per second are there in 1 kg of $^{238}$U? (*b*) How many curies is 1 kg of $^{238}$U (see Prob. 11)?

13 (*a*) What is the decay rate of the $^{14}$C in 1 g of carbon taken from a living tree? (The decay rate per gram is called the *specific activity*.) (*b*) The decay rate in 16 g of carbon taken from an ancient wood sample is 30 min$^{-1}$. What is the age of the sample?
*Ans.* (*a*) 15 min$^{-1}$; (*b*) $1.7 \times 10^4$ y

14 How many curies is 1 g of $^{90}$Sr?

15 (*a*) What nuclide alpha-decays to $^{238}$U? (*b*) To what nuclide does $^{135}$Xe beta-decay?
*Ans.* (*a*) $^{242}$Pu; (*b*) $^{135}$Cs

16 Which of the following nuclides are $e^-$ emitters and which are $e^+$ emitters: (*a*) $^{50}_{20}$Ca, (*b*) $^{88}_{40}$Zr, (*c*) $^{151}_{60}$Nd, (*d*) $^{189}_{80}$Hg?

17 Show that a free neutron can beta-decay into a proton and an electron but that a free proton cannot beta-decay into a neutron and a positron.

**REMARK** A free neutron beta-decays with a half-life of 12 min, whereas a free proton is stable. However, a proton inside a proton-rich nucleus can beta-decay.

18 *Electron capture* is a nuclear transformation in which a proton in a nucleus combines with an orbital electron to form a neutron. For example, a typical electron-capture reaction is $^{168}_{69}$Tm + $e^- \longrightarrow$ $^{168}_{68}$Er. Do nuclei that undergo electron capture lie above or below the band of stable nuclei in Fig. 20.3?

19 Determine the atomic number and mass number of the unknown nuclide in the following reactions?
(*a*) $^1_0n + ^{16}_8$O $\longrightarrow$ X
(*b*) $^4_2$He + $^{118}_{50}$Sn $\longrightarrow$ X + $^1_0n$
(*c*) $^1_1$H + $^{127}_{53}$I $\longrightarrow$ $^{50}_{21}$Sc + X
(*d*) $^{235}_{92}$U + $^1_0n \longrightarrow$ $^{107}_{43}$Tc + X + $5\,^1_0n$

*Ans.* (*a*) $^{17}$O; (*b*) $^{121}$Te; (*c*) $^{78}$As; (*d*) $^{124}$In

**REMARK** For the next four problems, use the table of isotopes in the "Handbook of Chemistry and Physics."

20 List the stable isotopes of tin.

21 Why are radioactive isotopes of oxygen and nitrogen seldom used as tracers in biological research?

22 Nuclei, such as $^{95}$Sr and $^{139}$Xe, which are the direct fragments of a fission event, undergo a sequence of beta decays until stable nuclei are formed. (a) Trace the sequence of decays that $^{95}$Sr and $^{139}$Xe each follows. (b) What is the stable nuclide formed from each fragment, and (c) what is the approximate half-life of each sequence?

23 Draw a chart similar to Fig. 20.5 starting with $^{232}$Th.

24 Approximately what percent of the mass of $^{235}$U is converted into energy by fission?

25 The average person in the United States consumes about $2.5 \times 10^{10}$ J of energy per year. How much $^{235}$U is required to generate this energy?  *Ans.* 0.29 g

26 Calculate the masses of $^3$He and $^3$H from Eq. 2.7 and the masses listed in Table 20.1.

27 Calculate the energy released in the "oxygen burning" reaction

$$^{16}_{8}O + {}^{16}_{8}O \longrightarrow {}^{28}_{14}Si + {}^{4}_{2}He$$

The mass of $^{28}$Si is 27.97693 u.

*Ans.* 9.58 MeV

**REMARK** Although the reaction in Prob. 27 is exothermic, it can occur only at temperatures of many billions of degrees because of the strong electrical repulsion between two oxygen nuclei. If temperatures this high were produced in an atomic explosion, all the oxygen in the atmosphere would convert to silicon. Fortunately, the temperature in an atomic explosion is only a few million degrees.

28 In a phototube with 10 electrodes, how many electrons reach the last electrode for every electron incident on the second electrode if the multiplication factor is 10?

29 Approximately how thick must a lead wall be to stop 97 percent of the 1-MeV gamma rays incident on it?  *Ans.* 4.45 cm

30 The intensity of a gamma ray beam is reduced to 25 percent of its initial value after passing through 0.84 cm of lead. What is the energy of the gamma rays?

31 A 50-keV (0.050-MeV) x-ray beam is used to take a chest x-ray. What is the ratio of the dose received by the chest skin to the dose received by the back skin? (Assume the chest is 10 cm thick.)  *Ans.* 8

32 For diagnostic purposes, radiopharmaceuticals are usually given in millicurie amounts. What is the mass of 50 millicuries of $^{99}$Tc$^m$?

33 A patient having a brain scan is injected with 20 millicuries of $^{99}$Tc$^m$. The energy of the gamma ray emitted when this nuclide decays to its ground state is 0.143 MeV. Assuming that half of the gamma rays escape from the body before interacting, what is the radiation dose received by a 60-kg patient? (*Hint:* Because of its short half-life, all the $^{99}$Tc$^m$ decays while still in the body.)  *Ans.* 0.44 rd

34 If the intensity of a 50-keV x-ray beam is 300 W/m$^2$, estimate the dose received by a patient during a 0.25-s exposure. (*Hint:* Take the body to be a 10-cm thick slab with the density of water.)

# BIBLIOGRAPHY

BEISER, ARTHUR: "Concepts of Modern Physics," McGraw-Hill Book Company, New York, 1963. A simple but quantitative treatment of nuclei and elementary particles.

EARLY, PAUL J., MUHAMMAD ABDEL RAZZAK, and D. BRUCE SODEE: "Textbook of Nuclear Medicine Technology," The C. V. Mosby Company, St. Louis, 1969. The basic principles of nuclear physics and their application to clinical practice for nuclear technologists.

GLASSTONE, SAMUEL: "Sourcebook on Atomic Energy," 3d ed., D. Van Nostrand Company, Inc., Princeton, N.J., 1968. The clearest, simplest, and most complete account of the basic principles of nuclear physics and the applications of nuclear physics to science, technology, and medicine.

JUNGK, ROBERT: "Brighter than a Thousand Suns," Harcourt, Brace, and Company, Inc., New York, 1958. Popular history of the development of the atomic bomb, including the history of Germany's atomic-bomb effort.

SMYTH, HENRY D.: "Atomic Energy for Military Purposes," Princeton University Press, Princeton, N.J., 1948. This is the "Smyth Report" written shortly after World War II as a semi-official history of the wartime effort to develop the atomic bomb.

WAGNER, HENRY, JR.: "Principles of Nuclear Medicine," W. B. Saunders Company, Philadelphia, 1968. Textbook of nuclear medicine for physicians and medical students.

WEAST, ROBERT C. (ed.): "Handbook of Chemistry and Physics," 51st ed., The Chemical Rubber Co., Cleveland, Ohio, 1970. This well-known reference book contains a 300-page table listing the properties of all the known nuclides. A new edition is published every year or so.

# appendixes

**Appendix I**

**POWERS OF TEN**

In algebra the expression $a^n$, where $n$ is an integer, means $a$ multiplied by itself $n$ times, or

$$a^n = \underbrace{a \cdot a \cdots a}_{n \text{ factors}}$$

This notation is useful in dealing with numbers in scientific work. For instance, we can write 1000 ($= 10 \cdot 10 \cdot 10$) as $10^3$, and 100,000 ($= 10 \cdot 10 \cdot 10 \cdot 10 \cdot 10$) as $10^5$. Furthermore, a number like 4500 can be written as $4.5 \times 10^3$, and 3,750,000 can be written as $3.75 \times 10^6$. A number expressed this way as a multiple of a power of 10 is said to be written in *scientific notation*.

There are three distinct advantages to scientific notation.

1 It enables one to express very large (and very small) numbers in a way that is simpler and more meaningful than using a lot of zeros. Avogadro's number (the number of molecules in a mole) is $6.02 \times 10^{23}$. It is clearly hopeless to use this number written out in decimal notation:

602,000,000,000,000,000,000,000

2 Scientific notation allows one to express the significant figures in

a large number unambiguously. When the mass of the sun is written as

$$1,971,000,000,000,000,000,000,000,000,000,000 \text{ g}$$

it is not clear whether any of the zeros after the 1 are significant. But when it is written as $1.971 \times 10^{33}$ g, it is clear that only the first four digits are significant. (The topic of significant figures is discussed in more detail in Sec. 1.3.)

3 Scientific notation greatly simplifies calculations that involve the multiplication and division of large (and small) numbers. You may recall the law of exponents, which says that the product of $a^n$ and $a^m$ is

$$a^n a^m = \underbrace{a \cdot a \cdots a}_{n \text{ factors}} \times \underbrace{a \cdot a \cdots a}_{m \text{ factors}}$$

$$= \underbrace{a \cdot a \cdots a}_{n + m \text{ factors}}$$

$$= a^{n+m}$$

That is, the exponent of the product is the sum of the exponents of the factors.

To apply this to a specific problem, let us try to calculate the number of atoms in the universe. This is neither as hard nor as preposterous as it sounds, since we are not interested in getting the answer to even one significant figure but only to within a factor of 10 or so. Of course we have to decide what is to be included in the universe. For the sake of this calculation the universe will be taken to include all the stars and galaxies visible from earth.

The sun is a fairly typical star with a mass of $2 \times 10^{33}$ g. It is composed mainly of hydrogen, which has an atomic mass of 1. This means that 1 mol of hydrogen has a mass of 1 g. Hence the sun contains $2 \times 10^{33}$ mol of hydrogen. But the number of atoms in a mole is just Avogadro's number, so the number of atoms in the sun is

$$(2 \times 10^{33})(6 \times 10^{23}) = 12 \times 10^{33} \times 10^{23}$$
$$= 12 \times 10^{56}$$

Of course, this number is not accurate to two significant figures, so it should be rounded off to just $10^{57}$.

There are approximately $10^{10}$ stars in an average galaxy and about $10^{11}$ galaxies in the universe. Therefore the number of atoms in the universe is approximately

$$10^{57} \times 10^{10} \times 10^{11} = 10^{78}$$

With this notation it is possible to express and handle arbitrarily large numbers, thus greatly expanding the realm of concepts that one can rationally think about. The very question of how many atoms there are in the universe is almost unthinkable without such a powerful arithmetical notation.

**REMARK** In his book "Sand Reckoner," Archimedes posed the question: How many grains of sand can the whole universe hold? The purpose in answering this question was to show how immense numbers could be handled. Archimedes had to develop his own number system, because the decimal system of notation was not known to the ancient Greeks.

A number bigger than $10^{78}$ seldom arises in physics, but much larger numbers can arise in biology. To understand how this is possible let us first answer the following question: How many distinct lines can be typed on a typewriter? Two lines are to be considered distinct if they differ by as much as one character in a given position. Let the typewriter have 80 spaces to the line and 100 characters on the keyboard. (A normal typewriter has only about 88 characters, but it is simpler to deal with 100.) Since each space can have any one of 100 possible characters in it, the total number of distinct lines is

$$\underbrace{100 \cdot 100 \cdot 100 \cdots 100}_{80 \text{ factors}} = 100^{80}$$

or $$\underbrace{(10 \cdot 10) \cdot (10 \cdot 10) \cdots (10 \cdot 10) = 10 \cdot 10 \cdots 10}_{160 \text{ factors}} = 10^{160}$$

The number of distinct lines that can be typed on a typewriter is immensely larger than the number of atoms in the universe. This means that all the lines that have ever been written in all the books in the world, together with all the lines that ever will be written, must be an infinitesimal fraction of the number of possible lines.

The implication of this for biology comes from the fact that each gene consists of a DNA molecule in which two purines [adenine (A) and guanine (G)] and two pyrimidines [cytosine (C) and thymine (T)] are arranged linearly, like characters in a line of type. In this case there are only four possible characters (A, G, C, T), but there are many hundreds of positions along the line. The total number of distinct genes is thus also immensely larger than the number of atoms in the universe. This means that since the beginning of life on earth, the evolutionary process has experimented with only a minute fraction of the number of possible genes.

Powers of 10 can also be used to divide very large numbers. The law of exponents says that the quotient of two powers $a^n$ and $a^m$ is

$$\frac{a^n}{a^m} = a^{n-m}$$

That is, the exponent of the quotient is the difference of the exponents of the numerator and denominator. For example, $10^8$ divided by $10^3$ is

$$\frac{10^8}{10^3} = \frac{10 \cdot 10 \cdot 10 \cdot 10 \cdot 10 \cdot \cancel{10} \cdot \cancel{10} \cdot \cancel{10}}{\cancel{10} \cdot \cancel{10} \cdot \cancel{10}}$$

$$= 10 \cdot 10 \cdot 10 \cdot 10 \cdot 10 = 10^{8-3} = 10^5$$

This rule implies that

$$\frac{10^n}{10^n} = 10^{n-n} = 10^0$$

But $10^n/10^n$ is also just 1, so that we have

$$1 = 10^0$$

This makes it easy to express fractions as negative exponents. For example,

$$\frac{1}{1,000,000} = \frac{10^0}{10^6} = 10^{0-6} = 10^{-6}$$

With negative exponents, very small numbers can be handled as easily as very large numbers. For instance, one can now ask for the mass of a hydrogen atom. All the necessary information is contained in this appendix. Since 1 g of hydrogen contains $6 \times 10^{23}$ atoms, each atom must have a mass of

$$\frac{1 \text{ g}}{6 \times 10^{23}} = \frac{1}{6} \times 10^{-23} \text{ g} = 0.16 \times 10^{-23} \text{ g}$$

**REMARK** While the rules for multiplying and dividing powers are clear, students sometimes have difficulty adding. For example, what is $10^7 + 10^4$? It certainly is not $10^{11}$! To discover what to do, it is best to write out the numbers in full:

$$10^7 + 10^4 = 10,000,000 + 10,000$$
$$= 10,010,000 = 1.001 \times 10^7$$

Often the answer is not wanted to this accuracy and so one has simply

$$10^7 + 10^4 = 10^7$$

That is, the largest exponent in a sum often dominates the others to such an extent that all the smaller exponents can be neglected.

## Problems

1 Express the following numbers in scientific notation:
  (a) 10,000        (b) 100,000,000
  (c) 700,000       (d) 36,400
                    *Ans.* (a) $10^4$; (b) $10^8$; (c) $7 \times 10^5$; (d) $3.64 \times 10^4$
2 Express the following numbers in conventional notation:
  (a) $10^7$            (b) $2 \times 10^3$
  (c) $4.76 \times 10^4$     (d) $1.4862 \times 10^3$
3 Express the following numbers in scientific notation:
  (a) 0.00479       (b) 173.28
  (c) 378,300       (d) 0.00000305
  *Ans.* (a) $4.79 \times 10^{-3}$; (b) $1.7328 \times 10^2$; (c) $3.783 \times 10^5$;
  (d) $3.05 \times 10^{-6}$
4 Express the following numbers in conventional notation:
  (a) $2 \times 10^{-3}$          (b) $1.46 \times 10^{-5}$
  (c) $76.254 \times 10^{-4}$     (d) $578.935 \times 10^2$
5 Calculate the following expressions:
  (a) $(1.3 \times 10^5) \times (3.5 \times 10^3)$      (b) $(2.4 \times 10^{14}) \times (5.0 \times 10^9)$

  (c) $\dfrac{8.4 \times 10^8}{2.1 \times 10^2}$        (d) $\dfrac{7.5 \times 10^{19}}{2.5 \times 10^{25}}$

      *Ans.* (a) $4.55 \times 10^8$; (b) $1.2 \times 10^{24}$; (c) $4.0 \times 10^6$; (d) $3.0 \times 10^{-6}$
6 Calculate the following expressions:
  (a) $(9.2 \times 10^{-3}) \times (4.1 \times 10^7)$       (b) $(3.9 \times 10^{-17}) \times (6.1 \times 10^{-5})$

  (c) $\dfrac{4.5 \times 10^{25}}{1.5 \times 10^{-10}}$        (d) $\dfrac{6.0 \times 10^{-17}}{1.2 \times 10^{13}}$

7 The following exercises are to be done with a slide rule:
  (a) $2.5 \times 3.2$      (b) $1.8 \times 7.7$
  (c) $4.3 \times 6.4$      (d) $5.8 \times 9.4$
                    *Ans.* (a) 8.0; (b) 13.9; (c) 27.5; (d) 54.5

8 The following exercises are to be done with a slide rule:

(a) $\dfrac{8.4}{3.5}$    (b) $\dfrac{5.2}{1.3}$    (c) $\dfrac{6.9}{9.7}$    (d) $\dfrac{2.8}{7.1}$

9 The following exercises are to be done with a slide rule:
(a) $2.55 \times 3.15$      (b) $1.84 \times 7.55$
(c) $4.33 \times 6.43$      (d) $6.23 \times 9.57$

*Ans.* (a) 8.03; (b) 13.89; (c) 27.8; (d) 59.6

10 The following exercises are to be done with a slide rule:

(a) $\dfrac{8.37}{3.45}$    (b) $\dfrac{5.15}{1.47}$    (c) $\dfrac{6.72}{9.64}$    (d) $\dfrac{2.73}{7.05}$

11 The following exercises are to be done with a slide rule:

(a) $\dfrac{3.15 \times 4.69}{6.05}$      (b) $\dfrac{7.86 \times 5.20}{9.15}$

(c) $\dfrac{6.17}{8.15 \times 1.05}$      (d) $\dfrac{9.05}{3.17 \times 5.71}$

*Ans.* (a) 2.44; (b) 4.47; (c) 0.721; (d) 0.500

12 The following exercises are to be done with a slide rule:

(a) $\dfrac{(6.07 \times 10^{13})(29.8)}{149}$      (b) $\dfrac{(183.25)(1.08 \times 10^{-17})}{6.15 \times 10^{16}}$

(c) $\dfrac{(7.96 \times 10^{7})(6.76 \times 10^{29})}{9.15 \times 10^{15}}$      (d) $\dfrac{(3.14)(6.47 \times 10^{9})^{2}}{4.81 \times 10^{-27}}$

13 Using powers of 10, calculate the number of seconds in a year to two significant figures.    *Ans.* $3.1 \times 10^{7}$ s

14 The speed of light is $3.0 \times 10^{8}$ m/s. How far does a pulse of light travel in 1 year (see Prob. 13)?

**REMARK**  The unit of length called a *light-year* is the distance a pulse of light travels in 1 year.

15 The visible universe is estimated to have a radius of $10^{10}$ light-years. (a) What is the volume of the universe in cubic meters? (See Prob. 14; the volume $V$ of a sphere of radius $r$ is $V = \frac{4}{3}\pi r^{3}$.) (b) On the average, how many atoms are there per cubic meter in the universe?

*Ans.* (a) $3.4 \times 10^{78}$ m³; (b) 0.3 atom/m³

16 Estimate the number of cells in the body of an 80-kg person. An average human cell has a linear size of about $2 \times 10^{-3}$ cm and is composed mostly of water. This means that a cubic centimeter of cells has a mass of 1 g.

## Appendix II

### ANGLES AND TRIANGLES

It is useful to think of the apex of an angle as coinciding with the center of a circle, as shown in Fig. II.1. Then, if the circumference of the circle is divided into equal divisions, the angle can be measured by the number of divisions included inside it. In Fig. II.1 the circle is divided into 16 equal parts and the angle includes 2.5 of them.

Of course in practice a circle is divided into 360 equal parts, called *degrees* (°). The number 360 is convenient because it has so many

FIGURE II.1
An angle coinciding with the center of a circle divided into 16 equal parts.

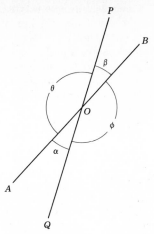

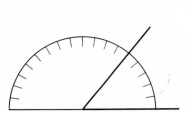

FIGURE II.2
A protractor measuring an angle.

FIGURE II.3
Two straight lines intersecting
at $O$.

divisors, but other than this it has no special significance. The division of the circle into 360 parts seems to be of very ancient origin, probably dating back to the Babylonians, who used a number system based on 60.

Angles are easily measured and drawn with a semicircular scale, called a *protractor*, whose circumference is divided into 180°. To measure an angle with it, you first place the center mark of the protractor over the apex of the angle (Fig. II.2). (The center mark is a point on the diameter of the protractor that marks the center of the circle.) Then the protractor is rotated until one side of the angle falls on the 0° division of the protractor. The division that coincides with the other side of the angle is the number of degrees in the angle.

**REMARK** Be sure you know where the center mark is on your protractor. A common mistake in using a protractor is to place the wrong point of the protractor over the apex of an angle. Another common mistake is to read the wrong scale. Some protractors have two scales, one starting with 0° from the left side of the protractor and one starting with 0° from the right side. The angle in Fig. II.2 is 50°, not 130°, because the scale that starts with 0° along one side of the angle must be used.

To draw a given angle, say 35°, you first draw a straight line and mark a point on it. Next, place the diameter of the protractor along the line with its center mark over the point. Then locate 35° on the circumference (being sure to use the correct scale) and mark a point there. Finally, draw a straight line between the two points. With a little practice you can learn to draw angles quickly and accurately.

You should be familiar with the elementary properties of angles and triangles. A few important theorems from geometry are given here for review.

**Theorem 1** *If two straight lines intersect, the opposite angles are equal.* In Fig. II.3 the two lines $AB$ and $QP$ intersect at $O$. The angles $\alpha$ and $\beta$ are opposite and therefore are equal. Likewise the angles $\theta$ and $\phi$ are equal.

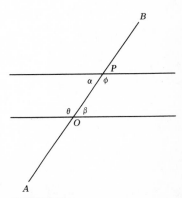

FIGURE II.4
A straight line intersecting two
parallel lines.

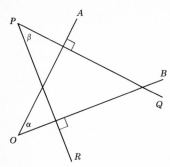

FIGURE II.5
Two angles with corresponding
sides perpendicular to each other.

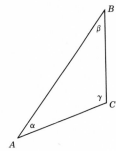

FIGURE II.6
The interior angles of a triangle.

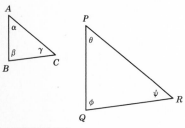

FIGURE II.7
Two similar triangles.

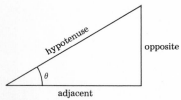

FIGURE II.8
A right triangle.

**Theorem 2** *If a line intersects two parallel lines, the opposite interior angles are equal.* In Fig. II.4 the line $AB$ intersects the two parallel lines at $O$ and $P$. Here $\alpha$ and $\beta$ are opposite interior angles and therefore are equal. Likewise the angles $\theta$ and $\phi$ are equal.

**Theorem 3** *If the corresponding sides of two angles are perpendicular to each other, the angles are equal.* In Fig. II.5 the side $AO$ of angle $\alpha$ is perpendicular to the side $QP$ of angle $\beta$, while the side $BO$ of $\alpha$ is perpendicular to the side $RP$ of $\beta$. Therefore the angles $\alpha$ and $\beta$ are equal. This theorem is used several times in this book.

**Theorem 4** *The sum of the interior angles of a triangle equals 180°.* In Fig. II.6 the interior angles of the triangle $ABC$ are $\alpha$, $\beta$, and $\gamma$. The sum of these angles is

$$\alpha + \beta + \gamma = 180°$$

**Theorem 5** *If the interior angles of two triangles are equal, the triangles are similar.* In Fig. II.7 the triangles $ABC$ and $PQR$ are similar because $\alpha = \theta$, $\beta = \phi$, and $\gamma = \psi$.

A *right triangle* is a triangle in which one of the angles is 90°. The sum of the other two angles must, by Theorem 4, equal $180 - 90° = 90°$. This means that if one of these two angles is given, the other angle is determined. For example, if one angle is 35°, the other must be 55°.

Consider the angle $\theta$ in the right triangle in Fig. II.8. The side across from $\theta$ is called the *opposite* side, whereas the side across from the right (90°) angle is called the *hypotenuse*. By Theorem 5 all right triangles with the same angle $\theta$ are similar, since the other angles are also equal. Consequently the ratio of corresponding sides is the same for all these triangles. For example, in all right triangles in which $\theta = 30°$, the ratio of the opposite side to the hypotenuse, opposite/hypotenuse, is 0.500. This ratio is called the *sine* of $\theta$, written $\sin \theta$. The sine can be calculated for any angle.

Ratios of other sides of a right triangle define other *trigonometric functions*. The two most important ones for our purposes are the sine and cosine. A third function, the tangent, will be used less often. These three functions are defined as follows:

$$\text{Sine} = \frac{\text{opposite}}{\text{hypotenuse}} \qquad \text{written } \sin \theta$$

$$\text{Cosine} = \frac{\text{adjacent}}{\text{hypotenuse}} \qquad \text{written } \cos \theta$$

$$\text{Tangent} = \frac{\text{opposite}}{\text{adjacent}} = \frac{\text{sine}}{\text{cosine}} \qquad \text{written } \tan \theta$$

The table of trigonometric functions inside the back cover gives the values of the sine, cosine, and tangent for angles from 0 to 90°. Note the small-angle approximation given for the sine and tangent.

A protractor and a table of sines and cosines provide alternative methods for solving problems involving triangles. These are the graphical and trigonometric methods discussed in Sec. 2.4. If you feel uncomfortable with trigonometry, you should use the graphical method. It has the advantages of being concrete and of helping you grasp the geometrical relations involved. The trigonometric method should be used only when you are confident that you can save time with it.

1 List all the divisors of 360.

2 Using a protractor, draw a right triangle with $\theta = 35°$. Determine sin 35° and cos 35° by measuring the sides of the triangle with a ruler. Compare your results with the values given in the table of trigonometric function inside the back cover.

3 Prove that $\sin \theta$ is equal to $\cos (90° - \theta)$. Check that the table of trigonometric functions satisfies this theorem.

4 Plot graphs of $\sin \theta$ and $\cos \theta$ against $\theta$. Plot at least 16 points between 0 and 360° for each graph and connect them with a smooth curve.

5 Find the length of the sides of a right triangle, given that the hypotenuse is 15 cm and that $\theta = 25°$.    *Ans.* 6.35 and 13.6 cm

6 In a 37° right triangle the opposite side is 2.4 cm. What are the lengths of the hypotenuse and the adjacent side?

7 The sides of a right triangle are 30 and 50 cm. (*a*) What is the angle opposite the 30-cm side? (*b*) What is the hypotenuse?
   *Ans.* (*a*) 31°; (*b*) 58.2 cm

8 A man starting from home walks 1.2 mi south and then 1.6 mi east to reach his office. (*a*) What is the straight-line distance between his home and office? (*b*) What angle does this line make with a north-south line?

9 In Fig. II.9 the angle $\theta$ is 23°. What is $\phi$?    *Ans.* 57°

10 What is the sum of the interior angles of a quadrilateral (four-sided) figure?

11 (*a*) Use the small-angle formula given in the table of trigonometric functions to calculate sin 3°. Compare the result with the value given in the table. (*b*) Compute sin 2.5°, sin 1.7°, and sin 0.3°.
   *Ans.* (*b*) 0.0435, 0.0296, and 0.00522

12 What are the angles whose sines are (*a*) 0.485, (*b*) 0.338, and (*c*) 0.025?

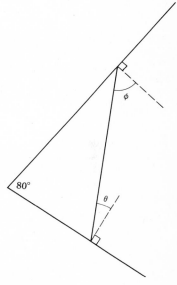

FIGURE II.9
Problem 9.

## Appendix III

## TORQUES ABOUT AN ARBITRARY POINT

According to Property 6 of force (Sec. 3.1), the sum of the torques acting on an object in equilibrium is zero. It was emphasized in Sec. 3.1 that all the torques must be taken about the same point but that this point is arbitrary. This proposition is demonstrated here for the special case in which all the forces on the object are parallel to each other.

Consider the object in Fig. III.1, which is in equilibrium under the action of the four forces $\mathbf{F}_1$, $\mathbf{F}_2$, $\mathbf{F}_3$, and $\mathbf{F}_4$. By Newton's first law (Property 5, Sec. 2.1), the sum of these forces is zero:

$$\mathbf{F}_1 + \mathbf{F}_2 + \mathbf{F}_3 + \mathbf{F}_4 = 0$$

Since these forces are parallel, their magnitudes satisfy the relation

$$F_1 - F_2 - F_3 + F_4 = 0 \qquad \text{III.1}$$

By Property 6 of force, the total torque $\tau_0$ about point $O$ is zero:

$$\tau_0 = d_1 F_1 - d_2 F_2 - d_3 F_3 + d_4 F_4 = 0 \qquad \text{III.2}$$

Here $d_1$, $d_2$, $d_3$, and $d_4$ are the perpendicular distances from $O$ to the lines of action of $\mathbf{F}_1$, $\mathbf{F}_2$, $\mathbf{F}_3$, and $\mathbf{F}_4$, respectively, as shown in Fig. III.1.

We want to show that, as a consequence of Eqs. III.1 and III.2, the sum of the torques is zero about any other point. For instance, the total torque $\tau_P$ about point $P$ in Fig. III.1 is

$$\tau_P = h_1 F_1 - h_2 F_2 - h_3 F_3 + h_4 F_4 \qquad \text{III.3}$$

where $h_1$, $h_2$, $h_3$, and $h_4$ are the perpendicular distances from $P$ to the lines of action of $\mathbf{F}_1$, $\mathbf{F}_2$, $\mathbf{F}_3$ and $\mathbf{F}_4$, respectively, as shown in Fig. III.1.

From Fig. III.1 it is clear that the distances from $P$ to the forces are each the distance $a$ longer than the distances from $O$ to the forces. That is, the $h$'s are related to the $d$'s by the relations,

$$h_1 = d_1 + a$$
$$h_2 = d_2 + a$$
$$h_3 = d_3 + a$$
$$h_4 = d_4 + a$$

When these relations are substituted into Eq. III.3, the total torque about $P$ becomes

$$\tau_P = (d_1 + a)F_1 - (d_2 + a)F_2 - (d_3 + a)F_3 + (d_4 + a)F_4$$

With a little algebra this can be rewritten as the sum of two terms:

$$\tau_P = (d_1 F_1 - d_2 F_2 - d_3 F_3 + d_4 F_4) + a(F_1 - F_2 - F_3 + F_4)$$

The first term is $\tau_0$, the total torque about $O$, which is zero according to Eq. III.2. The second term is $a$ times the total force, which is zero according to Eq. III.1. Therefore the total torque about $P$ is zero.

This proof obviously works for any number of parallel forces. The proof of the general case in which the forces are not parallel is more complicated, so it is not given here.

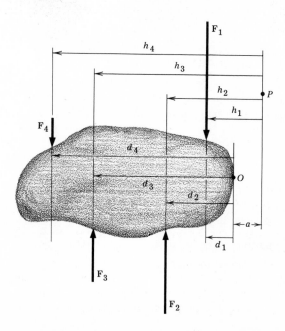

**FIGURE III.1**
An object in equilibrium under the action of four parallel forces.

## Appendix IV

## CENTRIPETAL ACCELERATION

An object moving at constant speed $v$ in a circle of radius $R$ has a centripetal acceleration **a**. In Sec. 4.2 it is stated without proof that the magnitude of **a** is $v^2/R$ and that **a** is directed toward the center of the circle. To prove these statements we must define acceleration more precisely than we did in Sec. 4.2.

Acceleration is a vector quantity that measures the change in velocity of an object. If an object has a velocity $\mathbf{v}_1$ at time $t_1$ and a velocity $\mathbf{v}_2$ at the later time $t_2$, the *average* acceleration $\mathbf{a}_\text{av}$ during this time interval is defined as

$$\mathbf{a}_\text{av} = \frac{\mathbf{v}_2 - \mathbf{v}_1}{t_2 - t_1} \qquad \text{IV.1}$$

The *instantaneous* acceleration, or what we have been calling simply the acceleration, is equal to the average acceleration during a very small time interval $t_2 - t_1$. There are ways to make the term "very small" more precise, but we need not go into these technicalities.

Consider an object moving at constant speed $v$ in a circle of radius $R$ (Fig. IV.1). At time $t_1$ the object is at point $A$ on the circle. The velocity $\mathbf{v}_1$ of the object at time $t_1$ is tangent to the circle at $A$. At the later time $t_2$, the object is at point $B$, having moved through an angle $\theta$. The velocity $\mathbf{v}_2$ at this time is tangent to the circle at $B$. We shall assume that $t_2 - t_1$ is small enough for Eq. IV.1 to be used to calculate the instantaneous acceleration.

First, the time interval $t_2 - t_1$ must be related to the angle $\theta$. Let $s$ be the distance between $A$ and $B$ measured along the circle. Then, since the object is moving with constant speed $v$, we have

$$t_2 - t_1 = \frac{s}{v} \qquad \text{IV.2}$$

The distance $s$ is some fraction of the total circumference $2\pi R$ of the circle. In fact, this fraction is just $\theta/360°$. (For example, if $\theta = 15°$, then $\theta/360° = \frac{1}{24}$, so that $s$ is $\frac{1}{24}$ the circumference.) In general, then, $s$ is related to $\theta$ and $R$ by

$$s = \frac{\theta}{360°} 2\pi R = \frac{2\pi R \theta}{360°} \qquad \text{IV.3}$$

Substituting this into Eq. IV.2, we get the time interval in terms of $\theta$, $R$, and $v$:

$$t_2 - t_1 = \frac{2\pi R \theta}{360° \, v} \qquad \text{IV.4}$$

Second, the difference in velocity $\mathbf{v}_2 - \mathbf{v}_1$ must be found. These velocity vectors are perpendicular to the two radii which form the sides of $\theta$, so (by Theorem 3 in Appendix II) the angle between $\mathbf{v}_1$ and $\mathbf{v}_2$ is also $\theta$. In Fig. IV.2 the vector $\mathbf{v}_1$ is replaced by $-\mathbf{v}_1$, and both it and $\mathbf{v}_2$ have been moved to the point $C$ midway between $A$ and $B$ for the purpose of vector addition. The sum of $\mathbf{v}_2$ and $-\mathbf{v}_1$ is, of course, $\mathbf{v}_2 - \mathbf{v}_1$, and from Fig. IV.2 this sum is seen to be in the direction of the line from $C$ to the center of the circle. This shows that the acceleration given by Eq. IV.1 is directed toward the center

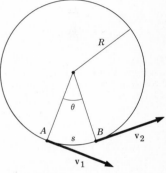

FIGURE IV.1
An object moving at constant speed $v$ in a circle of radius $R$. At each instant, the velocity **v** is tangent to the circle.

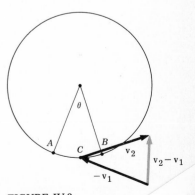

FIGURE IV.2
The vectors $-\mathbf{v}_1$ and $\mathbf{v}_2$ from a common point between $A$ and $B$

of the circle. (Since we are interested only in the case in which $\theta$ is very small, the three points $A$, $B$, and $C$ are nearly coincident.)

Third, the magnitude of $\mathbf{v}_2 - \mathbf{v}_1$ must be calculated. This magnitude will be denoted $\Delta v$. Figure IV.3 shows that $\Delta v$ can be thought of as the chord of a circle of radius $v$. The chord subtends the angle $\theta$. If $\theta$ is very small, the length of the chord is nearly equal to the arc length $s'$. This arc length is related to $\theta$ and $v$ by an equation analogous to Eq. IV.3, so

$$\Delta v = s' = \frac{2\pi v\theta}{360°} \qquad \text{IV.5}$$

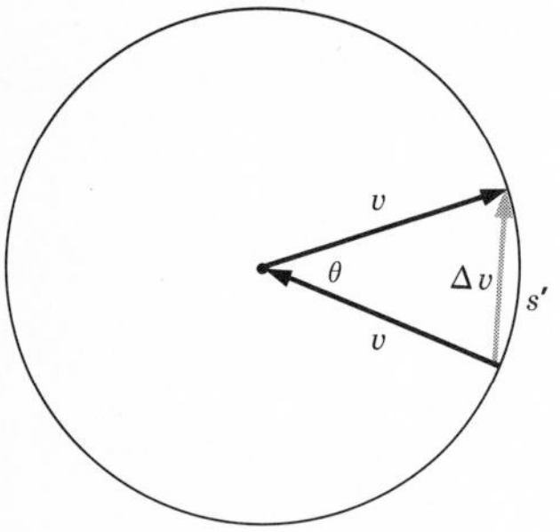

Finally, the magnitude of $\mathbf{a}$ is found by dividing the magnitude of $\mathbf{v}_2 - \mathbf{v}_1$, given in Eq. IV.5, by $t_2 - t_1$, given in Eq. IV.4. The result is

$$a = \frac{\Delta v}{t_2 - t_1} = \frac{2\pi v\theta/360°}{2\pi R\theta/360°v}$$

$$= \frac{v}{R/v} = \frac{v^2}{R} \qquad \text{IV.6}$$

and does not depend on $\theta$. This is true, of course, only when $\theta$ is so small that the chord in Fig. IV.3 is equal to the arc length. But it is only in this case that Eq. IV.1 gives the instantaneous acceleration. Therefore, Eq. IV.6 gives the magnitude of the acceleration exactly. We have already shown that, under these same conditions, the acceleration is directed, at each instant, from the position of the object to the center of the circle.

**FIGURE IV.3**
$\Delta v$ is the chord of an arc that subtends an angle $\theta$ on a circle of radius $v$.

## Appendix V

### THE SPEED OF A TRANSVERSE WAVE ON A CORD

It can be proved that a transverse wave of any shape propagates with the speed

$$\mathsf{v} = \sqrt{\frac{T}{\mu}} \qquad \text{V.1}$$

along a stretched cord (Sec. 12.2). Here $T$ is the tension in the cord, and $\mu$ is the mass per unit length of the cord. Although it is not possible to give the general proof here, we shall estimate the speed of the simple pulse shown in Fig. V.1. This calculation is instructive because it shows how Newton's second law applies to a mechanical wave.

The pulse in Fig. V.1 is assumed to be moving to the right with constant speed $\mathsf{v}$ along a very long cord. The entire pulse is of length $2l$, and it has a maximum displacement $h$. In the time

$$t = \frac{l}{\mathsf{v}} \qquad \text{V.2}$$

the pulse moves a distance $l$ along the cord, reaching the position shown by the dashed pulse. This requires that the segment of the cord between points $A$ and $B$ move up into the position between $A'$ and $B$ in the same time $t$. This segment starts from rest and is accelerated upward by the forces exerted on it by the remainder of the

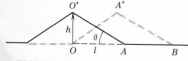

**FIGURE V.1**
A pulse moving to the right with constant speed $\mathsf{v}$ along a cord.

cord. If its average upward acceleration is $a$, the distance $d$ it travels in time $t$, according to Eq. 4.3, is

$$d = \tfrac{1}{2}at^2$$

Of course different parts of the segment travel different distances, but the average distance is $\tfrac{1}{2}h$. Putting this into the last equation, we get

$$\tfrac{1}{2}h = \tfrac{1}{2}at^2$$
or
$$h = at^2$$

Using Eq. V.2 for $t$, this becomes

$$h = a\left(\frac{l}{v}\right)^2$$

so that

$$v^2 = \frac{al^2}{h} \qquad\qquad \text{V.3}$$

This is the condition that determines $v$. It comes from the requirement that the segment of the cord between $A$ and $B$ move into the proper position in the proper time.

The acceleration of the segment is calculated from the total force acting on it. Figure V.2 shows the forces $\mathbf{F}_1$ and $\mathbf{F}_2$ on the segment at the instant the pulse reaches $A$. The magnitudes of $\mathbf{F}_1$ and $\mathbf{F}_2$ are both equal to the tension $T$ of the cord. The sum $\mathbf{S} = \mathbf{F}_1 + \mathbf{F}_2$ is shown in Fig. V.3. Recall that in Sec. 12.2 we said that the theory of mechanical waves is restricted to small displacements. In this case this means that $h$ must be much smaller than $l$ or that $\theta$ is a small angle. Under such a circumstance, $\mathbf{S}$ is directed nearly vertically, and so the triangle in Fig. V.3 is almost a right triangle. Thus to good approximation we have

$$\frac{S}{F_2} = \frac{S}{T} = \tan\theta$$
or
$$S = T\sin\theta$$

But from the triangle $AOO'$ in Fig. V.1 the tangent of $\theta$ is seen to be $h/l$, so that

$$S = T\frac{h}{l} \qquad\qquad \text{V.4}$$

The acceleration of the segment is given by Newton's second law,

$$a = \frac{S}{m}$$

where $S$ is the total force on the segment and $m$ is the mass of the segment. The mass is related to $\mu$, the mass per unit length, by $m = \mu l$. Therefore the acceleration is

$$a = \frac{S}{m} = \frac{Th/l}{\mu l} = \frac{Th}{\mu l^2} \qquad\qquad \text{V.5}$$

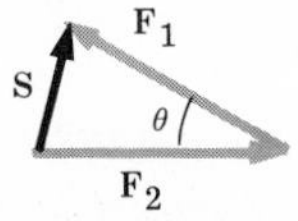

This is the acceleration of the segment at the instant the pulse reaches it. Once the segment moves upward, the forces on it change and its acceleration changes. However, for the purposes of this calculation,

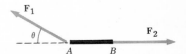

FIGURE V.2
The forces $\mathbf{F}_1$ and $\mathbf{F}_2$ on the segment $AB$ of the cord at the instant the pulse reaches it.

FIGURE V.3
The sum $\mathbf{S}$ of the forces $\mathbf{F}_1$ and $\mathbf{F}_2$. The angle $\theta$ is very small, so that $\mathbf{S}$ is nearly vertical.

we shall assume that the acceleration is given by Eq. V.5 throughout the motion of the segment.

When the value of the acceleration given by Eq. V.5 is substituted into Eq. V.3, we get

$$v^2 = \frac{(Th/\mu l^2)l^2}{h} = \frac{Th/\mu}{h} = \frac{T}{\mu}$$

so

$$v = \sqrt{\frac{T}{\mu}}$$

This is the correct expression for the speed as given by Eq. V.1.

This calculation is only an estimate of v. The final result agrees exactly with Eq. V.1 because the estimates of $d$ and $a$ were made with a bit of hindsight. Had other reasonable estimates been used, a result like $\sqrt{2T/\mu}$ or $\sqrt{T/2\mu}$, which differ from the exact value by a simple numerical constant, would have been obtained. The point is that it is possible to show that v is proportional to $\sqrt{T/\mu}$ from the requirement that the segment $AB$ must accelerate into position $A'B$ in the time it takes the wave to move the distance $l$. It is more difficult to prove rigorously that v is equal to $\sqrt{T/\mu}$.

**REMARK** The length $l$ and height $h$ of the pulse canceled out in the final expression for v. This is important because it demonstrates that the speed of a pulse does not depend on its size or shape.

## Appendix VI

### ENERGY IN A SOUND WAVE

Consider a pulse moving through a cylindrically shaped medium, as in Sec. 13.2. The cylinder has a cross-sectional area $A$, so that the volume of the pulse is $V = AL$, where $L$ is its length. The pressure $p$ in this volume is greater than the pressure $p_0$ in the rest of the medium.

A sound pulse is formed by compressing the air that originally occupied a volume $V_0 = AL_0$ at atmospheric pressure $p_0$ (Fig. VI.1), and the energy of the pulse is equal to the work required to form it. This work is equal to the average force $\bar{F}$ exerted while compressing the gas times the distance $d = L_0 - L$ through which the force moves:

$$W = \bar{F}(L_0 - L)$$
$$= \frac{\bar{F}(V_0 - V)}{A} \qquad \text{VI.1}$$

At the beginning of the compression there is a pressure $p$ on the outside of $V_0$ that exerts the force $F_1 = pA$ inward on the volume and a pressure $p_0$ inside $V_0$ that exerts the force $F_2 = p_0A$ outward. Thus to start, the total inward force is $(p - p_0)A$. As the gas is compressed, the internal pressure increases and the difference between the internal and external pressures decreases. At the end, the internal and external pressures are equal, so there is zero force compressing the pulse. The average force during compression then is

$$\bar{F} = \tfrac{1}{2}[(p - p_0)A + 0] = \tfrac{1}{2}(p - p_0)A$$

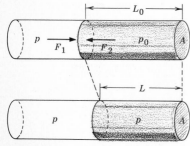

**FIGURE VI.1**
Air in the volume $V_0 = AL_0$ at the pressure $p_0$ is compressed to the volume $V = AL$ at the pressure $p$.

Using this in Eq. VI.1, we have

$$W = \tfrac{1}{2}(p - p_0)(V_0 - V) \qquad \text{VI.2}$$

If the compression takes place at constant temperature (isothermal compression), the pressures and volumes in Eq. VI.2 are related by the ideal-gas law:

$$p_0 V_0 = pV$$

By subtracting the quantity $p_0 V$ from both sides of this expression we get

$$p_0 V_0 - p_0 V = pV - p_0 V$$

or

$$V_0 - V = \frac{(p - p_0)V}{p_0}$$

When this is substituted into Eq. VI.2, we have

$$W = \frac{\tfrac{1}{2}V(p - p_0)^2}{p_0}$$

which is the energy in the pulse. It is proportional to the square of the difference between the pressure in the pulse and the undisturbed (atmospheric) pressure.

In a real sound wave, the compression is adiabatic, not isothermal. This modifies the expression for the energy by the factor $\gamma$, so that the energy is

$$E = \frac{\tfrac{1}{2}V(p - p_0)^2}{\gamma p_0} \qquad \text{VI.3}$$

However, the important point of the derivation, which is that the energy is proportional to the square of the pressure difference, is not changed.

Of special interest is the intensity $I$ of a wave, given by Eq. 13.12. In the case of the pulse, the energy passes through an area $A$ in the time $t = L/v$, so its intensity is

$$I = \frac{E}{At} = \frac{V(p - p_0)^2}{2\gamma p_0 AL/v} = \frac{v(p - p_0)^2}{2\gamma p_0}$$

$$= \frac{(p - p_0)^2}{2\rho v}$$

The last line is obtained by using Eq. 13.10 for the speed of sound in a gas.

The intensity of a sine wave is given by the same expression if $p - p_0$ is taken to be the pressure amplitude $A_p$ of the wave:

$$I = \frac{A_p{}^2}{2\rho v}$$

## Appendix VII

### THEORY OF NUCLEAR DECAY

The half-life $\tau$ of a radioactive nuclide is determined by measuring the decrease in the decay rate $R$ of a sample of the nuclide. The decay rate continuously decreases because the number $N$ of radioactive nuclei continuously decreases. A derivation of the mathematical relation between $R$, $N$, and $\tau$ is given here for students who are acquainted with calculus. Besides their importance for the study of radioactivity, the formulas derived here apply, with minor variations, to a wide variety of problems, such as the growth of a bacteria colony or the aging of a redwood forest.

Let $N(t)$ be the number of radioactive nuclei present in the sample at time $t$, and let $N(t + \Delta t)$ be the number present at the later time $t + \Delta t$. The difference

$$\Delta N = N(t) - N(t + \Delta t)$$

is the number that decayed during the time interval $\Delta t$. The decay rate during this interval is

$$R = \frac{\Delta N}{\Delta t} = -\frac{N(t + \Delta t) - N(t)}{\Delta t}$$

which, in the limit as $\Delta t$ goes to zero, becomes

$$R = \lim_{\Delta t \to 0} -\frac{\Delta N}{\Delta t} = -\frac{dN}{dt}$$

The minus sign is required because $N$ decreases as $t$ increases.

At any time $t$ the decay rate $R(t)$ is proportional to the number $N(t)$ of radioactive nuclei present. Thus we can write

$$R(t) = \lambda N(t) \qquad\qquad \text{VII.1}$$

or

$$-\frac{dN}{dt} = \lambda N \qquad\qquad \text{VII.2}$$

where $\lambda$ is the *decay constant*, a constant characteristic of each nuclide.

We want to solve Eq. VII.2 for $N(t)$. To do this we first divide both sides of the equation by $-N$

$$\frac{1}{N}\frac{dN}{dt} = -\lambda$$

and then integrate both sides with respect to time from 0 to $t$

$$\int_0^t \frac{1}{N}\frac{dN}{dt}\, dt = \int_0^t -\lambda\; dt$$

or

$$\int_{N(0)}^{N(t)} \frac{dN}{N} = -\lambda \int_0^t dt$$

Carrying out the indicated integration we get

$$\ln N \,\big|_{N(0)}^{N(t)} = -\lambda t \,\big|_0^t$$

which gives

$$\ln N(t) - \ln N(0) = -\lambda t$$

or

$$\ln \frac{N(t)}{N(0)} = -\lambda t$$

This equation is easily solved for $N(t)$ by exponentiating both sides.

The result is

$$\frac{N(t)}{N(0)} = e^{-\lambda t}$$

or

$$N(t) = N_0 e^{-\lambda t} \qquad \text{VII.3}$$

where $N_0 = N(0)$ is the number of radioactive nuclei present at time zero. This equation gives the number of radioactive nuclei remaining at time $t$ in terms of the number present at time zero and the decay constant.

The half-life $\tau$ is defined as the time in which half the nuclei decay. That is, by definition $N(\tau) = \frac{1}{2}N_0$, so setting $t$ equal to $\tau$ in Eq. VII.3, we get

$$\tfrac{1}{2}N_0 = N_0 e^{-\lambda \tau}$$

or

$$\tfrac{1}{2} = e^{-\lambda \tau}$$

Taking the natural logarithm of both sides, we get

$$\ln \tfrac{1}{2} = -\lambda \tau$$

or

$$\tau = -\frac{\ln \tfrac{1}{2}}{\lambda} = \frac{\ln 2}{\lambda} = \frac{0.693}{\lambda} \qquad \text{VII.4}$$

which is the relation between the decay constant $\lambda$ and the half-life $\tau$.

Equation 20.3 is obtained by substituting $\tau$ for $\lambda$ in Eq. VII.1:

$$R = \frac{0.693}{\tau} N$$

Substituting Eq. VII.3 into Eq. VII.1, we get

$$R(t) = \lambda N_0 e^{-\lambda t} \qquad \text{VII.5}$$

which gives the decay rate as a function of time. This is the expression we want, because $R(t)$ is an easily measured quantity. From Eq. VII.5 the logarithm of $R$ is

$$\ln R = \ln \lambda N_0 e^{-\lambda t} = \ln \lambda N_0 - \lambda t$$

Since $\ln \lambda N_0$ is a constant, a plot of $\ln R$ against $t$ is a straight line with slope $\lambda$. Thus the half-life of a radioactive nuclide can be determined by plotting measured values of $R$ against $t$ in this way.

## Appendix VIII

## THE METRIC SYSTEM

The metric system uses a single basic unit for each physical quantity, and multiples and fractions of this unit are formed by adding a prefix. For example, the metric unit of length is the meter (m); the millimeter (mm), centimeter (cm), and kilometer (km) are formed by adding the prefixes milli- (m), centi- (c), and kilo- (k) to it. The same prefixes are used for all physical quantities. Table VIII.1 lists the accepted prefixes and their abbreviations.

The following rules should be observed when using abbreviations of units:

1 Do not put a period after an abbreviation (except with inch). Thus the notation 7 km. is not correct, but 5 in. is.

TABLE VIII.I **Metric prefixes**

| Prefix | Abbreviation | Value | Prefix | Abbreviation | Value |
|--------|--------------|-------|--------|--------------|-------|
| tera-  | T | $10^{12}$ | milli- | m | $10^{-3}$ |
| giga-  | G | $10^{9}$  | micro- | $\mu$ | $10^{-6}$ |
| mega-  | M | $10^{6}$  | nano-  | n | $10^{-9}$ |
| kilo-  | k | $10^{3}$  | pico-  | p | $10^{-12}$ |
| deci-  | d | $10^{-1}$ | femto- | f | $10^{-15}$ |
| centi- | c | $10^{-2}$ | atto-  | a | $10^{-18}$ |

*2* Do not put an s at the end of an abbreviation. Thus, the notation 7 kms is not correct, but 7 km is.

*3* The abbreviation for the basic unit is a lowercase letter except when the unit is derived from a proper name. Thus, the symbol for the meter is m, whereas the symbol for the newton is N.

*4* A symbol with a prefix is treated as a new unit which can be raised to a power without using brackets. Thus, the notation $cm^3$ means $(10^{-2} \text{ m})^3 = 10^{-6} \text{ m}^3$, and not $10^{-2} \text{ m}^3$.

**REMARK**  The micrometer ($\mu$m) is the same as the micron ($\mu$). Scientific societies recommend the use of micrometer rather than micron, because micrometer is consistent with other metric notation. Nevertheless, the micron is still sometimes used, especially in the biological literature. Under no circumstances, however, should a hybrid symbol such as m$\mu$ be used; instead use nm.

**Appendix IX**

# GLOSSASY OF SYMBOLS USED IN THIS BOOK

Listed below are the most important symbols used in this book, together with the physical quantity they symbolize and the section in which the quantity is first defined. Symbols are in alphabetical order, with the Greek alphabet following the Latin alphabet.

| Symbol | Physical quantity | Section |
|--------|-------------------|---------|
| $a$ | Acceleration | 4.2 |
| $A$ | Amplitude | 12.4 |
| | Area | |
| | Mass number | 20.1 |
| $B$ | Bulk modulus | 10.2 |
| | Magnetic field | 18.1 |
| $c$ | Molar concentration | 9.4 |
| | Specific heat | 11.2 |
| | Speed of light | 14.1 |
| $C$ | Capacitance | 17.4 |
| | Heat capacity | 11.2 |
| $d$ | Distance | |
| $D$ | Disorder | 11.4 |
| | Distance | |
| $e$ | Efficiency | 6.1 |
| | Proton charge | 16.2 |
| $E$ | Electric field | 16.3 |
| | Energy | 5.5 |
| | Young's modulus | 10.2 |
| $E_m$ | Mechanical energy | 5.3 |
| $\mathcal{E}$ | emf | 17.1 |
| $f$ | Focal length | 15.1 |
| | Frequency | 12.4 |
| $F$ | Force | 2.1 |
| $g$ | Acceleration of gravity | 4.2 |
| $G$ | Gibbs free energy | 11.5 |
| | Gravitational constant | 5.4 |
| | Shear modulus | 10.2 |
| $h$ | Height | |
| | Planck's constant | 19.1 |
| $H$ | Enthalpy | 11.5 |
| $H_f$ | Heat of fusion | 10.1 |
| $H_s$ | Heat of sublimation | 10.1 |
| $H_v$ | Heat of vaporization | 9.1 |
| $I$ | Current | 17.1 |
| | Intensity | 11.3 |
| $k$ | Boltzmann's constant | 8.3 |
| $K$ | Compressibility | 10.2 |
| | Electric constant | 16.2 |
| | Kinetic energy | 5.2 |
| $L$ | Length | |
| $m$ | Magnification | 15.2 |
| | Mass | 4.3 |
| $M$ | Magnitude of molecular mass | 8.1 |
| | Mechanical advantage | 6.2 |
| $\mathfrak{M}$ | Molar mass | 81 |
| $n$ | Number | |
| $N$ | Avogadro's number | 8.1 |
| | Neutron number | 20.1 |
| | Number of particles | |
| $p$ | Pressure (absolute) | 7.1 |
| $\bar{p}$ | Pressure (gauge) | 7.3 |

| Symbol | Physical quantity | Section |
|---|---|---|
| $P$ | Power | 6.1 |
| $q$ | Charge | 16.2 |
| $Q$ | Fluid flow | 7.5 |
| | Heat | 11.2 |
| $r$ | Response factor | 13.4 |
| $R$ | Decay rate | 20.2 |
| | Electric resistance | 17.1 |
| | Fluid resistance | 7.5 |
| | Gas constant | 8.3 |
| $s$ | Object distance | 15.2 |
| $s'$ | Image distance | 15.2 |
| $S$ | Entropy | 11.5 |
| $t$ | Temperature (Celsius) | 8.2 |
| | Time | |
| $T$ | Temperature (absolute) | 8.2 |
| | Tension | 2.2 |
| $U$ | Potential energy | 5.3 |
| $v$ | Speed, velocity | 4.2 |
| $\vee$ | Wave speed | 14.1 |
| $V$ | Electric potential | 16.4 |
| | Volume | |
| $W$ | Work | 5.1 |
| $Z$ | Atomic number | 19.4 |
| $\alpha$ (alpha) | Angle | |
| $\beta$ (beta) | Angle | |
| | Intensity level | 11.3 |
| $\gamma$ (gamma) | Ratio of specific heats | 11.2 |
| | Shear strain | 10.2 |
| | Surface tension | 9.2 |
| $\Gamma$ (gamma) | Width of response curve | 14.4 |
| $\Delta$ (delta) | Change of | 10.2 |
| $\epsilon$ (epsilon) | Dielectric constant | 17.4 |
| | Normal strain | 10.2 |
| $\eta$ (eta) | Particle density | 8.1 |
| $\theta$ (theta) | Angle | |
| $\lambda$ (lambda) | Wavelength | 12.4 |
| $\mu$ (mu) | Mass per unit length | 12.2 |
| $\mu_k$ | Coefficient of kinetic friction | 2.2 |
| $\mu_0$ | Magnetic permeability | 18.2 |
| $\mu_s$ | Coefficient of static friction | 2.2 |
| $\nu$ (nu) | Viscosity | 7.5 |
| $\rho$ (rho) | Mass density | 7.2 |
| $\sigma$ (sigma) | Normal stress | 10.2 |
| $\tau$ (tau) | Characteristic time | 17.4 |
| | Half-life | 20.2 |
| | Period | 12.4 |
| | Shear stress | 10.2 |
| | Torque | 3.1 |
| $\psi$ (psi) | Quantum wave | 19.3 |

# Appendix X

## PHYSICAL CONSTANTS

| Name | Symbol | Value |
|------|--------|-------|
| Absolute zero | 0 K | $-273.15°C$ |
| Acceleration of gravity | $g$ | $9.807 \text{ m/s}^2$ |
| Avogadro's number | $N$ | $6.02 \times 10^{23}$ |
| Boltzmann's constant | $k$ | $1.381 \times 10^{-23} \text{ J/K}$ |
| Electric constant | $K$† | $8.988 \times 10^9 \text{ N-m}^2/\text{C}^2$ |
| Electron mass | $m_e$ | $9.110 \times 10^{-31} \text{ kg}$ |
| Gas constant | $R$ | $8.314 \text{ J/K}$ |
| Gravitational constant | $G$ | $6.673 \times 10^{-11} \text{ N-m}^2/\text{kg}^2$ |
| Magnetic constant (permeability) | $\mu_0$ | $4\pi \times 10^{-7} \text{ N/A}^2$ ‡ |
| Neutron mass | $m_n$ | $1.6749 \times 10^{-27} \text{ kg}$ |
| Planck's constant | $h$ | $6.626 \times 10^{-34} \text{ J-s}$ |
| Proton charge | $e$ | $1.602 \times 10^{-19} \text{ C}$ |
| Proton mass | $m_p$ | $1.6726 \times 10^{-27} \text{ kg}$ |
| Speed of light | $c$ | $2.9979 \times 10^8 \text{ m/s}$ |

† Usually written $1/4\pi\epsilon_0$.
‡ Exact.

# Appendix XI

## SOLAR-SYSTEM DATA

| Body | Mass, $10^{24}$ kg | Radius, $10^6$ m | Distance from sun, $10^{10}$ m |
|------|-------------------|------------------|-------------------------------|
| Sun | 1,970,000 | 696 | |
| Mercury | 0.328 | 2.57 | 5.8 |
| Venus | 4.83 | 6.31 | 10.8 |
| Earth | 5.983 | 6.378 | 14.9 |
| Mars | 0.634 | 3.43 | 22.8 |
| Jupiter | 1,880 | 71.8 | 77.8 |
| Saturn | 563 | 60.3 | 142.6 |
| Uranus | 86.1 | 26.7 | 286.9 |
| Neptune | 99.6 | 24.6 | 448.5 |
| Pluto | 6 | 6.3 | 590 |
| Moon | 0.0735 | 1.738 | 0.038† |

† Distance from earth.

# index

# table of the elements

The masses listed are based on $^{12}_{6}C = 12$ u. A value in parentheses is the mass number of the most stable (long-lived) of the known isotopes.

| Element | Symbol | Atomic Number Z | Average atomic mass |
|---|---|---|---|
| Actinium | Ac | 89 | (227) |
| Aluminum | Al | 13 | 26.9815 |
| Americium | Am | 95 | (243) |
| Antimony | Sb | 51 | 121.75 |
| Argon | Ar | 18 | 39.948 |
| Arsenic | As | 33 | 74.9216 |
| Astatine | At | 85 | (210) |
| Barium | Ba | 56 | 137.34 |
| Berkelium | Bk | 97 | (247) |
| Beryllium | Be | 4 | 9.0122 |
| Bismuth | Bi | 83 | 208.980 |
| Boron | B | 5 | 10.811 |
| Bromine | Br | 35 | 79.904 |
| Cadmium | Cd | 48 | 112.40 |
| Calcium | Ca | 20 | 40.08 |
| Californium | Cf | 98 | (251) |
| Carbon | C | 6 | 12.01115 |
| Cerium | Ce | 58 | 140.12 |
| Cesium | Cs | 55 | 132.905 |
| Chlorine | Cl | 17 | 35.453 |
| Chromium | Cr | 24 | 51.996 |
| Cobalt | Co | 27 | 58.9332 |
| Copper | Cu | 29 | 63.546 |
| Curium | Cm | 96 | (247) |
| Dysprosium | Dy | 66 | 162.50 |
| Einsteinium | Es | 99 | (254) |
| Erbium | Er | 68 | 167.26 |
| Europium | Eu | 63 | 151.96 |
| Fermium | Fm | 100 | (257) |
| Fluorine | F | 9 | 18.9984 |
| Francium | Fr | 87 | (223) |
| Gadolinium | Gd | 64 | 157.25 |
| Gallium | Ga | 31 | 69.72 |
| Germanium | Ge | 32 | 72.59 |
| Gold | Au | 79 | 196.967 |
| Hafnium | Hf | 72 | 178.49 |
| Helium | He | 2 | 4.0026 |
| Holmium | Ho | 67 | 164.930 |
| Hydrogen | H | 1 | 1.00797 |
| Indium | In | 49 | 114.82 |
| Iodine | I | 53 | 126.9044 |
| Iridium | Ir | 77 | 192.2 |
| Iron | Fe | 26 | 55.847 |
| Krypton | Kr | 36 | 83.80 |
| Lanthanum | La | 57 | 138.91 |
| Lawrencium | Lr | 103 | (257) |
| Lead | Pb | 82 | 207.19 |
| Lithium | Li | 3 | 6.939 |
| Lutetium | Lu | 71 | 174.97 |
| Magnesium | Mg | 12 | 24.312 |
| Manganese | Mn | 25 | 54.9380 |
| Mendelevium | Md | 101 | (256) |
| Mercury | Hg | 80 | 200.59 |
| Molybdenum | Mo | 42 | 95.94 |
| Neodymium | Nd | 60 | 144.24 |
| Neon | Ne | 10 | 20.183 |
| Neptunium | Np | 93 | (237) |
| Nickel | Ni | 28 | 58.71 |
| Niobium | Nb | 41 | 92.906 |
| Nitrogen | N | 7 | 14.0067 |
| Nobelium | No | 102 | (254) |
| Osmium | Os | 76 | 190.2 |
| Oxygen | O | 8 | 15.9994 |
| Palladium | Pd | 46 | 106.4 |
| Phosphorus | P | 15 | 30.9738 |
| Platinum | Pt | 78 | 195.09 |
| Plutonium | Pu | 94 | (244) |
| Polonium | Po | 84 | (209) |
| Potassium | K | 19 | 39.102 |
| Praseodymium | Pr | 59 | 140.907 |
| Promethium | Pm | 61 | (145) |
| Protactinium | Pa | 91 | (231) |
| Radium | Ra | 88 | (226) |
| Radon | Rn | 86 | 222 |
| Rhenium | Re | 75 | 186.2 |
| Rhodium | Rh | 45 | 102.905 |
| Rubidium | Rb | 37 | 85.47 |
| Ruthenium | Ru | 44 | 101.07 |
| Samarium | Sm | 62 | 150.35 |
| Scandium | Sc | 21 | 44.956 |
| Selenium | Se | 34 | 78.96 |
| Silicon | Si | 14 | 28.086 |
| Silver | Ag | 47 | 107.868 |
| Sodium | Na | 11 | 22.9898 |
| Strontium | Sr | 38 | 87.62 |
| Sulfur | S | 16 | 32.064 |
| Tantalum | Ta | 73 | 180.948 |
| Technetium | Tc | 43 | (97) |
| Tellurium | Te | 52 | 127.60 |
| Terbium | Tb | 65 | 158.924 |
| Thallium | Tl | 81 | 204.37 |
| Thorium | Th | 90 | 232.0381 |
| Thulium | Tm | 69 | 168.934 |
| Tin | Sn | 50 | 118.69 |
| Titanium | Ti | 22 | 47.90 |
| Tungsten | W | 74 | 183.85 |
| Uranium | U | 92 | 238.03 |
| Vanadium | V | 23 | 50.942 |
| Xenon | Xe | 54 | 131.30 |
| Ytterbium | Yb | 70 | 173.04 |
| Yttrium | Y | 39 | 88.905 |
| Zinc | Zn | 30 | 65.37 |
| Zirconium | Zr | 40 | 91.22 |